Lecture Notes in Mathematics

Edited by A. Dold and B. Eckmann

874

Abelian Group Theory

Proceedings of the Oberwolfach Conference,
January 12–17, 1981

Edited by R. Göbel and E. Walker

Springer-Verlag
Berlin Heidelberg New York 1981

Editors

Rüdiger Göbel
FB 6 – Mathematik, Universität Essen – Gesamthochschule –
Universitätsstr. 3, 4300 Essen 1, Federal Republic of Germany

Elbert Walker
Department of Mathematical Sciences, New Mexico State University
Las Cruces, New Mexico 88003, USA

AMS Subject Classifications (1980): 3 C xx, 3 E xx, 3 F xx, 13 E xx, 13 F xx, 13 G xx, 13 H xx, 16 A 46, 18 G xx, 20-04, 20 K xx, 22 B xx

ISBN 3-540-10855-6 Springer-Verlag Berlin Heidelberg New York
ISBN 0-387-10855-6 Springer-Verlag New York Heidelberg Berlin

Printed in Germany

Printing and binding: Beltz Offsetdruck, Hemsbach/Bergstr.
2141/3140-543210

A Memorial Tribute to REINHOLD BAER

PROFESSOR Dr.phil.,

Dr.rer.nat.h.c., Dr.rer.nat.h.c., Dr.Sc.h.c.

July 22, 1902 - October 22, 1979

Preface

A conference on Abelian Group Theory was held at the Mathematische Forschungsinstitut in Oberwolfach from January 12th to 17th, 1981. The conference brought together 39 Abelian group theorists from over the world who represented a wide range of interests in the subject. The interaction and cross-fertilization of ideas provided by this conference will no doubt stimulate the continuing development of Abelian Groups.

The first paper in this volume contains a short account of Reinhold Baer's contribution to Abelian Group Theory. The remaining papers are ones presented at this conference and a few others submitted by colleagues unable to attend. This collection reflects the current activity in the field and it is hoped that it will inspire others to participate actively in Abelian Group Theory.

Rüdiger Göbel
Essen
FRG

Elbert Walker
Las Cruces
USA

March, 1981

TABLE OF CONTENTS

LIST OF PARTICIPANTS

U. Albrecht	Department of Mathematics, New Mexico State University, Las Cruces, New Mexico 88003, USA and Fachbereich 6 - Mathematik, Universität Essen-Gesamthochschule, 4300 Essen, FRG
D. Beers	Department of Mathematics, New Mexico State University, Las Cruces, New Mexico 88003, USA and Wellesley College, Wellesley, Massachusetts, USA
C.-F. Bödigheimer	Mathematisches Institut, Universität Heidelberg, 6900 Heidelberg, FRG
K. Burkhard	Mathematisches Institut, Universität Würzburg, 8700 Würzburg, FRG
K. Benabdallah	Department of Mathematics and Statistics, University of Montreal, Montreal, Quebec, Canada
B. Charles	Université des Sciences et Techniques du Languedoc Institut de Mathématiques, Montpellier, France
Y. Cooper	Department of Mathematics, University of Texas of the Permian Basin, Odessa, Texas, USA
M. Dugas	Fachbereich 6 - Mathematik, Universität Essen-Gesamthochschule, 4300 Essen, FRG
P. Eklof	University of London, Bedford College and Department of Mathematics, University of California at Irvine, Irvine, California, USA
K. Faltings	Fachbereich Mathematik, Universität Kaiserslautern, 6750 Kaiserslautern, FRG
B. Franzen	Fachbereich 6 - Mathematik, Universität Essen - Gesamthochschule, 4300 Essen, FRG
U. Felgner	Mathematisches Institut, Universität Tübingen, 7400 Tübingen, FRG
L. Fuchs	Department of Mathematics, Tulane University, New Orleans, Louisiana 70118, USA
T. Giovanetti	Department of Mathematics, New Mexico State University, Las Cruces, New Mexico 88003, USA
R. Göbel	Fachbereich 6 - Mathematik, Universität Essen - Gesamthochschule, 4300 Essen, FRG
J. Hausen	Department of Mathematics, University of Houston, Central Campus, Houston, Texas, 77004, USA
M. Huber	Mathematisches Institut der Universität Freiburg, 7800 Freiburg, FRG

R. Hunter	Department of Mathematics, New Mexico State University, Las Cruces, New Mexico 88003, USA
F. Kiefer	Mathematisches Institut II, Freie Universität Berlin, 1 Berlin(West)
P. Kümmich	Mathematisches Institut II, Freie Universität Berlin, 1 Berlin(West)
H. Lenzing	Fachbereich Mathematik, Universität Paderborn - Gesamthochschule, 4790 Paderborn, FRG
W. Liebert	Mathematisches Institut, Technische Universität München, 8000 München, FRG
A. Mader	University of Hawaii, Honolulu, MI 96822, USA
C. Metelli	Seminario Matematico dell Universita Padova, Padova, Italy
R. Mines	Fachbereich 6 - Mathematik, Universität Essen - Gesamthochschule, 4300 Essen, FRG and Department of Mathematics, New Mexico State University, Las Cruces, New Mexico 88003, USA
J. Moore	Department of Mathematics, New Mexico State University, Las Cruces, New Mexico 88003, USA
O. Mutzbauer	Mathematisches Institut, Universität Würzburg, 8700 Würzburg, FRG
P. Plaumann	Mathematisches Institut Erlangen, Universität Erlangen, Bismarckstr. 1 1/2, 8520 Erlangen, FRG
L. Procházka	Matematicko-fyzikalni fakulta, Universita Karlova, Praha, CSSR
K.M. Rangaswamy	Department of Mathematical Sciences, University of Nevada, Las Vegas, Nevada 89154, USA
J.D. Rèid	Department of Mathematics, Wesleyan University Middletown, Conneticut, USA
F. Richman	Department of Mathematics, New Mexico State Universtity, Las Cruces, New Mexico 88003, USA
G. Sageev	Department of Mathematics, The Ohio State Universtity, Columbo, Ohio, USA
L. Salce	Seminario Matematico dell Universita Padova, Padova, Italy
S. Stock	Mathematisches Institut, Technische Universität München, 8000 München, FRG
R. Vergohsen	Fachbereich 6 - Mathematik, Universität Essen - Gesamthochschule, 4300 Essen, FRG
B. Wald	Fachbereich 6 - Mathematik, Universität Essen - Gesamthochschule, 4300 Essen, FRG

C. Walker — Department of Mathematics, New Mexico State University, Las Cruces, New Mexico 88003, USA

E.A. Walker — Department of Mathematics, New Mexico State University, Las Cruces, New Mexico 88003, USA

B. Zimmermann-Huisgen — Fachbereich Mathematik, Technische Universität München, Arcisstraße 21, 8000 München 2, FRG

Speakers of the joint sessions with a meeting on "Model-theory"

K.J. Devlin — Department of Mathematics, University of Lancaster, UK : "Combinatorial principles in set theory"

P. Eklof — University of London, Bedford College and Department of Mathematics, University of California at Irvine, Irvine, California, USA : "On the rank fo Ext".

F.D. Tall — Department of Mathematics, University of Toronto, Canada : "Martin's axiom"

Addresses of other contributors

D.M. Arnold — Department of Mathematics, New Mexico State University, Las Cruces, New Mexico 88003, USA

A. Birtz — Department of Mathematics and Statistics, University of Montreal, Montreal, Quebec, Canada

H. Bowman — Department of Mathematical Sciences, University of Nevada, Las Vegas, Nevada 89154, USA

P. Hill — Department of Mathematics, Auburn University, Auburn, Alabama 36849, USA

M. Höppner — Fachbereich Mathematik, Universität Paderborn - Gesamthochschule, 4790 Paderborn, FRG

J. Irwin — Department of Mathematics, Wayne State University, Detroit, Michigan, USA

E.L. Lady — Department of Mathematics University of Hawaii, Honolulu, MI 96822, USA

A. Mekler — Department of Mathematics Simon Fraser University, Burnaby, B.C., Canada

C.M. Ringel — Fakultät für Mathematik, Universität Bielefeld, 4800 Bielefeld 1, FRG

С.В.Рычков — Московская областъ, 142610, Орехево-Зуево, ул. Зелёная 26, ком.205,СССР

T. Snabb — Department of Mathematics, Wayne State University, Detroit, Michigan, USA

S. Shelah — Institute of Mathematics, The Hebrew University, Jerusalem, Israel

R.B. Warfield — Department of Mathematics, University of Seattle, Seattle, Washington, USA

P. Westphal — Fachbereich 6 - Mathematik, Universität Essen - Gesamthochschule, 4300 Essen, FRG

P. Zanardo — Seminario Matematico dell Universita Padova, Padova, Italy

REINHOLD BAER'S WORK ON ABELIAN GROUPS

L. Fuchs

In the mid 30's when Reinhold Baer entered the scene of abelian groups, the theory of infinite abelian groups was still in its infancy. The only advanced branch was the theory of p-groups. About a decade earlier, the pioneering papers of H. Prüfer were published, dealing with countable p-groups. In the early 30's, an astonishing structure theory was developed by H. Ulm and L. Zippin. Virtually nothing was known about the structure of torsion-free and mixed groups. Apart from sporadic examples of indecomposable groups of rank two, the only coherent theory, due to D. Derry, A. G. Kurosh and A. I. Malcev, was developed somewhat later to classify torsion-free groups of finite rank; its purpose was to provide examples of indecomposable groups of higher ranks. Mixed groups have not been studied at all, but thanks to F. Levi, an example of a non-splitting mixed group was on record.

The newly developed structure theory of countable p-groups had apparently a great impact on the young Baer. He immediately recognized the tremendous potentials of this theory, and more generally, of commutative groups, and set himself the goal to explore what commutativity can do for groups. The core of his achievements in abelian group theory is in a period of five years (the late 30's). In a short time, he changed both the shape and direction of the theory. He planted fresh ideas, in his numerous papers new methods surfaced and unexpected connections were discovered. He exploited the topics to such an extent that, for quite a while, only either

very hard or very easy problems were left to others.

His point of departure was the theory of p-groups, which opened the door to a new approach to mixed groups. He continued with developing a far-reaching theory of torsion-free groups, and went on to study relations between groups and their endomorphism rings, or automorphism groups, as well as the role of the lattice of subgroups in the group structure. This period culminates in his brilliant paper on injective modules and injective hulls. From the 40's on, his research was dominated by topics in non-commutative groups, but his commitment to the commutative case has never subsided: themes from abelian groups return again and again in his later papers, giving a new impetus to the subject. He even planned to publish a book on abelian groups; a good portion of the manuscript has already been completed, when he changed his mind, and instead he wrote a book on linear algebra and projective geometry (1952).

His continuous interest in and enthusiasm for commutative groups manifested in the research of his numerous doctoral students who chose to follow their master's footsteps in abelian groups. Undoubtedly, this was one of his favorite subjects, to which he attracted many of his best students through his extraordinarily stimulating and fascinating personality, his indefatigable energy and patience. But his real impact on abelian groups can not faithfully be assessed if, beyond his publications and the circle of his students, we fail to point out his tremendous influence on the mathematical public by his masterly presented talks and on all of us by his inspiring discussions.

In this survey, we plan to give a short summary of Reinhold Baer's contributions to infinite abelian group theory. In view of the abundance of the material to be covered, we can only try to trace some general lines. A variety of his theorems are included in monographs on the subject, so for further details the reader is referred to these books.

R. Baer's first result on abelian groups was published in a paper on non-commutative groups [1], dealing with a generalization of the center (the set of elements in a group which commute with every subgroup). Here he proves that a bounded abelian group is a direct sum of cyclic groups. He thus removed the countability hypothesis from one of Prüfer's famous theorems.

He immediately recognized the extraordinary importance of the Ulm-Zippin theory on countable p-groups. It was him who first gave non-trivial applications of the theory. In 1935, he investigates countable p-groups and shows inter alia that any two direct decompositions of a reduced p-group have isomorphic refinements if and only if the group is a direct sum of cyclic groups [2].

At the same time, he studies the common refinement property for direct decompositions [3]. Let A be an abelian group, R a subring of the endomorphism ring End A of A, and let R^0 denote the centralizer of R in End A. He proves the common refinement for direct decompositions of A into summands which are invariant under both R and R^0. Systematic use is made of the projections (which play a fundamental role in his later theory of direct decompositions).

In the same year, his third paper [4] is published on abelian groups. This is devoted to properties of elements in a p-group A which are invariant under automorphisms. He associates with $a \in A$ a finite sequence

$$I(a) = (s(a), s(pa), \ldots, s(p^{n(a)-1}a))$$

where $p^{n(a)}$ denotes the order of a and $s(a) = n(a) + h(a)$ with $h(a)$ standing for the height of a. Manifestly, this is the original version of the height-sequence (or indicator) of a, used in a modified form by I. Kaplansky and others later on. Though Baer deals primarily with direct sums of cyclics, his arguments extend easily to p-groups A without elements of infinite height. He shows that $a \in A$ can be embedded in a minimal direct summand of A, structurally characterized by $I(a)$, and in a form somewhat different from Kaplansky's, he classifies the characteristic subgroups of A.

The next year signals the publication of a most important paper [5], initiating the theory of mixed groups. At that stage of development, the most substantial question was to find out when a mixed group splits. He raises the question which can be reformulated as that of finding the torsion groups T (the torsion-free groups F) such that

$$(*) \qquad \mathrm{Ext}(F,T) = 0$$

for every torsion-free group F (every torsion group T), and goes on to ask for the general problem of characterizing pairs (T,F) of torsion and torsion-free groups satisfying (*). He gives a full answer to the first question (the sufficiency was proved simultaneously and independently by S. V. Fomin) by showing that these are precisely the direct sums of a divisible and a bounded group. In addition, he establishes several necessary conditions on the other two problems. This paper stimulated lot of research on torsion-free groups F satisfying (*) for all T (it was settled by P. Griffith only in 1969).

In 1937, he published three papers on totally different aspects of abelian groups. The most outstanding publication is [7] in which he lays down the foundation of the theory of torsion-free abelian groups, and at the same time he brings it to a fairly advanced level. His approach is entirely different from the Kurosh-Malcev ideas (which were developed about the same time). He concentrates on the rank 1 groups which are the fundamental building stones, and starts with their precise description. He introduces the most fundamental concepts, like characteristic and type of elements (his terminology was different), and uses the types in order to characterize the direct sums of rank one groups. The notion of separable group is his creation; except for their summands and endomorphism rings, almost all what we know about them today is already in his paper. In an example, he uses subgroups of the p-adic integers to construct indecomposable groups.

In the introduction of his paper [6], he states that his objective was to investigate relations between automorphisms and properties of the group itself. The set F(S) of all those elements of a p-group G is studied which are left invariant under all automorphisms of G leaving a subgroup S of G elementwise fixed. One of his results states that F(S) = S + F(0) if G is not reduced, while $F(S) = \overline{S} + F(0)$ if G has no elements of infinite height (here bar indicates the closure in the p-adic topology). In a remarkable proof, he succeeds in characterizing the center of the automorphism group by showing that (except for a very special case for $p = 2$) the center consists of multiplications by p-adic units or by integers mod p^r prime to p according as G is unbounded or is bounded by p^r with r minimal.

Already in this paper, he brings up the lattice of subgroups $L(G)$ of G, and investigates the relation between the automorphisms of G and the lattice-automorphisms of $L(G)$. At this time, he got deeply involved in the study of $L(G)$ in general, which he viewed as an additional source of information about G itself. In [9], he shows that in "most cases", the isomorphism of lattices of subgroups implies the isomorphism of the groups themselves. [8] is devoted to the question of dualism between two abelian groups G and G' : this is a bijective function $d: L(G) \to L(G')$ such that $S \cong G'/dS$ for every $S \in L(G)$. He settles this question by showing that such a dualism exists if and only if $G \cong G'$ and G is torsion whose p-components are finite.

His paper [11] is a true gem. It was published much before modules became so important. He proves nothing less than the existence of injective embeddings and a unique minimal one (in his setting, injective is a summand in everything containing it). His motivation came from abelian groups, but he noticed that his transfinite methods cover groups on which rings operate. He makes use of his famous criterion for injectivity which bears his name.

The interplay between group and ring theory is a source of enrichment for both disciplines. This is the theme of his remarkable study of endomorphism rings of bounded p-groups (in a slightly more general form, he considers modules over primary rings) [12]. His attention is focused on the ideal theory. He shows that (i) the groups are determined by their endomorphism rings and (ii) the endomorphism rings can be characterized ring theoretically. Occasionally, he assumes that the group contains two or three independent elements of maximal orders, but the results are valid without these hypotheses. Later on, (i) was generalized by

I. Kaplansky (1952) to arbitrary p-groups and (ii) by W. Liebert (1968) to separable p-groups.

From the beginning of his career, Baer was interested in the problem of group extensions. In one of his early publications (Math. Z. 38 (1934), 375-416), investigating extensions of abelian groups, he introduces the group of extensions; this turned out to play a crucial role in homological algebra developed by S. Eilenberg, S. Mac Lane, and others in the 50's. In his paper [13], he breaks with the tradition, and proposes a new equivalence relation between two (abelian) extensions, G and H, of a group S: call G and H of the same extension type if there exist homomorphisms $G \to H$ and $H \to G$ leaving S elementwise fixed. He succeeds in classifying the extension types of a given group S in the special case where G/S are torsion with trivial p-components for the primes p for which $G[p]$ is infinite.

His interest was attracted by the new ideas of homological algebra, and [14] is a beautiful illustration of the interplay between abelian groups and homological methods. The new feature was the study of endomorphisms of $\mathrm{Ext}(F,T)$ induced by endomorphisms of F or T. He obtains numerous information on the structure of Ext as an abelian group, and investigates $\mathrm{Ext}(P,T)$ where P is the group of sequences of integers and T is torsion.

Motivated by geometric considerations, in [15] he considers partitions π of an abelian group A; π is a set of subgroups $\neq 0$ of A such that every non-zero element of A belongs to exactly one member of π. Only elementary p-groups can have non-trivial partitions. He develops a theory for these partitions with a view to geometric applications.

In [16] those subgroups Γ of the automorphism group of a group A are considered for which 0 and A are the only admissible subgroups of A. If Γ is restricted to be locally finite, then A has to be an elementary p-group and the centralizer of Γ in the endomorphism ring of A turns out to be an absolute algebraic field of characteristic p. This paper is a beautiful example how ideas from commutative and non-commutative group theory can be combined.

As we close this survey, let us emphasize that the picture is necessarily incomplete: his contributions to other fields are frequently related to abelian groups too. His papers reflect the striking unity of mathematical ideas, and undoubtedly, they will continue to inspire mathematicians of new generations.

REFERENCES

[1] Der Kern, eine charakteristische Untergruppe, Compositio Math. 1 (1934), 254-283.

[2] The decomposition of enumerable, primary, abelian groups into direct summands, Quart. J. Math. Oxford 6 (1935), 217-221.

[3] The decomposition of abelian groups into direct summands, Quart. J. Math. Oxford 6 (1935), 222-232.

[4] Types of elements and characteristic subgroups of abelian groups, Proc. London Math. Soc. 39 (1935), 481-514.

[5] The subgroup of the elements of finite order of an abelian group, Ann. Math. 37 (1936), 766-781.

[6] Primary abelian groups and their automorphisms, Amer. J. Math. 59 (1937), 99-117.

[7] Abelian groups without elements of finite order, Duke Math. J. 3 (1937), 68-122.

[8] Dualism in abelian groups, Bull. Amer. Math. Soc. 43 (1937), 121-124.

[9] The significance of the system of subgroups for the structure of the group, Amer. J. Math. 61 (1939), 1-44.

[10] Duality and commutativity of groups, Duke Math. J. 5 (1939), 824-838.

[11] Abelian groups that are direct summands of every containing abelian group, Bull. Amer. Math. Soc. 46 (1940), 800-806.

[12] Automorphism rings of primary abelian operator groups, Ann. Math. 44 (1943), 192-227.

[13] Extension types of abelian groups, Amer. J. Math. 71 (1949), 461-490.

[14] Die Torsionsuntergrupe einer abelschen Gruppe, Math. Ann. 135 (1958), 219-234.

[15] Partitionen abelscher Gruppen, Arch Math. 14 (1963), 73-83.

[16] Irreducible groups of automorphisms of abelian groups, Pacific J. Math. 14 (1964), 385-406.

Pure Subgroups of Finite Rank Completely Decomposable Groups

David M. Arnold

R. Baer, in 1937, gave a complete set of invariants for finite rank completely decomposable groups. After such a promising start, the theory of the structure of finite rank torsion free groups has become stagnant. One of the difficulties has been the absence of a suitable intermediate class of groups, i.e. a class large enough to contain interesting examples, small enough that there is some hope of understanding the structure of groups in the class, and admitting enough different characterizations to provide a variety of techniques and reasonable problems.

This paper is devoted to the examination of a candidate for such a class, the class of pure subgroups of finite rank completely decomposable groups. Interest in this class stems from results of M.C.R. Butler [1] who proves that a torsion free group is a pure subgroup of a finite rank completely decomposable group iff it is the homomorphic image of a finite rank completely decomposable group and gives a characterization of these groups in terms of types and their associated fully invariant subgroups (Lemma 1.1).

Following Lady [4] and [5], a finite rank torsion free group is called a <u>Butler group</u> if it is a pure subgroup of a finite rank completely decomposable group. Butler groups are called quasi-essential groups by Koehler [1], purely finitely generated groups by Bican [1] and [2], and R-groups by Butler [1].

The class of Butler groups contains all rank-1 groups and is closed under finite direct sums, pure subgroups, torsion free homomorphic images, and quasi-isomorphism. If a Butler group A has a linearly ordered typeset (e.g. homogeneous or p-local) then A is completely decomposable (Butler [1]).

On the other hand, the class of Butler groups includes many of the known examples of "pathological" direct sum decompositions of finite rank torsion free groups (e.g. Fuchs [1], §90). If K is a finite dimensional Q-algebra then there is a Butler group with quasi-endomorphism algebra isomorphic to K (Brenner-Butler [1]). Moreover, if A is a finite rank torsion free group with finite typeset then there is a Butler group B, unique up to quasi-equality, with $B \subseteq A$, $A(\tau)/B(\tau)$ torsion for each type τ, and typeset(B) = typeset(A) (Theorem 1.9, due essentially to Koehler [1]).

Several characterizations of Butler groups, in addition to the characterization given by Butler [1], are given in Section 1 (Theorem 1.12 due to Bican [1], Theorem 1.13 due to Bican [2], and Theorem 1.10 which generalizes a result of Wang [1]).

Section 2 contains a definition and properties of the class of B_0-groups. Butler groups may be viewed as a generalization of almost completely decomposable groups with B_0-groups playing the role of completely decomposable groups (Corollary 2.5).

The notion of regulating subgroups of almost completely decomposable groups given by Lady [3] is generalized to the class of Butler groups in Section 3. The results of Lady [3], for almost completely decomposable groups, are generalized to a larger class of Butler groups in Section 4.

Finite rank torsion free rings with additive group a Butler group are considered in Section 5, e.g. if A is a Butler group then the additive group of the endomorphism ring of A modulo its nil radical is an almost completely decomposable group (Corollary 5.2).

Portions of Section 1 are expository in that some published results are reproved using techniques developed by Butler [1]. Generally, these techniques result in shorter and more conceptual proofs. Numerous examples of Butler groups and B_0-groups are given.

In summary, the class of Butler groups seems worthy of further study. To this end, several conjectures and problems are given in Section 6.

§0. Notation and Preliminaries

A type is an equivalence class of height sequences as defined by Fuchs [1]. If S is a finite set of types then $\inf(S)$ and $\sup(S)$ are types, where $\inf(S)$ and $\sup(S)$ are given by component minimums and maximums of height sequences representing types in S.

Suppose that A is a torsion free group and τ is a type. For $0 \neq a \in A$ let $\text{type}_A(a)$ be the type of the pure rank-1 subgroup of A generated by a. Define $A(\tau) = \{a\in A \mid \text{type}_A(a) \geq \tau\}$, a pure fully invariant subgroup of A, and $A^*(\tau)$ to be the subgroup of A generated by $\{a\in A \mid \text{type}_A(a) > \tau\}$. Note that $A^*(\tau) = \sum\{A(\sigma) \mid \sigma > \tau\}$. Let $A^*(\tau)_*$ denote the pure subgroup of A generated by $A^*(\tau)$. Define typeset(A) to be $\{\text{type}_A(a) \mid 0 \neq a \in A\}$.

Lemma 0.1 (Koehler [1]): Let A be a torsion free group of finite rank and assume that typeset(A) is finite. If T is a non-empty subset of typeset(A) and $\sigma = \inf(T)$ then there is a maximal Z-independent subset B of $A(\sigma)$ such that $\text{type}_A(b) = \sigma$ for each $b \in B$. In particular, $\sigma \in \text{typeset}(A)$.

Proof: Embed A in a Q-vector space V such that V/A is torsion. Define $S = \{\tau \in \text{typeset}(A) \mid \tau > \sigma\}$. If $\tau \in S$ then $QA(\tau) \subset QA(\sigma)$ since $A(\tau) \subset A(\sigma)$ and $A(\tau)$ is pure in A.

By induction on the cardinality of S there is a basis $(a_1, \ldots, a_r)$ of $QA(\sigma)$ such that $a_i \in A(\sigma)\setminus QA(\tau)$ for each $1 \leq i \leq r$ and each $\tau \in S$. For each i, $\text{type}_A(a_i) \geq \sigma$ and $\text{type}_A(a_i) \notin S$. Thus, $\text{type}_A(a_i) = \sigma$ for each i as desired.

Let A be a finite rank torsion free group. Then $QA = Q\otimes_Z A$ is a

Q-vector space and A may be embedded in QA with torsion cokernel. If B is another finite rank torsion free group then a quasi-homomorphism from A to B is an element of QHom(A,B) = $Q\otimes_Z$Hom(A,B). The groups A and B are quasi-isomorphic if there is f $\in$ QHom(A,B) and g $\in$ QHom(B,A) with fg = 1 and gf = 1. In this case f is a quasi-isomorphism. Note that A and B are quasi-isomorphic iff there is f $\in$ Hom(A,B) such that f is a monomorphism and B/f(A) is finite. The groups A and B are quasi-equal if QA = QB and there are non-zero integers m and n with mA $\subseteq$ B and nB $\subseteq$ A.

A finite rank torsion free group is completely decomposable if A is the direct sum of rank-1 groups and τ-homogeneous completely decomposable if A is the direct sum of rank-1 groups of type τ. A torsion free group A is strongly indecomposable if whenever A is quasi-isomorphic to a torsion free group B $\oplus$ C then B = 0 or C = 0.

§1. Butler Groups

Lemma 1.1 (Butler [1]): The following are equivalent for a finite rank torsion free group A:

(a) A is a Butler group;

(b) A is a homomorphic image of a finite rank completely decomposable group;

(c) typeset(A) is finite and for each type τ $A^*(\tau)_*/A^*(\tau)$ is finite and there is a τ-homogeneous completely decomposable group A_τ with $A(\tau) = A_\tau \oplus A^*(\tau)_*$.

Moreover, if A is a Butler group then $A(\sigma)/\sum\{A_\tau \mid \tau > \sigma\}$ is finite for each type σ.

An exact sequence $0 \to A \to B \to C \to 0$ of finite rank torsion free groups is balanced if $0 \to A(\tau) \to B(\tau) \to C(\tau) \to 0$ is exact for each type τ; equivalently if $c \in C$ then there is $b \in B$ with $b \to c$ and the p-height of b in B is equal to the p-height of c in C for each prime p (Fuchs [1], Lemma 86.4). Finite rank completely decomposable groups are balanced projectives (i.e., the groups projective relative to balanced exact sequences of finite rank torsion free groups, see Fuchs [1], Theorem 86.2)

Theorem 1.2: The class of Butler groups has enough balanced projectives and is closed under balanced extensions.

Proof. To show that there are enough balanced projectives, let A be a Butler group and τ a maximal type in typeset(A). Then $A^*(\tau) = 0$ and $C_\tau = A(\tau)$ is τ-homogeneous completely decomposable by Lemma 1.1. Now let $\tau \in$ typeset(A) and assume that for each $\tau < \sigma \in$ typeset(A) there is $C_\sigma \subseteq A$ with C_σ generated by pure rank-1 subgroups of A of type σ and that $A(\sigma) = \sum \{C_\delta | \delta \geq \sigma\}$. Now $A(\tau) = A_\tau \oplus A^*(\tau)_*$ where A_τ is τ-homogeneous completely decomposable and $A^*(\tau)_*/A^*(\tau)$ is finite. Write $A^*(\tau)_* = A^*(\tau) + A_1 + \ldots + A_n$ where each A_i is a pure rank-1 subgroup of A of type τ and define $C_\tau = A_\tau + A_1 + \ldots + A_n$. Then $A(\tau) = C_\tau + A^*(\tau) = C_\tau + \sum \{A(\sigma) | \sigma > \tau\} = C_\tau + \sum \{C_\delta | \delta > \tau\}$. Since typeset(A) is finite $A(\tau) = \sum \{C_\delta | \delta \geq \tau\}$ for each $\tau \in$ typeset(A) where C_δ is generated by pure rank-1 subgroups of A of type δ.

Define a finite rank completely decomposable group $D = \oplus_\tau D_\tau$ where $D_\tau = 0$ if $\tau \notin$ typeset(A) and D_τ is the direct sum of the rank-1 groups generating C_τ if $\tau \in$ typeset(A). Then there is an exact sequence $0 \to K \to D \to A \to 0$, noting that $\tau_0 = \inf\{\tau | \tau \in \text{typeset}(A)\} \in$ typeset(A) by Lemma 0.1 and $A(\tau_0) = A$. Moreover, the sequence is balanced exact since $D(\tau) = \oplus \{D_\sigma | \sigma \geq \tau\} \to A(\tau) = \sum \{C_\sigma | \sigma \geq \tau\}$ is onto for each $\tau \in$ typeset(A). Consequently, there are enough balanced projectives.

If $0 \to K \to C \to A \to 0$ is balanced exact and if K and A are

Butler groups then there are balanced exact sequences $0 \to X \to D \to K \to 0$ and $0 \to Y \to E \to A \to 0$ with D and E finite rank completely decomposable. Now D and E are balanced projectives so there is an epimorphism $D \oplus E \to C \to 0$. By Lemma 1.1 C is a Butler group.

Remark: The class of Butler groups is not closed under arbitrary pure extensions since there are rank-2 groups with infinite typesets. If A is a rank-2 Butler group and $0 \to B \to A \to C \to 0$ is a balanced exact sequence with $\mathrm{rank}(B) = \mathrm{rank}(C) = 1$ then $A \cong B \oplus C$ since rank-1 groups are balanced projectives. Thus if A is indecomposable then A has no proper balanced subgroups.

Let A be a Butler group and define the critical typeset of A, $T'(A)$, to be $\{\tau \mid A(\tau) \neq A^*(\tau)_*\}$. Then $T'(A) \subseteq \mathrm{typeset}(A)$ since $A(\tau) = A_\tau \oplus A^*(\tau)_*$, where A_τ is τ-homogeneous, and $\tau \in T'(A)$ iff $A_\tau \neq 0$. If A_1 and A_2 are two rank-1 groups of incomparable types τ_1 and τ_2 then $A = A_1 \oplus A_2$ is a Butler group with $T'(A) = \{\tau_1, \tau_2\} \neq \mathrm{typeset}(A) = \{\tau_0, \tau_1, \tau_2\}$ where $\tau_0 = \inf\{\tau_1, \tau_2\}$. On the other hand, typeset(A) is determined by $T'(A)$:

Theorem 1.3: If A is a Butler group then $\mathrm{typeset}(A) = \{\inf(S) \mid S \subseteq T'(A)\}$.

Proof. Since typeset(A) is finite, $\inf(S) \in \mathrm{typeset}(A)$ whenever $S \subseteq T'(A)$ (Lemma 0.1). If τ is a maximal type in typeset(A) then $\tau \in T'(A)$ since $A^*(\tau) = 0$. Now assume that $\tau \in \mathrm{typeset}(A)$ and that if $\tau < \sigma \in \mathrm{typeset}(A)$ then $\sigma = \inf(S_\sigma)$ for some $S_\sigma \subseteq T'(A)$. It suffices to suppose that $\tau \notin T'(A)$. Then $A(\tau)/A^*(\tau)$ is finite so there is $a \in A$ with $\mathrm{type}_A(a) = \tau$ and $a = \sum a_\sigma$ with $a_\sigma \in A_\sigma$ and $\sigma > \tau$. Hence $\tau \geq \inf\{\sigma \mid \tau < \sigma \in \mathrm{typeset}(A)\} \geq \tau$ and $\tau = \inf(S)$, where $S = \cup\{S_\sigma \mid \tau < \sigma \in \mathrm{typeset}(A)\}$. Since typeset(A) is finite the proof is complete.

Corollary 1.4 (Butler [1]). Let A be a Butler group. Then typeset(A) is linearly ordered iff $T'(A)$ is linearly ordered. In this case, typeset(A) = T'(A) and A is completely decomposable.

Proof: ($\rightarrow$) Note that $T'(A) \subseteq$ typeset(A).

($\leftarrow$) Apply Theorem 1.3 to see that typeset(A) = T'(A).

Let τ be a minimum type in typeset(A) = T'(A). Then $A = A_\tau \oplus A^*(\tau)_*$ with A_τ a τ-homogeneous completely decomposable group (Lemma 1.1). Now $A_\tau \neq 0$, since $\tau \in T'(A)$, and typeset$(A^*(\tau)_*)$ is linearly ordered. By induction on rank(A) $A^*(\tau)_*$, hence A, is completely decomposable.

The next two results are mild modifications of results of Lady [4] and Koehler [1], respectively.

Theorem 1.5: Suppose that A is a Butler group, B is a finite rank torsion free group and that $\varphi : QA \rightarrow QB$. Then $\varphi \in$ QHom(A,B) iff $\varphi(QA(\tau)) \subseteq QB(\tau)$ for each $\tau \in T'(A)$.

Proof. ($\rightarrow$) Write $\varphi = qf$ for $q \in Q$ and $f \in$ Hom(A,B) and observe that $f(A(\tau)) \subseteq B(\tau)$ for each type τ.

($\leftarrow$) Write $A(\tau) = A_\tau \oplus A^*(\tau)_*$ for each type τ. Then $A(\tau)/\Sigma\{A_\sigma | \sigma > \tau\}$ is finite as a consequence of Lemma 1.1. Thus $A(\tau)/\Sigma\{A(\sigma) | \tau < \sigma \in T'(A)\}$ is finite.

If τ is a maximal type in typeset(A) then $\tau \in T'(A)$ and $A(\tau)$ is τ-homogeneous completely decomposable. Therefore, $\varphi \in$ Hom(QA(τ), QB(τ)) = QHom(A(τ), B(τ)) (Warfield [1]). Now assume that $\tau \in$ typeset(A) and that $\varphi \in$ QHom(A(σ), B(σ)) for each $\tau < \sigma \in T'(A)$. Then QHom(A(τ), B(τ)) = QHom(A_τ, B(τ)) $\oplus$ QHom($A^*(\tau)_*$, B(τ)). But $\varphi \in$ QHom($A^*(\tau)_*$, B(τ)), since $A^*(\tau)_*/\Sigma\{A(\sigma) | \tau < \sigma \in T'(A)\}$ is finite, and $\varphi \in$ QHom(A_τ, B(τ)) = Hom(QA_τ, QB(τ)), as above, since A_τ is τ-homogeneous completely decomposable. Thus $\varphi \in$ QHom(A(τ), B(τ)) for each $\tau \in$ typeset(A). Letting $\tau = \inf\{\sigma | \sigma \in$ typeset(A)$\}$ completes the proof.

Corollary 1.6: Suppose that A and B are Butler groups. Then A and B are quasi-isomorphic iff $T'(A) = T'(B)$ and there is an isomorphism $\rho : QA \to QB$ with $\rho(QA(\tau)) = QB(\tau)$ for each $\tau \in T'(A)$.

The typeset of a Butler group A may also be recovered from the types of rank-1 groups generating A, due essentially to both Koehler [1] and Butler [1].

Theorem 1.7: Suppose that A is a Butler group with $A = A_1 + \ldots + A_n$ where each A_i is a rank-1 subgroup of A. If X is a pure rank-1 subgroup of A then $\text{type}(X) = \sup\{\inf\{\text{type}(A_i) \mid X \cap (\sum A_i) \neq 0\}\}$.

Proof. Let $C = A_1 \oplus \ldots \oplus A_n$ and $f : C \to A$ the canonical epimorphism. Then $f^{-1}(X)$ is a pure subgroup of C, hence a Butler group, so that $f^{-1}(X) = Y_1 + \ldots + Y_k$ with each Y_i a pure rank-1 subgroup of C (Lemma 1.1). If $f(Y_i) \neq 0$ and S_i is the smallest subset of $\{1, 2, \ldots, n\}$ with $Y_i \subseteq \oplus \{A_j \mid j \in S_i\}$ then type $(Y_i) = \inf\{A_j \mid j \in S_i\}$ and $X \cap \sum \{A_j \mid j \in S_i\} \neq 0$. On the other hand, if $X \cap \sum \{A_j \mid j \in S\} \neq 0$ then there is a pure rank-1 subgroup Y_S of C with $Y_S \subseteq (\oplus\{A_j \mid j \in S\}) \cap f^{-1}(X)$, and $f(Y_S) \neq 0$. Therefore, X is generated by $\{f(Y_S) \mid X \cap \sum\{A_j \mid j \in S\} \neq 0\}$, where each $f(Y_S) \neq 0$ and type $(Y_S) = \inf\{\text{type}(A_j) \mid j \in S\}$. Since $\text{rank}(X) = 1$, type $(X) = \sup\{\text{type}(Y_S)\}$ and the proof is complete.

If A is a Butler group then $T'(A)$ and $\text{rank}(A(\tau))$ are quasi-isomorphism invariants of A for each $\tau \in T'(A)$. The next example demonstrates that these invariants do not determine A up to quasi-isomorphism (compare Corollary 1.6).

Example 1.8: (a) There are strongly indecomposable Butler groups of rank > 1.

(b) There are rank-3 Butler groups A and B such that $T'(A) = T'(B)$, $A^*(\tau) = 0 = B^*(\tau)$ for each $\tau \in T'(A)$, $A(\tau) \cong B(\tau)$ for each $\tau \in T'(A)$ yet A and B are not quasi-isomorphic.

Proof. (a) Choose rank-1 groups A_1, A_2, A_3 of incomparable types τ_1, τ_2, τ_3, respectively, such that $\tau_0 = \inf\{\tau_i, \tau_j\}$ if $i \neq j$, and let $A = A_1 + A_2 + A_3 \subseteq QA_1 \oplus QA_2$, where $A_i \cap A_j = 0$ if $i \neq j$. Then A is a rank-2 Butler group. Since $0 \neq A_i \cap (A_j \oplus A_k) \neq A_i$ if $\{i,j,k\} = \{1,2,3\}$, it follows from Theorem 1.7 that typeset(A) = $\{\tau_0, \tau_1, \tau_2, \tau_3\}$. Thus A must be strongly indecomposable. Note that $T'(A) = \{\tau_1, \tau_2, \tau_3\}$.

(b) Choose rank-1 groups A_1, A_2, A_3, A_4 of incomparable types $\tau_1, \tau_2, \tau_3, \tau_4$ respectively such that $\tau_0 = \inf\{\tau_i, \tau_j\}$ whenever $i \neq j$. Define $A = A_1 \oplus (A_2+A_3+A_4)$ and $B = A_2 \oplus (A_1+A_3+A_4)$ where $A_1 + A_3 + A_4$ and $A_2 + A_3 + A_4$ are strongly indecomposable rank-2 Butler groups as constructed in (a). Then typeset(A) = $\{\tau_0, \tau_1, \tau_2, \tau_3, \tau_4\}$ = typeset (B). Moreover, $T'(A) = T'(B) = \{\tau_1, \tau_2, \tau_3, \tau_4\}$ and each τ_i is a maximal type in typeset(A). Finally, A and B are not quasi-isomorphic, otherwise the Krull-Schmidt theorem for quasi-isomorphism (Fuchs [1]) would imply that A_1 is quasi-isomorphic to A_2 which is impossible.

The next theorem is due essentially to Koehler [1].

Theorem 1.9: Let A be a finite rank torsion free group with typeset(A) finite. Then there is a Butler group B with $B \subseteq A$, $A(\tau)/B(\tau)$ torsion for each type τ, and typeset(B) = typeset(A). Moreover, B is unique up to quasi-equality.

Proof: For each $\tau \in$ typeset(A) choose B^τ, a direct sum of pure rank-1 subgroups of A of type τ with $A(\tau)/B^\tau$ torsion (Lemma 0.1). Define $B = \sum \{B^\tau \mid \tau \in \text{typeset}(A)\}$. Then B is a Butler group with typeset(A) $\subseteq$ typeset(B). Furthermore, $B^\tau \subseteq B(\tau) \subseteq A(\tau)$ so that

$A(\tau)/B(\tau)$ is torsion for each $\tau \in \text{typeset}(A)$. Thus $A(\tau)/B(\tau)$ is torsion for each type τ.

If τ is a maximal type in typeset(B) then $\tau \in \text{typeset}(A)$; otherwise $A(\tau)/B(\tau)$ torsion implies that there is $\tau < \sigma \in \text{typeset}(A) \subseteq \text{typeset}(B)$ which is impossible. Now assume that $\tau \in \text{typeset}(B)$ and that $\sigma \in \text{typeset}(A)$ for each $\tau < \sigma \in \text{typeset}(B)$. If $\tau = \inf\{\sigma \in \text{typeset}(B) \mid \sigma > \tau\}$ then $\tau \in \text{typeset}(A)$ (Lemma 0.1). Otherwise, $B^*(\tau) = B(\sigma)$ for some $\tau < \sigma \in \text{typeset}(B)$. Since B is a Butler group, $B(\tau) = B_\tau \oplus B(\sigma)$ for some $0 \neq B_\tau$. Now $B^\sigma \subseteq B(\sigma) \subseteq B \cap A(\sigma) \subseteq A(\sigma)$ with $A(\sigma)/B^\sigma$ torsion implies that $B(\sigma) = B \cap A(\sigma)$ since $B(\sigma)$ is pure in B. Also $A^*(\tau) = A(\sigma)$ by the induction hypothesis. Thus $0 \to B(\tau)/B^*(\tau) \to A(\tau)/A^*(\tau)$ is exact with $B(\tau)/B^*(\tau) \neq 0$. Hence $A(\tau)/A^*(\tau) \neq 0$ whence $\tau \in \text{typeset}(A)$. Since typeset(B) is finite, typeset(B) = typeset(A).

The uniqueness of B follows from Corollary 1.6 by letting φ be the identity on QB.

A subgroup B of a finite rank torsion free group A is a <u>regular subgroup of A</u> if $\text{type}_B(x) = \text{type}_A(x)$ for each $x \in B$. Note that pure subgroups of A are regular and if B is a subgroup of A with B_*/B finite then B is a regular subgroup of A.

<u>Theorem 1.10</u>: Let A be a torsion free group of finite rank. The following are equivalent:

(a) A is a Butler group;

(b) Typeset(A) is finite and B_*/B is finite for each regular subgroup B of A;

(c) Typeset(A) is finite and A/B is finite for each regular subgroup B of A with rank(A) = rank(B).

<u>Proof</u>: (a) $\to$ (b) Let B be a regular subgroup of A. Then B is a regular subgroup of B_* and B_* is a Butler group by Lemma 1.1.

Thus it suffices to assume that $B_* = A$, i.e., A/B is torsion. Suppose that τ is a maximal type in typeset(A). Since A/B is torsion and B is regular in A, typeset(A) = typeset(B). Thus $B(\tau) = B \cap A(\tau)$ is τ-homogeneous, $A(\tau)$ is τ-homogeneous completely decomposable and $A(\tau)/B(\tau)$ is torsion. Consequently, $B(\tau)$ is τ-homogeneous completely decomposable (Arnold-Lady [1]) so that $A(\tau)/B(\tau)$ is finite.

Now assume that $\tau \in$ typeset(A) and that $A(\sigma)/B(\sigma)$ is finite for each $\tau < \sigma \in$ typeset(A). Then $A^*(\tau)/B^*(\tau)$ is finite since $A^*(\tau) = \sum\{A(\sigma) \mid \tau < \sigma \in \text{typeset}(A)\}$ and typeset(A) is finite. Hence $A(\tau)/(A_\tau \oplus B^*(\tau))$ is finite where $A(\tau) = A_\tau \oplus A^*(\tau)_*$ and $A^*(\tau)_*/A^*(\tau)$ is finite by Lemma 1.1. Thus, $(A_\tau \cap B(\tau)) \oplus B^*(\tau) \subseteq B(\tau)$ and $(A_\tau \oplus B^*(\tau))/(A_\tau \cap B(\tau)) \oplus B^*(\tau)$ is finite since A_τ is τ-homogeneous completely decomposable, $A_\tau \cap B(\tau)$ is τ-homogeneous and $A_\tau/(A_\tau \cap B(\tau))$ is torsion (by the preceding remarks). Consequently, $A(\tau)/B(\tau)$ is finite for each $\tau \in$ typeset(A). In particular, A/B is finite since $A(\tau_0) = A$ and $B(\tau_0) = B$ where $\tau_0 = \inf\{\tau \mid \tau \in \text{typeset}(A)\}$.

(b) → (c) Clear.

(c) → (a) As a consequence of Theorem 1.9 there is a Butler group B such that $B \subseteq A$, A/B is torsion, and B is a regular subgroup of A. By (c), A/B is finite so that A is a Butler group by Lemma 1.1.

<u>Corollary 1.11 (Wang [1])</u>: Let A be a finite rank torsion free group such that typeset(A) is linearly ordered. The following are equivalent:

(a) A is completely decomposable;

(b) B_*/B is finite for each regular subgroup B of A;

(c) If B is a regular subgroup of A with rank(A) = rank(B) then A/B is finite.

Proof. Since A has finite rank and typeset(A) is linearly ordered, typeset(A) is finite. Now apply Theorem 1.10 and Corollary 1.4.

For a prime p let Z_p be the localization of Z at p (i.e., $Z_p = \{m/n \in Q \mid n$ is prime to $p\}$). If $S \subseteq \pi$, the set of all primes of Z, let $Z_S = \cap\{Z_p \mid p \in S\}$. For a finite rank torsion free group A define A_S to be $Z_S \otimes_Z A$ and regard A_S as a subgroup of $QA = Q\otimes_Z A$.

If A is a Butler group then A_S is a Butler group for each $S \subseteq \pi$ as a consequence of Lemma 1.1. Moreover, A is locally completely decomposable (i.e. A_p is completely decomposable for each prime p) by Corollary 1.4, noting that typeset(A_p) $\subseteq$ {type(Z_p), type(Q)}. Butler [1] gives an example of a locally completely decomposable group that is not a Butler group.

Theorem 1.12 (Bican [1]): Let A be a torsion free group of finite rank. Then A is a Butler-group iff $\Pi = S_1 \cup \ldots \cup S_n$ for some $1 \leq n \in Z$ such that A_S is completely decomposable with linearly ordered typeset for each $S \in \{S_1, S_2, \ldots, S_n\}$.

Proof. ($\rightarrow$) Write $A = A_1 + \ldots + A_m$ where each A_i is a pure rank-1 subgroup of A. Choose $0 \neq a_i \in A_i$ for each i. If p is a prime then there is a permutation α of $\{a_1, \ldots, a_m\}$ such that $h_p^A(\alpha a_1) \geq \ldots \geq h_p^A(\alpha a_m)$ where $h_p^A(x)$ = p-height of x in A. For each permutation α, let $S\alpha = \{p \mid h_p^A(\alpha a_1) \geq \ldots \geq h_p^A(\alpha a_m)\}$. Then $\{\alpha\}$ is finite and $\Pi = \cup_\alpha\{S\alpha\}$. Moreover, typeset($A_{S\alpha}$) is linearly ordered for each α so that each $A_{S\alpha}$ is completely decomposable by Corollary 1.4.

($\leftarrow$) If X and Y are subgroups of QA then there is an exact sequence $0 \rightarrow X\cap Y \rightarrow X\oplus Y \rightarrow X+Y \rightarrow 0$ is exact so that $X\cap Y$ is isomorphic to a pure subgroup of $X\oplus Y$. It now follows that

$A = \cap \{A_S \mid S \in \{S_1, S_2, \ldots, S_n\}\}$ is isomorphic to a pure subgroup of the completely decomposable group $\oplus_S A_S$. Thus A is a Butler group by Lemma 1.1.

A subgroup B of a finite rank torsion free group A is a <u>generalized regular subgroup of A</u> if for each $b \in B$ $h_p^B(b) = h_p^A(b)$ for all but a finite number of primes p. Note that regular subgroups of A are generalized regular subgroups of A.

<u>Theorem 1.13 (Bican [2])</u>: Let A be a finite rank torsion free group. Then A is a Butler group iff A is locally completely decomposable and if B is a generalized regular subgroup of A with A/B torsion then $(A/B)_p = 0$ for all but a finite number of p.

<u>Proof</u>. ($\rightarrow$) As noted above, A is locally completely decomposable. Write $A = A_1 + \ldots + A_n$ where each A_i is a pure rank-1 subgroup of A. Then $(A_i/A_i \cap B)_p = 0$ for all but a finite number of p. But $A_1/(A_1 \cap B) \oplus \ldots \oplus A_n/(A_n \cap B) \rightarrow A/B \rightarrow 0$ is an epimorphism so that $(A/B)_p = 0$ for all but a finite number of p.

($\leftarrow$) Let F be a maximal free subgroup of A and let $\{A_1, A_2, \ldots\}$ be an ordering of the set of pure rank-1 subgroups of A. For each m let $C_m = F + A_1 + \ldots + A_m$. If $C_m = A$ for some m then A is a Butler group.

Assume that A/C_m is infinite torsion for each m. Given m, let p be a prime and write $(A/C_m)_p = D \oplus T$ where D is p-divisible and T is finite. Then there is $n > m$ with $A(p^\infty)_p \subseteq (C_n)_p$, where $A(p^\infty)$ is the p-divisible subgroup of A. Thus $(A/C_n)_p$ is finite since A is locally completely decomposable. Consequently, $(A/C_m)_p \neq 0$ for infinitely many primes p else $A = C_k$ is a Butler group for some $k > m$ which is impossible. Choose an infinite set of primes $\{p_1, p_2, \ldots\}$ with $(A/C_i)_p \neq 0$ for

$p = p_i$. For each m choose $C_m \subseteq B_m \subseteq A$ such that A/B_m is a non-zero p_m-group and let $B = \cap_m B_m$. Then $(A/B)_p \neq 0$ for each $p = p_m$ and A/B is torsion since $F \subseteq B$. If $b \in B$ then $b \in A_m \subseteq \cap\{C_i \mid i \geq m\} \subseteq \cap\{B_i \mid i \geq m\}$ for some m. Thus $h_p^A(b) = h_p^B(b)$ for each prime $p \notin \{p_1, p_2, \ldots, p_{m-1}\}$, contradicting the hypotheses. Hence $A = C_m$ for each m and A is a Butler group.

§2. B_0-groups:

Let A be a Butler group and let τ be a type. Define the <u>Butler τ-invariant</u>, $b(A,\tau)$, to be the isomorphism class of the finite group $A^*(\tau)_*/A^*(\tau)$. Note that if A and B are isomorphic Butler groups then $b(A,\tau) = b(B,\tau)$; $b(A,\tau) = 0$ iff $A^*(\tau)$ is pure in A; if $\tau \notin \text{typeset}(A)$ then $A(\tau) = A^*(\tau)$ and $b(A,\tau) = 0$; and that $b(A,\tau) = b(A(\sigma),\tau)$ for each $\tau \geq \sigma$ (since $A(\sigma)(\tau) = A(\tau)$).

<u>Lemma 2.1:</u> If A and B are Butler groups and if τ is a type then $b(A \oplus B,\tau) = b(A,\tau) \oplus b(B,\tau)$.

<u>Proof:</u> A consequence of the fact that $(A\oplus B)(\tau) = A(\tau) \oplus B(\tau)$, $(A\oplus B)^*(\tau) = A^*(\tau) \oplus B^*(\tau)$, and $(A\oplus B)^*(\tau)_* = A^*(\tau)_* \oplus B^*(\tau)_*$.

A Butler group A is defined to be a B_0-group if $b(A,\tau) = 0$ for each type τ. B_0-groups are called torsionless groups by Butler [1]. As a consequence of Lemma 2.1, the class of B_0-groups is closed under finite direct sums and direct summands.

<u>Theorem 2.2:</u> Suppose that A is a Butler group with $A(\tau) = A_\tau \oplus A^*(\tau)_*$ for each type τ. The following are equivalent:

(a) A is a B_0-group;

(b) For each type σ, $A(\sigma) = \sum\{A_\tau \mid \tau \geq \sigma\}$;

(c) The sequence $0 \to K \to C \to A \to 0$ is balanced exact where $C =$

$\oplus_\tau A_\tau$ is finite rank completely decomposable and $K = \mathrm{Kernel}(C \to A)$.

Proof. (a) $\to$ (b) If τ is a maximal type in typeset(A) then $A(\tau) = A_\tau$ since $A^*(\tau) = 0$. Assume that $\tau \in$ typeset(A) and that $A(\sigma) = \Sigma\{A_\delta \mid \delta \geq \sigma\}$ for each $\tau > \sigma \in$ typeset(A). Then $A^*(\tau) = \Sigma\{A(\sigma) \mid \sigma > \tau\} = \Sigma\{A_\delta \mid \delta > \tau\}$. Since $b(A,\tau) = 0$, $A(\tau) = A_\tau \oplus A^*(\tau)$. Thus, $A(\tau) = \Sigma\{A_\delta \mid \delta \geq \tau\}$ for each $\tau \in$ typeset(A).

(b) $\to$ (c) Clear.

(c) $\to$ (a) If σ is a type then $C(\sigma) = \oplus\{A_\tau \mid \tau \geq \sigma\}$ so that $A(\sigma) = \Sigma\{A_\tau \mid \tau \geq \sigma\}$, $A^*(\sigma) = \Sigma\{A(\tau) \mid \tau > \sigma\} = \Sigma\{A_\delta \mid \delta > \sigma\}$ and $A(\sigma) = A_\sigma \oplus A^*(\sigma)$. Thus $b(A,\sigma) = 0$.

Corollary 2.3: (a) Suppose that A is a B_0-group, B is a finite rank torsion free group and that $\varphi : QA \to QB$. Then $\varphi \in \mathrm{Hom}(A,B)$ iff $\varphi(A(\tau)) \subseteq B(\tau)$ for each $\tau \in T'(A)$.

(b) Suppose that A and B are B_0-groups. Then A and B are isomorphic iff there is an isomorphism $\varphi : QA \to QB$ with $\varphi(A(\tau)) = B(\tau)$ for each $\tau \in T'(A)$.

Proof. As a consequence of Theorem 2.2, $A(\sigma) = \Sigma\{A(\tau) \mid \tau \geq \sigma, \tau \in T'(A)\}$ for each $\sigma \in$ typeset(A). The proof is now as in Theorem 1.5 and Corollary 1.6.

Remark: Let A_1 and A_2 be two rank-1 groups of incomparable type with a prime p such that $pA_1 \neq A_1$, $pA_2 \neq A_2$. Define $A = A_1 + A_2 + Z(a_1+a_2)/p \subseteq QA_1 \oplus QA_2$ and define $B = A_1 \oplus A_2$. Then B is a B_0-group, $B(\tau) = A(\tau)$ for each $\tau \in T'(A)$ yet A and B are not isomorphic since A is indecomposable. Thus the hypothesis that A and B are B_0-groups is necessary for Corollary 2.3.b.

Example 2.4: (a) There are strongly indecomposable B_0-groups of rank > 1.

(b) There are strongly indecomposable B_0-groups that are

quasi-isomorphic but not isomorphic.

Proof. Let $A = A_1 + A_2 + A_3$ be as constructed in Example 1.8.a.

(a) Note that A is a strongly indecomposable B_0-group since if $\tau_0 \neq \tau \in \text{typeset}(A)$ then $A^*(\tau) = 0$ while $A^*(\tau_0) = A_1 + A_2 + A_3 = A$. In fact, $E(A)$ is isomorphic to a subring of Q since if $0 \neq a_3 = a_1 \oplus a_2 \in (A_1 \oplus A_2) \cap A_3$ and $f \in E(A)$ then $f(a_i) = q_i a_i$ for some $q_i \in Q$ whence $q_1 = q_2 = q_3 = q$ and f is multiplication by $q \in Q$.

(b) Further assume that there is a prime q with $qA_i \neq A_i$ for $i = 1,2,3$. Then $A/qA \simeq Z/qZ \oplus Z/qZ$. Note that $A(\tau_i)/A_i$ is finite for $i = 1,2,3$, so assume that $A = A_1 + A_2 + A_3$ with $A_i = A(\tau_i)$ for $1 \leq i \leq 3$.

Define $B = qA_1 + qA_2 + A_3$. Then A/B is finite and $B = B(\tau_1) + B(\tau_2) + B(\tau_3)$ is a B_0-group, noting that typeset(B) = typeset(A). Assume that A and B are isomorphic say $f(A) = B$ for some $f \in E(A)$. Then $f = mu$ for some $m \in Z$ and u a unit of $E(A)$ and $qA \subseteq B = f(A) = mA \subseteq A$. Now $B \neq qA$ since $A_3 \subseteq B$ and $A_3 \neq qA_3$. Thus, $B = mA = A$ which is impossible since $B/qA \simeq Z/qZ$ while $A/qA \simeq Z/qZ \oplus Z/qZ$.

A finite rank torsion free group A is almost completely decomposable if A is quasi-isomorphic to a completely decomposable group. Note that almost completely decomposable groups are Butler groups.

Corollary 2.5: A group A is an almost completely decomposable B_0-group iff A is completely decomposable.

Proof: ($\leftarrow$) Write $A = A_1 \oplus \ldots \oplus A_n$ where each A_i has rank-1. If τ is a type then $A(\tau) = \oplus\{A_i \mid \text{type}(A_i) \geq \tau\}$ and $A^*(\tau) = \oplus\{A_i \mid \text{type}(A_i) > t\}$. Thus, $A^*(\tau)$ is a summand of A, hence pure in A.

($\rightarrow$) Write $A(\tau) = A_\tau \oplus A^*(\tau)$ where A_τ is a τ-homogeneous completely decomposable group for each type τ. Then $A = \Sigma_\tau A_\tau$ by Theorem 2.2. Let $C = C_1 \oplus \ldots \oplus C_n$ be a completely decomposable subgroup of finite index in A, where each C_i has rank-1. Then $C = \oplus_\tau C_\tau$ where $C_\tau = \oplus \{C_i \mid \text{type}(C_i) = \tau\}$ is τ-homogeneous completely decomposable. Since A/C is finite, $A(\tau)/C(\tau)$ and $A^*(\tau)/C^*(\tau)$ are finite for each type τ. Thus $\text{rank}(A_\tau) = \text{rank}(C_\tau)$ for each type τ since $A(\tau) = A_\tau \oplus A^*(\tau) \supseteq C(\tau) = C_\tau \oplus C^*(\tau)$. Consequently, $\text{rank}(A) = \text{rank}(C) = \sum_\tau \text{rank}(C_\tau) = \sum_\tau \text{rank}(A_\tau)$ whence $A = \oplus_\tau A_\tau$ is completely decomposable.

Remark: There are well known examples of almost completely decomposable groups that are not completely decomposable (e.g., Fuchs [1]). Consequently, the class of B_0-groups is not closed under quasi-isomorphism.

Two finite rank torsion free groups A and B are nearly isomorphic if for each $0 \neq n \in Z$ there is a monomorphism $f_n : A \to B$ such that $[A : f_n(B)]$ is finite and relatively prime to n. If A and B are nearly isomorphic then $A \oplus A' \cong B \oplus B$ for some A' nearly isomorphic to B (Lady [2]). Warfield has proved that A and B are nearly isomorphic iff there is a positive integer n with $A^n \cong B^n$ where A^n is the direct sum of n copies of A (see Arnold [3]). Consequently, if A and B are nearly isomorphic Butler groups then $b(A,\tau) = b(B,\tau)$ for each type τ (Lemma 2.1). In particular, the class of B_0-groups is closed under near isomorphism.

A B_0-group A is an N-group if whenever B is a finite rank torsion free group nearly isomorphic to A then B is isomorphic to A.

Theorem 2.6: Suppose that A is a Butler group, A is quasi-isomorphic to an N-group, and that B is a finite rank torsion free group. Then the following are equivalent:

(a) $A \oplus D \cong B \oplus D$ for some finite rank torsion free group D;

(b) A is nearly isomorphic to B;

(c) $A \oplus C \cong B \oplus C$ for some N-group C quasi-isomorphic to A.

Proof: (a) → (b) Lady [2].

(b) → (c) Let C be a subgroup of finite index in A such that C is an N-group. There is a monomorphism $f : B \to A$ such that $[A : f(B)]$ is finite and relatively prime to $[A:C]$. It follows that $A \oplus D \cong B \oplus C$ for some D. Thus, D is nearly isomorphic to C since A is nearly isomorphic to B (Lady [2]). Since C is an N-group, $D \cong C$ and $A \oplus C \cong B \oplus C$.

(c) → (a) Clear.

It is conjectured that every B_0-group is an N-group (see Section 6). The class of N-groups contains each B_0-group A such that (i) A is completely decomposable; (ii) $pA = A$ for all but a finite number of primes p (Lady [2]); or (iii) A is strongly indecomposable and there is a type τ such that rank $(A(\tau)/A^*(\tau)) = 1$ (since in this case $E(A)/NE(A)$ is isomorphic to a subring of $E(A(\tau)/A^*(\tau)) \subseteq Q$, where $NE(A)$ = nil radical of $E(A)$, so that $E(A)/NE(A)$ is a principal ideal domain, see Arnold [2] or [3]). In particular, the consequences of Theorem 2.6 are true if A is almost completely decomposable, first proved by Lady [3]. Note that by (iii), the B_0-groups constructed in Example 2.4 are N-groups.

§3. Regulating Subgroups of Butler Groups

A subgroup B of a Butler group A is a regulating subgroup of A if $B = \sum_\tau A_\tau$, where $A(\tau) = A_\tau \oplus A^*(\tau)_*$ for each type τ. By Lemma 1.1, A/B is finite and B is a Butler group. Note that A_τ need not be unique. However, if $A(\tau) = A'_\tau \oplus A^*(\tau)_*$ then $A_\tau \cong A'_\tau$.

Proposition 3.1: (a) If C is a regulating subgroup of A and if C

$\subseteq B \subseteq A$ then C is a regulating subgroup of B.

(b) If C is a regulating subgroup of B and if B is a regulating subgroup of A then C is a regulating subgroup of A.

(c) If $B = \sum_\tau A_\tau$ is a regulating subgroup of A and if σ is a type then $\sum\{A_\tau \mid \tau \geq \sigma\}$ is a regulating subgroup of $A(\sigma)$ and $\sum\{A_\tau \mid \tau > \sigma\}$ is a regulating subgroup of $A^*(\sigma)$ and of $A^*(\sigma)_*$.

(d) If A is a Butler group, σ is a type, and $B' = \sum_\tau A'_\tau$ is a regulating subgroup of $A(\sigma)$ then $B = B' + \sum\{A_\tau \mid \tau \not\geq \sigma\}$ is a regulating subgroup of A, where $A(\tau) = A_\tau \oplus A^*(\tau)_*$ for each $\tau \not\geq \sigma$.

Proof. (a) Suppose that $C = \sum_\tau A_\tau$ where $A(\tau) = A_\tau \oplus A^*(\tau)_*$ for each τ.. Then $B(\tau) = B(\tau) \cap A(\tau) = A_\tau \oplus (A^*(\tau)_* \cap B(\tau)) = A_\tau \oplus B^*(\tau)_*$, since $A_\tau \subseteq C(\tau) \subseteq B(\tau)$, for each τ. Hence C is a regulating subgroup of B.

(b) Write $C = \sum_\tau B_\tau$ and $B = \sum_\tau A_\tau$ where $B(\tau) = B_\tau \oplus B^*(\tau)_*$ and $A(\tau) = A_\tau \oplus A^*(\tau)_*$ for each type τ. Then $A(\tau) = B_\tau \oplus A^*(\tau)_*$ since $A^*(\tau)/B^*(\tau)$ finite implies that $B_\tau \cap A^*(\tau)_* = 0$ and $A_\tau \subseteq B(\tau) \subseteq B_\tau \oplus A^*(\tau)_* \subseteq A(\tau)$ implies that $A(\tau) = B_\tau + A^*(\tau)_*$. Thus C is a regulating subgroup of A.

(c) If $\tau \in \text{typeset}(A(\sigma))$ then $\sigma \leq \tau$ and $A(\sigma)(\tau) = A(\tau) = A_\tau \oplus A^*(\tau)_* = A_\tau \oplus A(\sigma)^*(\tau)_*$. Thus, $\sum\{A_\tau \mid \tau \geq \sigma\}$ is a regulating subgroup of $A(\sigma)$. Let $C = A^*(\sigma)_*$. Then $C = C^*(\sigma)_*$ and if $\tau > \sigma$ then $C(\tau) = A(\tau) = A_\tau \oplus A^*(\tau)_* = A_\tau \oplus C^*(\tau)_*$. Hence $\sum\{A_\tau \mid \tau > \sigma\}$ is a regulating subgroup of C. Also $\sum\{A_\tau \mid \tau > \sigma\} \subseteq A^*(\sigma)$ is a regulating subgroup of $A^*(\sigma)$ by (a).

(d) is clear.

Corollary 3.2: Suppose that A is a Butler group. The following are equivalent:

(a) A is a B_0-group;

(b) $A(\sigma)$ is a unique regulating subgroup of $A(\sigma)$ for each $\sigma \in \text{typeset}(A)$;

(c) $A(\sigma)$ is a regulating subgroup of $A(\sigma)$ for each $\sigma \in$ typeset(A).

Proof: (a) → (b) Apply Proposition 3.1 and Theorem 2.2.

(b) → (c) Clear.

(c) → (a) Let $\sigma = \inf\{\tau \mid \tau \in \text{typeset}(A)\}$. Then $\sigma \in$ typeset(A) by Lemma 1.1 and Lemma 0.1. Thus, $A = A(\sigma)$ is a regulating subgroup of A, say $A = \sum_\tau A_\tau$ where $A(\tau) = A_\tau \oplus A^*(\tau)_*$ for each τ. Now $A = A_\sigma \oplus A^*(\sigma)$, where $A^*(\sigma) = \sum\{A_\tau \mid \tau > \sigma\}$, so that $b(A,\sigma) = 0$. Let $\sigma \in$ typeset(A). Then $A(\sigma)$ is a regulating subgroup of $A(\sigma)$ and $\sigma = \inf\{\tau \mid \tau \in \text{typeset}(A(\sigma))\}$. By the preceding remarks, $b(A(\sigma),\sigma) = 0$. But $b(A(\sigma),\sigma) = b(A,\sigma)$ so that $b(A,\sigma) = 0$ for each $\sigma \in$ typeset(A). Therefore A is a B_0-group.

Corollary 3.3: If A is an almost completely decomposable group and if B is a regulating subgroup of A then B is completely decomposable.

Proof. In view of Proposition 3.1.a, B is a regulating subgroup of B. Thus B is completely decomposable, as in the proof of Corollary 2.5.

If A is a B_0-group then A is a unique regulating subgroup of A by Corollary 3.2. The converse is not true in general:

Example 3.4: There is a rank-3 Butler group A such that A is a unique regulating subgroup of A but A is not a B_0-group. However, there is a subgroup B of finite index in A such that B is a B_0-group.

Proof. Let $V = Qv_1 \oplus Qv_2 \oplus Qv_3$ be a vector space and $\{p_1, p_2, p_3, p_4, p_5\}$ be a set of distinct primes. Define $v_4 = v_1 + v_2 + p_2v_3$ so that each 3-element subset of $\{v_1, v_2, v_3, v_4\}$ is a basis of V. Let $A = A_1 + A_2 + A_3 + A_4 \subseteq V$ where $A_1 = \langle v_1/p_3{}^\infty p_4{}^\infty\rangle$, $A_2 = \langle v_2/p_3{}^\infty p_5{}^\infty\rangle$, $A_3 = \langle v_3/p_1{}^\infty\rangle$, and $A_4 = \langle v_4/p_2{}^\infty\rangle$ (i.e., A_1 is the subgroup of V generated by $\{v_1/p_3^ip_4^j \mid i,j = 0,1,2, \ldots\})$.

As a consequence of Theorem 1.7, typeset $(A) = \{\tau_0, \tau, \tau_1, \tau_2, \tau_3, \tau_4\}$ where $\tau_i = \text{type}(A_i)$ for $1 \leq i \leq 4$; $\tau_0 = \text{type}(Z) = \inf\{\tau_3,\tau_4\}$; and $\tau = \inf\{\tau_1,\tau_2\} > \tau_0$. Now $A_i = A(\tau_i)$, since A_i is pure in A, and $A^*(\tau_i) = 0$ for $1 \leq i \leq 4$. Also $A^*(\tau) = A(\tau_1) + A(\tau_2) = A_1 \oplus A_2$ has rank 2 and $\text{rank}(A(\tau)) < \text{rank}(A) = 3$ so that $A(\tau) = A^*(\tau)_*$. Moreover, $A^*(\tau_0) = A$ from which it follows that A is a unique regulating subgroup of A.

Note that $v_4/p_2 = x + v_3$ where $x = (v_1+v_2)/p \in A(\tau)\setminus A^*(\tau)$ so that $A^*(\tau)$ is not pure in A and $b(A,\tau) \neq 0$. Thus, A is not a B_0-group.

Let $B = A_1 + A_2 + p_2A_3 + A_4 \subseteq A$ and note that A/B is finite. Now $A(\tau_i) = B(\tau_i) = A_i$ for $i = 1, 2, 4$, $B^*(\tau_i) = 0$ for $i = 1,2,3,4$, and $B^*(\tau_0) = B$. Furthermore, $B(\tau_3) = B \cap A(\tau_3) = B \cap A_3 = p_2A_3 + (A_3 \cap (A_1+A_2+A_4))$ and it is easy to see that $A_3 \cap (A_1+A_2+A_4) \subseteq p_2A_3$. Thus, $B(\tau_3) = p_2A_3$.

Note that $B/(A_1\oplus A_2) \cong (p_2A_3\oplus A_4)/(p_2A_3\oplus A_4) \cap (A_1\oplus A_2)$. If $x = p_2b_3 + b_4 \in p_2A_3 \oplus A_4$ with $mx = b_1 + b_2 \in A_1 \oplus A_2$ and $b_4 = (k/p_2^i)v_4 \in A_4$ for some $k \in Z$ then $m(p_2b_3) + m(k/p_2^i)p_2v_3 = 0$, (recalling that $v_4 = v_1 + v_2 + p_2v_3$). Thus $p_2b_3 + (k/p_2^i)p_2v_3 = 0$ so that $k/p_2^i \in Z$, since p_2v_3 has p_2-height $= 0$ in B. Therefore, $x \in A_1 \oplus A_2$, $A_1 \oplus A_2 = B^*(\tau)$ is pure in B and B is a B_0-group as desired.

§4. B_1-Groups:

A Butler group A is defined to be a B_1-group if whenever $C = \Sigma_\tau C_\tau$ is a subgroup of finite index in A with $C(\tau)/(C_\tau\oplus C^*(\tau)_*)$ finite for each type τ then there is a B_0-group D with $C \subseteq D \subseteq A$. The class of B_1-groups includes all B_0-groups and all almost completely decomposable groups (see proof of Corollary 2.5).

<u>Lemma 4.1</u>: Suppose that A is a B_1-group.

(a) Each regulating subgroup of A is a B_0-group.

(b) If σ is a type then $A(\sigma)$ is a B_1-group.

<u>Proof</u>. (a) If B is a regulating subgroup of A then $B \subseteq D \subseteq A$ for some B_0-group D. Thus B is a regulating subgroup of D (Proposition 3.1.a) so that $B = D$ is a B_0-group (Corollary 3.2).

(b) Let $C = \Sigma_\tau C_\tau$ be a subgroup of finite index in $A(\sigma)$ such that $C(\tau)/(C_\tau \oplus C^*(\tau)_*)$ is finite for each type τ. Define $C' = C + \{A_\delta | \delta > \sigma\}$ where $A(\delta) = A_\delta \oplus A^*(\delta)_*$ for each δ and note that $C_\tau = 0$ if $\tau > \sigma$. Since A is a B_1-group there is a B_0-group D with $C' \subseteq D \subseteq A$. Thus $C \subseteq C'(\sigma) \subseteq D(\sigma) \subseteq A(\sigma)$ where $D(\sigma)$ is a B_0-group as a consequence of Corollary 3.2. Thus, $A(\sigma)$ is a B_1-group.

<u>Lemma 4.2</u>: Assume that A is a Butler group such that each regulating subgroup of A is a B_0-group. If B and C are regulating subgroups of A then $[A:B] = [A:C]$.

<u>Proof</u>. Write $B = \Sigma_\tau B_\tau$ and $C = \Sigma_\tau C_\tau$ where $A(\tau) = B_\tau \oplus A^*(\tau)_* = C_\tau \oplus A^*(\tau)_*$ for each type τ. Assume that $B \neq C$ and choose a σ with $B_\sigma \neq C_\sigma$. Define $D = C_\sigma + \Sigma\{B_\tau | \tau \neq \sigma\}$, a regulating subgroup of A. By induction on the cardinality of $\{\tau | B_\tau \neq C_\tau\}$ it suffices to prove that $[A:B] = [A:D]$.

Define $E = B + D$. Then $E = B + C_\sigma = B + E(\sigma)$ and $E = D + B_\sigma = D + E(\sigma)$. Thus, $[E:B] = [E(\sigma):B \cap E(\sigma)] = [E(\sigma):B(\sigma)]$ and $[E:D] = [E(\sigma):D(\sigma)]$. Therefore, it is sufficient to prove that $[A(\sigma):B(\sigma)] = [A(\sigma):D(\sigma)]$, in which case $[A(\sigma):E(\sigma][E(\sigma):B(\sigma)] = [A(\sigma):E(\sigma)][E(\sigma):D(\sigma)]$, $[E:B] = [E:D]$, and $[A:B] = [A:E][E:B] = [A:E][E:D] = [A:D]$.

Now $[A(\sigma):B(\sigma)] = [B_\sigma \oplus A^*(\sigma)_*:B_\sigma \oplus B^*(\sigma)] = [A^*(\sigma)_*:B^*(\sigma)]$ and $[A(\sigma):D(\sigma)] = [C_\sigma \oplus A^*(\sigma)_*:C_\sigma \oplus D^*(\sigma)] = [A^*(\sigma)_*:D^*(\sigma)]$ since B and D are B_0-groups. But each regulating subgroup of $A^*(\sigma)_*$ is a B_0-group and $B^*(\sigma)$, $D^*(\sigma)$ are regulating subgroups of $A^*(\sigma)_*$ as a consequence of Proposition 3.1 and Theorem 2.2. Also $\mathrm{rank}(A) > \mathrm{rank}(A^*(\sigma)_*)$ by the choice of σ. By induction on rank(A) the proof is complete.

<u>Theorem 4.3:</u> Suppose that A is a B_1-group and that B is a subgroup of finite index in A. Then B is a regulating subgroup of A iff $[A:B] = \min\{[A:C] \mid C \subseteq A$ and C is a B_0-group$\}$. In this case, if C is a B_0-group with $C \subseteq A$ then $[A:B]$ divides $[A:C]$.

<u>Proof:</u> In view of Lemma 4.1 and Lemma 4.2, it is sufficient to prove that if A/C is finite where C is a B_0-group but not a regulating subgroup of A then there is a B_0-group $D' \subseteq A$ such that $[A:D']$ is a proper divisor of $[A:C]$.

There is a maximal $\sigma \in T'(C) = T'(A)$ such that $C(\sigma)$ is not a regulating subgroup of $A(\sigma)$: otherwise $C = \sum\{C(\sigma) \mid \sigma \in T'(C)\}$ is a regulating subgroup of A as an application of Theorem 2.2 and Proposition 3.1.

Write $A(\sigma) = A_\sigma \oplus A^*(\sigma)_*$ and define $D = A_\sigma + \sum\{C_\tau \mid \tau \neq \sigma\}$ where $C(\tau) = C_\tau \oplus C^*(\tau)$ for each $\tau \neq \sigma$. It suffices to prove that $[A:D]$ is a proper divisor of $[A:C]$: in which case there is a B_0-group D' with $D \subseteq D' \subseteq A$ so that $[A:D']$ is a proper divisor of $[A:C]$ as needed.

By the choice of σ, $C^*(\sigma) = \sum\{C_\tau \mid \tau > \sigma\}$ is a regulating subgroup of $A^*(\sigma)_*$ and $C^*(\sigma) \subseteq D^*(\sigma) \subseteq (D')^*(\sigma)$ where D' is a B_0-group containing D. Thus $C^*(\sigma)$ is a regulating subgroup of $(D')^*(\sigma)$ (Proposition 3.1.a) whence $C^*(\sigma) = (D')^*(\sigma)$ (Corollary 3.2). Therefore, $C^*(\sigma) = D^*(\sigma)$ is pure in D', hence in D.

Define $E = C + D$. Then $E = C + A_\sigma = C + E(\sigma)$ and $E = D + C_\sigma = D + E(\sigma)$ so that $[E:C] = [E(\sigma):C(\sigma)]$ and $[E:D] = [E(\sigma):D(\sigma)]$. As in the proof of Lemma 4.2, it is sufficient to prove that $[A(\sigma):D(\sigma)]$ is a proper divisor of $[A(\sigma):C(\sigma)]$; in which case $[A:D]$ is a proper divisor of $[A:C]$. But $[A(\sigma):D(\sigma)] = [A^*(\sigma)_*:D^*(\sigma)]$ is a proper divisor of $[A(\sigma):C(\sigma)]$ since $C^*(\sigma) = D^*(\sigma)$ and $C(\sigma) = C_\sigma \oplus C^*(\sigma)$ but $A(\sigma) \neq C_\sigma \oplus A^*(\sigma)_*$ by the choice of σ.

Corollary 4.4 (Lady [3]): Assume that A is an almost completely decomposable group and that B is a subgroup of finite index in A. Then B is a regulating subgroup of A iff $[A:B] = \min\{[A:C] \mid C \subseteq A$ and C is completely decomposable$\}$. In this case, if $C \subseteq A$ with C completely decomposable then $[A:B]$ divides $[A:C]$.

Proof. Apply Corollary 2.5 and Theorem 4.3.

Remark: Lady [3] gives an example of an almost completely decomposable group A with regulating subgroups B and C such that $[A:B] = [A:C] = p^2$, $A/B \simeq Z/pZ \oplus Z/pZ$ and $A/C \simeq Z/p^2Z$. In particular, A/B and A/C are not isomorphic.

If A is a B_1-group define $i(A) = [A:B]$ where B is a regulating subgroup of A. By Theorem 4.3, $i(A)$ is well-defined.

Corollary 4.5: Suppose that A is a B_1-group. The following are equivalent:

(a) A is a B_0-group;

(b) A is a unique regulating subgroup of A;

(c) A is a regulating subgroup of A;

(d) $i(A) = 1$.

Corollary 4.6: If A and D are B_1-groups and if A and D are nearly isomorphic then $i(A) = i(D)$.

Proof. Choose a monomorphism $f : A \to D$ such that $[D:f(A)]$ is relatively prime to $i(D)$. If B is a regulating subgroup of A then B, hence $f(B)$, is a B_0-group. Thus $i(D)$ divides $[D:f(B)]$ by Theorem 4.3. But $[D:f(B)] = [D:f(A)][f(A):f(B)] = [D:f(A)]i(A)$ so that i(D) divides i(A). Symmetrically, i(A) divides i(D).

§5. Endomorphism Rings of Butler Groups

A ring R with identity is a finite rank torsion free ring if R^+, the additive group of R, is finite rank torsion free. In this case, R may be regarded as a subring of $QR = Q\otimes_Z R$. Let JQR be the Jacobson radical of QR and define $NR = (JQR) \cap R$. Since QR is a finite dimensional Q-algebra, NR is a nilpotent ideal of R. Also $(NR)^+$ is a pure subgroup of R^+.

Theorem 5.1: Let R be a finite rank torsion free ring.

(a) R^+ is a Butler group iff $(NR)^+$ is a Butler group and $(R/NR)^+$ is almost completely decomposable.

(b) If $(R/NR)^+$ is a Butler group and $Q(R/NR)$ is a simple algebra then $(R/NR)^+$ is homogeneous completely decomposable.

Proof: First of all, R/NR is quasi-equal to a ring product $R_1 \times \ldots \times R_n$ where each QR_i is a simple algebra (Beaumont-Pierce [1]). If S_i = Center (R_i) then S_i is a domain and R is a finitely generated S_i-module (Pierce [1]). Thus, R_i is quasi-equal to a finitely generated free S_i-module. Since S_i is a domain, S_i is homogeneous with idempotent type = type of the identity of S_i.

(a) (→) Since $(NR)^+$ is pure in R^+, $(NR)^+$ and $(R/NR)^+$ are Butler groups (Lemma 1.1). If QR_i is a simple algebra and S_i = Center (R_i) then S_i^+ is a homogeneous Butler group (by the preceding remarks) hence completely decomposable (Corollary 1.4). Again by the preceding remarks, $(R/NR)^+$ is almost completely decomposable.

(←) By Beaumont-Pierce [1], R^+ is quasi-isomorphic to $(NR)^+ \oplus (R/NR)^+$ so that R^+ is a Butler group.

(b) is a consequence of the proof of (a) ($\rightarrow$) since R quasi-equal to a free S-module implies that R^+ is quasi-isomorphic, hence isomorphic to a homogeneous completely decomposable group.

<u>Remark</u>: If A is a Butler group with $\text{Hom}(A,Z) = 0$ then $R = E(Z\oplus A)$ is a finite rank torsion free ring with $A \cong \text{Hom}(Z,A) \subseteq (NR)^+$.

<u>Corollary 5.2</u>: If A is a Butler group then $(E(A)/NE(A))^+$ is almost completely decomposable, Furthermore, if $Q(E(A)/NE(A)) \cong QE(A)/JQE(A)$ is a simple algebra then $(E(A)/NE(A))^+$ is homogeneous completely decomposable.

<u>Proof</u>. Butler [1] proves that if A is a Butler group then $E(A)^+$ is a Butler group: if C_1 and C_2 are completely decomposable of finite rank such that $\pi : C_1 \rightarrow A \rightarrow 0$ is an epimorphism and A is a pure subgroup of C_2 then $\phi : E(A)^+ \rightarrow \text{Hom}(C_1,C_2)$, defined by $\phi(f) = f\pi$, is a monomorphism with pure image whence $E(A)^+$ is isomorphic to a pure subgroup of the completely decomposable group $\text{Hom}(C_1,C_2)$.

Now apply Theorem 5.1.

§6. Conjectures and Problems

<u>Conjecture 6.1 (Butler [1])</u>: Every Butler group contains a B_0-group as a subgroup of finite index.

Note that Conjecture 6.1 is true for almost completely decomposable groups and the groups constructed in Example 3.4.

<u>Conjecture 6.2</u>: If A is a B_0-group and if B is a finite rank torsion free group nearly isomorphic to A then $B \cong A$ (see remarks following Theorem 2.6).

Problem 6.3: Find a "useful" complete set of isomorphism (quasi-isomorphism, near-isomorphism) invariants for B_0-groups.

Such a set of invariants is likely to include typeset (or critical typeset), rank($A(\tau)$) for each τ, and rank($A(\tau)/A^*(\tau)$). If A is completely decomposable then the critical typeset of A and rank($A(\tau)/A^*(\tau)$) for each τ is a complete set of isomorphism invariants for A.

Problem 6.4: Prove or disprove that every Butler group is a balanced subgroup of a finite rank completely decomposable group. Compute balanced projective dimensions of Butler groups.

Problem 6.5: Determine special ring theoretic properties of $E(A)$ the endomorphism ring of a B_0-group A. For example, if A is completely decomposable and $NE(A)$ is the nil radical of $E(A)$ then $E(A)/NE(A)$ is a product of matrix rings over subrings of Q. Is it true that if A is a B_0-group then $E(A)/NE(A)$ is a product of maximal orders?

Problem 6.6: Compute the group and ring structure of K_0(Butler), the Grothendieck group of Butler groups modulo split exact sequences.

If T is the torsion subgroup of K_0(Butler) then $T = \{[A]-[B] \mid A$ nearly isomorphic to $B\}$ and K_0(Butler)/T is a free abelian group. (a consequence of results of Lady [2]). Note that if both Conjecture 6.1 and 6.2 are true then $T = 0$ (Theorem 2.6). In particular, K_0(almost completely decomposable) is a free abelian group. Let S be a set of at most three primes and let R_s be the class of Butler groups divisible by all primes not in S. Cruddis [1] has characterized all indecomposable R_s-groups and proved that $K_0(R_s)$ is free (also see Arnold [1]).

Define QButler to be the category of Butler groups with morphism sets $Q\otimes_Z \mathrm{Hom}(A,B)$. The ring structure of K_0(QButler) is considered by

Lady [5]. Lady [4] also proves that certain full subcategories of QButler are equivalent to full subcategories of finitely generated modules over artinian rings and that certain full subcategories of Butler are equivalent to full subcategories of finitely generated modules over hereditary Noetherian rings.

Problem 6.7: Extend the known results about Butler groups to valuated groups and mixed groups of finite torsion free rank. The valuated group analog of Butler groups is the class of valuated subgroups of valuated finite direct sums of torsion free cyclic groups (see Richman-Walker [1]). The mixed group analog of Butler groups is the class of mixed groups A such that there is an exact sequence $0 \to B \to A \to T \to 0$ where B is a Butler group, B is a nice subgroup of A, and T is totally projective (see Hunter-Richman [1]). Note that $\mathrm{rank}(A(\tau)/A^*(\tau)_*)$ is analogous to the Warfield invariants for mixed groups (e.g. Hunter-Richman [1]).

List of References

Arnold, D. M.

[1] A class of pure subgroups of completely decomposable abelian groups, Proc. Amer. Math. Soc. 41 (1973), 37-44.

[2] Genera and decompositions of torsion free modules, Springer-Verlag Lecture Notes #616 (1977), 197-218.

[3] Finite rank torsion free abelian groups and subrings of finite dimensional Q-algebras, Lecture Notes, 220p, preprint.

Arnold, D. M., and Lady, L.

[1] Endomorphism rings and direct sums of torsion free abelian groups, Trans. Amer. Math. Soc., 211 (1975), 225-237.

Baer, R.

[1] Abelian groups without elements of finite order, Duke Math. J. 3 (1937), 68-122.

Beaumont, R. A., and Pierce, R. S.

[1] Torsion-free rings, Illinois J. Math. 5 (1961), 61-98.

Bican, L.

[1] Purely finitely generated Abelian groups, Comment. Math. Univ. Carolinae 11 (1970), 1-8.

[2] Splitting in Abelian groups, Czech. Math. J. 28 (103) (1978).

Brenner, S., and Butler, M. C. R.

[1] Endomorphism rings of vector spaces and torsion free abelian groups, J. London Math. Soc. 40 (1965), 183-187.

Butler, M. C. R.

[1] A class of torsion-free abelian groups of finite rank, Proc. London Math. Soc. 15 (1965), 680-698.

Cruddis, T. B.

[1] On a class of torsion free abelian groups Proc. London Math. Soc. 21 (1970), 243-276.

Fuchs, L.

[1] Infinite Abelian Groups, Vol. II, Academic Press, 1973.

Hunter, R. and Richman, F.

[1] Global Warfield groups, preprint.

Koehler, J.

[1] The type set of a torsion-free group of finite rank, Illinois J. Math. 9 (1965), 66-86.

Lady, E.L.

[1] Summands of finite rank torsion free abelian groups, J. Alg. 32 (1974), 51-52.

[2] Nearly isomorphic torsion free abelian groups, J. Alg. 35 (1975), 235-238.

[3] Almost completely decomposable torsion free abelian groups, Proc. A.M.S. 45 (1974), 41-47.

[4] Extension of scalars for torsion free modules over Dedekind domains, Symposia Mathematica XXIII (1979), 287-305.

[5] Grothendieck rings for certain categories of quasi-homomorphisms of torsion free modules over Dedekind domains, preprint.

Pierce, R. S.

[1] Subrings of simple algebras, Michigan Math. J. 7 (1960), 241-243.

Richman, F. and Walker, E.

[1] Valuated groups, J. Algebra 56 (1979) 145-167.

Wang, J. S. P.

[1] On completely decomposable groups, Proc. Amer. Math. Soc. 15 (1964), 184-186.

Warfield, R. B., Jr.

[1] Homomorphisms and duality for torsion-free groups, Math. Z. 107 (1968), 189-200.

ON SPECIAL BALANCED SUPGROUPS OF TORSIONFREE SEPARABLE ABELIAN GROUPS

H. Bowman and K. M. Rangaswamy

The concept of a strongly balanced subgroup introduced in [5] seems to have a close relationship with torsionfree separable abelian groups. This is explored here. The classical theorem of L. Fuchs [3] on summands of a torsionfree separable abelian group A is extended to the strongly balanced subgroups of A when the typeset of A satisfies the maximum condition. The connection between a special type of balanced subgroups and the Butler groups [2] is also investigated.

All the groups that we consider here are assumed to be abelian. For the general notation, terminology and results we refer to [3]. A group is said to have finite rank if it contains a finitely generated essential subgroup.

We say the short exact sequence

$$0 \to B \xrightarrow{\alpha} A \xrightarrow{\beta} C \to 0$$

is

(i) <u>strongly isotype</u> if, to each $b \in B$, there is a homomorphism $f: A \to B$ such that $f\alpha(b) = b$

(ii) <u>strongly nice</u> if, to each $c \in C$, there is a homomorphism $g: C \to A$ such that $\beta g(c) = c$, and

(iii) <u>strongly balanced</u> if it is both strongly isotype and strongly nice.

A subgroup B of a group A is said to be strongly balanced if the exact sequence $0 \to B \xrightarrow{i} A \xrightarrow{\eta} A/B \to 0$ is strongly balanced where i is the inclusion map and η is the natural map. Clearly a strongly balanced subgroup is balanced, but the converse is not true as it is clear from the balanced exact sequence $0 \to B \to F \to G \to 0$ where G is torsionfree homogeneous indecomposable of type $(0,0,\ldots,0,\ldots)$ and F is free. Actually a strongly nice subgroup is already balanced. For this reason the strongly nice exact sequences were called strongly balanced in [4]. In [5] it was pointed out that the torsion part A_t of a separable mixed group A is strongly balanced.

<u>Proposition 1</u>: An exact sequence $0 \to B \overset{\alpha}{\to} A \overset{\beta}{\to} C \to 0$ is strongly balanced if and only if for each finite subset $\{b_1,\ldots,b_n\}$ of B there is an h: $A \to B$ satisfying $h\alpha(b_i) = b_i$, $i = 1,\ldots,n$ and for each finite subset $\{c_i,\ldots, c_m\}$ of C there is an $h':C \to A$ such that $\beta h'(c_j) = c_j$, $j = 1,\ldots,m$.

<u>Proof</u>: By induction Suppose for each non-empty subset X of B with r elements where $r < n$, there is a $\gamma:A \to B$ such that $\gamma\alpha|X$ = identity. Let f: $A \to B$ satisfy $f\alpha(b_i) = b_i$, $i = 1,\ldots,n-1$ and let $f': A \to B$ satisfy $f'\alpha(b_n - f\alpha(b_n)) = b_n - f\alpha(b_n)$. Then $h = f' + f - f'\alpha f$ satisfies $h\alpha(b_i) = b_i$, $i = 1,\ldots,n$. A similar induction argument proves the existence of $h': C \to A$ such that $\beta h'(c_j) = c_j$, $j = 1,\ldots,m$.

<u>Proposition 2</u>: If $0 \to B \to A \to C \to 0$ is strongly balanced exact, then the following induced sequences are strongly balanced exact: $0 \to B_t \to A_t \to C_t \to 0$, $0 \to B/B_t \to A/A_t \to C/C_t \to 0$ and $0 \to D(B) \to D(A) \to D(C) \to 0$, where X_t is the torsion part and D(X) is the divisible part of the group X.

<u>Proof</u>: Straightforward.

The next proposition indicates how finite rank summands of B and C "lift" to summands of A.

<u>Proposition 3</u>: Suppose $0 \to B \overset{\alpha}{\to} A \overset{\beta}{\to} C \to 0$ is strongly balanced exact. If B_1 and C_1 are finite rank summands respectively of B and C, then there is a decomposition of A: $A = B_1' \oplus C_1' \oplus H$ with $B_1 \cong \alpha(B_1) = B_1'$, $C_1' \cong \beta(C_1') = C_1$. Moreover, $B = B_1 \oplus B_2$, $C = C_1 \oplus C_2$ such that the sequence $0 \to B_2 \overset{\alpha'}{\to} H \overset{\beta'}{\to} C_2 \to 0$ is strongly balanced exact, where $\alpha' = \alpha|B_2$ and $\beta' = \beta|H$.

<u>Proof</u>: In view of Proposition 2, we may assume that the groups are all reduced. Let $C_1 = T_1 \oplus F_1$ and $B_1 = T_1' \oplus F_1'$, where T_1 and T_1' are respectively the torsion parts of C_1 and B_1 and are clearly finite. Let X and Y be maximal independent subsets respectively of F_1 and F_1'. If f: $A \to B$ is a homomorphism satisfying $f\alpha$ is identity on both T_1' and Y and if g: $C \to A$ is such that βg is identity on both T_1 and X, then $f\alpha$ and βg act as identity respectively on B_1 and C_1. Let $C = C_1 \oplus C_2$. Then $A = A_1 \oplus C_1'$, where $C_1' = g(C_1)$ and $A_1 = \beta^{-1}(C_2)$ and $A_1 = B_1' \oplus H$ where $B_1' = \alpha(B_1)$. Now $\alpha(B) = B_1' \oplus (\alpha(B) \cap H)$. If we write $B = B_1 \oplus B_2$ where $\alpha(B) \cap H = \alpha(B_2)$, then the sequence $0 \to B_2 \overset{\alpha'}{\to} H \overset{\beta'}{\to} C_2 \to 0$ is strongly balanced exact, where $\alpha' = \alpha|B_2$ and $\beta' = \beta|H$.

It appears that the strongly balanced subgroups are to separable groups what the direct summands are to completely decomposable groups. For instance, if C is torsionfree completely decomposable, then, for each type τ, $C(\tau)$ and $C^*(\tau)$ are direct summands of C. For separable groups we have:

Proposition 4: Let A be torsionfree separable. Then for each type τ, the exact sequences

$$0 \to A(\tau) \overset{i}{\to} A \overset{\eta}{\to} A/A(\tau) \to 0$$

$$0 \to A^*(\tau) \to A \to A/A^*(\tau) \to 0$$

are strongly balanced.

Proof: Let us identify $A(\tau)$ with $i(A(\tau))$. Let $a \in A(\tau)$. If $a \in D$, where D is completely decomposable and $A = D \oplus E$, then $a \in D \cap A(\tau) = D(\tau)$, a summand of D and hence of A. Then the projection $h: A \to D(\tau)$ satisfies $hi(a) = a$. Let $\bar{b} = b + A(\tau) \in A/A(\tau)$. If $b \in G$ where $A = G \oplus H$ and G is completely decomposable of finite rank, then $A/A(\tau) = \eta(G) \oplus \eta(H)$, $G = G(\tau) \oplus G'$ and $G' \simeq \eta(G') = \eta(G)$. Using the projection $A/A(\tau) \to \eta(G)$ followed by the isomorphism of $\eta(G)$ to G' one gets a $g: A/A(\tau) \to A$ which satisfies that $\eta g(\bar{b}) = \bar{b}$.

A similar argument shows that $A^*(\tau)$ is strongly balanced in A.

Proposition 5: If $0 \to B \to A \to C \to 0$ is strongly balanced exact and B, C are both separable, then A is separable.

Proof: Follows from Proposition 3.

A natural question is whether the converse of the above proposition holds, namely, must a strongly balanced subgroup of a separable torsionfree group be again separable? The classical theorem of L. Fuchs [3] says that the direct summands of torsionfree separable groups are again separable. We shall extend this to strongly balanced subgroups of A when the typeset $\mathcal{T}(A)$ satisfies the maximum condition.

THEOREM 6. Suppose $0 \to B \overset{\xi}{\to} A \overset{\eta}{\to} C \to 0$ is strongly balanced exact and A is torsionfree separable with its typeset $\mathcal{T}(A)$ satisfying the maximum condition. Then B and C are also separable.

In order to prove Theorem 6, we need the following:

Lemma 7: If $0 \to B \overset{\xi}{\to} A \overset{\eta}{\to} C \to 0$ is a strongly balanced exact sequence of torsionfree groups, then, for each type τ, all the rows of the following commutative diagram are strongly balanced exact. If furthermore A is separable, then all the columns are also strongly balanced exact:

$$\begin{array}{ccccccccc}
 & & 0 & & 0 & & 0 & & \\
 & & \downarrow & & \downarrow & & \downarrow & & \\
0 & \to & B(\tau) & \xrightarrow{\xi'} & A(\tau) & \xrightarrow{\eta'} & C(\tau) & \to & 0 \\
 & & \downarrow i'' & & \downarrow i & & \downarrow i' & & \\
0 & \to & B & \xrightarrow{\xi} & A & \xrightarrow{\eta} & C & \to & 0 \\
 & & \downarrow \delta'' & & \downarrow \delta & & \downarrow \delta' & & \\
0 & \to & B/B(\tau) & \xrightarrow{\bar\xi} & A/A(\tau) & \xrightarrow{\bar\eta} & C/C(\tau) & \to & 0 \\
 & & \downarrow & & \downarrow & & \downarrow & & \\
 & & 0 & & 0 & & 0 & &
\end{array} \qquad \ldots\ldots \quad (1)$$

Proof: Let $\bar{b} = \delta''(b) \in B/B(\tau)$. If $g: A \to B$ satisfies $g\xi(b) = b$, then δ'' vanishes on $A(\tau)$ and hence induces a $\bar{g}: A/A(\tau) \to B/B(\tau)$ such that $\bar{g}\delta = \delta''g$. It is then easy to see that $\bar{g}\,\bar{\xi}\,(\bar{b}) = \bar{b}$. Likewise, if $\bar{c} = \delta'(c) \in C/C(\tau)$ and $f: C \to A$ satisfies $\eta f(c) = c$, then δf vanishes on $C(\tau)$ and the induced homomorphism $\bar{f}: C/C(\tau) \to A/A(\tau)$ satisfies $\bar{\eta}\bar{f}\,(\bar{c}) = \bar{c}$. Thus the third row is strongly balanced.

Suppose A is separable. Let $\bar{c} = \delta'(c) \in C/C(\tau)$ and let $\bar{f}: C/C(\tau) \to A/A(\tau)$ be as in the preceding paragraph. By Proposition 4, there is an $h: A/A(\tau) \to A$ satisfying $\delta h(\bar{f}(\bar{c})) = \bar{f}(\bar{c})$. Then $\alpha = \eta h\bar{f}: C/C(\tau) \to C$ satisfies $\delta'\alpha(\bar{c}) = \bar{c}$. Likewise, suppose $x \in C(\tau)$ and $f: C \to A$ satisfies $\eta f i'(x) = i'(x)$, then there is $f': C(\tau) \to A(\tau)$ such that $fi' = if'$. Clearly $i'\eta'f'(x) = i'(x)$ and since i' is (1-1), $\eta'f'(x) = x$. Since $A(\tau)$ is strongly balanced, there is a $g: A \to A(\tau)$ such that $gi(f'(x)) = f'(x)$. Then $\alpha' = \eta'gf: C \to C(\tau)$ satisfies $\alpha'i'(x)$. Thus the third column is strongly balanced.

A similar diagram chase argument using the fact that the middle column is strongly balanced shows that the first column is also strongly balanced.

Lemma 8: If $0 \to B \xrightarrow{\xi} A \to C \to 0$ is strongly balanced exact sequence of torsionfree groups and if A is separable, then for each type τ, $\xi(B) \cap A^*(\tau) = \xi B^*(\tau)$ and $C^*(\tau)$ is pure in $C(\tau)$, and the diagram (1)* obtained from (1) by replacing $B(\tau)$, $A(\tau)$, $C(\tau)$ respectively by $B^*(\tau)$, $A^*(\tau)$, $C^*(\tau)$ has all its rows and columns strongly balanced. In particular, $(C/C^*(\tau))(\tau) = C(\tau)/C^*(\tau)$ and $(B/B^*(\tau)(\tau) = B(\tau)/B^*(\tau)$.

Lemma 9: Suppose $0 \to B \to A \to C \to 0$ is strongly balanced exact and A is torsionfree separable. Then, for each type τ, the rows and the columns of the following commutative diagram are strongly balanced exact:

$$\begin{array}{ccccccccc}
 & & 0 & & 0 & & 0 & & \\
 & & \downarrow & & \downarrow & & \downarrow & & \\
0 & \to & B(\tau)/B^*(\tau) & \to & A(\tau)/A^*(\tau) & \to & C(\tau)/C^*(\tau) & \to & 0 \\
 & & \downarrow & & \downarrow & & \downarrow & & \\
0 & \to & B/B^*(\tau) & \to & A/A^*(\tau) & \to & C/C^*(\tau) & \to & 0 \\
 & & \downarrow & & \downarrow & & \downarrow & & \\
0 & \to & B/B(\tau) & \to & A/A(\tau) & \to & C/C(\tau) & \to & 0 \\
 & & \downarrow & & \downarrow & & \downarrow & & \\
 & & 0 & & 0 & & 0 & &
\end{array}$$

Proof: Use Lemmas 7 and 8.

Proof of Theorem 6. First we shall prove that B is separable. For convenience, we identify B with $\xi(B)$ and consider it as a subgroup of A. Observe that if $B(\tau)$ is separable for each type τ in the typeset $T(B)$, then B is separable, since, by Lemma 7, $B(\tau)$ is strongly balanced in B and so, by Proposition 3, the finite rank summands of $B(\tau)$ map under the inclusion map to direct summands of B. Suppose $\sigma \in T(B)$ is maximal with respect to the property that $B(\sigma)$ is not separable. Clearly σ is not a maximal element of $T(B)$ since, otherwise, $B(\sigma) \simeq B(\sigma)/B^*(\sigma)$ which is homogeneous separable by Lemma 9. Thus $B^*(\sigma) \neq 0$. Let $b \in B^*(\sigma)$ so that $b = b_1 + \ldots + b_n$ with $b_i \in B(\sigma_i)$ and $\sigma_i > \sigma$, for $i = 1, \ldots, n$. We may also assume, without loss of generality, that the σ_i's are pairwise incomparable. By the separability of $B(\sigma_1)$, $b_1 \in D_1$ which is a finite rank completely decomposable summand of $B(\sigma_1)$ and hence, by Lemma 7 and Proposition 3, a summand of B and also of A: $A = D_1 \oplus E_1$, $B = D_1 \oplus B_1$, with $B = B \cap E_1$ and B_1 strongly balanced in E_1. Write $b_i = x_i + b_i'$ with $x_i \in D_1$ and $b_i' \in B_1(\sigma_i)$, $i = 2,\ldots,n$. Again, by the separability of $B_1(\sigma_2)$, $b_2' \in D_2$, a finite rank completely decomposable summand of $B_1(\sigma_2)$ and hence of E_1: $E_1 = D_2 \oplus E_2$, $A = D_1 \oplus D_2 \oplus E_2$, $B = D_1 \oplus D_2 \oplus B_2$ where $B_2 = B \cap E_2$ is strongly balanced in E_2 and $b_1, b_2 \in D_1 \oplus D_2$. Proceeding like this and removing, at each step, a finite rank completely decomposable summand of A, we get at the nth step a decomposition: $A = D_1 \oplus \ldots \oplus D_n \oplus E_n$, $B = D_1 \oplus \ldots \oplus D_n \oplus B_n$ and $b = b_1 + \ldots + b_n \in D_1 \oplus \ldots \oplus D_n$, a completely decomposable summand contained in $B^*(\sigma)$. Thus $B^*(\sigma)$ is separable. By Lemma 8,

$$0 \to B^*(\sigma) \to B(\sigma) \to B(\sigma)/B^*(\sigma) \to 0$$

is strongly balanced and $B(\sigma)/B^*(\sigma)$ is homegeneous separable.

Hence, by Proposition 5, $B(\sigma)$ is separable too, a contradiction. This proves that B is separable.

A similar argument shows that C is also separable. This completes the proof of Theorem 6.

Carol Walker [6] asked if a balanced subgroup of a completely decomposable group is again completely decomposable. One would wonder if this holds at least for the strongly balanced subgroups of completely decomposable groups. The following proposition gives a partial answer:

Proposition 10: If A is torsionfree completely decomposable with $T(A)$ finite, then any strongly balanced subgroup B of A is also completely decomposable.

Proof: Apply induction on $|T(A)|$. Let τ be a maximal element in $T(A)$. Then $B(\tau)$, being pure in the homogeneous completely decomposable group $A(\tau)$, is itself completely decomposable. By Lemma 7, $B/B(\tau) \cong (B + A(\tau))/A(\tau)$, is strongly balanced in $A/A(\tau)$ and hence, by induction, $B/B(\tau)$ is completely decomposable. Since, by Lemma 7 $0 \to B(\tau) \to B \to B/B(\tau) \to 0$ is strongly balanced exact, it splits. Thus B is completely decomposable.

Remark: Note that $C = A/B$ need not be completely decomposable as it is clear from the strongly balanced exact sequence $0 \to B \to F \to \Pi Z \to 0$, where F is free abelian and Z is the additive group of integers.

It is well known [3] that to each torsionfree group A there is a balanced exact sequence $0 \to B \to C \to A \to 0$ with C completely decomposable. If, further, A is separable, then clearly this sequence is strongly nice. Can it actually be strongly balanced? The next proposition answers this in the affirmative when $T(A)$ is finite. Moreover, it shows that B is completely decomposable too.

Proposition 11: Suppose A is torsionfree separable with $T(A)$ finite. If $0 \to B \xrightarrow{\xi} C \to A \to 0$ is balanced exact and C is completely decomposable, then B is completely decomposable and the sequence is strongly balanced exact.

Proof: By induction on $|T(A)|$. Let τ be a maximal type in $T(A)$. Identifying B with $\xi(B)$, we shall consider B as a subgroup of C. Since, for all σ,

$$0 \to B(\sigma) \to C(\sigma) \to A(\sigma) \to 0 \text{ is exact, } C^*(\tau) \subset B.$$

Factoring out the summand $C^*(\tau)$, we may assume that $C(\tau)$ is homogeneous so that $B(\tau)$ is a direct summand of $C(\tau)$. Consider the following commutative diagram with exact rows and columns, in which the vertical maps are all natural:

$$\begin{array}{ccccccccc} 0 \to & B & \to & C & \to & A & \to 0 \\ & \downarrow & & \downarrow & & \downarrow & \\ 0 \to & B/B(\tau) & \to & C/C(\tau) & \to & A/A(\tau) & \to 0 \\ & \downarrow & & \downarrow & & \downarrow & \\ & 0 & & 0 & & 0 & \end{array}$$

Now the bottom row is balanced exact, since the third column is strongly nice and the top row is balanced. Then, by induction, it is strongly balanced and $B/B(\tau)$ is completely decomposable. Since $B(\tau)$ is a summand of C, B is then completely decomposable. For the same reason, B is strongly balanced in C. Hence the result.

If A is a torsionfree group and B is strongly balanced in A, then, for each type τ, $B \cap \langle A^*(\tau) \rangle_* = \langle B^*(\tau) \rangle_*$, where $\langle S^*(\tau) \rangle_*$ denotes the pure subgroup generated by $S^*(\tau)$. A balanced subgroup may not have this property as the following example shows:

<u>Example</u>: Let A be a torsionfree strongly indecomposable group of rank 2 with typeset $= \{\tau_0, \tau_1, \tau_2, \tau_3\}$ where $\tau_0 = \inf\{\tau_1, \tau_2, \tau_3\}$ and the τ_i, $i = 1,2,3$ are pairwise non-comparable. Furthermore, let $A = R_1 a_1 + R_2 a_2 + R_3 a_3$ with $\chi(a_i) = \tau(R_i) = \tau_i$, $i = 1,2,3$ and $a_3 = a_1 + a_2$. (See, for eg., R. A. Beaumont and R. S. Pierce, Torsionfree groups of rank 2, Memoirs of Amer. Math. Soc. No. 38(1961) or D. Arnold's unpublished Lecture Notes on torsionfree groups of finite rank.)
If $C = R_1 x_1 \oplus R_2 x_2 \oplus R_3 x_3$ with $\chi(a_i) = \tau(R_i) = \tau_i$, $i = 1,2,3$ then the map $\Sigma r_i x_i \to \Sigma r_i a_i$ is a balanced epimorphism $C \to A$. The kernel B is a rank 1 group containing $x_1 + x_2 - x_3$ and hence has type τ_0. Clearly $B \cap C^*(\tau_0) \neq \langle B^*(\tau_0) \rangle_*$ since $C = C^*(\tau_0)$ and $B^*(\tau_0) = 0$.

<u>Definition</u>: A balanced subgroup B of a torsionfree group A is said to be <u>*-balanced</u> if, for each type τ, $B \cap \langle A^*(\tau) \rangle_* = \langle B^*(\tau) \rangle_*$. An exact sequence $0 \to B \xrightarrow{\alpha} A \to C \to 0$ is *-balanced if $\alpha(B)$ is a *-balanced subgroup of A.

We wish to investigate the connection between the *-balanced subgroups of completely decomposable groups and the Butler groups [2] which are pure subgroups of finite rank torsionfree completely decomposable groups. D. Arnold [1] and L. Fuchs (unpublished) noted that, to each Butler group A, there is a balanced exact sequence $0 \to B \to C \to A \to 0$ where C is completely decomposable of finite rank. When can this be *-balanced? This is answered below. Here, by an almost completely decomposable group we mean a finite rank torsionfree group containing a completely decomposable subgroup of finite index. Clearly, these groups are Butler groups.

<u>Proposition 12</u>: If A is torsionfree almost completely decomposable, then any balanced exact sequence $0 \to B \xrightarrow{i} C \to A \to 0$ with C completely decomposable is *-balanced.

Proof: Let us identify B with i(B) and consider it as a subgroup of C and that $A = C/B$. Let $S \supset B$ be such that $S' = S/B$ is a completely decomposable subgroup of A with $[A: S'] = n$, say. Then $[C: S] = n$ and S splits, $S = B \oplus T$ with $T \simeq S'$. Since C/S is finite, any $x \in S$ has the same type in S as in C. Hence $S \cap C(\tau) = S(\tau)$, for all types τ. Let $x \in C^*(\tau)$, so that $x = x_1 + \dots x_k$, where $x_i \in C(\sigma_i)$, $\sigma_i > \tau$, $i = 1,\dots,k$. Then $nx_i \in S \cap C(\sigma_i) = S(\sigma_i)$, $i = 1,\dots,k$ and so $nx \in S^*(\tau)$. Thus $S \cap C^*(\tau) = \langle S^*(\tau) \rangle_*$, for all types τ. Then B, being a summand of S, is $*$-balanced in C.

The next result explores the connection between the $*$-balanced subgroups and the B_0 groups, which are the Butler groups A with the property that $A^*(\tau)$ is pure in A for all types τ (see [1]).

Proposition 13: Let C be torsionfree completely decomposable group of finite rank. If B is $*$-balanced in C and if C/B is a B_0-group, then B is a direct summand of C.

Proof: Let τ be a maximal element in the typeset of B, so that $B(\tau)$ is homogeneous of type τ and we get $B \cap C^*(\tau) = \langle B^*(\tau) \rangle_* = 0$. Since C/B is a B_0-group and B is balanced, $(B(\tau) + C^*(\tau))/B(\tau) = (B/C)^*(\tau)$ is pure in B/C and so $B(\tau) \oplus C^*(\tau)$ is pure and hence a summand of $C(\tau)$. (This fact was noted in [1]). Thus $C = B(\tau) \oplus C_1$ and $B = B(\tau) \oplus B_1$, where $B_1 = B \cap C_1$. Now B_1 is $*$-balanced in C_1 and the induction on the typeset gives that B_1 and hence B is a direct summand.

From Propositions 12 and 13 we get

Corollary 14 ([1]): If a B_0-group is almost completely decomposable then it is completely decomposable.

REFERENCES

[1] D. Arnold, Pure subgroups of finite rank completely decomposable torsionfree abelian groups. (This publication)

[2] M.C.R. Butler, A class of torsionfree abelian groups of finite rank, Proc. London Math. Soc. (3) 15 (1965), 680 - 698.

[3] L. Fuchs, Infinite Abelian Groups, Vol. II, Pure and Appl. Math., Vol. 36, Academic Press, New York, 1973.

[4] K. M. Rangaswamy, An aspect of purity and its dualisation in abelian groups and modules, Symposia Math. 23 (1979), 307 - 320.

[5] K. M. Rangaswamy, The theory of separable mixed abelian groups (to appear).

[6] C. L. Walker, Projective classes of completely decomposable abelian groups, Arch. Math. 23 (1972), 581 - 588.

Abelian Groups Finitely Generated Over Their Endomorphism Rings

J. D. Reid[1]

1. Introduction. This paper concerns the study of those abelian groups that are finitely generated over their endomorphism rings (or, as we shall say, finitely E-generated groups). While a torsion group is finitely E-generated if and only if it is bounded, this class of groups is otherwise quite extensive. Thus, if R is any ring (with identity) and M is any finitely generated R-module, then the underlying group, M^+, is finitely E-generated. This is true in particular of the additive group, R^+, of R itself. In a slightly different spirit, any abelian group that has Z as a summand is actually cyclic as a module over its endomorphism ring, hence is finitely E-generated. It is our intention, however, to concentrate on torsion free groups of finite rank that are finitely E-generated so the word group should be interpreted from now on to mean torsion free abelian group of finite rank.

By way of setting a context we point out first the fact that there has been a considerable amount of recent work done studying groups G viewed as modules over their endomorphism rings E. Groups that are projective, injective, quasi-projective, quasi-pure injective, pure injective, flat, etc. over their endomorphism rings have been and are being studied, but to our knowledge no one has yet studied the class of finitely E-generated groups, as such, in a systematic way. We feel that, in this context, such a program is a natural one. Moreover, these groups have begun to surface in the investigation of various problems. Thus any torsion free group of finite rank that is projective over its endomorphism ring is finitely E-generated ([2]). The theory of strongly homogeneous groups recently investigated by Arnold ([1]) can be made to depend on the finitely E-generated case and when this is done one obtains a very natural, and more general, exposition of this theory (cf. [9]). As a final example, the additive structure of fractional ideals of (arbitrary) subrings of algebraic number fields is, as we shall see in Section 3, precisely that of the irreducible groups that are finitely E-generated.

To a considerable extent our work here is based on [10], a few results from which are summarized in Section 2 for the reader's convenience, and also to establish some useful notation. In addition, on two occasions we make use of a very basic result of Beaumont and Pierce ([4]) - the analogue for torsion free rings of the Wedderburn Principal Theorem. We assume familiarity with the notions of quasi-isomorphisms, quasi-decompositions, etc. as set forth in these two references, or in the book of Fuchs [6], which is also our general reference. Finally, one remark about notation. A dot placed above the usual symbol indicates the corresponding

[1] Partially supported by the National Science Foundation, Grant No. MCS8004456.

quasi-concept. Thus, for example, $G \doteq H$ means that G is quasi-equal to H.

Each section contains its own brief introduction, to which we refer the reader for an idea of the contents of the paper. We find it interesting that everything seems to reduce, in a quite explicit way, to irreducible groups (definition recalled below). We might remark also that we regard this paper as laying foundations and we intend to develop some of these topics in greater detail, and with applications, elsewhere.

2. Representing Irreducible Groups. Quasi-isomorphisms, quasi-decompositions, etc. refer to isomorphism and decompositions respectively in the category whose objects are torsion free abelian groups and in which the morphisms from A to B are the elements of $Q \otimes \mathrm{Hom}(A, B)$. From this point of view, perhaps the simplest groups are the irreducible groups:

Definition [10]. We say that the group G is irreducible if $Q \otimes G$ is an irreducible module over $Q \otimes E$.

In this definition, and in what follows, E is the endomorphism ring of G. Clearly G is irreducible if and only if G has no proper pure fully invariant subgroup. We recall the following

Theorem 0 [10]. For a torsion free group G of finite rank, the following are equivalent:

1) G is irreducible.
2) $G \doteq H^m$ with H irreducible and strongly indecomposable.
3) $Q \otimes E = D_m$, the $m \times m$ matrices over D, where D is a division ring and $m[D:Q] = \mathrm{rank}\, G$.

It follows that G is irreducible and strongly indecomposable if and only if $Q \otimes E$ is a division algebra whose dimension over Q is the rank of G. Suppose that G is such a group and choose $x \in G$, $x \neq 0$. Then Ex is a fully invariant subgroup of G, hence is full in G in the (standard) sense that the quotient is torsion. Thus for $y \in G$ there exist $\alpha \in E$ and $n \in Z$, $n \neq 0$, such that $ny = \alpha x$. We may therefore define a map $\hat{x}$, easily seen to be well defined, from G to $Q \otimes E$ by $\hat{x}(y) = \alpha/n$. Thus, $\hat{x}$ is defined by the condition

$$\hat{x}(y)x = y, \quad y \in G.$$

It is clear that $\hat{x} \in \mathrm{Hom}_E(G, Q \otimes E)$ and is monic. Moreover, if $w \in G$, $w \neq 0$, then $\hat{x}$ and $\hat{w}$ are related by

$$\hat{x}(w)\hat{w} = \hat{x}.$$

This yields $\hat{x}(G) \doteq \hat{w}(G)$ so this E-submodule of $Q \otimes E$ is unique up to quasi-equality. Clearly $E \subseteq \hat{x}(G)$ and E is the set of all elements of $Q \otimes E$ that take $\hat{x}(G)$ into itself under left multiplication.

Conversely, if D is any division algebra over Q and G is a subgroup of D such that the ring

$$R = \{\alpha \in D \mid \alpha G \subseteq G\}$$

is full in D (i.e., contains a Q-basis of D), then G is irreducible ([10]); indeed any R-submodule of D is full in D. Such a group G of course may fail to be strongly indecomposable and is in fact strongly indecomposable if and only if the endomorphisms of G are given by left multiplication by elements of R. This follows easily from Theorem 0. Now let

$$S = \{\alpha \in D \mid G\alpha \subseteq G\}.$$

The elements of S induce endomorphisms of G so if G is strongly indecomposable, right multiplication by an element of S is just left multiplication by some element of R. It follows that S is the intersection of R with the center of D. We summarize these remarks in the following, which we will sometimes refer to as the Representation Theorem.

<u>Theorem 1</u> (Representation Theorem). If G is strongly indecomposable and irreducible with endomorphism ring E, then $D = Q \otimes E$ is a division ring and G is E-isomorphic to an E-submodule of D containing E. Conversely any strongly indecomposable subgroup G of a division algebra D over Q with full left order $R = \{\alpha \in D \mid \alpha G \subseteq G\}$ is irreducible and R is its endomorphism ring. For such subgroups G of D, the ring $S = \{\alpha \in D \mid G\alpha \subseteq G\}$ is the intersection of R with the center of D.

3. <u>Finitely E-generated Irreducible Groups</u>.

We get more properly into our subject in this section by characterizing first the finitely E-generated irreducible groups. For this we need

<u>Definition</u> [11]. We say that the group G is <u>strongly irreducible</u> if G is quasi-equal to each of its non-zero fully invariant subgroups; equivalently for groups of finite rank, each non-zero fully invariant subgroup of G has finite index in G.

<u>Theorem 2</u>. Let G be irreducible. Then G is finitely E-generated if and only if G is strongly irreducible. In this case, $Q \otimes E$ is isomorphic to a matrix algebra over a field.

<u>Proof</u>: From the structure theorem for irreducible groups (Theorem 0), the group G is quasi-equal to a group H^m where H is strongly indecomposable and irreducible. It is easy to see that G is strongly irreducible if and only if H is, and that G is finitely E-generated if and only if H is. Moreover the division ring D in Theorem 0 is just the algebra of quasi-endomorphisms of H.

It follows from these remarks that it suffices to show that a strongly indecomposable irreducible group, H, is strongly irreducible if and only if it is finitely

E-generated and that in this case its division algebra D of quasi-endomorphisms is commutative. Furthermore, by the representation theorem we may assume that $R \subseteq H \subseteq D$ and every endomorphism of H is given by multiplication by elements of R. Assume then that H is strongly irreducible. Then R is a fully invariant subgroup of H so that $nH \subseteq R$ for some integer n, $n \neq 0$. It is clear from this that H is finitely generated over R, hence is finitely E-generated. Now given $x, y \in H$ we have $x(ny) = (nx)y \in Ry \subseteq G$ so that $ny \in S = \{\alpha \in D \mid G\alpha \in G\}$. Since y is arbitrary in G we have $nG \subseteq S$ so that S is a full subring of D. But, as we have seen, S is commutative and it follows that D is commutative as well.

Conversely if H is irreducible, strongly indecomposable and finitely E-generated, generated over its endomorphism ring by $x_1, \ldots, x_k$, say, then these same elements generate H over R. Since R is full in H there exists an integer n, $n \neq 0$, such that $nx_i \in R$ for each $i = 1, \ldots, k$. Clearly then $nH \subseteq R$. Now if L is any fully invariant subgroup of H then nL is a left ideal of R, non-zero if L is non-zero. Since R is a full subring of a division ring, it is easy to see, and standard by now, that $mR \subseteq nL$ for some integer $m, m \neq 0$. Hence $mnH \subseteq L$. Thus H is strongly irreducible.

These strongly irreducible groups were introduced in [11] where it was established that they were, among the strongly indecomposable groups, exactly the class of groups that admit non-nilpotent (associative) ring structures. It had been observed earlier ([3], [10]), though not in these terms, that the additive groups of subrings of algebraic number fields are strongly irreducible. The additive groups of these rings - or, better, of fractional ideals over subrings (with identity) of algebraic number fields - were characterized by Beaumont and Pierce in [3]. We can give, from our point of view, a different and very simple characterization of these groups. Recall that if R is a subring of the algebraic number field F, then a <u>fractional ideal</u> over R is a non-zero additive subgroup of F that is finitely generated over R.

<u>Theorem 3</u>. A torsion free group of finite rank is isomorphic to the additive group of a fractional ideal of a subring of an algebraic number field if and only if it is irreducible and finitely E-generated.

<u>Proof</u>: It is obvious that fractional ideals have the indicated additive structure. Conversely suppose that G is irreducible and finitely E-generated. Then, according to our previous results, we have $nG \subseteq H^m \subseteq G$ for some integers $n \neq 0$ and $m \geq 1$, and some strongly indecomposable strongly irreducible group H. Moreover the algebra of quasi-endomorphisms of H is a field F and we may assume that $R \subseteq H \subseteq F$ where R is a subring of F and H is a finitely generated R-module. Thus H is a fractional ideal of R, which proves the theorem in the strongly in-

decomposable case. In the general case it suffices to work with H^m itself since if $nG \subseteq H^m \subseteq G$ and H^m is a fractional ideal for the ring S, then G is a fractional ideal for the subring $nS + Z$ of S.

To treat the case of H^m, then, we use the fact that $R \subseteq H \subseteq F$ with H finitely generated over R. Let K be any extension of F of degree m over F. Since R is full in F we may, following Beaumont and Pierce now, choose an F-basis $1 = x_1, x_2, \ldots, x_m$ of K such that $x_i x_j = \Sigma c_{ijk} x_k$ with $c_{ijk} \in R$. Then $S = \Sigma R x_j$ is a full subring of K and $\Sigma H x_j$, which is isomorphic to H^m, is a fractional ideal for S.

Among the fractional ideals of subrings of algebraic number fields of course are the subrings themselves. There is a result - to our mind a beautiful result - of Beaumont and Pierce that says that the subrings of algebraic number fields cluster around the integrally closed rings. These authors used their theory of q.d. invariants in establishing this result. Since we need to use this result in the next section we give here a simple direct proof, which may therefore be of some interest.

Theorem 4 (Beaumont and Pierce [3]). Every quasi-equality class of full subrings of an algebraic number field contains a unique integrally closed ring - the common integral closure of every ring in the class.

Proof: We assume our rings R, S contain 1. (In any event, adjunction of the identity would not change the quasi-equality class.) Note that if $S \doteq R$ then $R \subseteq RS \doteq R$. Thus if R and S are integrally closed and quasi-equal then they are each quasi-equal to RS. But the ring RS is finitely generated as module over R (and over S) and integral closure now forces $R = RS = S$. Hence there is at most one integrally closed ring in each quasi-equality class. On the other hand, if R is any full subring of the algebraic number field F and J is the ring of integers in F, then J is finitely generated as Z-module so $R \doteq RJ$. But $J \subseteq RJ$ so, as is well known, RJ is integrally closed.

4. Split Irreducible Groups. Invariants.

In this section we construct invariants for a class of groups that includes the irreducible finitely E-generated groups. These invariants determine the group up to quasi-isomorphism and are analogous to the types of rank 1 groups. They are equivalent to the types in the rank 1 case.

First, let G be a full subgroup of the algebraic number field F with full "order" $R = \{\alpha \in F \mid \alpha G \subseteq G\}$. By Theorem 4, R is quasi-equal to a unique integrally closed ring R' and then G is quasi-equal to the R'-module $R'G$. Hence we lose nothing up to quasi-isomorphism by assuming that R is integrally closed, hence Dedekind, to begin with and that the ring, J, of integers of F is contained in R. Now for each prime ideal P of J, denote the P-adic valuation on F by v_P and put

$$v_P(G) = \inf\{v_P(x) \mid x \in G\}.$$

Then we denote by $v(G)$ the sequence $\langle v_P(G)\rangle$, indexed by the primes of J, whose entry at P is $v_P(G)$. It is easy to check that $-\infty \leq v(G) \leq 0$ holds for almost all P. Ribenboim [12] has shown that there is a bijection between functions f from the primes of J to $Z \cup \{-\infty\}$, satisfying $f(P) \leq 0$ for almost all P, and J-submodules of F. In particular, G can be recovered from $v(G)$:

(*) .... $G = \{x \in F \mid v_P(x) \geq v_P(G)\}$.

Our interest, however, is not so much in G as J-module, but rather as Z-module. In addition we are viewing G as a representative of a quasi-equality class, so we proceed a bit further. Given a function f, as above, and an automorphism $s : x \longrightarrow x^s$ of the field F, s maps J onto itself and permutes the primes of J in finite batches. We define f^s by $f^s(P) = f(P^s)$. If f and g are two such functions we define $f \sim g$ provided that $f(P) = g(P)$ for almost all primes P and for all P for which either value is $-\infty$. It is easy to see that automorphisms of F are compatible with this relation so that if [f] denotes the equivalence class of f, we may define $[f]^s = [f^s]$ for an automorphism s. We observe also that for quasi-equal groups G and G', we have $v(G) \sim v(G')$.

Definition. Let G be an irreducible group with endomorphism ring E. We will say that G is split if $Q \otimes E$ is a matrix algebra over a field F. The field F will be called a field of definition of G.

We view the field of definition as being defined only up to isomorphism (which it certainly is) since quasi-isomorphic split groups have isomorphic fields of definition which we will sometimes want to identify with each other. As we have seen, irreducible finitely E-generated groups are split but split groups need not be finitely E-generated as, for example, the rank 1 groups show. It follows from Theorem 0 that an irreducible group G is split if and only if $G \doteq H^m$ where H is strongly indecomposable and the algebra of quasi-endomorphisms of H is a field, namely a field of definition, F, of G. By the representation theorem we may assume $R \subseteq H \subseteq F$ where R is a subring of F and multiplication by elements of R gives the action of the endomorphisms of H. Then we can compute $[v(H)]$ as described above.

Definition. Let G be a split irreducible group and let H be a strongly indecomposable quasi-summand of G. Then the equivalence class $[v(H)]$ is called the invariant for G. We will denote the invariant for G by inv(G).

It is clear that the invariant for G is well defined and that quasi-isomorphic split irreducible groups have the same invariant. We can now prove

Theorem 5. Let G and G' be split irreducible groups then G and G' are quasi-isomorphic if and only if they have equal ranks, a common field of definition F and, for some automorphism s of F, $\mathrm{inv}(G)^s = \mathrm{inv}(G')$.

Proof: It suffices to prove the theorem in case G and G' are strongly indecomposable since any irreducible group is determined up to quasi-isomorphism by its rank and the quasi-isomorphism class of its strongly indecomposable summands. Hence we suppose first that G and G' are strongly indecomposable, irreducible and split, and that they are quasi-isomorphic. Then it is obvious that they have equal ranks and a common field of definition. Moreover, in this case, $\mathrm{inv}(G) = [v(G)]$ and $\mathrm{inv}(G') = [v(G')]$ and since these are invariant under quasi-equality, we may assume that G and G' are actually isomorphic under a map φ, say.

We now choose $x \in G$, $x \neq 0$, and put $z = \varphi(x) \in G'$. These provide embeddings of G and G' into F as described in the representation theorem. We denote these maps by σ and ρ respectively (rather than by $\hat{x}$, $\hat{z}$ as before). The endomorphisms of G correspond to multiplication of elements of $\sigma(G)$ by elements of a certain subring R of F and, similarly, the endomorphism ring of G' is isomorphic to a subring R' of F. Hence R and R' are isomorphic rings and since F is the field of fractions of each of these, this isomorphism is induced by some automorphism s of F. Finally, for any automorphism s of F, prime ideal P of the ring of integers of F and element $a \in F$, we have $v_{P^s}(a^s) = v_P(a)$. Since $\rho(G') = \sigma(G)^s$, this yields $\mathrm{inv}(G)^s = \mathrm{inv}(G')$ as required.

Conversely assume that G and G' have equal ranks, field of definition F, and that $\mathrm{inv}(G)^s = \mathrm{inv}(G')$ for some automorphism s of F. Since $\mathrm{inv}(G)^s = \mathrm{inv}(G^s)$ and G^s is isomorphic to G, we may restrict ourselves to the identity automorphism, i.e., $\mathrm{inv}(G) = \mathrm{inv}(G')$. Furthermore, we will identify G with $\sigma(G)$, G' with $\rho(G')$ and thus assume the situation $R \subseteq G \subseteq F$, $R' \subseteq G' \subseteq F$ as usual.

Now let $P_1, \ldots, P_t$ be the finite set of primes for which $v_P(G) \neq v_P(G')$. Then neither value is infinite and we can choose an integer $n \neq 0$ such that

$$v_{P_i}(n) \geq v_{P_i}(G') - v_{P_i}(G),\ i = 1, \ldots, t .$$

Since $v_P(nG) = v_P(n) + v_P(G)$ for any P we now have $v_P(nG) \geq v_P(G')$ for all P. Hence, by (*), $nG \subseteq G'$. Similarly $n'G' \subseteq G$ for some integer n'. This proves the theorem.

We intend to study split irreducibles in detail elsewhere, since as we noted above, they need not be finitely E-generated and that is our topic here. However we might note in passing that there are examples of split irreducible groups that are not finitely E-generated other than the rank 1 groups. Indeed let F be any algebraic number field with ring of integers J. Then by a result of Zassenhaus [15] (see also Butler [5]) there is a group G with rank equal to $[F : Q]$ and with J as endomorphism ring. Clearly G is split irreducible and in fact has F as field of definition. However G is not finitely E-generated for then it would be quasi-equal to J and hence quasi-decomposable, contrary to the fact that its endomorphism ring has no zero divisors. Incidentally, it is also the case, as follows easily from results in [3], that every algebraic number field occurs as field of definition for some split irreducible group that is finitely E-generated.

5. Strongly Indecomposable Groups. In this section we introduce an invariant for finitely E-generated groups which we hope will prove of interest. Our immediate use for this idea is in identifying the strongly indecomposable finitely E-generated groups. This identification also follows from the results of the next section, but we believe that the direct approach here has its merits. In any case, it serves as an illustration of the usefulness of the invariant.

Definition. Let G be a torsion free group that is finitely E-generated. Denote by $\rho(G)$ the least integer n such that there exists a group G' that is quasi-isomorphic to G and that has a set of n generators over its endomorphism ring.

In making this definition we note that the property of being finitely E-generated is a quasi-isomorphism invariant. It follows from results in [2] that if G has finite rank and is projective over its endomorphism ring then $\rho(G)$ is defined and equal to 1. It is not hard to show that, in general, $\rho(G)$ is less than or equal to the rank of G for any finitely E-generated group G.

Lemma. Let G be finitely E-generated with $\{g_1, \dots, g_n\}$ a set of generators, and put

$$L = \{(\lambda_1, \dots, \lambda_n) \in E^n \mid \Sigma \lambda_j g_j = 0\}.$$

Suppose $(\lambda_1, \dots, \lambda_n) \in L$ with some λ_i monic. Then $G \doteq \sum_{j \neq i} Eg_j$.

Proof: To be definite, suppose that λ_1 is monic. Then λ_1 is a quasi-automorphism - i.e., there exists $\lambda \in E$ such that $\lambda\lambda_1 = \lambda_1\lambda = t$, t a non-zero integer. Then

$$0 = \lambda_1 g_\lambda + \dots + \lambda_n g_n$$

yields

$$0 = tg_1 + \dots + \lambda\lambda_n g_n$$

so that $tg_1 \in \sum_{j \neq 1} Eg_j$. Clearly then $tG \subseteq \sum_{j \neq 1} Eg_j$.

We use this now to determine the strongly indecomposable finitely E-generated groups. We will state the result in terms of what are called E-rings. These rings were named and studied by Schultz ([13], see also [14]). Their theory, in the finite rank torsion free case, is intimately entwined with the theory of those groups that are projective over their endomorphism rings ([7], [2]). We recall the definition: A ring, associative and with an identity, is an E-ring if the left regular representation is an isomorphism of the ring onto the ring of all endomorphisms of its underlying additive group.

Theorem 6. The following are equivalent for the group G.

(1) G is strongly indecomposable and finitely E-generated.

(2) G is quasi-isomorphic to the additive group of a strongly indecomposable E-ring.

(3) G is strongly indecomposable and strongly irreducible.

If these conditions hold then $\rho(G) = 1$ and the E-ring in question is the endomorphism ring of G.

Proof: Suppose G is strongly indecomposable and finitely E-generated and let $\rho(G) = n$. Then up to quasi-isomorphism we may assume that G has n generators over its endomorphism ring E and that any group quasi-isomorphic to G needs at least n generators over its endomorphism ring. If $\{g_1, \ldots, g_n\}$ generate G and L is defined as in the lemma above, then we have an exact sequence of E-modules

$$0 \longrightarrow L \longrightarrow E^n \xrightarrow{\varphi} G \longrightarrow 0$$

where $\varphi(\alpha_1, \ldots, \alpha_n) = \Sigma\, \alpha_j g_j$. From the lemma and the fact that $\rho(G) = n$ it follows that, if $(\lambda_1, \ldots, \lambda_n) \in L$ then no λ_j is monic. Since G is strongly indecomposable, every endomorphism of G is either monic or is nilpotent ([10]). Thus if N denotes the nil radical of E, then $L \subseteq N^n$.

Now by a fundamental result of Beaumont and Pierce [4], N is a quasi-summand of E so for some S we have $E \doteq N \oplus S$ (group direct sum). Therefore $E^n \doteq N^n \oplus S^n$ and since $L \subseteq N^n$ we now have

$$G \approx E^n/L \doteq S^n \oplus N^n/L .$$

But G is strongly indecomposable and $S \neq 0$, so we conclude that $n = 1$ and $L = N$. In particular, $\rho(G) = 1$. Again invoking strong indecomposability of G, it follows that E/N is a full subring of a division ring, hence, additively, is an irreducible group ([10]). Since G is isomorphic to E/N, G is irreducible. This in turn implies that $N = 0$ after all so G is isomorphic to E under φ. This yields the fact that E is an E-ring and we have shown that (1) implies both (2) and (3). It is obvious that (2) implies (1) and by Theorem 2, (3) implies (1). This completes the proof.

This theorem yields the fact that any finitely E-generated group G with $\rho(G) > 1$ has non trivial quasi-decompositions. We have not encountered any finitely E-generated groups that are not quasi-decomposable into a direct sum of groups with $\rho = 1$. Along these lines we have the following:

Theorem 7. Let G be finitely E-generated. Then $\rho(G) \leq$ rank G and equality holds if and only if G is quasi-equal to a direct sum of rank 1 groups of incomposable idempotent types.

Proof: Let $\rho(G) = n$. Since all the properties mentioned in the theorem are invariant under quasi-equality, we may assume that G has a system $\{g_1, \ldots, g_n\}$ of n generators over its endomorphism ring E. Consider the system of pure subgroups

$$\langle Eg_1\rangle_* \subseteq \langle Eg_1 + Eg_2\rangle_* \subseteq \ldots \subseteq \langle Eg_1 + \ldots + Eg_n\rangle_* = G.$$

(Here, $\langle X\rangle_*$ denotes the pure subgroup generated by the set X.) If $g_{i+1} \in \langle Eg_i + \ldots + Eg_n\rangle_*$ then $\{g_1, \ldots, g_i, g_{i+2}, \ldots, g_n\}$ generate a full E-submodule of G. But it is clear that any full E-submodule of a finitely E-generated group is quasi-equal to G and we would violate $\rho(G) = n$. Hence all the terms in the sequence of pure subgroups above are distinct. This shows that $\rho(G) \leq$ rank G.

Now if $\rho(G)$ = rank G then the fact that all terms in the sequence above are distinct implies in particular that $\langle Eg_1\rangle_*$ has rank 1, and $Eg_1 \not\subseteq \langle Eg_2 + \ldots + Eg_n\rangle_*$. Since it is clear that $G = Eg_1 + \langle Eg_2 + \ldots + Eg_n\rangle_*$ we have in fact $G = Eg_1 \oplus \langle Eg_2 + \ldots + Eg_n\rangle_*$. These two summands being E-submodules, this is a module decomposition. Therefore $\langle Eg_2 + \ldots + Eg_n\rangle_* = H$, say, is an image of G under an E-map, so H is finitely E-generated. Moreover every endomorphism of H is induced by one of G so, since $\rho(G) = n$ we have $\rho(H) \geq n - 1$. On the other hand, rank H = n - 1 so we conclude that $\rho(H) = n - 1$ = rank H. By induction, H is quasi-equal to a direct sum of rank 1 groups of incomparable types and it is clear that these types are incomparable with the type of Eg_1 as well. The converse is clear.

6. Structure of Finitely E-generated Groups. In this section we obtain the general structure theory of finitely E-generated groups. For the first result we need the full force of the result of Beaumont and Pierce [4] mentioned above. This theorem states that any torsion free ring of finite rank is quasi-equal, as additive group, to a direct sum of its nilradical N and a subring S, where $Q \otimes S$ is a semi-simple algebra. The following may be viewed in a sense as an extension of this theorem.

Theorem 8. Let G be finitely E-generated with endomorphism ring E and let N be the nilradical of E. Let $\langle NG\rangle_*$ be the pure subgroup of G generated by NG. Then there is a quasi-decomposition

$$G \doteq \langle NG\rangle_* \oplus H$$

where H is a subgroup of G isomorphic to $\overline{G} = G/\langle NG\rangle_*$. The endomorphism ring of $\overline{G}$ is quasi-equal to the ring of endomorphisms induced on $\overline{G}$ by E, and has nilradical zero. Finally, $\overline{G}$ is finitely E-generated.

Proof: We write $E \doteq N \oplus S$, as described above, and view G as S-module. Then $\bar{G}$ is also an S-module and it is clear that $\bar{G}$ is finitely generated over S. Moreover, since S has nilradical zero it is easy to see that S is Noetherian. Hence, by a standard change of rings formula we may identify $Q \otimes \mathrm{Hom}_S(\bar{G}, S)$ and $\mathrm{Hom}_{Q\otimes S}(Q \otimes \bar{G}, Q \otimes S)$. Now $Q \otimes S$ is semi-simple Artinian so its module $Q \otimes \bar{G}$ is projective and clearly is finitely generated. Thus we have a finite dual basis $\{\psi_i, w_i\}$ for $Q \otimes \bar{G}$: i.e., $\psi_i \in \mathrm{Hom}_{Q\otimes S}(Q \otimes \bar{G}, Q \otimes S)$, $w_i \in Q \otimes \bar{G}$ and for any $w \in Q \otimes \bar{G}$ we have $w = \Sigma \psi_i(w) w_i$. Now there exist $m_i \in Z$ such that $m_i w_i \in \bar{G}$ and, by change of rings, there exist $\ell_i \in Z$ such that $\ell_i \psi_i \in \mathrm{Hom}_S(\bar{G}, S)$. Hence there exists a fixed $k \in Z$ such that $kw_i = x_i \in \bar{G}$ and $k\psi_i = \varphi_i \in \mathrm{Hom}_S(\bar{G}, S)$ for all i. Moreover, we have, for each $x \in \bar{G}$,

$$k^2 x = \Sigma k \psi_i(x)(kw_i) = \Sigma \varphi_i(x) x_i .$$

Thus $\bar{G}$ is "almost projective" over S and, just as in the case of projective modules, it follows that the S-exact sequence

$$0 \longrightarrow \langle NG \rangle_* \longrightarrow G \xrightarrow{\nu} \bar{G} \longrightarrow 0$$

almost splits in the sense that there exists an S-map $\bar{\varphi} : \bar{G} \longrightarrow G$ such that $\nu\bar{\varphi}$ is multiplication by k on $\bar{G}$. Taking $H = \bar{\varphi}(\bar{G})$ we then have $G \doteq \langle NG \rangle_* \oplus H$ as desired. The proofs of the remaining statements are routine.

In the next theorem we will get a satisfactory description of the term H (equivalently, $\bar{G}$) in the above decomposition. On the other hand, little can be said in general about the quasi-summand $\langle NG \rangle_*$. For example, let $G = Z \oplus M$ where M is strongly indecomposable and, say, not finitely E-generated. It is easy to see that, if N is the nilradical of the endomorphism ring of G then $NG = \langle NG \rangle_* = M$. It follows in particular that $\langle NG \rangle_*$ need not be finitely E-generated even in the case where G is E-cyclic, as here. At any rate, we have

Theorem 9. If G is finitely E-generated and its endomorphism ring E has nilradical zero then G is quasi-equal to a direct sum of strongly irreducible groups G_i such that $\mathrm{Hom}(G_i, G_j) = 0$ if $i \neq j$. Conversely any group with this structure is finitely E-generated and its endomorphism ring has nilradical zero.

Proof: Let G be finitely generated over E and suppose E has nilradical zero. It is clear that if $mG \subseteq A \oplus B \subseteq G$ with A and B fully invariant in G then $\mathrm{Hom}(A, B) = 0$, and A and B are finitely generated over E, hence over their own endomorphism rings. Therefore we may assume that G is strongly indecomposable over E, in the obvious sense, and we must show then that G is strongly irreducible. Since G is finitely E-generated it suffices, by Theorem 2, to show that G is irreducible.

Now since the nilradical of E is zero, the algebra $Q \otimes E$ is semi-simple and $Q \otimes E$ contains no central idempotents since G does not quasi-decompose as E-module. Thus $Q \otimes E$ is simple and its module $Q \otimes G$ is completely reducible.

However, by change of rings, $Q \otimes G$ is indecomposable over $Q \otimes E$. We conclude that $Q \otimes G$ is an irreducible $Q \otimes E$ module so that, by definition, G is an irreducible group.

The converse is clear.

REFERENCES

1. D. M. Arnold, Strongly homogeneous torsion free abelian groups of finite rank, Proc. A. M. S., 56 (1976), 67-72.
2. D. M. Arnold, R. S. Pierce, J. D. Reid, C. Vinsonhaler, W. Wickless, Torsion free abelian groups of finite rank projective as modules over their endomorphism rings, J. Algebra, to appear.
3. R. A. Beaumont and R. S. Pierce, Subrings of algebraic number fields, Acta. Sci. Math. Szeged., 22 (1961), 202-216.
4. ______________________, Torsion free rings, Ill. J. Math., 5 (1961), 61-98.
5. M. C. R. Butler, On locally free torsion-free rings of finite rank, J. London Math. Soc., 43 (1968), 297-300.
6. L. Fuchs, Infinite Abelian Groups, Vol. II, Academic Press, New York, 1973.
7. G. P. Niedzwicki and J. D. Reid, Torsion free abelian groups cyclic projective over their endomorphism rings, to appear.
8. R. S. Pierce, Subrings of simple algebras, Michigan Math. J., 7 (1960).
9. J. D. Reid, Abelian groups cyclic over their endomorphism rings, to appear.
10. ________, On the ring of quasi-endomorphisms of a torsion free group, Topics in Abelian Groups (Proc. Sympos. New Mexico State University, 1962), Scott, Foresman, Chicago, Ill., 1963, pp. 51-68.
11. ________, On rings on groups, Pac. J. Math., 53 (1974), 229-237.
12. P. Ribenboim, Modules sur un anneau de Dedekind, Summa Brasiliensis Math., 3 (1952), 21-36.
13. P. Schultz, The endomorphism ring of the additive group of a ring, J. Austral. Math. Soc., 15 (1973) 60-69.
14. P. Schultz and R. Bowshell, Unital rings whose additive endomorphisms commute, Math. Ann., 228 (1977), 197-214.
15. H. Zassenhaus, Orders as endomorphism rings of modules of the same rank, J. London Math. Soc., 42 (1967), 180-182.

RELATIONS BETWEEN HOM, EXT, AND TENSOR PRODUCT FOR CERTAIN CATEGORIES OF MODULES OVER DEDEKIND DOMAINS

E. L. Lady[1]

Tensor products of finite rank torsion free abelian groups are poorly understood. Computing $G \otimes H$ seems to be at least as complicated as computing $\mathrm{Hom}(G,H)$. In fact, in [4] these two problems were shown to be equivalent if G and H are both p-local and have a splitting field K which is a finite extension of the rationals or if they are both T-Butler groups for some set T of mutually incomparable idempotent types. In these two cases we have $\mathrm{Hom}(G,FH) \approx F(G \otimes H)$, where F is a contravariant functor which is almost a duality on a suitable category containing both G and H. In this paper, we further elaborate this relationship by giving an explicit formula for the kernel of the canonical natural epimorphism from $G \otimes H$ onto $F^2(G \otimes H)$. Furthermore, we show that there is a natural isomorphism between the divisible subgroup of $G \otimes H$ and $\mathrm{QHom}(AH,G)$, where AH is the Arnold dual of H. The latter result holds for much larger categories than those considered in [4], for instance for the category of all finite rank modules over a discrete valuation ring.

Let W be a dedekind domain and Q denote its quotient field. We consider a family of finite rank dedekind domains $W_1, \ldots, W_n$ containing W such that for each prime ideal p of W, p-rank $\Pi W_i = 1$. Then each W_i is a strongly indecomposable W-module, $\mathrm{End}\, W_i \approx W_i$, and $\mathrm{Hom}(W_i,W_j) = 0$ for $i \neq j$. We write $I = \Pi W_i$. Then $\mathrm{End}_W I \approx I$. We fix a W-homomorphism $\tau: I \to Q$ such that $\tau(W_i) \neq 0$ for all i. We let C be the category of homomorphisms of W-modules G such that the reduced quotient of G is isomorphic to a pure submodule of I^r for some r.

In particular, if W is a discrete valuation ring, then $n = 1$ and C is the category of homomorphisms of modules having QW_1 as a splitting field. On the other hand, if $QW_i = Q$ for all i, then C is the category of T-Butler modules, where T is the set of types $t(W_i)$. We let $\mathrm{Ext}_C(G,H)$ be the submodule of $\mathrm{Ext}^1_W(G,H)$ determined by classes of extensions $0 \to H \to X \to G \to 0$ such that G, H, and X belong to C. We let $\mathrm{QExt}_C(G,H)$ denote the quotient of $\mathrm{Ext}_C(G,H)$

[1]This research was supported by NSF Grant MCS7801705.

modulo the submodule of quasi-split extensions. It is well known [6] that the latter submodule is simply the torsion submodule of $\mathrm{Ext}_C(G,H)$. Since C is closed under quasi-isomorphism, it follows that this is also the torsion submodule of $\mathrm{Ext}^1_W(G,H)$. Since $\mathrm{Ext}_C(G,H)$ is divisible, $Q\mathrm{Ext}_C(G,H) \approx Q \otimes \mathrm{Ext}_C(G,H)$. We let $F = \mathrm{Hom}(\_,C^+Q)$, where C^+Q is a module to be defined below.

The indecomposable pure injective modules in C are Q and the various W_i (compare [2, Lemma 3.4, p. 300]). We let $i(G)$ denote the maximal pure injective submodule of G and $d(G)$ denote its divisible submodule. We say that G is <u>reduced</u> if $d(G) = 0$ and <u>strongly reduced</u> if $i(G) = 0$. We write $G * H = (G\otimes H)/d(G\otimes H)$. We write $w_i = \mathrm{rank}\, W_i$, and for any G, W_i-rank $G = \mathrm{rank}_{W_i}\mathrm{Hom}(G,W_i) = (\mathrm{rank\,Hom}(G,W_i))/w_i$.

<u>THEOREM 1</u>. For any modules G and H in C, $G \otimes FH \approx F\mathrm{Hom}(G,H) \oplus Q\mathrm{Ext}_C(G,H) \oplus \Pi W_i^{k_i}$, where $k_i = \mathrm{length}\ \mathrm{Ext}^1_W(G,H)[p]$ for any prime ideal p such that $pW_i \neq W_i$. In particular, $i(G\otimes FH) \approx Q\mathrm{Ext}_C(G,H) \oplus \Pi W_i^{k_i}$. Furthermore, there are natural isomorphisms $d(G\otimes FH) \approx \mathrm{Hom}(\mathrm{Ext}_C(G,H),Q) \approx Q\mathrm{Hom}(C^-H,G)$.

For Theorem 2, we remove the restriction that the torsion free dedekind domains W_i have finite rank over W. Thus, for instance, if W is local, we can choose $I = W_1$ to be the completion of W, so that C becomes the category of all finite rank torsion free W-modules. In this context, the functors C^-, C^+, and F are no longer available. However, there does exist a pure exact contravariant functor A on the category of quasi-homomorphisms of modules in C which is called Arnold duality, is essentially independent of the category C, and has the property that A^2 is naturally quasi-isomorphic to the identity functor.

<u>THEOREM 2</u>. For any modules G and H in C, $d(G\otimes H) \approx Q\mathrm{Hom}(AH,G)$.

1. <u>The Coxeter Functors and F.</u>

<u>Proposition 1.1</u>. A finite rank torsion free W-module belongs to C if and only if each $W_i * G$ is a projective W_i-module.

<u>Proof</u>. We may suppose G reduced, in which case the canonical map $G \to \Pi W_i * G$ is a monomorphism with pure image. Thus the stated condition is sufficient. Conversely, a comparison of p-ranks shows that

the multiplication map $W_i * W_i \to W_i$ must be an isomorphism. So if G is pure in $I^k = \Pi W_i{}^k$, then so is $W_i * G$. But $\mathrm{Hom}(W_i * G, W_j) = 0$ for $i \neq j$, so $W_i * G$ is a pure submodule of $W_i{}^k$ and hence is a projective W_i-module [2, Lemma 1.3(3)].

<u>Lemma 1.2</u>. If G and H are in C, then the map $\varphi \mapsto (\tau \otimes H)\varphi$ is an isomorphism

$$\mathrm{QHom}_I(I * G, I \otimes H) \to \mathrm{Hom}_W(I * G, QH).$$

<u>Proof</u>. Since I is the ring-theoretic product of the W_i, we have

$$\mathrm{Hom}_I(I * G, I \otimes H) \approx \Pi \mathrm{Hom}_{W_i}(W_i * G, W_i \otimes H),$$

and likewise $\mathrm{Hom}_W(I * G, QH) \approx \Pi \mathrm{Hom}_W(W_i * G, QH)$. Thus it suffices to see that if τ_i is the restriction of τ to W_i, then $\varphi \mapsto (\tau_i \otimes H)\varphi$ is an isomorphism from $\mathrm{QHom}_{W_i}(W_i * G, W_i \otimes H)$ to $\mathrm{Hom}_W(W_i * G, QH)$. In fact, it is monic since Image φ is a W_i-module and $\tau_i \otimes H$ contains no nontrivial W_i-module in its kernel. And since $W_i * G$ is a free W_i-module, we see that

$$\mathrm{rank}\ \mathrm{Hom}_{W_i}(W_i * G, W_i \otimes H) = w_i(W_i\text{-rank } G)(\mathrm{rank}\ H) = \mathrm{rank}\ \mathrm{Hom}_W(W_i * G, QH)$$

so that $\varphi \mapsto (\tau_i \otimes H)\varphi$ is an isomorphism.

Now consider the following commutative diagram with exact rows:

$$\begin{array}{ccccccc} 0 \longrightarrow & Q(I * G) & \xrightarrow{\varphi} & Q(I \otimes H) & \longrightarrow & Q(I * H) & \longrightarrow 0 \\ & \uparrow & & \downarrow \tau & & & \\ G \longrightarrow & I * G & \xrightarrow{\psi} & QH & \longrightarrow & 0. & \end{array}$$

The left-hand map in the bottom row is the canonical one and is a monomorphism if and only if G is reduced. The left-hand vertical map is the inclusion and is I-linear. The right-hand map in the top row is the canonical one induced by the definition of $I * H$ as $(I \otimes H)/d(I \otimes H)$ and the top row is required to the I-linear. The vertical map marked τ is actually $Q(\tau \otimes H)$. Thus we must have $Q(I * G) \approx d(I \otimes H)$ and $QH \approx (I * G)/G$ (which is necessarily divisible, as a computation of p-ranks shows).

Now if G is given, then the bottom sequence is determined and the map φ is uniquely determined by Lemma 1.2. Furthermore, φ

is a monomorphism because ψ contains no nontrivial I-submodule of $I * G$ in its kernel [2, Lemma 3.5]. $Q(I * H)$ is then determined as the cokernel of φ and the restriction of the cokernel map to QH yields a map $QH \to Q(I * H)$. Then $I * H$ and therefore also H are uniquely determined up to quasi-equality [2, Lemma 1.2, Theorem 3.8]. We write $H = C^- G$.

Conversely, suppose H given. Then the top sequence is determined and so $I * G$ is determined up to quasi-equality and ψ is determined by the commutativity of the diagram. We then define $G = \operatorname{Ker} \psi$ and write $G = C^+ H$.

One then easily checks that C^- and C^+ are in fact endofunctors on the category of quasihomorphisms of objects in C, that C^- is left adjoint to C^+, that $C^+C^-G \approx G$ if G has no pure injective quasisummand, and that $C^-C^+G \approx G$ if G has no pure projective quasisummand.

In particular, we notice the following exact sequences:

$$(*) \qquad 0 \longrightarrow C^+Q \longrightarrow \Pi W_i \xrightarrow{\ \tau\ } Q \longrightarrow 0$$

$$(**) \qquad G \longrightarrow I * G \longrightarrow QC^-G \longrightarrow 0$$

<u>Lemma 1.3</u>. If p is a prime ideal such that $pW_i \neq W_i$, then for any G belonging to C, $W_i\text{-rank } G = p\text{-rank } G$.

<u>Proof</u>. $W_i\text{-rank } G = \operatorname{rank}_{W_i} \operatorname{Hom}(G, W_i) = \operatorname{rank}_{W_i} \operatorname{Hom}(W_i * G, W_i)$. Now $W_i * G$ is projective W_i-module by Lemma 1.1, and so $W_i * G \sim W_i^k$ for some k. Since $p\text{-rank } W_i = 1$, $k = p\text{-rank } W_i * G = p\text{-rank } G$. Since $\operatorname{rank}_{W_i} W_i = 1$, the lemma follows.

<u>Proposition 1.4</u>. The functor $F = \operatorname{Hom}(\_, C^+Q)$ has the following properties:

1) For all G, FG is strongly reduced.
2) If $0 \to G \to H \to M \to 0$ is exact and M is strongly reduced, then $0 \to FM \to FH \to FG \to 0$ is exact.
3) The canonical map $G \to F^2G$ is a split epimorphism and its kernel is $i(G)$.
4) For any G, $\operatorname{rank} FG = \operatorname{rank} C^-G$. If G is strongly reduced, then for all i, $W_i\text{-rank } FG = W_i\text{-rank } G$.

<u>Proof</u>. [4, Theorem 1].

2. Hom, Ext, and Tensor Product.

Proposition 2.1. For any G belonging to C,

$$i(G) \approx Q\mathrm{Ext}_C(G,C^+Q) \oplus \Pi W_j^{k_j},$$

where k_j = length $\mathrm{Ext}^1_W(G,C^+Q)[p]$ for any prime ideal p such that $pW_j \neq W_j$.

Proof. This follows from three facts:

1) $\mathrm{Ext}_C(Q,C^+Q) \approx Q$.

2) $\mathrm{Ext}_C(W_j,C^+Q) \approx (QW_j)/W_j$.

3) $\mathrm{Ext}_C(G,C^+Q) = 0$ if G is strongly reduced.

Also note that $\mathrm{Ext}^1_W(G,C^+Q)[p] = \mathrm{Ext}_C(G,C^+Q)[p]$. Now 1) follows by applying $\mathrm{Hom}(Q,\_)$ to the short exact sequence (*) and 2) follows from applying $\mathrm{Hom}(W_i,\_)$ to the same sequence and noting that $\mathrm{Hom}(W_j,C^+Q) = 0$, $\mathrm{Ext}_C(W_j,I) = 0$, $\mathrm{Hom}(W_j,I) \approx W_j$ and $\mathrm{Hom}(W_j,Q) \approx QW_j$. Finally 3) follows by applying $\mathrm{Hom}(\_,C^+Q) = F$ to (**) by Proposition 1.4. (Recall that $G \to I * G$ is monic if G is reduced.)

Proposition 2.2. There is a natural isomorphism $\mathrm{Ext}_C(G \otimes FH,C^+Q) \approx \mathrm{Ext}_C(G,H)$.

Proof. By applying $\mathrm{Hom}(FH,\_)$ to (*) and noting that $\mathrm{Ext}_C(FH,C^+Q) = 0$ by Proposition 2.1, we get

$$0 \longrightarrow F^2H \longrightarrow \mathrm{Hom}(FH,I) \longrightarrow \mathrm{Hom}(FH,Q) \longrightarrow 0.$$

Applying $\mathrm{Hom}(G,\_)$ gives

$$0 \to \mathrm{Hom}(G,F^2H) \to \mathrm{Hom}(G,\mathrm{Hom}(FH,I)) \to \mathrm{Hom}(G,\mathrm{Hom}(FH,Q)) \to \mathrm{Ext}_C(G,F^2H) \to 0.$$

On the other hand, applying $\mathrm{Hom}(G\otimes FH,\_)$ to (*) yields

$$0 \to \mathrm{Hom}(G\otimes FH,C^+Q) \to \mathrm{Hom}(G\otimes FH,I) \to \mathrm{Hom}(G\otimes FH,Q) \to \mathrm{Ext}\ (G\otimes FH,C^+Q) \to 0.$$

Comparing these two exact sequences and using the natural isomorphism $\mathrm{Hom}(G\otimes FH,\_) \approx \mathrm{Hom}(G,\mathrm{Hom}(FH,\_))$ yields $\mathrm{Ext}_C(G\otimes FH,\_) \approx \mathrm{Ext}_C(G,F^2H)$. But since $H \approx F^2H \oplus i(H)$, $\mathrm{Ext}_C(G,F^2H) \approx \mathrm{Ext}_C(G,H)$.

Proof of Theorem 1. $F(G\otimes FH) \approx \mathrm{Hom}(G\otimes FH,C^+Q) \approx \mathrm{Hom}(G,F^2H)$. Thus $G \otimes FH \approx F\mathrm{Hom}(G,F^2H) \oplus E$ and by Propositions 2.1 and 2.2,

$E = i(G\otimes FH) \approx QExt_C(G\otimes FH, C^+Q) \oplus \Pi W_j^{k_j}$, where

$$k_j = \text{length } Ext_C(G\otimes FH, C^+Q)[p] = \text{length } Ext_C(G,H)[p]$$

and $QExt_C(G\otimes FH, C^+Q) \approx QExt_C(G,H)$. Furthermore, it follows easily from [1, Proposition 1, p. 76] that there is a natural isomorphism $QExt_C(G,H) \approx Hom(Hom(C^-H,G),Q)$. By vector space duality, then, $d(G\otimes FH) \approx Hom(Q, G\otimes FH) \approx Hom(QExt_C(G\otimes FH, C^+Q),Q) \approx Hom(QExt_C(G,H),Q)$. Finally, $Hom(G,H) \approx Hom(G,F^2H) \oplus E'$, where $E' = Hom(G,i(H))$ is pure injective, so that $FHom(G,H) \approx FHom(G,F^2H)$.

Recall that a module G is pure pre-projective if $G \approx C^{-r}P$ for some pure projective module P and some r and that H is pure pre-injective if $H \approx C^{+s}E$ for some pure injective module E and some s. In most cases, the pure pre-projective and pure pre-injective modules are the only ones we have a real systematic understanding of. For these modules, Theorem 1 yields some explicit calculations.

Corollary 2.3. For any module H and any r,

1) $d(C^{+r}Q\otimes FH) \approx QExt_C(C^{+r}Q,H) \approx QExt_C(FH,C^{-r+1}V) \approx Hom(C^{-r-1}H,Q)$;

2) $d(C^{+r}W_i\otimes FH) \approx QExt_C(C^{+r}W_i,H) \approx QExt_C(FH,FC^{+r}W_i) \approx QHom(C^{-r-1}H,W_i)$.

Corollary 2.4. If G is pure pre-projective and H has no pure pre-injective quasi-summand, then $G\otimes H$ is reduced.

Proof. We may suppose G strongly indecomposable. Since H is strongly reduced, $H \approx F^2H$. Then $d(G\otimes H) \approx d(G\otimes F^2H) \approx QExt_C(G,FH) \approx Hom(Hom(C^-FH,G),Q) \approx Hom(Hom(C^{+r-1}FH,C^{+r}G),Q) = 0$ for large r.

Proposition 2.5. [5, Lemma 2.2, p. 279][7, Theorem 2, p. 147]. Let $r = \text{rank } G$, $s = \text{rank } H$, $b_i = W_i\text{-rank } G$, $c_i = W_i\text{-rank } H$, and $w_i = \text{rank } W_i$. Then

$$\text{rank } Hom(G,H) - \text{rank } Ext_C(G,H) = rs + \Sigma w_i b_i c_i - r\Sigma w_i c_i \quad \text{and}$$

$$W_i\text{-rank } Hom(G,H) + \text{length } Ext^1_W(G,H)[p_i] = b_i c_i,$$

where p_i is a prime ideal of W such that $p_iW_i \neq W_i$.

Proof. Clearly the result is valid if $H = Q$ or $H = W_i$ and is valid for $H_1 \oplus H_2$ if valid for H_1 and H_2. Thus we may suppose H strongly reduced. We then get

$$0 \to H \to \Pi W_i * H \to QC^- H \to 0$$

which yields

$$0 \to \mathrm{Hom}(G,H) \to \Pi\mathrm{Hom}(G,W_i * H) \to \mathrm{Hom}(G,QC^- H) \to \mathrm{Ext}_C(G,H) \to 0$$

since $W_i * H$ is pure injective. Now $W_i * H \approx W_i^{c_i}$, so rank $\mathrm{Hom}(G,W_i * H) = w_i c_i (W_i\text{-rank } G) = w_i b_i c_i$ and rank $\mathrm{Hom}(G,QC^- H) = r(\Sigma w_i c_i - s)$. The formula for ranks follows. The formula for W_i-ranks follows from the same sequence, since W_i-rank $\mathrm{Hom}(G,QC^- H) = 0$, W_i-rank $(\Pi\mathrm{Hom}(G,W_j * H)) = b_i c_i$.

3. Arnold Duality. If H is strongly reduced, then there is a natural isomorphism of functors $d(\_\otimes H) \approx Q\mathrm{Hom}(C^- FH,\_)$. It follows that $C^- FH$ is independent of the category C, as long as H is strongly reduced in that category. We now define a contravariant functor A called Arnold duality which is completely independent of C and has the property that $d(\_\otimes H) \approx Q\mathrm{Hom}(AH,\_)$ for all H.

In this section, we remove the hypothesis that the dedekind domains W_i have finite W-rank.

Definition. If G is a module in C, we define AG as follows: $QAG = \mathrm{Hom}(G,Q)$, and $d(I\otimes AG) = Q\mathrm{Hom}(G,I)$, where we make the identifications

$$Q\mathrm{Hom}(G,I) \subseteq \mathrm{Hom}(G,QI) = QI \otimes \mathrm{Hom}(G,Q) = QI \otimes AG.$$

This determines the canonical map $Q(I\otimes AG) \to Q(I * AG)$ and thus determines $I * AG$, and hence AG, up to quasi-isomorphism. It is readily seen that A is a pure-exact contravariant functor on the category of quasi-homomorphisms of modules in C, that $A^2 \sim 1$, and that rank AG = rank G and p-rank AG = rank G - p-rank G for all p. (Note that if W is local, the definition here is essentially identical to that given in [3].)

Proposition 3.1. If rank $I < \infty$, then the functors A and $C^- F$ have isomorphic restrictions to the category of quasi-homomorphisms of strongly reduced modules in C.

Proof. The contruction of $C^{-}FG$ is based on the following diagram:

$$\begin{array}{ccccccccc} & & 0 & \to & QHom(G,I) & \xrightarrow{\varphi} & I \otimes Hom(G,Q) & \to & Q(I * C^{-}FG) \to 0 \\ & & & & \uparrow & & \downarrow \tau & & \\ 0 & \to & FG & \to & Hom(G,I) & \xrightarrow{\tau} & Hom(G,Q) & \longrightarrow & 0 \end{array}$$

Furthermore, if we identify $I \otimes Hom(G,Q) = Hom(G,QI)$, then φ becomes the inclusion map. But then the top sequence is the one determining $I * AG$ and the proposition follows.

Now consider the set of all torsion free W-algebras $I = \Pi_1^n W_i$ (with n variable) such that each W_i is a dedekind domain and p-rank $I = 1$ for each prime ideal p of W. This set is directed, where we say $I \leq I'$ if I is isomorphic to a pure subalgebra of I'. Each such W-algebra I determines a category $C(I)$. In the remainder of the paper, we let C be the direct union of all such categories.

Proposition 3.2. The functor A is compatible with the inclusions $C(I) \subseteq C(I')$ and hence can be defined on the entire category C.

Proof. This follows by noting that if $G \in C(I)$ and I is isomorphic to a pure subalgebra of I', then $QHom(G,I) = Hom(G,QI) \cap QHom(G,I')$.

The proof of Theorem 2 now follows directly from the following proposition and the fact that $H \sim A^2H$.

Proposition 3.3. Define $\theta: d(G \otimes AH) \to Hom(H,QG)$ by setting $\theta(\Sigma g_i \otimes \varphi_i)(h) = \Sigma \varphi_i(h) g_i$ where $g_i \in G$, $h \in H$, and $\varphi_i \in Hom(H,Q) = QAH$. Then θ is a monomorphism whose image is $QHom(H,G)$.

Proof. First, consider the case $G = I$. (The fact that rank I may be infinite is unimportant.) In this case, $\theta: d(I \otimes AH) \to Hom(H,QI)$ is nothing but the restriction of the canonical identification $I \otimes Hom(H,Q) \approx Hom(H,QI)$, so that the proposition simply restates the definition of AH. Second, if $G = Q^m$, where m is a possibly infinite cardinal, then $\theta: Q^m \otimes Hom(H,Q) \to Hom(H,Q^m)$ is a standard vector space isomorphism. (Note that $QHom(H,Q^m) = Hom(H,Q^m) = Hom(QH,Q^m)$.) Finally, note that the functors $QHom(H,\_)$ and $d(\_ \otimes AH) = Hom(Q, \_ \otimes AH)$ are both left exact. Thus the proposition follows for arbitrary G in $C(I)$ by considering the sequence $0 \to G \to I * G \to Q^m \to 0$ and remembering that $I * G$ is a finitely generated projective I-module.

REFERENCES

[1] M. C. R. Butler, The construction of almost split sequences I, Proc. London Math. Soc. (3) 40(1980), 72-86.

[2] E. L. Lady, Extension of scalars for torsion free modules over dedekind domains, Symposia Math. 23(1979), 287-305.

[3] E. L. Lady, Splitting fields for torsion free modules over discrete valuation rings I, J. Algebra 49(1977), 261-275.

[4] E. L. Lady, Splitting fields for torsion free modules over discrete valuation rings III, J. Algebra 66(1980), 307-320.

[5] C. M. Ringel, Representations of K-species and bimodules, J. Algebra 41(1976), 269-302.

[6] C. P. Walker, Properties of Ext and quasi-splitting of abelian groups, Acta Math. Acad. Sci. Hungar. 15(1964), 157-160.

[7] R. B. Warfield, Jr., Extensions of torsion free abelian groups of finite rank, Arch. Math. 23(1972), 145-150.

ON DIRECT DECOMPOSITIONS OF TORSION FREE ABELIAN GROUPS OF RANK 4

Khalid Benabdallah and Otto Mutzbauer

Torsion free abelian groups of rank 3, having non-isomorphic direct decompositions into direct indecomposable summands are almost completely decomposable [1;4.1] or [2;Lemma 1]. The same doesn't hold for torsion free abelian groups of rank 4.

Arnold and Lady proved in [1;4.1], that all direct decompositions of a torsion free abelian group G of rank 4, which is not almost completely decomposable, into indecomposable summands are isomorphic unless $G = A \oplus B$ where either $\text{rk}\, A = 1$, $\text{rk}\, B = 3$ and B is quasi-isomorphic to $A \oplus X$, where X is strongly indecomposable, or A, B quasi-isomorphic of rank 2. In both cases there are groups having non-isomorphic decompositions. The number of pairwise non-isomorphic decompositions will be calculated for a special class of groups.

The notation of Fuchs [3] will be used.

<u>Example 1</u> Let P,Q,R,S be disjoint infinite sets of primes and let t be a prime not contained in $P \cup Q \cup R \cup S$. Let for instance
$P^{-1}a = \{p^{-1}a \mid p \in P\}$. Define the torsion free abelian groups of rank 1, 2 and 4 respectively $A = \langle P^{-1}a\rangle \cong B = \langle P^{-1}b\rangle$,
$C = \langle Q^{-1}c, R^{-1}d, S^{-1}(c+d)\rangle$, $G = A \oplus \langle B \oplus C, t^{-1}(b+c)\rangle$.
Let $1 \leq u < t$, $a' = ua + \beta b$, $b' = ta + \delta b$ where β, δ are integers satisfying $u\delta - t\beta = 1$. Since $a = \delta a' - \beta b'$, $b = -ta' + ub'$ we get
$A \oplus B = A' \oplus B'$ if we define $A' = \langle P^{-1}a'\rangle$ and $B' = \langle P^{-1}b'\rangle$
consequently $G = A' \oplus \langle B' \oplus C, t^{-1}(ub' + c)\rangle$.
Let us assume an isomorphism φ between the indecomposable summands of rank 3. By the incomparability of the charcteristics defined by the sets P,Q,R,S respectively we have $\varphi b = \pm b'$, $\varphi c = \pm c$, $\varphi d = \pm d$ and consequently $\varphi\,[t^{-1}(b+c)] = \pm t^{-1}(b' + c)$ contradicting the assumed isomorphism if $t > 3$.

This group G has $\frac{t-1}{2}$ pairwise non-isomorphic direct decompositions and is not almost completely decomposable.

Example 2 Let P, Q, R be disjoint infinite sets of primes and let s,t be different primes not contained in $P \cup Q \cup R$. Let be

$G = \langle P^{-1}x_1, Q^{-1}x_2, R^{-1}(x_1+x_2) \rangle \oplus \langle P^{-1}x_3, Q^{-1}x_4, R^{-1}(x_3+x_4), (st)^{-1}x_3 \rangle$

and $y_1 = sx_1+ax_3$, $y_2 = sx_2+ax_4$, $y_3 = tx_1+bx_3$, $y_4 = tx_2+bx_4$ where a, b are integers satisfying $sb - ta = 1$. Define

$H = \langle P^{-1}y_1, Q^{-1}y_2, R^{-1}(y_1+y_2), s^{-1}y_1 \rangle \oplus \langle P^{-1}y_3, Q^{-1}y_4, R^{-1}(y_3+y_4), t^{-1}y_3 \rangle$.

Obviously $H \leq G$, moreover $H = G$ because

$\langle P^{-1}x_1 \rangle = \langle P^{-1}(by_1-ay_3) \rangle \leq H$ etc. and $(st)^{-1}x_3 = s^{-1}y_1 - t^{-1}y_3$.

Assume $\varphi: \langle P^{-1}x_1, Q^{-1}x_2, R^{-1}(x_1+x_2) \rangle \to \langle P^{-1}y_1, Q^{-1}y_2, R^{-1}(y_1+y_2), s^{-1}y_1 \rangle$ to be an isomorphism. Then by the incomparabilities of the characteristics defined by the different sets of primes we have $\varphi x_2 = \pm y_2$, $\varphi(x_1+x_2) = \pm y_1 \pm y_2$ consequently $\varphi x_1 = \pm y_1$ contradicting the fact $h_s(x_1) = o \neq 1 = h_s(y_1)$, and the two decompositions of G are proved to be not isomorphic. The group G has only these two non-isomorphic decompositions.

We use a description of torsion free abelian groups of rank 2 developed in [4]. Let $\pi = \sum_{i=o}^{\infty} \pi_i p^i \in \mathbb{Q}_p^*$ be a p-adic integer in standard expansion, i.e. π_i integers $0 \leq \pi_i < p$. $\mathbb{K}_p^*$ denotes the field of p-adic numbers. $\pi^{(n)} = \sum_{i=o}^{n-1} \pi_i p^i$ is the (n-1)-th partial sum of π.

Let G be a torsion free abelian group of rank 2 and u,v independent elements of G. Using the notations and theorems of [3; §93]

$\mathbb{Q}_p^* \otimes G$ is a p-adic module and

$\mathbb{Q}_p^* \otimes G)/[\mathbb{Q}_p^*(1 \otimes u) \oplus \mathbb{Q}_p^*(1 \otimes v)] \cong \mathbb{Z}(p^{n_p}) \oplus \mathbb{Z}(p^{m_p}) \cong [G/\langle u,v \rangle]_p$ where

$0 \leq m_p \leq n_p \leq \infty$ and $\mathbb{Z}(p^\infty)$ denotes the group of Prüfer. There is a p-adic integer ρ_p such that either

$$\mathbb{Q}_p^* \otimes G = p^{-n_p}\mathbb{Q}_p^*[(1 \otimes v) + \rho_p^{(n_p-m_p)}(1 \otimes u)] + p^{-m_p}\mathbb{Q}_p^* \otimes \langle u,v \rangle \quad \text{or}$$

$$\mathbb{Q}_p^* \otimes G = p^{-n_p}\mathbb{Q}_p^*[(1 \otimes u) + \rho_p^{(n_p-m_p)}(1 \otimes v)] + p^{-m_p}\mathbb{Q}_p^* \otimes \langle u, v \rangle \quad \text{where}$$

$\mathbb{K}_p^* = p^{-\infty}\mathbb{Q}_p^*$ and $\rho^{(\infty)} = \rho$. Define a <u>characteristic</u>

$\mathbb{M} = \{m_p, n_p, \pi_p \mid p \text{ prime}\}$ where $\pi_p = \rho_p$ in the first case and $\pi_p = \rho_p^{-1}$

($\infty = 0^{-1}$) in the second case. The group G is given up to isomorphism by $\mathbb{M}$ [4;93.2] and the characteristic $\mathbb{M}$ of $G = \langle u,v \mid \mathbb{M} \rangle$ is uniquely determined if u and v are fixed [4;Lemma 1].

<u>Lemma 1</u> Let G and H be quasi-isomorphic torsion free abelian groups of rank 2. Let $\varphi: H \to G$ be an embedding monomorphism. If $\varphi H \leq G$ is an embedding of minimal index then there is a natural number n, such that $G/\varphi H \cong \mathbb{Z}(n)$ cyclic of order n and for all primes p dividing n $|G/pG| = p^2$, i.e. the p-rank of G is 2.

<u>Proof.</u> Let φH be of minimal index n in G, $G/\varphi H \cong \oplus_{p \in P} [\mathbb{Z}(p^{n_p}) \oplus \mathbb{Z}(p^{m_p})]$ where P is a finite set of primes and m_p, n_p are integers with $0 \leq m_p \leq n_p < \infty$. There are three possibilities for the quotient G/pG namely 0, $\mathbb{Z}(p)$ or $\mathbb{Z}(p) \oplus \mathbb{Z}(p)$. If $G/pG = 0$, i.e. $G = pG$, then $G/\varphi H$ is a finite p-divisible group implying $m_p = n_p = 0$. If $G/pG \cong \mathbb{Z}(p)$ then $\varphi H/p(\varphi H) \cong \mathbb{Z}(p)$ because $G/\varphi H$ is finite and there is a free subgroup F of φH such that $\varphi H/F \cong \mathbb{Z}(p^\infty)$ and $(G/\varphi H)_p$ is finite cyclic implying $m_p = 0$.

Let $r = \prod_{p \in P} p^{n_p}$, $s = \prod_{p \in Q} p^{n_p - m_p}$ where $Q = \{p \in P \mid G/pG \cong \mathbb{Z}(p)\}$.

Then $rG \leq \varphi H$ and $s(\varphi H) \leq rG$ with $\varphi H/rG \cong rG/s(\varphi H) \cong \mathbb{Z}(s)$. Using $s(\varphi H) \cong H$, $rG \cong G$ and the minimality of n we get $s = n$, i.e. $\prod_{p \in Q} p^{n_p - m_p} = \prod_{p \in P} p^{n_p + m_p}$ and $m_p = 0$ for all $p \in P$, $m_p = n_p = 0$ for all p if $|G/pG| \neq p^2$, proving the lemma.

<u>Definition</u> A torsion free abelian group G of rank 2 is said to be quadratic if the quasi-endomorphism ring $\mathbb{Q} \otimes \operatorname{End} G$ is a quadratic number field.

<u>Definition</u> A characteristic $\mathbb{M} = \{m_p, n_p, \pi_p \mid p \text{ prime}\}$ is said to be quadratic if there are rationals

$x, y \in \mathbb{Q}$, $\sqrt{x^2 - y} \notin \mathbb{Q}$ such that $\sum_p n_p - m_p = \infty$ and for almost all primes, especially for all primes p with $n_p = \infty$, $m_p < \infty$,

$$\pi_p^2 + 2x\pi_p + y \in p^{n_p - m_p} \mathbb{Q}_p^*$$

where $p^\infty \mathbb{Q}_p^* = \{0\}$ and $\infty - \infty = 0$.

Lemma 2 A torsion free abelian group of rank 2 is quadratic if and only if every describing characteristic is quadratic.

Proof. A quadratic group G has an irreducible endomorphism η, i.e. there are rationals x,y such that $\sqrt{x^2-y} \notin \mathbb{Q}$ and $\eta^2 + 2x\eta + y = 0$. Let u,v be independent elements of $G = \langle u,v \mid \mathbb{M} \rangle$ then every matrix describing η relative to u and v has the form $\begin{pmatrix} d-2xb & b \\ -yb & d \end{pmatrix}$ with rationals b,d and $\eta u = (d-2xb)u + bv$, $\eta v = -ybu + dv$, and $v + \pi_p u$ (or $u+\pi_p^{-1}v$) is an eigenvector of η by its height. This implies for almost all primes and especially for all primes p with $n_p = \infty$, $m_p < \infty$:

$$\eta(v+\pi_p u) = (d+b\pi_p)\, v + (d\pi_p - 2xb\pi_p - yb)u$$

$$\equiv (d+b\pi_p)\ (v + \pi_p u) \mod p^{-m_p}\mathbb{Q}_p^* \otimes \langle u,v \rangle$$

such that $d\,\pi_p - 2\,x\,b\,\pi_p - yb - (d + b\pi_p)\pi_p \in p^{n_p-m_p}\mathbb{Q}_p^*$,

i.e. $\pi_p^2 + 2x\pi_p + y \in p^{n_p-m_p}\mathbb{Q}_p^*$. Moreover a quadratic group is always strongly indecomposable [6;7.1] such that the assumtion $\sum_p n_p - m_p < \infty$, i.e. $G/(\langle u \rangle_* \oplus \langle v \rangle_*)$ finite by $\pi_p = 0$ if $n_p = m_p$, would lead to an immediate contradiction; and the characteristic $\mathbb{M}$ is quadratic.

Conversely, let the characteristic $\mathbb{M}$ be quadratic and $G = \langle u,v|\mathbb{M} \rangle$. There is a natural number b such that $\eta u = -2xbu + bv$, $\eta v = -ybu$ defines an endomorphism because

$$\eta(v+\pi_p u) = b\pi_p v - (2xb\pi_p + yb)u \equiv b\pi_p(v+\pi_p u) \mod p^{-m_p}\mathbb{Q}_p^*,$$

using $\pi_p^2 + 2x\pi_p + y \in p^{n_p-m_p}\mathbb{Q}_p^*$ and π_p unit in $\mathbb{Q}_p^*$ for almost all p with $n_p \neq m_p$. This endomorphism is irreducible and a group with quadratic characteristic is strongly indecomposable [5], consequently the group G is quadratic [6; 7.1].

Theorem 3 Let $G = A \oplus B$ be a torsion free abelian group of rank 4 and A, B quasi-isomorphic and strongly indecomposable homogeneous not quadratic group of rank 2. If $\varphi: A \to B$ is an embedding of minimal index k and $r = |\{p \mid p \text{ prime dividing } k\}|$, then 2^{r-1} is the number of pairwise non-isomorphic decompositions of G. And a complete set of non-isomorphic decompositions is given by

$$H_m = \langle \langle y_1,y_2 \mid \mathbb{M} \rangle, m^{-1}y_1 \rangle \oplus \langle \langle y_3,y_4 \mid \mathbb{M} \rangle, mk^{-1}y_3 \rangle$$

where $A \cong \langle y_1,y_2 \mid \mathbb{M} \rangle$, m is a natural number dividing k such that m and $m^{-1}k$ are relatively prime.

Proof. [1; 4.1] settles all cases except of A and B quasi-isomorphic homogeneous strongly indecomposable not quadratic groups of rank 2. Assume $G = C \oplus D$ to be a second decomposition of G which is not isomorphic to the first one. Let φ, ψ, ρ be embeddings of A in B, C and D of minimal index respectively. Let $\mathbb{M}$ be a characteristic such that

$$A = \langle x_1,x_2 \mid \mathbb{M} \rangle \quad , \quad B = \langle \varphi A \; , \; k^{-1}x_3 \rangle$$

$$C = \langle \psi A \; , \; l^{-1}(\alpha y_1+\beta y_2) \rangle , \quad D = \langle \rho A \; , \; m^{-1}(\gamma y_3+\delta y_4) \rangle$$

where $\varphi x_1 = x_3$, $\varphi x_2 = x_4$, $\psi x_1 = y_1$, $\psi x_2 = y_2$, $\rho x_1 = y_3$, $\rho x_2 = y_4$,

and $\alpha,\beta,\gamma,\delta$ can be assumed to be integers ($0 \le \alpha,\beta < l$, $0 \le \gamma,\delta < m$) such that the pairs α,β and γ,δ are relatively prime implied by the minimality of the indices and by multiplication modulo l, m respectively. If the prime p divides klm, then $h_p(x_1)$, $h_p(x_2)$ are finite by lemma 1 and it can be assumed $m_p = n_p = 0$. Moreover the special form of B is no restriction, this can be proved using an unimodular transformation of x_1, x_2.

$$F^* = \mathbb{Q}_p^* \otimes \langle x_1,x_2,x_3,x_4 \rangle = \mathbb{Q}_p^* \otimes \langle y_1,y_2,y_3,y_4 \rangle \text{ holds if p doesn't}$$

divide klm and then (identifying x_i and $1\otimes x_i$ etc.)

$$\mathbb{Q}_p^* \otimes G = \mathbb{Q}_p^* \otimes (A \oplus B) = p^{-n_p}[\mathbb{Q}_p^*(x_2+\pi_p x_1) \oplus \mathbb{Q}_p^*(x_4+\pi_p x_3)] + p^{-m_p}F^*$$
$$= p^{-n_p}[\mathbb{Q}_p^*(y_2+\pi_p y_1) \oplus \mathbb{Q}_p^*(y_4+\pi_p y_3)] + p^{-m_p}F^*$$

if $\pi_p \in \mathbb{Q}_p^*$ and analogously in the case $\pi_p \notin \mathbb{Q}_p^*$. There exist p-adic integers $a_p,b_p,c_p,d_p \in \mathbb{Q}_p^*$ where $a_pd_p - b_pc_p$ is a unit such that

$$y_2+\pi_p y_1 \equiv a_p(x_2+\pi_p x_1) + b_p(x_4+\pi_p x_3)$$

$$y_4+\pi_p y_3 \equiv c_p(x_2+\pi_p x_1) + d_p(x_4+\pi_p x_3)$$

modulo $p^{-m_p}F^*$. Substituting $y_i = \sum_{j=1}^{4} a_{ij}x_j$ $(1\le i\le 4)$ where (a_{ij}) is a regular matrix with rational entries a_{ij}, we get for instance

$a_{21} + \pi_p a_{11} - a_p \pi_p$, $a_{22} + \pi_p a_{12} - a_p \in p^{n_p - m_p} \mathbb{Q}_p^*$ or in a matrix equation, using the notation $(a_{ij}) = \begin{pmatrix} X_a & X_b \\ X_c & X_d \end{pmatrix}$ where X_a, X_b, X_c, X_d are 2×2 matrices: $(\pi_p, 1)(X_a - a_p E) \in p^{n_p - m_p}(\mathbb{Q}_p^* \oplus \mathbb{Q}_p^*)$ etc., and $(\pi_p, 1)$ is eigenvector of X_a with eigenvalue a_p modulo $p^{n_p - m_p}\mathbb{Q}_p^*$.

If for instance X_a is a singular matrix then a non-trivial linear combination of y_1 and y_2 is contained in B and $B \cap C \neq 0$. Let $\tau : G \to D$ be the projection with kernel C, then $\tau B \cong (B+C)/C \cong B/(B \cap C) \lesssim D$ and τ is a quasi-endomorphism of B with kernel not equal 0. This contradicts the homogeneity of B [6; 7.1]. Therefore all possible intersections of the four summands are 0 and X_a, X_b, X_c, X_d are regular.

If for instance a_p is not a unit for infinitely many primes p with $m_p \neq n_p$ then, using first that all corresponding π_p are p-adic integers

$$a_{11}a_{22} - a_{12}a_{21} = a_{11}(a_{22} + \pi_p a_{12}) - a_{12}(a_{21} + \pi_p a_{11}) \in p\mathbb{Q}_p^* .$$

This contradicts the regularity of X_a. The case that infinitely many π_p are not contained in $\mathbb{Q}_p^*$ leads to the same result. Likewise we have that for almost all p with $m_p \neq n_p$ hold: a_p, b_p, c_p, d_p are units in $\mathbb{Q}_p^*$. This implies for almost all primes p with $m_p \neq n_p$ for instance:

$$a_{12}\pi_p^2 + (a_{22} - a_{11})\pi_p - a_{21} \in p^{n_p - m_p}\mathbb{Q}_p^* .$$

Consequently we have $a_{12} = a_{21} = 0$, $a_{11} = a_{22} \neq 0$ because A is not quadratic and by the regularity of X_a. The matrices X_b, X_c and X_d have also this diagonal form by the same reasons and $(a_{ij}) = \begin{pmatrix} aE & bE \\ cE & dE \end{pmatrix}$ with rationals a, b, c, d, $ad-bc \in \mathbb{Q}/\{0\}$. The case that infinitely many π_p are not contained in $\mathbb{Q}_p^*$ leads to the same result. If the prime p divides the denominator of a or b then it divides x_1 and x_2 by $y_1 = ax_1 + bx_3$, $y_2 = ax_2 + bx_4$ and therefore p is no divisor of klm. If the prime p divides the numerators of a and b then again y_1 and y_2 are divisible by p such that p doesn't divide klm. And there are rationals λ, μ being units in $\mathbb{Q}_p^*$ for all primes p dividing klm, satisfying $a = \lambda a', b = \lambda b'$, $c = \mu c', d = \mu d'$; a', b', c', d' integers and the pairs a', b' and c', d' are relatively prime respectively.

We have $(a_{ij})^{-1} = (ad-bc)^{-1}\begin{pmatrix} dE & -bE \\ -cE & aE \end{pmatrix}$, i.e.

$x_1 = (ad-bc)^{-1}(dy_1-by_3)$, $x_2 = (ad-bc)^{-1}(dy_2-by_4)$,

$x_3 = (ad-bc)^{-1}(-cy_1+ay_3)$, $x_4 = (ad-bc)^{-1}(-cy_2+ay_4)$, and the greatest common divisor of k and $a'd'-b'c'$ divides all y_i using that the pairs a', b' and c', d' are relatively prime respectively. Consequently $a'd'-b'c'$ and klm are relatively prime. l divides

$$\alpha y_1 + \beta y_2 = a(\alpha x_1 + \beta x_2) + b(\alpha x_3 + \beta x_4)$$

therefore l divides a' and $\alpha x_3+\beta x_4$, but the greatest common divisor of β and k is just maximal divisor of $\alpha x_3+\beta x_4$ which is relatively prime to all divisors caused by M only. This implies that l divides β and k such that $\alpha = 1, \beta = 0$. Similarly $\gamma = 1$, $\delta = 0$ and m divides c' and k. The greatest common divisor of m and l divides therefore a', c' and k , i.e. divides $a'd'-b'c'$ and k which are relatively prime, and m and l are relatively prime. Now k divides

$$x_3 = (ad-bc)^{-1}(-cy_1+ay_3) = (ad-bc)^{-1}lm[-(m^{-1}c)(l^{-1}y_1)+(l^{-1}a)(m^{-1}y_3)]$$

and k = lm because a prime dividing $(ml)^{-1}k$ divides $m^{-1}c'$ and $l^{-1}a'$ consequently a' and c' which contradicts the fact that k and $a'd'-b'c'$ are relatively prime. If the index k is minimal then l and m are minimal too. Now we have proved that the given direct sums contain all decompositions up to isomorphism, i.e. the given number is an upper bound. It remains to prove that all these direct sums are in fact decompositions of $G = A \oplus B$. Define

$y_1 = m^{-1}kx_1+bx_3$, $y_2 = m^{-1}kx_2+bx_4$, $y_3 = mx_1+dx_3$, $y_4 = mx_2+dx_4$

where m divides k, m and $m^{-1}k$ are relatively prime and b, d are integers satisfying $m^{-1}kd-mb = 1$. Then for all primes p , relatively prime to k:

$$p^{-n}{}^{p}Q_p^*[(1\otimes y_2) + \pi_p(1\otimes y_1)] + p^{-m}{}^{p}Q_p^*\otimes\langle y_1,y_2,y_3,y_4\rangle = p^{n}{}^{p}Q_p^*\{m^{-1}k[(1\otimes x_2)$$
$$+ \pi_p(1\otimes x_1)] + b[(1\otimes x_4) + \pi_p(1\otimes x_3)]\} + p^{-m}{}^{p}Q_p^*\otimes\langle x_1,x_2,x_3,x_4\rangle \quad \text{by}$$

$\langle y_1,y_2,y_3,y_4\rangle = \langle x_1,x_2,x_3,x_4\rangle$ because the transformation is unimodular. Analogously in the second case of these modules. This module is contained in $Q_p^* \otimes (A \oplus B) = Q_p^* \otimes G$. Moreover $mk^{-1}y_1$, $m^{-1}y_3 \in G$ and by [3; 93.2] holds $H_m \leq G$. Again by $\langle x_1,x_2,x_3,x_4 \rangle = \langle y_1,y_2,y_3,y_4 \rangle = U$ we have $H_m/U = G/U$ implying $H_m = G$. This proves the theorem.

Remark A torsion free abelian group $G = A \oplus B$ of rank 4 where A and B are strongly indecomposable has only isomorphic decompositions except of some special cases namely A quasi-isomorphic to B. But even in this case non-isomorphic decompositions can be impossible, for instance if the minimal index k in theorem 3 is a power of a prime.

REFERENCES

[1] D. Arnold and L. Lady, Endomorphism rings and direct sums of torsion free abelian groups, Transact. Amer. Math. Soc., 211 (1975), 225 - 237.

[2] R. Burkhardt and O. Mutzbauer, Decompositions of torsion free abelian groups of rank 3, Arch. Math., (1981).

[3] L. Fuchs, Infinite abelian groups I + II, Academic Press, (197o, 1973).

[4] O. Mutzbauer, Klassifizierung torsionsfreier abelscher Gruppen des Ranges 2, Rend. Sem. Mat. Univ. Padova, 55 (1976), 195 - 2o8.

[5] O. Mutzbauer, Zerlegbarkeitskriterien für Invarianten torsionsfreier abelscher Gruppen des Ranges 2, Czech. Math. J., 29(1o4), (1979), 337 - 339.

[6] J.D. Reid, On the ring of quasi-endomorphisms of a torsion-free group, Topics in abelian groups, 51-68 (Chicago, 1963).

HYPER-INDECOMPOSABLE GROUPS

by

K. *Benabdallah** *et* A. *Birtz*

An indecomposable torsion free group is said to be hyper-indecomposable if all proper subgroups between its divisible hull and itself are indecomposable. On closer examination, hyper-indecomposable groups turn out to be nothing but reduced cohesive groups. We recall that a torsion free group G is cohesive if and only if G/K is divisible for every non-zero pure subgroup K of G. Cohesive groups were studied and examples of such groups were given for all ranks up to $2^{\aleph_0}$ by D.W. DUBOIS in [4]. We arrived at the notion of hyper-indecomposable groups via the concept of p-indicators of pairs of elements of torsion free groups which was introduced in [2]. The p-indicator of a pair of elements is a triple, one of whose components is a p-adic number. If this p-adic number is irrational (in the sense that it does not belong to the set of rational numbers that are included in the p-adic field) for every pair of independant elements in a group G, G is said to be a p-irrational group. This notion leads to yet another characterization of reduced cohesive groups and helps to conceptualize the largely computational constructions given in [4].

All groups considered here are abelian torsion free groups. For notation and terminology we follow the standard in [5], however, what we call p-indicator here is an entirely different concept.

1. Preliminaries

In [2] we developped the notion of a p-indicator of a pair of elements in a group. For convenience, we recall here some definitions and results without proof from [2] and proceed to the definition of p-irrational groups.

Definition 1.1. Let (a,b) be a pair of elements of a group G, the p-indicator $I_p(a,b)$ of the pair (a,b) is a triple:

$$I_p(a,b) = (\eta_p, M_p, \beta_p)$$

where $\eta_p \in K_p$ the field of p-adic numbers and $M_p, \beta_p \in Z \cup \{\infty\}$. These numbers are obtained as follows:

i) β_p is the p-height of b in G (denoted also by β).

---

* The work of the first author was partially supported by the C.R.S.N.G. of Canada grant no A5591.

ii) $\eta_p = \lim_i p^{\alpha-\beta} n_i$, where $\alpha = h_p(\alpha)$, and n_i is the unique integer (if it exists) which satisfies $h_p(a+p^{\alpha-\beta} n_i b) \geq i+\alpha$, and $0 \leq n_i < p^i$, $i \in Z^+$.

iii) The n_i's exist either for all $i \in Z^+$ or up to a certain $\ell \in Z^+$. M_p is defined by $M_p = \ell+\alpha$, where $\ell \in Z^+ \cup \{\infty\}$.

Remark. If α or β is infinite we take $\eta_p = 0$. n_i is a partial sum of $p^{\beta-\alpha}\eta_p$. If we write $n_{i+1} = n_i + s_i p^i$, $0 \leq s_i < p$, for all $i < \ell$, then $\eta_p = p^{\alpha-\beta} \sum_{i=0}^{\ell} s_i p^i$ and $n_{i+1} = \sum_{j=0}^{i} s_i p^i$.

Then p-indicator of (a,b) allows for a useful description of the generators of the smallest p-pure subgroup of G containing a and b . We call this subgroup the p-pure enveloppe of a,b in G and denote it by $\langle a,b\rangle_p$.

Lemma 1.2. Let $a,b,G,\alpha,\beta,n_i,s_i,\ell$, be as in the preceding discussion and set:

$$x_i = p^{-i-\alpha}(a+n_i p^{\alpha-\beta} b) , \quad 0 \leq i \leq \ell .$$

then: $\langle a,b\rangle_p = \langle \{x_i\}_{i=0}^{\ell} , p^{-\beta} b\rangle$.

Moreover: $x_i = px_{i+1} - s_i p^{-\beta} b$.

We also need the following result from [2].

Proposition 1.3. Let (a,b) be a pair of linearly independant elements of G , then $\langle a,b\rangle_p$ contains a non-zero element of infinite p-height if and only if $\eta_p \in Q$ and either M_p or β_p is infinite.

Definition 1.4. Let (a,b) and (a',b') be two pairs of elements of G . We say that these pairs are linked if they generate the same pure subgroup in G and this pure subgroup is of rank 2.

Clearly if (a,b) and (a',b') are linked pairs then {a,b} as well as {a',b'} is linearly independant and there exist uniquely determined rational numbers A_1,A_2,B_1,B_2 such that

$$\text{(I)} \qquad \begin{aligned} a &= A_1 a' + B_1 b' \\ b &= A_2 a' + B_2 b' . \end{aligned}$$

The following proposition will clarify the relation between the p-indicators of linked pairs in a special case of interest to us for the rest of this article. The general case is presented in [3] lemma 4.1.

Proposition 1.5. Let (a,b) and (a',b') be linked pairs of elements of a group G and let $I_p(a,b) = (\eta,M,\beta)$ and $I_p(a',b') = (\eta',M',\beta')$. Suppose further that

$\eta \notin Q$. Then $\alpha, \beta, \alpha', \beta'$ are finite and

(i) $\alpha' = h_p(a') = s + \beta - d$

(ii) $\beta' = h_p(b') = r + \beta - d$

(iii) $\eta' = (B_1+\eta B_2)/(A_1+\eta A_2) \notin Q$.

where s,r,d are the values at p of $B_1+\eta B_2$, $A_1+\eta A_2$, $A_1B_2 - B_1A_2$ respectively.

Proof. From (I), letting $D = A_1B_2 - A_2B_1$ we have

$$a' = (B_2a-B_1b)/D$$

and

$$b' = (-A_2a+A_1b)/D .$$

Now, using Theorem 2.7 of [2] we have

$$\alpha' = h_p(a') = \min\{v_p(B_2/D) + M \, , \, \beta + v_p((B_1+\eta B_2)/D)\}$$

but $M = \infty$ since $\eta \notin Q$. Therefore:

$$\alpha' = \beta + s - d .$$

Similarly, $\beta' = r + \beta - d$.

Again, since $\eta \notin Q$, r and s are finite and there exists a positive integer k such that for all $i \geq k$ we have both:

$$r = v_p(A_1+n_ip^{\alpha-\beta}A_2)$$

and

$$s = v_p(B_1+n_ip^{\alpha-\beta}B_2)$$

where n_i is the i-1-th partial sum of η .

Replacing a and b from equation (I), we have:

$$\text{(II)} \qquad a + n_ip^{\alpha-\beta}b = (A_1+n_ip^{\alpha-\beta}A_2)a' + (B_1+n_ip^{\alpha-\beta}B_2)b' .$$

Now, given $j \in Z^+$, we can choose i such that the p-height of the right hand side of equation (II) is larger or equal to $r + \alpha' + j$, but $r + \alpha' = h_p(A_1+n_ip^{\alpha-\beta}A_2)a')$. Thus by Lemma 1.1 of [2] we see that there exist an n_j' , such that both:

$$h_p(a'+p^{\alpha'-\beta'}n_j'b') \geq \alpha' + j \text{ , and } 0 \leq n_j' < p^j \text{ , } j \in Z^+ .$$

Furthermore, dividing the right hand side of (II) by $A_1 + n_ip^{\alpha-\beta}A_2$, we see that:

$$(B_1+n_ip^{\alpha-\beta}B_2)/(A_1+n_ip^{\alpha-\beta}A_2) \equiv p^{\alpha'-\beta'}n_j' \text{ , } (p^{j+\alpha'-\beta'}) .$$

This congruence is to be taken in K_p . Now taking the limits in K_p of both sides we obtain:

$$\eta' = (B_1 + \eta B_2)/(A_1 + \eta A_2) .$$

Clearly $\eta' \notin Q$.

We can now introduce an interresting class of groups.

Definition 1.6. A group of rank ≥ 2 is said to be a p-irrationnal group (abreviated: a p-i-group) if for the prime p and every pair (a,b) of independant elements of G , $\eta_p(a,b) \notin Q$. If G is a p-i-group for all primes p we say that it is an irrational group or an i-group.

From proposition 1.5, we see that a rank -2 group is a p-i-group whenever it contains a pair (a,b) with $\eta_p(a,b) \notin Q$. From proposition 1.3 we note that a p-i-group contains no elements of infinite p-height. The next result shows that p-i-groups have a strong indecomposability property.

Proposition 1.7. Let G be a p-i-group then any p-pure subgroup of G is indecomposable. In particular G itself is undecomposable.

Proof. Without loss of generality we may assume that G is of rank two. Suppose $G = A \oplus B$, and let $a \neq 0$, $a \in A$ and $b \neq 0$, $b \in B$, then: $\langle a,b\rangle_p = \langle a\rangle_p \oplus \langle b\rangle_p$. Let x_i be as in lemma 1.2 then, since $\eta_p(a,b) \notin Q$, we have:
$p^{\alpha+i}x_i = a + n_i p^{\alpha-\beta} b \in \langle a,b\rangle_p$ for all $i < \infty$. Therefore, $h_p(a) \geq \alpha + i$, that is to say $h_p(a) = \infty$. Now using proposition 1.3 we must have $\eta_p(a,b) \in Q$. This is clearly a contradiction.

2. *Hyper-indecomposable groups*

Let D be a divisible hull of a group G . We say that G is hyper-indecomposable if G and all proper subgroup of D containing G are indecomposable. As we mentionned in the introduction a group is hyper-indecomposable if and only if it is a reduced cohesive group. We establish here this and several other characterizations of these groups. We give first a group of characterizations which could be described as "abstract" in that they do not imply (at least not directly) a way to construct non-trivial exemples of such groups. Finally, we establish a "concrete" characterization which helps to understand the constructions given in [4].

Théorème 2.1. Let G be a reduced group and D its divisible hull. Then the following properties are equivalent.

(a) G is hyper-indecomposable

(b) $G + R = D$ for any divisible subgroup R of D with $R \neq 0$.

(c) G is cohesive.

Proof. (a) implies (b) obvious. (b) implies (c): Let K be a pure subgroup of G, $K \neq 0$ and let R be the divisible hull of K in D. Since K is pure in G we have $G \cap R = K$. Therefore G/K is isomorphic to $(G+R)/R$ which is torsion free divisible. (c) implies (b). Let R be a non-zero divisible subgroup of D then $G \cap R$ is a pure subgroup of G therefore $G/G \cap R$ is torsion free divisible. Now by the isomorphism theorems $(G+R)/R$ is torsion free divisible and as such it is a summand of D/R. But $D/G + R$ is a torsion group. Therefore $G + R = D$. (b) implies (a): Let H be a subgroup of D containing G and suppose that $H = A \oplus B$ where both A and B are non-zero. Let R be the divisible hull of A in D. Then: $H + R \supset G + R = D$. Therefore $A \oplus B = D$ and B is divisible. It follows that:

$$H \supset G + B = D \text{ that is to say } H = D.$$

Before we pass on to the main result, we give a somewhat amusing characterization of hyper-indecomposable groups of rank ≥ 3.

Theorem 2.2. Let G be a reduced group of rank ≥ 3. Then G is hyper-indecomposable if and only if the sum of any two pure subgroups of G with non-zero intersection is again a pure subgroup of G.

Proof. If G is hyper-indecomposable and H,K are two pure subgroups of G such that $H \cap K \neq 0$ then $H \cap K$ is pure in G and $G/H \cap K$ is divisible. Therefore $(H+K)/H \cap K$ is divisible and $H + K$ is pure in G. Conversely, let M be any rank 1 pure subgroup of G, then G/M is a group if rank ≥ 2. G/M has the property that the sum of any two pure subgroups is again a pure subgroup. Such groups must be either divisible or rank 1 groups. (See [1]). Therefore G/M is divisible and G is cohesive. Thus G is hyper-indecomposable.

We turn now to our main result: the characterization of hyper-indecomposable groups in terms of p-i-groups. We need the following lemma:

Lemma 2.3. Let G be a rank two p-i-group, D its divisible hull and $G \subset H \subset D$. Then if H is decomposable it must be p-divisible.

Proof. Let (a,b) be a pair of elements in G and $\eta = \eta_p(a,b) \notin Q$. Now $M_p(a,b) = \infty$ in G therefore $M_p(a,b)$ is also infinite when computed in H. Moreover if H is decomposable $\eta_p(a,b)$ computed in H must be rational as can be deduced from proposition 1.5 and proposition 1.7. Therefore, from proposition 1.3 we see that H contains a non-zero element of infinite p-height in H. Without loss of generality we may suppose that b is of infinite p-height in H. Thus $H = \langle p^{-\infty}b\rangle^* \oplus K$. Let $a' \neq 0$, $a' \in K \cap G$ then $\eta_p(a',b) \notin Q$ when computed in G and

$$\langle a',b\rangle_p^G \subset \langle a',b\rangle_p^H = \langle p^{-\infty}b\rangle \oplus \langle p^{-\alpha'}a'\rangle = M$$

where $\alpha' = h_p^H(a')$. Now from lemma 1.2, for every $i < \infty$, $x_i \in M$ and $p^{i+\alpha}x_i = a' + n_i p^{\alpha-\beta}b \in M$.

Therefore $\alpha' = h_p^H(a') \geq i + \alpha$ for all i .

That is to say $h_p^H(\alpha') = \infty$ and H is p-divisible.

Theorem 2.4. Let G be a reduced group then G is hyper-indecomposable if and only if G is a p-i-group for every p for which $pG \neq G$.

Proof. Let G be hyper-indecomposable and suppose that for some p , $pG \neq G$. If G is not a p-i-group then by porposition 1.3 it contains an element of infinite p-height different from zero. Let a be that element and let $K = \langle a \rangle_*$ the pure enveloppe of $\langle a \rangle$ in G . K is then p-divisible. Now, from Theorem 2.2 G/K is divisible, therefore G is p-divisible. This is a contradiction. Conversely if G is a p-i-group for every p such that $pG \neq G$. Let H be a subgroup of D containing G . Suppose that $H = A \oplus B$ where $A \neq 0 \neq B$. Let $a \neq 0$, $a \in A$ and $b \neq 0$, $b \in B$ then $\langle a,b \rangle_p^H \supset \langle ma,nb \rangle_p^G = X$ for some $m,n \in Z$. Now $K = \langle a,b \rangle_p^H = \langle a \rangle_p^H \oplus \langle b \rangle_p^H$ then K is between X and its divisible enveloppe in D . Furthermore $\eta_p(ma,nb)$ is the same in X and in G , therefore $\eta_p(ma,nb) \notin Q$. So that by lemma 2.3, K is p-divisible. It follows that a and b are of infinite p-height. This implies that H is p-divisible for every p such that $pG \neq G$. Now let q be such that $qG = G$ then the q-primary component of D/G is null. This means that all subgroups of D/G and in particular H/G are q-divisible. Therefore H itself is q-divisible. We conclude that H is divisible and $H = D$. That is to say G is hyper-indecomposable.

An immediate consequence of this result is that irrational groups are hyper-indecomposable. Now using lemma 1.2 and proposition 1.5 one can immediately construct a rank two hyper-indecomposable group. For higher ranks the constructions involve essentially set theoretical methods and the basic idea in the rank two case. For a different development from that of [4] we refer to [3].

References

[1] K. BENABDALLAH and A. BIRTZ: "Abelian groups where sums of pure subgroups are pure". Comment. Math. Univ. St.Pauli XXVII-2(1978),91-95.

[2] K. BENABDALLAH and A. BIRTZ: "p-pure enveloppes of pairs in torsion free abelian groups". Comment. Math. Univ. St.Pauli XXVIII-1(1979), 107-113.

[3] A. BIRTZ: "Nouveaux invariants pour les groupes abeliens sans torsion de rang deux et applications à divers problèmes de structure". Ph.D. dissertation. Université de Montreal, 1980.

[4] D.W. DUBOIS: "Cohesive groups and p-adic integers". Publ. Math. Debrecen 12 (1965), 51-58.

[5] L. FUCHS: "Infinite Abelian groups", Academic press, New York. vol. 36-1 and 36-2 (1973).

ON A PAPER OF I.FLEISCHER

Luigi Salce and Paolo Zanardo[1]

1. The last theorem of Kaplansky's paper [5], which develops the theory of modules over Dedekind and valuation domains, states that a finitely generated module over an almost maximal valuation domain R is a direct sum of cyclic modules. The proof is based on the fact that, if M' is a cyclic module with minimal annihilator of a torsion R-module M, then M' is pure in M ;unfortunately this fact is not generally true if R is not archimedean, as we will show by an easy counterexample. Matlis gave a correct proof of Kaplansky's theorem ([6,Prop. 14]) by means of a more careful argument.

In the paper [1] , Fleischer rederived and completed the results of Kaplansky by a unified theory of modules over Prüfer domains. Unfortunately he fell into a trap similar to the one which occured in Kaplansky : in fact he stated ([1,Prop.8]), without giving a detailed proof, that every submodule of a finitely generated module over an almost maximal valuation domain is a direct sum of rank one modules. Probably Fleischer based his proof on the fact (see [1,Lemma 3]) that a rank one dominating submodule of a torsion module is pure (for the definition of "rank" and "dominating submodule" see the next section); but a cyclic submodule with minimal annihilator is a rank one dominating submodule, therefore the same counterexample used for contradicting Kaplansky's argument shows that Lemma 3 of Fleischer's paper is

---

[1]Lavoro eseguito nell'ambito dei Gruppi di Ricerca Matematica del CNR.

false, without assuming the archimediecity of the ring.

The main goal of this paper is to give a proof of the result stated by Fleisher, by using a technique which resembles the one used by Matlis in [6,Prop.14] . In a forthcoming paper by L.Fuchs and the first author a more constructive proof will be given in the case of an archimedean valuation domain.

In the secon section we briefly discuss the definition of the rank of a module, and the notions of dominating submodule and archimedean ideal. In the third section we prove the main theorem and we show, by a counterexample, that it does not hold if the finiteness condition is dropped.

2. In the following R will always denote a valuation domain, i.e. a domain such that its ideals, or equivalently the submodules of its field of quotients Q, are totally ordered by inclusion ; P denotes the maximal ideal of R and v the value map of R. For general reference about valuation domains and related topics see [4] . Recall that R is an almost maximal valuation domain, briefly denoted by AMVD, if every proper quotient of R, or equivalently of its field of quotients Q, is linearly compact in the discrete topology. There are many characterizations for an AMVD, for which we refer to [5], [6], [3] and [9]: the ones which we shall use are that, for any submodule I of Q , Q/I is injective and that E(Q/I) is uniserial. A module is uniserial if its submodules are totally ordered by inclusion. Obviously submodules and quotients of an uniserial module are still uniserial, therefore every ry R-module of the form J/I, with $I \subsetneq J \subsetneq Q$, is uniserial ; if R is an AMVD, every uniserial R-module U is of this form, as is easily seen

by looking at the injective envelope E(U) , which has to be indecomposable, hence $E(U) \cong Q/I$ for some $I \subseteq Q$ (see [6]) .

We shall examine now the concept of rank r(M) of an R-module M. If M is torsion free, there are no problems in defining r(M) as the dimension of $M \otimes_R Q$ as a vector space over Q ; if M is an arbitrary R-module, Fleischer defines r(M) as the minimum r(F), where F ranges over the family of torsion free modules with M as epimorphic image.

Note that if R fails to be an AMVD, then E(Q/I) is not uniserial for a nonzero ideal I, therefore $r(E(Q/I)) > 1$; but if R is an AMVD , modules of finite rank have a more agreeable behaviour with respect to their injective envelope.

<u>Lemma 1</u> . Let R be an AMVD and M a torsion R-module. Then r(M) = n , where n is a positive integer, if and only if E(M) is the direct sum of n nonzero indecomposable injective modules.

<u>Proof</u>. It is easy to see, by induction on n, that every quotient of Q^n (the direct sum of n copies of Q) is the direct sum of at most n nonzero indecomposable injective modules. If r(M)=n , then M = F/K with F torsion free and $E(F) \cong F \otimes_R Q \cong Q^n$; therefore $E(M) \subseteq Q^n/K$ and the preceding remark shows that $E(M) \cong Q/I_1 \oplus \dots \oplus Q/I_m$ with $m \leq n$ and the I_i's ideals of R. If however $m < n$, then M is an epimorphic image of a torsion free module of rank m , which is absurd.

The preceding lemma shows that a torsion module over an AMVD R is of finite rank if and only if it is a submodule of a finite direct sum of torsion uniserial modules and that an R-module has rank one if and only if it is uniserial.

Recall that a submodule M' of an R-module M is pure in M if

$$rM \cap M' = rM' \quad \forall r \in R .$$

In order to find a rank one pure submodule in a torsion R-module M , Fleischer introduced the concept of dominating submodule : the submodule M' of M is said to be dominating in M if, for every $x \in M$, there exists an $y \in M'$ such that Ann $y \subseteq$ Ann x (recall that, if $z \in M$, then Ann $z = \{r \in R : rz=0\}$ and Ann $M = \bigcap_{x \in M}$ Ann x). The following lemma clarifies the relation between dominating submodules and submodules with minimal annihilator.

Lemma 2. Let M' be a submodule of a torsion R-module M . Then :

1) If M' is dominating in M, then Ann M'= Ann M.

2) If M' is cyclic and Ann M'= Ann M , then M' is dominating.

Proof. 1) If Ann $M \subsetneq$ Ann M', there exists an $r \in R$ such that $rM' = 0$ and $rx \neq 0$ for some $x \in M$; then Ann $x \subsetneq$ Ann M', therefore no $y \in M'$ satisfies the condition Ann $y \subseteq$ Ann x.

2) If $M' = \langle y \rangle$, then Ann M'= Ann y ; if $x \in M$, then Ann y = Ann M implies that Ann $y \leq$ Ann x.

The second claim in lemma 2 does not hold for arbitrary uniserial modules M'; for instance, let R be a non discrete archimedean valuation domain, M = R/aR and M'= P/aR , with aR a proper principal ideal ; then Ann M'= aR = Ann M but M' is clearly not dominating in M.

Lemma 3 in [1] asserts that a dominating rank one submodule of a torsion R-module is pure ; this is not generally true, as example 4 shows ; to get a correct result we need an extra-condition.

Recall that a nonzero ideal I of R is archimedean if rI = I, with

$r \in R$, implies that r is a unit. This definition goes back to Matlis [6] ; archimedean ideals are also implicitely used by Nishi [7] and Shores and Lewis [8] in studying endomorphism rings of indecomposable injective and uniserial R-modules. A detailed analysis of this property can be found in the forthcoming paper by Fuchs and the first author; it is quite obvious that a nonzero principal ideal is archimedean , while a nonzero prime ideal different from P is not archimedean.

<u>Lemma 3</u>. Let U be a dominating submodule of a torsion R-module M such that $U \cong J/I$, with $0 \neq I \subsetneq J \subsetneq Q$; if I is archimedean, then U is pure in M .

<u>Proof</u>. (See [1,Lemma 3]) We identify U with J/I. Let $0 \neq rx \in U$, where $r \in R$, $x \in M$; let $y \in U$ be such that $\mathrm{Ann}\, y \subsetneq \mathrm{Ann}\, x$, and let $rx = a + I$, $y = b + I$, with $a, b \in J$. We have the equalities :

$$(1) \quad \mathrm{Ann}\, y = \mathrm{Ann}(b + I) = b^{-1}I$$

$$(2) \quad r^{-1}\mathrm{Ann}\, x = \mathrm{Ann}\, rx = \mathrm{Ann}(a + I) = a^{-1}I \ ;$$

(2) implies that $\mathrm{Ann}\, x = ra^{-1}I$, therefore by (1) we get that $b^{-1}I \subsetneq ra^{-1}I$, or equivalently $I \subsetneq bra^{-1}I$. If $bra^{-1} \in R$, then the archimedeicity of I implies that bra^{-1} is a unit, therefore $a \in rbR$ and

$$rx = a + I \in r\langle b + I \rangle \subsetneq rU \ .$$

If $bra^{-1} \notin R$, then br is not a multiple of a, hence the reverse inclusion holds : $a = tbr$, for some $t \in R$; then

$$rx = a + I = r(tb + I) \in rU \ .$$

Therefore U is pure in M .

We will show now that the archimedeicity hypothesis in Lemma 3 cannot be avoided.

Example 4. Let R be a valuation domain with a nonzero prime ideal P' different from P. If $r \in P \setminus P'$, then r is not a unit and $rP' = P'$. Then rR/P' is a cyclic submodule of R/P' and $\mathrm{Ann}(rR/P') = \mathrm{Ann}(r + P') = r^{-1}P' = P' = \mathrm{Ann}(R/P')$; lemma 2 shows that rR/P' is a dominating submodule of R/P' ; but it is not pure, because

$$r(R/P') \cap (rR/P') = rR/P' \supsetneq r(rR/P') = r^2R/P' .$$

3. We shall prove Fleischer's [1,Prop.8] in the following more general form.

Theorem 5. A torsion module M of rank n over an AMVD R is a direct sum of n uniserial modules.

Proof. The case n = 1 is trivial. By induction on n, we assume the claim true for n-1, if n > 1. Let $E(M) = \bigoplus_{1\le i\le n} U_i$, with $U_i = Q/K_i$ and $M_i = M \cap U_i$ for every i = 1,..,n. Note that $M_i \neq 0$ for all i , otherwise M can be embedded into the $\bigoplus_{j\neq i} U_j$, and this implies that $r(M) \le n-1$. Henceforth we shall assume that $M_i = I_i/K_i$, with $K_i \subsetneq I_i \subseteq Q$ for every i. We shall prove that there exists an index j such that M_j is pure in M ; from this fact it is easy to conclude : M_j is pure injective, because it is linearly compact in the discrete topology (see [10]) therefore $M = M_j \oplus N$, with N canonically embedded into the $\bigoplus_{i\neq j} U_i$, and again the induction hypothesis ensures that N is a direct sum of n-1 uniserial modules.

Assume then, by way of contradiction, that no one of the M_j's is pure ; hence for every j there exist elements $m_j \in M$ and $r_j \in R$ such that

$$0 \neq r_j m_j \in (r_j M \cap M_j) \setminus rM_j .$$

Note that $r_j m_j \in M_j$ ensures that r_j annihilates all the coordinates of

m_j with index different from j ; moreover $r_j m_j \notin rM_j$ implies that the coordinate of m_j of index j is not in M_j . Thus there exists a set of n elements of M :

$$A = \{m_j : j = 1,\dots,n\}$$

(with $m_j = (a_i^j + K_i)_i$, $1 \leq i \leq n$, $a_i^j + K_i \in U_i$ for all i), satisfying the following condition :

(*) if $1 \leq j \leq n$, there exists an $r_j \in R$ such that :

$$0 \neq r_j m_j \in M_j \ ; \ a_j^j + K_j \notin M_j \ .$$

Condition (*) is equivalent to :

$$r_j a_j^j \in I_j \setminus K_j \ ; \ r_j a_i^j \in K_i \text{ if } i \neq j \ ; \ a_j^j \notin I_j \ .$$

Let now :

$$l(A) = \inf\{ k \in N : \{m_{j(1)},\dots,m_{j(k)}\} \subseteq M \cap (U_{j(1)} \ \dots \ U_{j(k)})\}$$

where the k indices $j(i)$'s are obviously all different.

Note that $l(A) > 1$, because $m_j \notin M_j = M \cap U_j$ for every j . Let assume now that A has minimal $k = l(A)$ among the sets of n elements of M satisfying property (*). Without loss of generality we can also assume that :

$$\{m_1, m_2, \dots, m_k\} \subseteq M \cap (U_1 \ \dots \ U_k).$$

The rest of the proof is based on the following

<u>Claim</u>. There exists an $h \in \{1,\dots,k\}$ such that, if $1 \leq j \leq k$, $a_h^j = s_j a_h^h$ for some $s_j \in R$.

Let assume, by way of contradiction, that for every $j \leq k$, there exists an index $i(j)$ (obviously different from j) such that :

$$a_j^j = s_{i(j)} \ a_j^{i(j)}$$

with $s_{i(j)} \in P$ (i.e. $v(s_{i(j)}) > 0$). Then $0 \neq r_j m_j$ implies that :

$$r_j s_{i(j)} a_j^{i(j)} = r_j a_j^j \notin K_j \ ,$$

therefore $v(r_j s_{i(j)}) < v(r_{i(j)})$, because $r_{i(j)} a_j^{i(j)} \in K_j$. If

$$v(r_t) = \max \{v(r_i) : 1 \leq i \leq k\} ,$$

then $v(r_t) < v(s_{i(t)} r_t) < v(r_{i(t)})$, which is absurd .

Note the resemblance of the previous claim with the argument used in [6,Prop. 14] . We are now ready to conclude the proof, by constructing a set A' of n elements of M satisfying property (*) and such that $l(A') < k$, contradicting the minimality of $l(A)$.

Without loss of generality we can assume h=1 in the claim . If $1 \leq j \leq n$, let

$$m'_j = \begin{cases} m_j & \text{if } j=1 \text{ or } k+1 \leq j \leq n \\ s_j m_1 - m_j & \text{if } 2 \leq j \leq k \end{cases} .$$

The set $A' = \{m'_j : 1 \leq j \leq n\}$ is obviously contained in M . Note that

$$(3) \qquad v(r_j s_j) > v(r_1) \qquad \text{if } 2 \leq j \leq k$$

because $r_j s_j a_1^1 = r_j a_1^j \in K_1$ and $r_1 a_1^1 \notin K_1$. Let now, for $2 \leq j \leq k$,

$$m'_j = (b_i^j + K_i)_i \qquad (1 \leq i \leq n)$$

with $b_i^j + K_i \in U_i$ for every i . Then, if $2 \leq j \leq k$

$$b_i^j = s_j a_1^1 - a_1^j = 0$$

and, if $k+1 \leq i \leq n$, $b_i^j = 0$ because $m_j \in U_1 \oplus \ldots \oplus U_k$.

Hence $\{m'_2, \ldots, m'_k\} \subseteq M \cap (U_2 \oplus \ldots \oplus U_k)$ and $l(A') < k$. To get the desired contradiction we have only to show that A' satisfies property (*) . It is enough to verify this for an index j such that $2 \leq j \leq k$. Now we have :

$$r_j b_j^j = r_j (s_j a_j^1 - a_j^j) \in I_j \setminus K_j$$

because of (3) and $j \neq 1$. If $i \neq j$, then

$$r_j b_i^j = r_j (s_j a_i^1 - a_i^j) \in K_i$$

because $r_j a_i^j \in K_i$ and (3) implies $r_j s_j a_i^1 \in K_i$, except for the case

$i=1$, for which $b_1^j = 0$. Finally

$$b_j^j = s_j a_j^1 - a_j^j \notin I_j$$

because otherwise $a_j^j \notin I_j$ implies $v(s_j a_j^1) = v(a_j^j)$, and $r_j a_j^j \notin K_j$ implies $r_j s_j a_j^1 \notin K_j$, contradicting (3) and $r_1 a_j^1 \in K_j$.

If R is an archimedean AMVD, it is possible to say which of the M_i's of the proof of theorem 5 is pure in M , by simply comparing them and their annihilators ; but the proof in that case uses quite different arguments. We show now by an example that it is not possible , in the general case , to decide which of the M_i's is pure , by just looking at their properties .

Example 6. Let R be a non archimedean AMVD and I a nonzero prime ideal of R different from P . Let $c \in I$ and $K = cP$; if $x \in P \setminus I$, consider the submodule M of $U_1 \oplus U_2$, where $U_i = R/K$ for i=1,2 , defined by

$$M = \langle (1 + K, x + K) ; (a + K, 0) : a \in I; (0, b + K) : b \in I \rangle.$$

Let $y \in M_1 = M \cap U_1$; then $y = r(1 + K , x + K) + s(a + K , 0) + t(0 , b + K)$, with $r,s,t \in R$, $a,b \in I$ and $rx + tb \in K \subseteq I$; but $b \in I$ implies that $rx \in I$. Thus $r \in I$ because I is prime, hence $r1 + sa \in I$ and this obviously implies that $y \in I/K \oplus \{0\}$. Let $y' = r'(1 + K , x + K) + s'(a' + K , 0) + t'(0 , b' + K)$, with $r',s',t' \in R$, $a',b' \in I$ and $r'1 + s'a' \in K \subseteq I$. Now $a' \in I$ implies that $r' \in I$, thus $r'x + t'b' \in I$ and $y' \in \{0\} \oplus I/K$. It follows that M_1 and M_2 are both isomorphic to I/K , but their behaveiour with respect to the purity in M is different ; one of them has to be pure in M , by theorem 5 , but M_1 cannot be pure : in fact $c(1 + K , x + K) \in M_1$ because $c \in I$ and $cx \in K$, but it is not 0 , because $c \notin K$. Moreover, if $(y + K , 0) \in M_1$, then $c(y + K,$

0) = 0 , because $y \in I$ and $cy \in K$.

We conclude this paper with an example of a module of infinite rank which is pure in a direct sum of uniserial modules and which fails to be a direct sum of uniserial modules .

Example 7. Let R be an AMVD with divisibility group G(R) isomorphic to a dense subgroup G of the additive group of the real numbers (for these notions we refer to [4]). Let $g \in G$, $g > 0$ and , for every $q \in G$ such that $0 < q \leq g$, let $r_q \in R$ such that $v(r_q) = q$. Let H_g be the R-module generated by the elements :

$$\{ x \, ; \, x_q : 0 < q < g \}$$

subjected to the relations :

$$\mathrm{Ann}\, x = P \, ; \, r_q x_q = x \quad \text{for every } q \, .$$

The construction of H_g is similar to that of the Prüfer group $H_{\omega+1}$ (see [2, pag.150]), the simplest example of an abelian p-group with nonzero elements of infiniti height . If we adjoin to H_g a generator y , subjected to the relation :

$$r_g y = x \, ,$$

then it is easy to verify that $H_g + \langle y \rangle$ is a direct sum of cyclic modules and $H_g + P\langle y \rangle$ is a direct sum of uniserial modules , in which H_g is pure . We shall prove that H_g is not a direct sum of uniserial modules . Assume , by way of contradiction , that $H_g = \bigoplus_{i \in I} U_i$, with U_i uniserial for all i . $\langle x \rangle$ is simple, therefore $\langle x \rangle \subseteq U_j$ for some $j \in I$. The module $H_g/\langle x \rangle$ is trivially a direct sum of cyclics , hence every U_i is cyclic. But this is impossible, because $x \in \bigcap_{q<g} r_q H_g$ and $x \notin r_g H_g$, which cannot happen in a direct sum of cyclic modules .

REFERENCES

[1] FLEISCHER,I. "Modules of finite rank over Prüfer rings" Annals Math. 65 , 2 (1957) 250-254 .

[2] FUCHS,L. "Infinite Abelian Groups" Academic Press ; London , New York 1971 .

[3] GILL,D.T. "Almost maximal valuation rings" J.London Math.Soc. 4 , 1 (1971) 140-146 .

[4] GILMER,R. "Multiplicative Ideal Theory" Marcel Dekker , New York 1972 .

[5] KAPLANSKY,I. "Modules over Dedekind rings and valuation rings" Trans.Amer.Math.Soc. 72 (1952) 327-340 .

[6] MATLIS,E. "Injective modules over Prüfer rings" Nagoya Math.J. 15 (1959) 57-69 .

[7] NISHI,M. "On the ring of endomorphisms of an indecomposable injective module over a Prüfer ring" Hiroshima Math.J. 2 (1972) 271-283 .

[8] SHORES,T.S. and LEWIS,W.T. "Serial modules and endomorphism rings" Duke Math.J. 41 , 4 (1974) 889-909 .

[9] VAMOS,P. "Classical rings" J.Alg. 34 (1975) 114-129 .

[10] WARFIELD,R.B.Jr. "Purity and algebraic compactness for modules" Pacific J.Math. 28 , 3 (1969) 699-719 .

Weak compactness and the structure of Ext(A,$\mathbb{Z}$) *

by

G. Sageev and S. Shelah

In this paper we address a conjecture of the second author, namely that in L, for any cardinal $\kappa \geq \aleph_1$ and any prescription of cardinals $\lambda p \leq \kappa^+$ to the primes p, there exists an abelian group A for which $\nu_0(A) = \kappa^+$ = the rank of the torsion free part of Ext(A,$\mathbb{Z}$), and $\lambda p = \nu_p(A)$ = the rank of the p-part of Ext(A,$\mathbb{Z}$).

It is well known that these cardinals characterize the divisible group Ext(A,$\mathbb{Z}$) for torsion free A. The conjecture is false for countable A, where it has been shown by C. Jensen [11] that $\nu_p(A)$ is either finite or $2^{\aleph_0}$ and $\nu_p(A) \leq \nu_0(A)$. Similarly Hulanicki [8,9] has shown that for divisible abelian groups which admit a compact topology $\nu_p(A) \leq \nu_0(A)$ and $\nu_p(A)$ is finite or of the form 2^λ, λ infinite. However we have shown that the conjecture is true for $\kappa = \aleph_1 = |A|$ under ZFC + GCH alone, see [12], and Eklof and Huber this volume, [4]. Just using the fact that $\mathrm{Ext}_p(\oplus A_i,\mathbb{Z}) = \Pi \mathrm{Ext}_p(A_i,\mathbb{Z})$ it is now easy to see that for any cardinal $\kappa \geq \aleph_1$ and successor cardinals $\lambda p \leq \kappa^+$ there exists an abelian group for which $|A| = \kappa$ and $\nu_p(A) = \lambda p$.

The question remains whether we can have $\nu_p(A) = |A|$ or $\nu_p(A)$ singular or $\nu_p(A)$ inaccessible. We show that the conjecture is not true in all generality by proving that

<u>THEOREM</u>. If A is a torsion free abelian group of weakly compact cardinality κ and $\nu_p(A) \geq \kappa$, then $\nu_p(A) = 2^\kappa$.

Since weak compactness is consistent with V = L, provided it is consistent with ZFC, the above theorem displays some restriction of the conjecture.

There are a number of equivalent definitions of weak compactness [1,10]; a suitable one for a non logician is:

*This work has been partially supported by The National Science Foundation (grant No. 710646).

<u>DEFINITION</u>. (i) A cardinal κ has the <u>tree property</u> iff, for every tree, T, of height κ with levels of cardinality $<\kappa$ has a branch of length κ. (ii) κ is <u>weakly compact</u> iff it is inaccessible and has the tree property.

Weak compactness was originally introduced in relation to compactness of certain infinitary languages which in turn can be readily related to the following equivalent property for inaccessible κ:

Any κ-complete filter D in a κ-complete field B of subsets of κ can be extended to a κ-complete prime filter in B.

As for the tree property: $\aleph_0$ has the tree property by Köenigs lemma and it is also a well known result of Aronszajn that $\aleph_1$ does not have the tree property; moreover singular cardinals, and with GCH, also successor to regular cardinals do not have the tree property.

The treatment of $\mathrm{Ext}_p(A,\mathbb{Z})$ is based on the following theorem.

<u>DEFINITION</u>. Let $H: \mathrm{Hom}(A,\mathbb{Z}) \to \mathrm{Hom}(A,\mathbb{Z}/p\mathbb{Z})$ be the natural homomorphism defined by:

$$[H(h)](x) = h(x)/p\mathbb{Z}\ ,\ h \in \mathrm{Hom}(A,\mathbb{Z})\ ,\ x \in A,\ p \text{ a prime.}$$

<u>THEOREM</u>. For abelian torsion free A

$$\mathrm{Ext}_p(A,\mathbb{Z}) \cong \mathrm{Hom}(A,\mathbb{Z}/p\mathbb{Z})/H[\mathrm{Hom}\ A,\mathbb{Z}]\ .$$

<u>Proof</u>: The exact sequence $0 \to p\mathbb{Z} \overset{\alpha}{\to} \mathbb{Z} \overset{\beta}{\to} \mathbb{Z}/p\mathbb{Z} \to 0$, α the identity embedding, β natural, induces the long exact sequence

$$0 \to \mathrm{Hom}(A,p\mathbb{Z}) \to \mathrm{Hom}(A,\mathbb{Z}) \to \mathrm{Hom}(A,\mathbb{Z}/p\mathbb{Z})$$
$$\overset{E_*}{\to} \mathrm{Ext}(A,p\mathbb{Z}) \overset{\alpha_*}{\to} \mathrm{Ext}(A,\mathbb{Z}) \overset{\beta_*}{\to} \mathrm{Ext}(A,\mathbb{Z}/p\mathbb{Z}) \to 0$$

(see Fuchs [6]). Since the sequence is exact,

$$J = \mathrm{Hom}(A,\mathbb{Z}/p\mathbb{Z})/H[\mathrm{Hom}(A,\mathbb{Z})] \cong \mathrm{Ker}(A_*) = \mathrm{Im}(E_*)$$

$\mathbb{Z}$, $p\mathbb{Z}$ are isomorphic; hence also $\mathrm{Ext}(A,\mathbb{Z})$, $\mathrm{Ext}(A,p\mathbb{Z})$; in particular elements of order p of $\mathrm{Ext}(A,\mathbb{Z})$, are represented by elements of order p in $\mathrm{Ext}(A,p\mathbb{Z})$. All elements of J are of order p. Hence it suffices to show that all extension $E \in \mathrm{Ext}(A,p\mathbb{Z})$ of order p are mapped to 0 by α_*. Let $E \in \mathrm{Ext}(A,p\mathbb{Z})$, $pE = 0$, be represented by a factor set $f: A \times A \to p\mathbb{Z}$. Thus for some function $g: A \to p\mathbb{Z}$ with $g(o) = 0$, $pf(x,y) = g(x) + g(y) - g(x+y) \in \mathrm{Trans}(A,\mathbb{Z})$, $\forall\ x,y \in A$.

Since α is an injection, $\alpha_*(E)$ can be represented by the same f. Now since $A, \mathbb{Z}$ are torsion free, there is a unique $g': A \to \mathbb{Z}$ such that $pg'(x) = g(x)$, $\forall\, x \in A$. Therefore $f(x,y) = g'(x) + g'(y) - g'(x+y)$, hence also $\alpha^*(E) = 0$. □

<u>THEOREM (ZFC)</u>. If G is a group of weakly compact cardinality, κ, for which $\nu_p(G) \geq \kappa$, then $\nu_p(G) = 2^\kappa$.

<u>Proof</u>: Let G be a group of weakly compact cardinality κ with $\nu_p(G) \geq \kappa$. We shall show that $\nu_p(G) = 2^\kappa$. This is done by constructing a filtration $\langle G_\alpha : \alpha \leq \kappa \rangle$ of G, and a tree of homomorphisms $h_\eta : G_\alpha \to \mathbb{Z}/p\mathbb{Z}$, $\eta \in {}^\alpha 2$, ordered by inclusion ($\eta \subset \eta' \Rightarrow h_\eta \subset h_{\eta'}$) and continuous, i.e. if $C = \{h_{\eta_\beta} : \beta < \alpha,\ h_{\eta_\beta} \subseteq \mathrm{Hom}(G_\beta, \mathbb{Z}/p\mathbb{Z})\}$ is a chain with α limit and $\eta = \bigcup_{\alpha<\beta} \eta_\beta$, then $h_\eta = \bigcup_{\beta<\alpha} h_{\eta_\beta}$. We will construct the tree T, such that at each level α, $\{h_\eta : \eta \in {}^\alpha 2\} = T_\alpha$ are independent homomorphisms mod $H[\mathrm{Hom}(G_\alpha, \mathbb{Z})]$, where H is the operation described above. It is easy to see that this property will be preserved at limit ordinals.

We first exploit the tree property of κ to obtain the following lemma:

<u>LEMMA</u>. If $D \subset \mathrm{Hom}(G, \mathbb{Z}/p\mathbb{Z})$ are independent mod $H[\mathrm{Hom}(G,\mathbb{Z})]$ and $|D| < \kappa$, then for any subgroup $G' \subseteq G$, $|G'| < \kappa$ there exists a subgroup G'', $G' < G'' < G$, $|G''| < \kappa$ such that $\{h \restriction G'' : h \in D\}$ are independent mod $H[\mathrm{Hom}(G,\mathbb{Z})]$.

<u>Proof</u>: If not, then there exists a continuous strictly increasing sequence of subgroups of G, $\langle G_\alpha \rangle_{\alpha<\kappa}$ with $|G_\alpha| < \kappa$ and $G_0 = G'$, $G_\kappa = G$, such that $\forall \alpha < \kappa$, $h', h'' \in D, h' \neq h''$ and $h' \restriction G_\alpha = h'' \restriction G_\alpha \bmod (H[\mathrm{Hom}(G_\alpha, \mathbb{Z})])$.

<u>Notation</u>. Henceforth we denote for $h', h'' \in \mathrm{Hom}(G, \mathbb{Z}/p\mathbb{Z})$, $h' \equiv_\alpha h''$ iff $h' \equiv h'' \bmod (H[\mathrm{Hom}(G\ ,\mathbb{Z})])$.

Now for $\beta < \alpha$, $h' \restriction G_\alpha \equiv_\alpha h'' \restriction G_\alpha \Rightarrow h' \restriction G_\beta \equiv_\beta h'' \restriction G_\beta$. Moreover if $h \in \mathrm{Hom}(G_\alpha, \mathbb{Z})$ is such that $h' \restriction G_\alpha = h'' \restriction G_\alpha + h/p\mathbb{Z}$, then $h' \restriction G_\beta = h'' \restriction G_\beta + h \restriction G_\beta/p\mathbb{Z}$. Clearly if $h', h'' \in D$, $\alpha < \kappa$ and $h' \restriction G_\alpha \not\equiv h'' \restriction G_\alpha$, then $h' \restriction G_\beta \not\equiv_\beta h'' \restriction G_\beta, \forall \beta$, $\alpha \leq \beta \leq \kappa$. Therefore,

$\exists h^*, h^{**} \in D$ such that $\forall \alpha < \kappa$ $h^* \upharpoonright G_\alpha \equiv_\alpha h^{**} \upharpoonright G_\alpha$. For these fixed h^*, h^{**} let $K_\alpha = \{h \in \mathrm{Hom}(G_\alpha, \mathbb{Z}) : h^* \upharpoonright G_\alpha = h^{**} \upharpoonright G_\alpha + h/p\mathbb{Z}\}$

Clearly $\{h \upharpoonright G_\beta : h \in K_\alpha\} \subseteq K_\beta$, $\beta < \alpha$ and $|K_\alpha| < \kappa$. If $K = K_\alpha$ and K is partially ordered by extension, $\prec$, then $\langle K, \prec\rangle$ is a tree of height κ with levels of cardinality less than κ. Thus by weak compactness there exists a branch b of length κ, $b = \{h_\alpha : \alpha < \kappa\}$. Then $\bigcup_{\alpha<\kappa} h_\alpha = h \in \mathrm{Hom}(G, \mathbb{Z})$ and $h^* = h^{**} + h/p\mathbb{Z}$.

This is a contradiction. □

<u>Construction of the filtration and respective tree, T , of homomorphisms.</u>

For $\alpha = 0$, we set $G_0 = 0$, $T_0 = \emptyset$. For $\alpha = 1$, we choose $G_1 \subset G$ of cardinality $<\kappa$ such that $\exists\, h_0, h_1 \in \mathrm{Hom}(G_1, \mathbb{Z}/p\mathbb{Z})$ which are independent mod $H[\mathrm{Hom}(G_1, \mathbb{Z})]$. Such h_0, h_1 exist by the lemma. We set $T_1 = \{h_0, h_1\}$. For α limit we just take unions, i.e. $G_\alpha = \bigcup_{\beta<\alpha} G_\beta$, and $T_\alpha = \{\bigcup_{\eta \in b} h_\eta : b$ is a branch through $\bigcup_{\beta<\alpha} T_\beta\}$. For $\alpha = \beta + 1$ successor, we first choose $G_{\beta+1} > G_\beta$ such that

$\nu_p(G_{\beta+1}) > (2^{|G_\beta|})^+$. This is again possible by our lemma. For every

$\eta \in {}^\beta 2$ let h_{η_0} be any extension of h_η in $\mathrm{Hom}(G_\alpha, \mathbb{Z}/p\mathbb{Z})$. These will be independent mod $H[\mathrm{Hom}(G_\alpha, \mathbb{Z})]$ since the $\{h_\eta : \eta \in {}^\beta 2\}$ are independent mod $H[\mathrm{Hom}(G^\beta, \mathbb{Z})]$. We must choose the h_{η_1} so that they

are all similarly independent. By our choice of $G_{\beta+1}$ we can find a

family of $(2^{|G_\beta|})^+$ homomorphisms containing $\{h_{\eta_0} : \eta \in {}^\beta 2\}$ which are independent mod $H[\mathrm{Hom}(G_\alpha, \mathbb{Z})]$. Again from our cardinality assumptions, we can choose from these, $|G_\beta|$ distinct disjoint pairs of homomorphisms (h', h'') such that $h' \upharpoonright G_\beta = h'' \upharpoonright G_\beta$. Thus we can assign from these to every $\eta \in {}^\beta 2$ a distinct pair (h'_η, h''_η) and set $h_{\eta_1} = h_{\eta_0} + (h'_\eta - h''_\eta)$. Thus we clearly have

$h_{\eta_0} \upharpoonright G_\beta = h_{\eta_1} \upharpoonright G_\beta = h_\eta$: and the $(T_\alpha =)\ \{h_{\eta_0}, h_{\eta_1} : \eta \in {}^\beta 2\}$ are in-

dependent mod $H[\mathrm{Hom}(G_{\beta+1}, \mathbb{Z})]$.

Since $G = G_\kappa$, $T = T_\kappa \subseteq \mathrm{Hom}(G, \mathbb{Z}/p\mathbb{Z}) - H[\mathrm{Hom}(G, \mathbb{Z})]$ and $|T_\kappa| = 2^\kappa$, we have $\nu_p(G) = 2^\kappa$. □

REFERENCES

1. Drake, F. R., Set Theory, North Holland Publishing Co., (1974).

2. Eklof, P., Methods of Logic in Abelian Group Theory, in: Abelian Group Theory, Springer Verlag Lecture Notes 616 (1977).

3. Eklof, P. and Huber, M., Abelian Group Extensions and the Axiom of Constructibility, Math. Helv. 54, 440-457 (1979).

4. Eklof, P. and Huber, M., On the p-ranks of Ext(A,G), Assuming CH, This volume (1981).

5. Fuchs, L., Infinite Abelian Groups, Vol. I, Academic Press, New York (1970).

6. Fuchs, L., Infinite Abelian Groups, Vol. II, Academic Press, New York (1973).

7. Hiller, H., Huber, M., Shelah, S., The Structure of Ext(A,$\mathbb{Z}$) and V = L , Math. Zeitschr., 162, 39-50(1978).

8. Hulanicki, A., Algebraic Characterization of Abelian Divisible Groups which Admit Compact Topologies, Fund. Math. 44, 192-197 (1957).

9. Hulanicki, A., Algebraic Structure of Compact Abelian Groups, Bull. Acad. Polon. Sci. Sér. Sci. Math. Astronom. Phys. 6, 71-73 (1958).

10. Jech, T., Set Theory, Academic Press, New York (1978).

11. Jensen, C., Les Foncteurs Dérives de Lim et Leurs Applications en Theorie des Modules. Lecture Notes in Mathematics 254, Berlin-Heidelberg-New York, Springer (1972).

12. Sageev, G. and Shelah, S., On the Structure of Ext(A,$\mathbb{Z}$) in ZFC^+, submitted to the Journal of Symbolic Logic (1980).

13. Shelah, S., Whitehead groups may not be free even assuming CH, I, Israel J. Math. 28, 193-204 (1977).

14. Shelah, S., Whitehead groups may not be free even assuming CH, II, Israel J. Math., in Press.

15. Shelah, S., On Uncountable Abelian Groups, Israel J. Math. 32, 311-330 (1979).

16. Shelah, S., Consistency of Ext(G,ℤ) = Q, submitted to I.J.M.

ON THE p-RANKS OF EXT(A,G), ASSUMING CH

Paul C. Eklof[1] and Martin Huber[2]

1. Introduction. Given Abelian groups A and G with A torsion-free, Problem 39 in [F] asks for the determination of the (torsion-free) rank and the p-ranks of the divisible group $\mathrm{Ext}(A,G)$. For A countable and $G = Z$ it was observed by Chase and by Jensen, respectively, that these invariants are subject to certain restrictions ([C3, Lemma 1.4], [J, Théorème 2.7]). In [HW] Jensen's result has been extended to the case that G is any reduced rational group. For uncountable A and G satisfying suitable cardinality conditions, the present authors proved that under various set-theoretic hypotheses there are similar restrictions on the rank of $\mathrm{Ext}(A,G)$ [EH2] (thus improving results in [HHS] and [EH1]). In contrast to this, recent results of Shelah [Sh] and Sageev [SS] show that assuming CH (the Continuum Hypothesis) the p-ranks of $\mathrm{Ext}(A,Z)$ can be arbitrary. More precisely, they established the following theorem.

THEOREM 1.1. Assume CH. Let χ be any function assigning to each prime p a cardinal $\chi(p) \leq \aleph_1$ or $\chi(p) = 2^{\aleph_1}$. Then there exists an $\aleph_1$-free group A of cardinality $\aleph_1$ such that $r_o(\mathrm{Ext}(A,Z)) = 2^{\aleph_1}$ and for every p, $r_p(\mathrm{Ext}(A,Z)) = \chi(p)$. (By $r_o(H)$ $[r_p(H)]$ we denote the rank [the p-rank] of the Abelian group H.)

It is the aim of this note to give an alternate proof of this theorem which avoids much of the elaborate combinatorial machinery of the proof of Sageev and Shelah and makes use instead of ideas of Chase involving function topologies on groups [C2]. Chase used these ideas (in [C3]) to construct - assuming CH - an $\aleph_1$-free group A of cardinality $\aleph_1$ such that $\mathrm{Ext}(A,Z)$ is non-zero and torsion-free; so our construction is really a generalization of his, which combines his

1. Partially supported by NSF Grant No. MCS76-12014 A01
2. This work was done while the second author was a Fellow of the "Schweizerische Nationalfonds"

ideas with ideas from [SS]. In addition, we shall extend Theorem 1.1 to the case of Ext(A,G) where G is any reduced rational group (Theorem 5.1).

In Section 2 we gather some auxiliary results including two theorems from [C2] concerning subgroups of dual groups which are dense with respect to a certain topology. Section 3 is devoted to the proof of two key lemmas which provide the essential step in the inductive process by which the group A in question is built. This process is carried out in Section 4. In the final section we consider the case of Ext(A,G) where G is reduced rational. With a view to this latter result, we will, in the earlier sections, work in the category of R-modules where R is a principal ideal domain (PID).

2. Auxiliary results. Throughout this paper, R will be a PID with quotient field $Q \neq R$. We shall use the following terminology and notations. Given an R-module A we shall write $A^* = \mathrm{Hom}_R(A,R)$, the dual module of A. If $f: A \to B$ is an R-homomorphism, the induced homomorphism: $B^* \to A^*$ is denoted by f^*. For a prime p in R we define as usual, $pA = \{pa \mid a \in A\}$ and $A[p] = \{a \in A \mid pa = 0\}$; we recall that the p-rank of A is the dimension of the R/pR-vector space A[p]. The canonical epimorphism $R \to R/pR$ induces a natural map: $A^* \to \mathrm{Hom}_R(A,R/pR)$, denoted by Φ_A^p. If $f \in A^*$ we sometimes will write f/p instead of $\Phi_A^p(f)$; $f \equiv g$ (p) shall mean $f/p = g/p$.

The following lemma is certainly known; we include a proof for completeness. We shall write $\mathrm{Ext}_R(A,R)$ instead of $\mathrm{Ext}_R^1(A,R)$.

LEMMA 2.1. For any R-module A,

$$r_p(\mathrm{Ext}_R(A,R)) = \dim_{R/pR} \mathrm{Coker}\, \Phi_A^p.$$

Proof. The short exact sequence

$$0 \longrightarrow R \xrightarrow{\ p\ } R \longrightarrow R/pR \longrightarrow 0$$

(p denoting multiplication by p) gives rise to an exact sequence

$$A^* \xrightarrow{\Phi_A^p} \mathrm{Hom}_R(A,R/pR) \longrightarrow \mathrm{Ext}_R(A,R) \xrightarrow{p_*} \mathrm{Ext}_R(A,R).$$

Therefore we have $\mathrm{Coker}\ \Phi_A^p \cong \mathrm{Ker}\ p_*$. However, p_* is nothing but multiplication by p. Hence $\mathrm{Ker}\ p_* = \mathrm{Ext}_R(A,R)[p]$, and the lemma follows. □

Next we list (without proof) some simple but useful observations. If B is a submodule of a torsion-free R-module A, the pure closure of B in A will be denoted by B_*.

(2.2) Suppose that B is a submodule of A such that A/B is uniquely p-divisible. Let $h \in \mathrm{Hom}_R(B,R/pR)$ and $f \in B^*$. Then (i) there is exactly one $h' \in \mathrm{Hom}_R(A,R/pR)$ such that $h'|B = h$; (ii) there is at most one $f' \in A^*$ such that $f'|B = f$. □

(2.3) Let A be torsion-free and let B be a submodule of A such that A/B is a Q-vector space. Let $\iota : B \hookrightarrow A$ denote the inclusion and let F be a submodule of A^*. Then ι^* maps F_* onto the pure closure of $\iota^*(F)$ in B^*. □

(2.4) Let A be a free R-module, let p, q be different primes; let $h_o \in \mathrm{Hom}_R(A,R/pR)$ and $h_1 \in \mathrm{Hom}_R(A,R/qR)$. Then there is an $f \in A^*$ such that $f/p = h_o$ and $f/q = h_1$. □

Next we briefly recall some terminology and results from [C2]. Given an R-module A, its dual module A^* may be equipped with the A-<u>topology</u> i.e., the Hausdorff linear topology which is given by taking as neighborhoods of 0 the submodules

$$U_X = \{f \in A^* \mid f(X) = 0\},$$

where X ranges over all finite subsets of A. In the sequel, when we say that a submodule F of A^* is dense in A^*, we shall always mean that F is dense in A^* with respect to the A-topology.

<u>THEOREM</u> 2.5. (cf. [C2, Theorem 1.4]) A submodule F of A^* is dense in A^* iff for every $a \in A$ which generates a pure submodule of A there exists $f \in F$ such that $f(a) = 1$. □

<u>THEOREM</u> 2.6. Let A be a free R-module of countably infinite rank and let F be a pure submodule of A^* of countable rank (hence F is free). Then F is dense in A^* iff there exist a basis $\{a_n | n<\omega\}$ of A and a basis $\{f_n | n<\omega\}$ of F such that $f_m(a_n) = \delta_{mn}$.

<u>Proof</u>. Chase [C2, Theorem 3.2] proved that if F is dense then the two bases exist. That the converse holds is easily seen, using the characterization of Theorem 2.5. □

3. <u>The key lemmas</u>. From now on let R always denote a <u>countable</u> PID. As before let Q be its quotient field, and if p is a prime, let R_p denote the localization of R at p, a discrete valuation ring. The proof of Theorem 1.1 is based on the following two lemmas, the first of which is similar to Lemma 4.3 of [C3].

<u>LEMMA</u> 3.1. Let p be a prime of R. Let B be a free R-module of countably infinite rank and let F be a countable dense pure submodule of B^*. Then there is a short exact sequence of R-modules

$$0 \to B \to A \to R_p \to 0$$

such that A is free and every $f \in F$ extends to a homomorphism on A.

<u>Proof</u>. The lemma is trivial if R is a discrete valuation ring. Otherwise, let $\{p_n | n<\zeta\}$ be an enumeration of the primes in R such that $p_o = p$. By Theorem 2.6 there exist a basis $\{b_n | n<\omega\}$ of B and a basis $\{f_n | n<\omega\}$ of F such that $f_n(b_m) = \delta_{nm}$. We define A to be generated by B and $\{y_n | n<\omega\}$ with relations

$$r_n\, y_{n+1} = y_n - b_n \quad (n<\omega)$$

where $r_n = \prod_{i=1}^{n} p_i$. (If ζ is finite we let $r_n = r_{\zeta-1}$ for $n \geq \zeta$). Clearly A is an extension of B by R_p, and A is free with basis $\{y_n | n<\omega\}$. Let

$$\phi_n(y_m) = \begin{cases} \delta_{nm} & \text{if } n \leq m; \\ r_m \phi_n(y_{m+1}) & \text{working downwards if } n > m. \end{cases}$$

This defines a homomorphism $\phi_n \in A^*$ which agrees on B with f_n. Since the f_n's generate F, the lemma is proved. □

If A is an R-module and $a_i \in A, i < \kappa$, we shall write $\langle a_i | i<\kappa \rangle$ for the submodule generated by $\{a_i | i<\kappa\}$ (although this notation is not standard in module theory).

<u>LEMMA</u> 3.2. Let R be any countable PID and let p be a prime of R. Let B be a free R-module of countably infinite rank, let F be a countable dense pure submodule of B^*, and suppose that $g \in B^*$ is such that $g/p \notin \Phi_B^p(F)$. Then there exists an extension

$$0 \to B \to A \to Q \to 0$$

such that

(a) A is free;

(b) each $f \in F$ extends to $f' \in A^*$;

(c) $\{f' \in A^* | (f'|B) \in F\}$ is a dense submodule of A^*;

(d) g does not extend to a homomorphism on A.

<u>Proof</u>. Let $\{p_n | n<\zeta\}$ enumerate the primes in R; suppose $p_0 = p$; and let $\{r_n | n<\omega\}$ be an enumeration of R. Let $\{b_n | n<\omega\}$ and $\{f_n | n<\omega\}$ be bases of B and of F, respectively, such that $f_n(b_m) = \delta_{nm}$ (which exist by 2.6). By induction on n we define a permutation $\sigma : \omega \to \omega$ and new bases $\{a_n | n<\omega\}$ of B and $\{\phi_n | n<\omega\}$

of F satisfying the following five conditions.

(i) For all n, $\phi_n = f_{\sigma(n)}$.

(ii) If $n \not\equiv 0\ (3)$, $a_n = b_{\sigma(n)}$.

(iii) If $n \equiv 0\ (3)$, $\langle a_n, a_{n+1}\rangle = \langle b_{\sigma(n)}, b_{\sigma(n+1)}\rangle$.

(iv) $\phi_n(a_m) = \delta_{nm}$ if $m \geq n$.

For each $n < \omega$, let A_n be the R-module generated by B and $\{y_i \mid i \leq n\}$ with relations

$$s_i\, y_{i+1} = y_i - a_i \quad (i<n)$$

where $s_i = \prod_{j=0}^{i} p_j$. (If ζ is finite we let $s_i = s_{\zeta-1}$ for $i \geq \zeta$.) Let $g^{(k)}$ denote the unique extension of g to a homomorphism on A_{3k} such that $g^{(k)}(y_o) = r_k$ (if it exists at all). We require also:

(v) for all k, if $g^{(k)}$ exists then $g^{(k)}(y_{3k} - a_{3k}) \not\equiv 0\ (p)$.

Suppose that $\{a_i \mid i<n\}$ and $\{\phi_i \mid i<n\}$ have been defined. We first consider the case where $n \equiv 2\ (3)$. Let $\sigma(n)$ be the least m such that $m \notin \{\sigma(i) \mid i<n\}$. (This will insure that σ is onto.) Let $a_n = b_{\sigma(n)}$ and $\phi_n = f_{\sigma(n)}$, so of course (i), (ii) and (iv) are satisfied. Now suppose that $n \equiv 0\ (3)$, say $n = 3k$. By (i) we have

$$\langle b_m \mid m \notin \{\sigma(i) \mid i<n\}\rangle = \bigcap_{i<n} \operatorname{Ker} \phi_i .$$

Therefore, since $g/p \notin \langle \phi_i/p \mid i<n\rangle$, there exists $m \notin \{\sigma(i) \mid i<n\}$ such that $g(b_m) \not\equiv 0\ (p)$. Let $\sigma(n)$ be some t such that $t \notin \{\sigma(i) \mid i<n\} \cup \{m\}$, let $\phi_n = f_t$, and define

$$a_n = \begin{cases} b_m + b_t & \text{if } g^{(k)}(y_n - b_m - b_t) \not\equiv 0\ (p); \\ 2b_m + b_t & \text{otherwise.} \end{cases}$$

Furthermore, let $\sigma(n+1) = m$, $\phi_{n+1} = f_m$ and $a_{n+1} = b_m$. Then clearly (i), (ii) and (iii) hold, (iv) is readily verified, whereas (v) holds because

$$g^{(k)}(y_n-b_m-b_t) - g^{(k)}(y_n-2b_m-b_t) = g(b_m) \not\equiv 0 \ (p)$$

(if $g^{(k)}$ exists).

Now let $A = \bigcup_{n<\omega} A_n$ i.e., $A = \langle B, y_n \mid s_n y_{n+1} = y_n - a_n, n<\omega\rangle$. Clearly A is an extension of B by Q, and since $\{a_n \mid n<\omega\}$ is a basis of B, A is a free R-module with basis $\{y_n \mid n<\omega\}$. Moreover, it follows from (v) that g does not extend to a homomorphism on A. So it remains to show that F extends to a dense submodule of A^*. Let

$$\psi_n(y_m) = \begin{cases} \delta_{nm} & \text{if } m \geq n; \\ s_m\psi_n(y_{m+1}) + \phi_n(a_m) & \text{if } m < n. \end{cases}$$

By (iv) this defines for each $n < \omega$ an R-homomorphism $\psi_n : A \to R$ which extends ϕ_n. Thus every $f \in F$ extends to a homomorphism on A. Finally to see that $F' = \langle \psi_n \mid n<\omega\rangle$ is dense in A^*, by Theorem 2.6 it suffices for each n to find $\theta_n \in F'$ such that $\theta_n(y_m) = \delta_{nm}$ for all $m < \omega$. But by the definition of the ψ_n's there is such a θ_n of the form $\Sigma_{i=0}^n c_i^n \psi_i$. This completes our proof. □

4. Proof of Theorem 1.1. Let R be a countable PID and let χ be a function assigning to each prime p in R a cardinal $\chi(p) \leq \aleph_1$ or $\chi(p) = 2^{\aleph_1}$. We observe that it suffices for each prime p to construct an $\aleph_1$-free R-module A^p of cardinality $\aleph_1$ such that $r_o(\mathrm{Ext}_R(A^p,R)) = 2^{\aleph_1}$, $r_p(\mathrm{Ext}_R(A^p,R)) = \chi(p)$ and $\mathrm{Ext}_R(A^p,R)$ is q-torsion-free for every prime q with $q \neq p$. For then we let $A = \oplus_p A^p$ which is clearly $\aleph_1$-free and the desired rank conditions hold because

$$\mathrm{Ext}_R(A,R) \cong \prod_p \mathrm{Ext}_R(A^p,R).$$

Thus we fix a prime p and let $\kappa = \chi(p)$. We distinguish two cases: $\kappa = 2^{\aleph_1}$ and $\kappa \leq \aleph_1$. Let us consider the first case and assume first that R is a discrete valuation ring with single prime p. Let $\{A_\nu | \nu < \omega_1\}$ be a continuous increasing chain of countable free R-modules such that for each ν, $A_{\nu+1}/A_\nu$ is an indecomposable torsion-free R-module of rank 2. (Such a module exists because R, being countable, is not complete: see [K, Theorem 19]). Let $A = \bigcup_{\nu<\omega_1} A_\nu$. (From here on we omit the superscript p). We claim that

(i) $|A^*| \leq 2^{\aleph_0}$.

(ii) $\dim_{R/pR}\mathrm{Hom}_R(A,R/pR) = 2^{\aleph_1}$.

Since $A_{\nu+1}/A_\nu$ is indecomposable, it is clear that $(A_{\nu+1}/A_\nu)^* = 0$. But then by induction we have $(A_\nu/A_o)^* = 0$ for all ν. Hence $(A/A_o)^* = 0$, so the natural map $A^* \to A_0^*$ is injective, proving (i). To prove (ii) we consider the exact sequence

$$0 \longrightarrow \mathrm{Hom}_R(A_{\nu+1}/A_\nu,R/pR) \longrightarrow \mathrm{Hom}_R(A_{\nu+1},R/pR) \xrightarrow{\varepsilon} \mathrm{Hom}_R(A_\nu,R/pR).$$

We note that ε is onto because $A_{\nu+1}/A_\nu$ is torsion-free. On the other hand, we have $\mathrm{Hom}_R(A_{\nu+1}/A_\nu,R/pR) \neq 0$. So at each step every homomorphism: $A_\nu \to R/pR$ has at least 2 extensions to $A_{\nu+1}$, and (ii) follows.

By hypothesis $2^{\aleph_0} < 2^{\aleph_1}$, so we deduce from (i), (ii) and Lemma 2.1 that $r_p\mathrm{Ext}_R(A,R) = 2^{\aleph_1}$. Moreover if the invariant Γ is defined for R-modules as in [EH2, II.1] for Abelian groups, we have $\Gamma(A) = 1$ and hence $r_o(\mathrm{Ext}_R(A,R)) = 2^{\aleph_1}$ by Theorem 2.11 of [EH2].

Now suppose that $\kappa = 2^{\aleph_1}$ and that R has at least two different primes. We enumerate the primes $\{p_n | n < \zeta\}$ in R such that $p_o = p$. Let A_o be a free R-module of countably infinite rank. For each $n < \zeta$, $\mathrm{Hom}_R(A_o,R/p_nR)$ is an R/p_nR-vector space of dimension $2^{\aleph_0}$. As we assume CH we may enumerate

$$\mathrm{Hom}_R(A_o, R/p_nR) = \{{}^nh_\nu \mid \nu < \omega_1\}.$$

We shall define a continuous increasing chain $\{A_\nu \mid \nu < \omega_1\}$ of countable free R-modules and homomorphisms $f_\nu \in A^*_{\nu+1}$, $\nu < \omega_1$, such that the following conditions hold:

(1) for all $\nu < \omega_1$, $A_{\nu+1}/A_\nu \cong R_p$;

(Thus by 2.2 for every $f \in A^*_\nu$ [resp. for every $h \in \mathrm{Hom}_R(A_\nu, R/p_nR)$, $0 < n < \zeta$], $\nu \le \rho \le \omega_1$, there is at most one [exactly one] extension to an R-homomorphism on A_ρ, denoted by $f^{(\rho)}$ [respectively $h^{(\rho)}$].)

(2) if $\nu = \zeta\delta + n$ where $0 < n < \zeta$, $f_\nu/p_n = {}^nh_\delta^{(\nu+1)}$;

(3) for all $\rho > \nu$, $f_\nu^{(\rho)}$ exists.

Suppose that A_μ has been defined for all $\mu < \nu$. If ν is a limit ordinal we let $A_\nu = \bigcup_{\mu<\nu} A_\mu$. Then A_ν is free by Pontryagin's criterion, and (1) - (3) continue to hold. Suppose that ν is a successor, say $\nu = \alpha + 1$. In this case let F be a countable dense pure submodule of A^*_α which contains $\{f_\mu^{(\alpha)} \mid \mu < \alpha\}$. Since F is countable, we may apply Lemma 3.1 which provides a free R-module $A_{\alpha+1}$ of countable rank such that (1) and (3) hold. Furthermore, we define $f_\alpha \in A^*_{\alpha+1}$ such that (2) holds which is possible because $A_{\alpha+1}$ is free. Finally, let $A = A_{\omega_1} = \bigcup_{\nu<\omega_1} A_\nu$. We claim that for every n with $0 < n < \zeta$, $\Phi_A^{p_n} : A^* \to \mathrm{Hom}_R(A, R/p_nR)$ is onto. So let $h : A \to R/p_nR$ be any R-homomorphism and let $\delta < \omega_1$ such that $h|A_o = {}^nh_\delta$. If we set $\nu = \zeta\delta + n$, we have $f_\nu/p_n = {}^nh_\delta^{(\nu+1)}$ (by (2)), and f_ν extends to $f \in A^*_{\omega_1}$ (by (3)). Therefore, by uniqueness $f/p_n = h$, so $\Phi_A^{p_n}$ is onto, and hence by Lemma 2.1 $\mathrm{Ext}_R(A,R)$ has no p_n-torsion. This is true for every n with $0 < n < \zeta$, so it remains to verify that $r_p(\mathrm{Ext}_R(A,R)) = r_o(\mathrm{Ext}_R(A,R)) = 2^{\aleph_1}$. But this can be done as in the case that R is a discrete valuation ring, using the facts that $|A^*| \leq 2^{\aleph_0}$, $\dim_{R/pR}\mathrm{Hom}_R(A,R/pR) = 2^{\aleph_1}$, and $\Gamma(A) = 1$.

We now turn to the case where $\kappa \leq \aleph_1$. We shall assume that $\kappa > 0$ since the proof in the case $\kappa = 0$ is considerably simpler and in fact was given by Chase [C3, Theorem 4.4]. Let $\{p_n | n < \zeta\}$ enumerate the primes; let $p_o = p$. Let A_ω be a free module of rank $\aleph_o$. Choose a countable dense pure submodule F of A^*_ω and let $\{f_n | n < \omega\}$ and $\{a_n | n < \omega\}$ be as in 2.6. Then it is easy to see that $\{f_n/p | n < \omega\}$ is independent in $\mathrm{Hom}(A_\omega, R/pR)$ so we may write

$$\mathrm{Hom}_R(A_\omega, R/pR) = H_o \oplus \langle {}^o h_\nu | \nu < \omega_1 \rangle,$$

where $\dim_{R/pR} H_o = \kappa$, $\{{}^o h_\nu | \nu < \omega_1\}$ is independent, and for all $\nu < \omega$, $f_\nu/p = {}^o h_\nu$. Let

$$E \overset{\mathrm{def}}{=} \{g \in A^*_\omega | g/p \in H_o - \{0\}\}.$$

Enumerate $E = \{g_\nu | \omega \leq \nu < \omega_1\}$ and for each n with $0 < n < \zeta$,

$$\mathrm{Hom}_R(A_\omega, R/p_n R) = \{{}^n h_\nu | \nu < \omega_1\}$$

such that $f_\nu/p_n = {}^n h_\delta$ if $\nu = \zeta\delta + n < \omega$ (where $\delta \geq 0$). This time we shall define a continuous increasing chain $\{A_\nu | \nu < \omega_1\}$ of countable free R-modules and homomorphisms $f_\nu \in A^*_{\nu+1}$, $\nu < \omega_1$, such that

(1) for all $\nu \geq \omega$, $A_{\nu+1}/A_\nu \cong Q$;

(This insures uniqueness of extension of homomorphisms; we will use the same notation as in the first case.)

(2) for all ν, $f_\nu/p = {}^o h_\nu^{(\nu+1)}$;

(3) if $\nu = \zeta\delta + n$ and $0 < n < \zeta$, $f_\nu/p_n = {}^n h_\delta^{(\nu+1)}$;

(4) for all $\rho > \nu$, $f_\nu^{(\rho)}$ exists;

(5) for all $\nu \geq \omega$, $F_\nu = \langle f_\mu^{(\nu)} | \mu < \nu \rangle_*$ is a dense submodule of A^*_ν;

(6) for all $\nu \geq \omega$, $g_\nu^{(\nu+1)}$ does not exist.

For $\nu < \omega$, let $A_\nu = A_\omega$ and let f_ν be as chosen above. So, by definition, conditions (1) - (6) are satisfied for $\nu \leq \omega$. Now let $\nu > \omega$ and suppose A_μ, f_μ have been defined for $\mu < \nu$ such that (1) - (6) hold. If ν is a limit ordinal, let $A_\nu = \bigcup_{\mu<\nu} A_\mu$. Then (1) - (4) and (6) continue to hold. As to (5) we have to verify that $F_\nu = \langle f_\mu^{(\nu)} | \mu < \nu \rangle_*$ is a dense submodule of A_ν^*. Let $\underline{a}$ generate a pure submodule of A_ν; then $\underline{a} \in A_\mu$ for some $\mu < \nu$ and $\langle \underline{a} \rangle$ is pure in A_μ. Since F_μ is dense in A_μ^*, by Theorem 2.5 there exists $f \in F_\mu$ with $f(\underline{a}) = 1$. But then by (4) and by (2.3) $f^{(\nu)}$ exists and $f^{(\nu)} \in F_\nu$, and $f^{(\nu)}(\underline{a}) = 1$. Therefore F_ν is dense in A_ν^*.

Now suppose that ν is a successor, say $\nu = \alpha + 1$. By (5) F_α is a dense submodule of A_α^*. We wish to apply Lemma 3.2 with $B = A_\alpha$, $F = F_\alpha$ and $g = g_\alpha^{(\alpha)}$ (if the latter exists, otherwise a simplified version of 3.2 suffices). So we need to show that $g/p \neq f/p$ for any $f \in F_\alpha$. If $f \in F_\alpha$ by definition we have $xf = \sum_{i=1}^k x_i f_{\mu_i}^{(\alpha)}$ for some $\mu_i < \alpha$ with $x \neq 0$ where we may assume that $\langle x, x_1, \cdots, x_k \rangle = R$. Now if $g/p = f/p$, it follows that $\bar{x}\, g_\alpha/p = \sum_{i=1}^k \bar{x}_i\, {}^o h_{\mu_i}$ (where $\bar{x} = x + pR$). However, by definition we have $g_\alpha/p \in H_o - \{0\}$ and $H_o \cap \langle {}^o h_\mu | \mu < \alpha \rangle = 0$; hence p divides x. On the other hand, the ${}^o h_\mu$'s are independent, so p divides x_i for $i = 1, \cdots, k$. But this contradicts our assumption that $\langle x, x_1, \cdots, x_k \rangle = R$. Hence, as desired, $g/p \neq f/p$ for any $f \in F_\alpha$. Now Lemma 3.2 provides an extension

$$0 \longrightarrow A_\alpha \xrightarrow{\iota} A_{\alpha+1} \longrightarrow Q \longrightarrow 0$$

such that $A_{\alpha+1}$ is free, every $f \in F_\alpha$ extends to $f^{(\alpha+1)} \in A_{\alpha+1}^*$ such that $F' = \{f^{(\alpha+1)} | f \in F_\alpha\}$ is dense in $A_{\alpha+1}^*$, but $g = g_\alpha^{(\alpha)}$ does not extend. Hence conditions (1), (4) and (6) are satisfied. Furthermore, by (2.4) we can choose $f_\alpha \in A_{\alpha+1}^*$ such that (2) and

(3) hold. So it remains to verify (5) for $F_{\alpha+1}$. By (2.3) the map $\iota^* : A^*_{\alpha+1} \to A^*_\alpha$ sends $F_{\alpha+1}$ onto a pure submodule of A_α which of course contains $\{f^{(\alpha)}_\mu \mid \mu < \alpha\}$ and hence F_α. Therefore $F_{\alpha+1}$ contains F', so $F_{\alpha+1}$ is dense in $A^*_{\alpha+1}$ as required.

Finally, let $A = \bigcup_{\nu<\omega_1} A_\nu$. As in the second case one verifies that $\mathrm{Ext}_R(A,R)$ has no p_n-torsion for $0 < n < \zeta$ (using (3) and (4)) and that $r_o(\mathrm{Ext}_R(A,R)) = 2^{\aleph_1}$ (using (1)). On the other hand, (2), (4) and (6) imply that Coker $\Phi^p_A \cong H_o$ and hence has dimension κ. Therefore by Lemma 2.1 $r_p(\mathrm{Ext}_R(A,R)) = \kappa$. This completes the proof of the second case and hence of Theorem 1.1. □

<u>Remark</u>. The structure of the torsion submodule of $\mathrm{Ext}_R(A,R)$ yields some information about the structure of A itself. For example, if A is an $\aleph_1$-free R-module then $\mathrm{Ext}_R(A,R)$ is torsion-free iff A is separable and coseparable iff A is coseparable iff A is <u>finitely projective</u> i.e., A is projective w.r.t. exact sequences

$$0 \to L \to M \to N \to 0$$

where N is finitely generated (cf. [C1, Theorem 4.2], [G, Theorem 2.3], [Hi, Theorem 8], [Ha, Lemma 2.6]).

If A is an $\aleph_1$-free R-module such that for some prime p, $\mathrm{Ext}_R(A,R)$ has no p-torsion, then by [C1, Theorem 4.1] A is <u>torsionless</u> i.e., for every $a \in A - \{0\}$ there exists $f \in A^*$ so that $f(a) \neq 0$. If A is an $\aleph_1$-free R-module such that for some p, $r_p(\mathrm{Ext}_R(A,R))$ is countable, the methods of [C1] show that A is an extension of a free R-module of finite rank by a torsionless R-module.

5. <u>The torsion of</u> Ext(A,G) <u>for a rational group</u> G. In this final section we extend Theorem 1.1 as follows. Recall that an Abelian group G is termed <u>rational</u> if it is isomorphic to a subgroup of the rationals. As usual, we omit the subscript on Hom and Ext if the ground ring is $\mathbb{Z}$.

THEOREM 5.1. Assume CH. Let G be any reduced rational group and let χ be as in Theorem 1.1. Then there exists a torsion-free group A of cardinality $\aleph_1$ such that $r_o(\mathrm{Ext}(A,G)) = 2^{\aleph_1}$ and for every prime p for which $pG \neq G$, $r_p(\mathrm{Ext}(A,G)) = \chi(p)$.

In the proof of this theorem we shall apply the well known fact that given a (left) R-module B, the functors $-\otimes_R B$ and $\mathrm{Hom}(B,-)$ are adjoint to each other. In other words, there is a natural isomorphism

$$\eta_C^A : \mathrm{Hom}(A\otimes_R B,C) \xrightarrow{\sim} \mathrm{Hom}_R(A,\mathrm{Hom}(B,C)).$$

Given $f \in \mathrm{Hom}(A\otimes_R B,C)$, recall that $\eta_C^A(f) : A \to \mathrm{Hom}(B,C)$ associates to each $a \in A$ the homomorphism taking b to $f(a\otimes b)$.

Proof of Theorem 5.1. Let $R = \mathrm{Hom}(G,G)$, which is isomorphic to the subring of $\mathbb{Q}$ characterized by the property that for every (rational) prime p, $pR \neq R$ iff $pG \neq G$. By the result of Section 4 there is an $\aleph_1$-free R-module A' of cardinality $\aleph_1$ such that $r_o(\mathrm{Ext}_R(A',R)) = 2^{\aleph_1}$ and $r_p(\mathrm{Ext}_R(A',R)) = \chi(p)$ if $pR \neq R$. We claim that for $A = A'\otimes_R G$ the conclusion of the theorem holds. By a variant of Lemma 2.1, $r_p(\mathrm{Ext}(A,G))$ is the dimension of the $\mathbb{Z}/p\mathbb{Z}$-vector space

$$\mathrm{Coker}(\psi : \mathrm{Hom}(A,G) \to \mathrm{Hom}(A,G/pG))$$

where ψ denotes the natural map. So concerning the p-rank it suffices to prove that

$$\mathrm{Coker}\ \psi \cong \mathrm{Coker}(\Phi_{A'}^p : \mathrm{Hom}_R(A',R) \to \mathrm{Hom}_R(A',R/pR)).$$

To this end we consider the commutative diagram

$$\begin{array}{ccc} \mathrm{Hom}(A'\otimes_R G,G) & \xrightarrow{\psi} & \mathrm{Hom}(A'\otimes_R G,G/pG) \\ \downarrow \eta_G^{A'} & & \downarrow \eta_{G/pG}^{A'} \\ \mathrm{Hom}_R(A',\mathrm{Hom}(G,G)) & \xrightarrow{\psi'} & \mathrm{Hom}_R(A',\mathrm{Hom}(G,G/pG)) \end{array}$$

where ψ' is the natural map. We note that $\mathrm{Hom}(G,G/pG)$ is the cokernel of the mutiplication map

$$\mathrm{Hom}(G,G) \xrightarrow{p} \mathrm{Hom}(G,G)$$

because G/pG is cyclic. Therefore ψ' may be identified with $\Phi^p_{A'}$, and hence $\mathrm{Coker}\ \psi \cong \mathrm{Coker}\ \psi' \cong \mathrm{Coker}\ \Phi^p_{A'}$, as desired.

As to the torsion-free rank we recall that by definition A' is the union of a continuous increasing chain $\{A'_\nu | \nu < \omega_1\}$ of countable R-submodules such that for each ν, $A'_{\nu+1}/A'_\nu$ is torsion-free but not free. Therefore, if $A_\nu = A'_\nu \otimes_R G$ then $\{A_\nu | \nu < \omega_1\}$ is a continuous increasing chain with union A, and for each ν, $A'_{\nu+1}/A'_\nu$ can be embedded into $A_{\nu+1}/A_\nu$ since R is an R-submodule of G. By [EH2, Theorem 2.16] it follows that for each ν, $\mathrm{Ext}(A_{\nu+1}/A_\nu, G) \neq 0$. Therefore we have $\Gamma_G(A) = 1$ and hence $r_o(\mathrm{Ext}(A,G)) = 2^{\aleph_1}$ by Theorem 2.11 of [EH2]. (For the definition of Γ_G see [EH2, II.1].) □

<u>Remarks</u>. 1) In Theorem 5.1 if G is isomorphic to a subring of $\mathbb{Q}$ then $G \cong \mathrm{Hom}(G,G) = R$, hence the group A considered in the above proof is an $\aleph_1$-free R-module. If G is not a ring A is still homogeneous of the same type as G, because it is the union of a chain of countable pure subgroups $\{A_\nu | \nu < \omega_1\}$ each of which is isomorphic to $G^{(\omega)}$, the direct sum of countably many copies of G.

2) Analyzing the above proof we see that we only need the following three properties of a rational group G:

(i) $\dim_{Z/pZ} G/pG \leq 1$ for every prime p;

(ii) $R = \mathrm{Hom}(G,G)$ is a countable PID which is not a field;

(iii) G contains an R-submodule isomorphic to R.

We also used the fact that every rational prime p such that $pG \neq G$ is also a prime in R; but this follows because $pG \neq G$ implies

$pR \neq R$ and, by (i), $\dim_{Z/pZ} R/pR \leq 1$. It is easy to see that a group G satisfying (ii) is indecomposable and reduced. Moreover, it follows from results of Murley [M] that if G is an indecomposable torsion-free reduced group of finite rank which satisfies (i), then (ii) and (iii) also hold. (Indeed, (ii) is just Corollary 7 of [M], and (iii) holds because by Theorem 4 of [M], G is a torsion-free R-module). We conclude that for such groups G , Theorem 5.1 is still valid.

REFERENCES

[C1] Chase, S.: Locally free modules and a problem of Whitehead, Illinois J. Math. 6 (1962), 682-699.

[C2] Chase, S.: Function topologies on Abelian groups, Illinois J. Math. 7 (1963), 593-608.

[C3] Chase, S.: On group extensions and a problem of J.H.C. Whitehead, 173-193 in Topics in Abelian Groups, Scott, Foresman and Co., Chicago 1963.

[EH1] Eklof, P. and Huber, M.: Abelian group extensions and the Axiom of Constructibility, Comment. Math. Helv. 54 (1979), 440-457.

[EH2] Eklof, P. and Huber M.: On the rank of Ext, Math. Z. 174 (1980), 159-185.

[F] Fuchs, L.: Infinite Abelian Groups, Vol. I, Academic Press, New York 1970.

[G] Griffith, P.: Separability of torsion-free groups and a problem of J.H.C. Whitehead, Illinois J. Math. 12 (1968), 654-659.

[Ha] Hausen, J.: On generalizations of projectivity for modules over Dedekind domains, J. Austral. Math. Soc., to appear.

[HHS] Hiller, H., Huber, M. and Shelah, S.: The structure of Ext(A,Z) and V = L , Math. Z. 162 (1978), 39-50.

[Hi] Hiremath, V. A.: Finitely projective modules over a Dedekind domain, J. Aus. Math. Soc. 26 (1978), 330-336.

[HW] Huber, M. and Warfield, Jr., R.: On the torsion subgroup of Ext(A,G) , Arch. Math. (Basel) 32 (1979), 5-9.

[J] Jensen, C.: Les Foncteures Dérivés de lim et leurs Applications en Theorie des Modules. Lecture Notes in Math. No. 254, Springer-Verlag, Berlin 1972.

[K] Kaplansky, I.: Infinite Abelian Groups, rev. ed., Univ. of Michigan Press, Ann Arbor 1969.

[M] Murley, C.: The classification of certain classes of torsion-free Abelian groups, Pac. J. Math. 40 (1972), 647-665.

[SS] Sageev, G. and Shelah, S.: On the structure of Ext(A,Z) in L. Preprint 1980.

[Sh] Shelah, S.: On the structure of Ext(G,Z) assuming V = L. Preprint 1979.

A CARDINAL-DETERMINED PROJECTIVITY CONDITION FOR ABELIAN GROUPS AND MODULES

Jutta Hausen[1]

1. INTRODUCTION. A module M over an associative ring R is said to be κ-projective, κ a cardinal, if M is projective relative to all short exact sequences

$$0 \to A \to B \to C \to 0$$

of R-modules such that C has a generating set of cardinality less than κ. Conditions are given for a module over a left hereditary ring to be κ-projective. The κ-projectivity, κ infinite, of an abelian group G is related to the structure of Ext(G,Z). The reduced $\aleph_o$-projective groups are precisely the $\aleph_o$-coseparable ones. Using results of Griffith [8] and Chase [3] it follows that, for abelian groups, $\aleph_o$-coseparability implies $\aleph_o$-separability. For κ uncountable, κ-projective abelian groups are shown to be Whitehead groups. An application of these concepts leads to a new characterization of left hereditary rings.

Throughout, R is an associative ring with identity element and all modules are unital left R-modules. Every group will be abelian, and we shall write mappings to the right. Z will denote the ring of integers.

2. κ-PROJECTIVE MODULES. Throughout, κ denotes a cardinal. Following Eklof and Huber [6], an R-module M is said to be κ-generated if M has a generating set of cardinality strictly less then κ. Standard arguments, together with [1, p. 186, 16.10], prove

LEMMA 2.1. (i) Direct sums and direct summands of κ-projective modules are κ-projective.

(ii) Every κ-generated κ-projective module is projective.

It is well known that a ring R is left hereditary if and only if submodules of projective (left) R-modules are projective [2; p. 14]. Note that the submodule N' in the following lemma is a direct summand.

LEMMA 2.2. If M is a κ-projective module over a left hereditary ring R, then every submodule N of M with κ-generated quotient M/N contains a submodule N' of M such that M/N' is projective; if submodules of κ-generated modules are κ-generated, then N' can be chosen such that M/N' is κ-generated and projective.

Proof. Let $N \leq M$ with M/N κ-generated, let F be a κ-generated free

---

[1] This research was partially supported by a University of Houston Central Campus Research Enabling Grant.

R-module and let $\alpha: F \twoheadrightarrow M/N$ be an epimorphism. Then there exists a homomorphism $\psi: M \to F$ making the diagram

$$\begin{array}{ccc} & & M \\ & \psi \swarrow & \downarrow \pi \\ F & \overset{\alpha}{\twoheadrightarrow} & M/N \end{array}$$

commutative, where π is the canonical map. Since R is left hereditary, $M\psi \leq F$ is projective (and κ-generated under the additional hypothesis). Let $N' = \ker \psi$.

The following result holds with and without the parenthetical statement.

PROPOSITION 2.3. Let R be a left hereditary ring and let κ be a cardinal such that submodules of κ-generated R-modules are κ-generated. Then an R-module M is κ-projective if and only if every submodule N of M with κ-generated quotient M/N contains a submodule N' such that M/N' is projective (and κ-generated).

Proof. Only the "if" remains to be established. Let $\alpha: B \twoheadrightarrow C$ an epimorphism of R-modules where C is κ-generated and let $\phi: M \to C$ be a homomorphism. By hypothesis, $M/\ker \phi \simeq M\phi \leq C$ is κ-generated. It follows that $M = K \oplus P$ for some submodules K and P of M with $K \leq \ker \phi$ and $P \simeq M/K$ projective. Thus, there exists a homomorphism $\eta: P \to B$ such that $\eta\alpha = \phi|P$. Extend η to a homomorphism $\psi: M \to B$ by defining $\psi|K = 0$.

Suppose κ is an infinite cardinal and R is a Dedekind domain. Then submodules of κ-generated R-modules are κ-generated [11], and Proposition 2.3 is a characterization of the κ-projective R-modules. This extends Hiremath's results on finitely projective modules [10] which are just the $\aleph_o$-projective ones in our terminology.

3. κ-PROJECTIVE ABELIAN GROUPS. Following Griffith [9] and Nunke [12], we call a group G κ-<u>coseparable</u> if G is torsion-free, its subgroups of rank less than κ are free, and every subgroup H of G with κ-generated quotient G/H contains a direct summand H' of G such that G/H' is κ-generated. Pontryagin's theorem, of course, implies that $\aleph_o$-coseparable groups are $\aleph_1$-free. From Proposition 2.3 we obtain

LEMMA 3.1. Every κ-coseparable abelian group is κ-projective.

The following lemma will be of importance.

LEMMA 3.2. If G is an $\aleph_o$-projective abelian group then $\operatorname{Ext}(G,Z)$ is torsion-free.

Proof. Let n be a non-zero integer. The exactness of the sequence

$$0 \to Z \xrightarrow{n\cdot 1} Z \xrightarrow{\alpha} Z/nZ \to 0$$

implies the exactness of the induced sequence

$$\operatorname{Hom}(G,Z) \xrightarrow{\alpha_*} \operatorname{Hom}(G,Z/nZ) \xrightarrow{\gamma} \operatorname{Ext}(G,Z) \xrightarrow{\beta} \operatorname{Ext}(G,Z)$$

where $\beta = n\cdot 1$ is the multiplication by n [7, p. 222, 52.1]. Since M is $\aleph_o$-projective, α_* is surjective which implies $\gamma = 0$ and $\ker \beta = 0$ as desired.

Following Nunke [12] and Griffith [8], we call the group G κ-<u>separable</u> if every κ-generated subgroup of G is contained in a κ-generated direct summand of G. The $\aleph_o$-coseparable $\aleph_o$-separable groups then are precisely the locally free coseparable ones of Griffith [8]. Using a deep result of Chase [3], Griffith has proved that a reduced abelian group G is both $\aleph_o$-coseparable and $\aleph_o$-separable if and only if $\mathrm{Ext}(G,Z)$ is torsion-free [8, p. 655, 2.3]. Thus we have

THEOREM 3.3. The following properties of the reduced abelian group G are equivalent.

(i) G is $\aleph_o$-projective.

(ii) G is $\aleph_o$-coseparable.

(iii) G is $\aleph_o$-coseparable and $\aleph_o$-separable.

(iv) $\mathrm{Ext}(G,Z)$ is torsion-free.

COROLLARY 3.4. Every $\aleph_o$-coseparable abelian group is $\aleph_o$-separable.

It follows from Hiremath [10] that an abelian group G is $\aleph_o$-projective if and only if the reduced part of G has this property. Thus, $\aleph_o$-projective groups need not be free. However, the existence of non-free reduced $\aleph_o$-projective groups is not so certain. Assuming the Continuum Hypothesis, Chase has constructed a non-free reduced abelian group G such that $\mathrm{Ext}(G,Z)$ is torsion-free [4, p. 191, 4.4]. Obviously, every Whitehead group is $\aleph_o$-projective; denying the Continuum Hypothesis (and assuming Martin's Axiom), a non-free reduced $\aleph_o$-projective abelian group (which is not a Whitehead group) has been constructed by Shelah [14, p. 328, 3.4]. In contrast to Corollary 3.4, for $\kappa \geq \aleph_1$, however, κ-coseparability does not imply κ-separability (cf. Eklof [5], pp. 70, 74, 7.5 and 8.3).

For uncountable cardinals we have the following result.

LEMMA 3.5. If G is a κ-projective abelian group where κ is an uncountable cardinal then $\mathrm{Ext}(G,Z) = 0$.

Proof. Let

$$0 \to \langle x\rangle \xrightarrow{i} A \xrightarrow{\eta} G \to 0$$

be a representative of an equivalence class in $\mathrm{Ext}(G,Z)$. It follows from Lemma 2.1 that G is torsion-free which implies A torsion-free. Let X be a maximal linearly independent subset of A containing x and let B be the pure subgroup of A generated by $X - \{x\}$. Then A/B is torsion-free of rank one, hence countable, and so is $A\eta/B\eta = G/B\eta$. By Proposition 2.3, $G = H \oplus F$ with $H \leq B\eta$ and F free. One verifies that $A = (\eta^{-1}(H) \cap B) \oplus \eta^{-1}(F)$. Since $\eta^{-1}(F)/\langle x\rangle = \eta^{-1}(F)/\ker \eta \cong F$ is free, $\langle x\rangle$ is a direct summand of $\eta^{-1}(F)$ and the given sequence splits.

By Lemma 3.5, every κ-projective abelian group, $\kappa \geq \aleph_1$, is a Whitehead group. Assuming the Axiom of Constructibility, Shelah has shown that every Whitehead group is free [13]; on the other hand, assuming Martin's Axiom and denying the Continuum Hypothesis, Shelah recently has constructed a non-free $\aleph_1$-coseparable group [14, p. 312, 1.1] which, by Lemma 3.1, must be $\aleph_1$-projective. Thus, the existence of non-free $\aleph_1$-projective abelian groups is independent of and consistent with our standard set theory.

4. A CHARACTERIZATION OF LEFT HEREDITARY RINGS. By definition, a ring R is left hereditary if every left ideal of R is a projective R-module. It is well known that this is equivalent to the property that submodules of projective left R-modules are projective [2, p. 14]. We have the following generalization.

THEOREM 4.1. Let R be a ring with identity and let λ be a cardinal such that every left ideal of R is a λ-generated R-module. Then the following conditions are equivalent.

(i) R is left hereditary.

(ii) Submodules of λ-projective left R-modules with λ-generated quotients are λ-projective.

(iii) For every cardinal κ, submodules of κ-projective left R-modules with κ-generated quotients are κ-projective.

Proof. (i) $\Rightarrow$ (iii): Let N be a submodule of the κ-projective left R-module M such that M/N is κ-projective. By Lemma 2.2, $M = N' \oplus P$ for some submodules N' and P of M with $N' \leq N$ and P projective. Hence $N = N \cap M = N' \oplus (N \cap P)$. Since P is projective, so is $N \cap P$, and N is κ-projective by Lemma 2.1, (i).

(ii) $\Rightarrow$ (i): Let L be a left ideal of R. Since R/L is λ-generated, L is λ-projective. Since L is λ-generated, Lemma 2.1, (ii), implies that L is projective, completing the proof.

REFERENCES

1. F. W. Anderson and K. R. Fuller, "Rings and Categeries of Modules", Springer-Verlag, New York 1974.

2. H. Cartan and S. Eilenberg, "Homological Algebra", Princeton University Press, Princeton 1956.

3. S. Chase, Locally free modules and a problem of Whitehead, Illinois J. Math. 6(1962), 682-699.

4. S. Chase, On group extensions and a problem of J. H. C. Whitehead, Topics in Abelian groups, ed. by J. M. Irwin and E. A. Walker, Scott, Foresman, Chicago, 1963, pp. 173-193.

5. P. C. Eklof, Set theoretic methods in homological algebra and abelian groups, Proceedings, 18e session du Séminaire de mathématiques supérieures, Groupes abéliens, modules et sujets connexes, Les Presses de l'Université de Montréal, Montréal, 1980, pp. 7-117.

6. P. C. Eklof and M. Huber, Abelian group extensions and the axiom of constructibility, Comment. Math. Helvetici 54 (1979), 440-457.

7. L. Fuchs, "Infinite Abelian Groups", Vol. I, Academic Press, New York, 1970.

8. P. Griffith, Separability of torsion free groups and a problem of J. H. C. Whitehead, Illinois J. Math. 12 (1968), 654-659.

9. P. Griffith, "Infinite Abelian Groups", University of Chicago Press, Chicago, 1970.

10. V. A. Hiremath, Finitely projective modules over a Dedekind domain, J. Austral Math. Soc. (Series A) 26 (1978), 330-336.

11. I. Kaplansky, Modules over Dedekind rings and valuation rings, Trans. Amer. Math. Soc. 72 (1952), 327-340.

12. R. J. Nunke, Whitehead's Problem, Abelian Group Theory, edited by D. Arnold, R. Hunter and E. Walker, Lecture Notes in Mathematics, Vol. 616, Springer-Verlag, Berlin, 1977, pp. 240-250.

13. S. Shelah, Infinite abelian groups, Whitehead problem and some constructions, Israel J. Math. 18 (1974), 243-256.

14. S. Shelah, On uncountable abelian groups, Israel J. Math. 32 (1979), 311-330.

$\aleph_1$-SEPARABLE GROUPS OF MIXED TYPE

Alan H. Mekler*

By "group" I will mean "torsion-free abelian group". Groups which are almost free have been extensively studied. In this paper some groups which are almost completely decomposable will be constructed. Recall that a group is completely decomposable, if it is the direct sum of rank 1 groups. For $a \in G$, let $\chi(a)$ denote the characteristic of a and $\tau(a)$ the type (cf. [F], pp. 108-109). Throughout χ (respectively τ) will always stand for a characteristic (respectively type). A group is homogeneous if every element has the same type. So a free group is a completely decomposable group of type $(0,0,\ldots)$. For κ an infinite cardinal, G is κ-separable if every subset of cardinality $< \kappa$ is contained in a completely decomposable direct summand of G. Some non-trivial κ-separable groups exist.

Theorem 1. (1) [G] There is an $\aleph_1$-separable group of cardinality $\aleph_1$ which is not completely decomposable.

(2) [M] For all $n \geq 1$, there is an $\aleph_n$-separable group of cardinality $\aleph_n$ which is not completely decomposable.

(3) [M] Assume $V = L$. For all regular non-weakly compact cardinals κ, there is a κ-separable group of cardinality κ which is not completely decomposable.

All the groups above are homogeneous of type $(0,0,\ldots)$. One can produce a homogeneous κ-separable group of another type by tensoring one of the groups

---

*Financial support for this paper was furnished by the Ministerium für Wissenschaft und Forschung des Landes Nordrhein-Westfalen under the title Überabzählbare abelsche Gruppen and the National Sciences and Engineering Research Council of Canada under Grant # U0075

given above by a rank 1 group of the desired type. (Some care must be taken. See Lemma 3 and the remark following Theorem 4 for details.) I will construct some κ-separable groups which are not the direct sum of homogeneous groups. (Only the case $\kappa = \aleph_1$ will be done in detail.)

I wish to thank Lazlo Fuchs for suggesting this problem to me. He pointed out that tensoring produces homogeneous groups of new types. After I had proven Theorem 5, I learned that he had used a similar technique to construct separable groups.

Before constructing the promised groups, note that there is a condition which forces a separable group to be the direct sum of homogeneous groups. Recall that characteristics are partially ordered by: $(k_1,\ldots,k_n,\ldots) \geq (\ell_1,\ldots,\ell_n,\ldots)$ iff $k_n \geq \ell_n$ for all n. This ordering extends to types. Types τ_1 and τ_2 are <u>incomparable</u> if neither $\tau_1 \geq \tau_2$ nor $\tau_2 \geq \tau_1$.

<u>Proposition 2</u>. If A is a separable group and the different types of rank 1 summands are pairwise incomparable, the A is the direct sum of homogeneous groups.

<u>Proof</u>. Let $\{\tau_i \mid i \in I\}$ be the set of types of rank 1 summands. Let $A(\tau)$ denote $\{\kappa \in A \mid \tau(x) \geq \tau\}$. Since the types are pairwise incomparable, $A(\tau_i)$ is homogeneous for all i. If G is a completely decomposable direct summand of A, then $G = \bigoplus_{i\in I} G(\tau_i) = \bigoplus_{i\in I} G \cap A(\tau_i)$. So $A = \bigoplus_{i\in I} A(\tau_i)$.

Before showing this proposition is the only non-trivial condition on types which forces an $\aleph_1$-separable group to be a direct sum of homogeneous groups, we need to know more about $\aleph_1$-separable groups. Let A be a group of cardinality $\aleph_1$. Then $\{A_\nu \mid \nu < \omega_1\}$ is an ω_1-<u>filtration</u> of A, if it is an increasing chain of subgroups of A such that: $A = \bigcup_{\nu<\omega_1} A_\nu$; $|A_\nu| < \aleph_1$; and if λ is a limit ordinal $A_\lambda = \bigcup_{\nu<\lambda} A_\nu$. A is $\aleph_1$-separable iff there is an ω_1-filtration $\{A_\nu \mid \nu < \omega_1\}$ of A by completely decomposable groups so that for all ν, $A_{\nu+1}$

is a direct summand of A. For such a filtration, let $S = \{\nu \mid A_\nu$ is not a direct summand of $A\}$. Then A is completely decomposable iff S is not stationary (i.e., there is some closed unbounded set $C \subseteq \omega_1$ which is disjoint from S). See [M] or [E] for definitions and proofs. Given a stationary set $S \subseteq \omega_1$ which consists of limit ordinals and a map $\phi : S \to$ countable torsion-free groups, it is possible ([M]) to construct an $\aleph_1$-separable group A with an ω_1-filtration $\{A_\nu \mid \nu < \omega_1\}$ such that : each A_ν is free; $A_{\nu+1}$ is a direct summand of A, for all ν ; and $A_{\nu+1}/A_\nu \cong \phi(\nu)$, for all $\nu \in S$. (A variant of this construction will be used in Theorem 7.).

Lemma 3. Suppose R is a rank 1 group and $R \not\cong Q$. Then there is a $\aleph_1$-separable group A of cardinality $\aleph_1$ and type $(0,0,\ldots)$ such that $A \otimes R$ is not completely decomposable.

Proof. Choose A with an ω_1-filtration $\{A_\nu \mid \nu < \omega_1\}$ so that for a stationary set $S \subseteq \omega_1$: if $\nu \in S$, then $(A_{\nu+1}/A_\nu) \otimes R$ is not a direct sum of copies of R. Now $\{A_\nu \otimes R \mid \nu \subset \omega_1\}$ is an ω_1-filtration of $A \otimes R$. For $\nu \in S$, $A_\nu \otimes R$ is not a direct summand of $A \otimes R$. Otherwise $(A_{\nu+1}/A_\nu) \otimes R$ would be isomorphic to a direct summand of $A_{\nu+1} \otimes R$. Hence it would be a direct sum of copies of R. □

Theorem 4. There are homogeneous $\aleph_1$-separable groups of cardinality $\aleph_1$ of all types other than $(\infty,\infty,\ldots)$ which are not completely decomposable.

Remark. Lemma 3 and Theorem 4 (also Theorems 5 and 6 below), extend to $\aleph_n$-separable groups. Assuming $V = L$, they extend to κ-separable groups for all regular non-weakly compact cardinals κ.

In the proof of lemma 3, if no such stationary set exists then $A \otimes R$ is completely decomposable.

Theorem 5. Suppose $\tau_1 \geq \tau_2$ are types and $\tau_1 \neq (\infty,\infty,\ldots)$. There is an

$\aleph_1$-separable group H such that: for any $x \in H$, $\tau(x) = \tau_1$, or $\tau(x) = \tau_2$; and H is not the direct sum of homogeneous groups.

Proof. For simplicity assume the first coordinate of τ_1 is 0 (i.e., not ∞). Let R be a rank 1 group whose type is $(0,\infty,\infty,\ldots)$. Choose A an $\aleph_1$-separable group of cardinality $\aleph_1$ and type $(0,0,\ldots)$ so that $A \otimes R$ is not completely decomposable. Let

$$0 \to K \to F \xrightarrow{\varphi} A \to 0$$

be a free resolution of A where $|F| = \aleph_1$. Let R_2 be a rank 1 group of type τ_2 and R_1 a rank 1 group such that $R_1 \otimes R_2$ is of type τ_1. Consider the following pushout

$$\begin{array}{ccccccccc} 0 \to & K \otimes R_2 & \to & F \otimes R_2 & \to & A \otimes R_2 & \to 0 \\ & \downarrow & & \downarrow & & \| & \\ 0 \to & K \otimes R_2 \otimes R_1 & \to & H & \to & A \otimes R_2 & \to 0 \end{array}$$

I claim that H is the desired group. Let $\{A_\nu \mid \nu < \omega_1\}$ be an ω_1-filtration of A by free groups so that $A_{\nu+1}$ is a direct summand of A for all ν. Assume $F = \bigcup_{\nu<\omega_1} F_\nu$ where each F_ν is a countable direct summand of F and $\varphi(F_\nu) = A_\nu$. Let $K_\nu = F_\nu \cap K$. Note that: $K_\nu \otimes R_2$ is a direct summand of $F_\nu \otimes R_2$ (either by noting F_ν/K_ν is free or by using [F] 86.5). Let H_ν be the subgroup of H generated by $F_\nu \otimes R_2$ and $K_\nu \otimes R_2 \otimes R_1$ (more precisely by their images in H). Now $H = \bigcup_{\nu<\omega_1} H_\nu$. Also for any ν, H_ν is completely decomposable (as $K_\nu \otimes R_2$ is a direct summand of $F_\nu \otimes R_2$). To see that H is $\aleph_1$-separable, let $\alpha = \nu + 1$. Choose G so that $F_\alpha + K/K \oplus G/K = F/K$. This is possible by the choice of the filtration. Now K_α is a direct summand of F and so of G. Let G' be a complementary summand of K_α. Let $K' = G' \cap K$. So $F = F_\alpha \oplus G'$ and $K = K_\alpha \oplus K'$. Let H' be the pushout of

$$\begin{array}{ccccccccc} 0 & \to & K' \otimes R_2 & \to & G' \otimes R_2 & \to & (G'/K') \otimes R_2 & \to & 0 \\ & & \downarrow & & \downarrow & & \| & & \\ 0 & \to & K' \otimes R_2 \otimes R_1 & \to & H' & \to & (G'/K') \otimes R_2 & \to & 0 \end{array}$$

Then $H = H_\alpha \oplus H'$.

It remains to show H is not the direct sum of homogeneous groups. Suppose not, then $H = K' \otimes R_2 \otimes R_1 \oplus G$ where $G \cong A \otimes R_2$. Consider $H \otimes R \cong F \otimes R$. So $G \otimes R$ is a completely decomposable group. This contradicts the choice of A , as $A \otimes R_2 \otimes R \cong A \otimes R$ which is not completely decomposable.

The same idea can be used to construct other $\aleph_1$-separable groups.

<u>Theorem 6</u>. There is an $\aleph_1$-separable group which is not the direct sum of groups in which the number of different types of elements is finite.

<u>Proof</u>. Let $\{\tau_n | n < \omega\}$ be a set of types all of which are greater than $(0,0,\ldots)$ such that: the first coordinate of each type is 0; and if $\tau_m > \tau_n$, then $m < n$. Let K,F and R be as in the proof of Theorem 5 . In [M] it is shown that $K = \bigcup_{n<\omega} K_n$ where for all n $K_{n+1} \supset K_n$ and K_n is a direct summand of F .

Let $F_0 = F$ and $G_0 = K_0$. Inductively choose F_{n+1} so that $G_n \oplus F_{n+1} = F_n$. Let $G_{n+1} = F_{n+1} \cap K_{n+1}$. So $K_{n+1} = G_n \oplus K_n$. Also $K = \bigoplus_{n<\omega} G_n$. Let $G = \bigoplus_{n<\omega} G_n \otimes R_n$, where R_n is a rank 1 group of type τ_n . Let H be the group obtained from the following pushout diagram

$$\begin{array}{ccccccccc} 0 & \to & K & \to & F & \to & A & \to & 0 \\ & & \downarrow & & \downarrow & & \| & & \\ 0 & \to & G & \to & H & \to & A & \to & 0 \end{array} .$$

As in the proof of Theorem 5, one can show H is $\aleph_1$-separable and not the direct sum of homogeneous groups.

Note that for any n , $\langle K_n \rangle_*$ (the pure closure of K_n) is a direct summand of H. Since $\bigoplus_{m>n} G_m \subseteq F_{n+1}$, $\bigoplus_{m>n} G_m \otimes R_m$ and F_{n+1} generate a complementary summand of $< K_n >_*$. As $H(\tau_m) \subseteq < K_m >_*$, $H(\tau_m)$ is a direct summand of H. To complete the proof, it suffices to show: if H_1 is direct summand of H and the number of different types of elements of H_1 is finite, then H_1 is the direct sum of homogeneous groups. Suppose $H = H_1 \oplus H_2$ and H_1 is a counterexample to the statement above. Assume the number of types of elements of H_1 is minimal. Let τ be maximal among the types of elements of H_1. So for some m, $\tau = \tau_m$. Since $H(\tau_m) = H_1(\tau_m) \oplus H_2(\tau_m)$ and $H(\tau_m)$ is a direct summand of H, $H_1(\tau_m)$ is a direct summand of H_1.

In the construction above by varying the choice of the τ_n's one can construct groups where the types are totally ordered or groups which are not the direct sum of groups whose types are totally ordered.

Each group H constructed so far fails to be the direct sum of homogeneous groups for the same reason. There is some type τ such that there are no subgroups H_1 and H_2 such that: $H = H_1 \oplus H_2$; H_1 is homogeneous of type τ; and H_2 has no rank 1 summand of type τ. The next examples show there are other ways for an $\aleph_1$-separable group to fail to be the direct sum of homogeneous groups.

Theorem 7. There is an $\aleph_1$-separable group A of cardinality $\aleph_1$-separable group A of cardinality $\aleph_1$ which is not the direct sum of homogeneous groups such that: for any type τ, $A = A' \oplus A''$ where A' is completely decomposable and A'' has no element of type τ.

Proof. Choose $\{\chi_n | n < \omega\}$ an ascending sequence of characteristics which are pairwise different as types. The group A will be a subgroup of

$$\bigoplus_{\alpha < \omega_1} \mathbb{Q}x_\alpha \oplus \bigoplus_{\lambda < \omega_1} \mathbb{Q}y_\lambda .$$

For each limit ordinal λ choose $\eta_\lambda : \omega \to \lambda$ such that: $\eta_\lambda(n) < \eta_\lambda(n)$, if $n < m$; $\bigcup_{n<\omega} \eta_\lambda(n) = \lambda$; and for all n, $\eta_\lambda(n) = \gamma + n$ for some limit ordinal γ. ($\{\eta_\lambda | \lambda$ a limit ordinal$\}$ is a ladder system for ω_1. cf. [E] chap. 10). Define

$y_{n\lambda}$ by induction. Let $y_{0\lambda} = y_\lambda$ and $y_{n+1\,\lambda} = \dfrac{y_{n\lambda} - x_{\eta_\lambda(n)}}{n!}$.
Let A be the smallest group containing $\{x_\alpha | \alpha < \omega_1\} \cup \{y_{n\lambda} | n < \omega_1$ and λ a limit ordinal $< \omega_1\}$ such that: $\chi(x_\alpha) = \chi_n$, if $\alpha = \gamma + n$ for some limit ordinal γ; and $\chi(y_{n\lambda}) = \chi_n$, for all λ and n.

If $\nu < \omega_1$, let $A_\nu = \langle \{x_\alpha | \alpha < \nu\} \cup \{y_\alpha \in A | \alpha < \nu\} \rangle_*$. I will omit proving each A_ν is completely decomposable (cf. [E] p. 99 or [M] 2.4. Some work is necessary to see the kinship between A and [M] 2.4). To see A is $\aleph_1$-separable consider $A_{\nu+1}$. For each limit ordinal $\lambda > \nu$, let n_λ be the least number such that $\eta_\lambda(n_\lambda) > \nu$. Let $B = \langle \{x_\alpha | \alpha > \nu\} \cup \{y_{n_\lambda\,\lambda} | \lambda > \nu$ and λ a limit ordinal$\} \rangle_*$. I claim $A = A_{\nu+1} \oplus B$. It is clear $A_{\nu+1} \cap B = 0$. All that remains to show is $\langle y_{n\lambda} \rangle_* \subseteq A_{\nu+1} \oplus B$ for all n and limit ordinals $\lambda < \omega_1$. Suppose $\lambda > \nu$ and $n_\lambda = m + 1$, I will show why $\langle y_{m\lambda} \rangle_*$ is in the group. The general result will then be obvious. First note $y_{m\lambda} = m! y_{m+1\lambda} + x_{\eta_\lambda(m)}$. Also if $k | y_{m\lambda}$ then $k | y_{m+1\lambda}$ and $k | x_{\eta_\lambda(m)}$, as $\chi_m = \chi(y_{m\lambda}) = \chi(x_{\eta_\lambda(m)}) < \chi(y_{m+1\lambda})$.

If λ is a limit ordinal $A_{\lambda+1}/A_\lambda$ has a non-zero divisible subgroup. So A is not completely decomposable.

Also for any n

$$A = \bigoplus_{i \in I_1} \langle x_i \rangle_* \oplus \ldots \oplus \bigoplus_{i \in I_m} \langle x_i \rangle_* .$$

$$\oplus \langle \{x_{\alpha+m+1} | \alpha < \omega_1\} \cup \{y_{m+1\lambda} \in A\} \rangle_*$$

where $I_k = \{\alpha | \alpha = \gamma + k$ for some limit ordinal $\gamma\}$. The last part of the theorem has been verified. Note for each m, $A/A(\tau_m)$ is completely decomposable (τ_m is the type associated with χ_m). So any homogeneous summand is completely decomposable. Since A is not completely decomposable, A is not the direct sum of homogeneous groups.

Each of the groups constructed so far has involved only countably many types. It is possible to modify the construction above to include uncountably many types.

Theorem 8. There is an $\aleph_1$-separable group A of cardinality $\aleph_1$ which is not completely decomposable. If B is a direct summand of A and the types of elements of B come from a countable set, then B is completely decomposable.

Proof. Choose $\{\chi_\alpha | \alpha < \omega_1\}$ such that $\tau_\alpha < \tau_\nu$ is $\alpha < \nu$, where τ_α is the type associated with χ_α. Again we will define A as a subgroup of

$$\oplus_{\alpha < \omega_1} \mathbb{Q} x_\alpha \oplus_\lambda \oplus_{\omega_1} \mathbb{Q} y_\lambda .$$

Let $\{\eta_\lambda | \lambda$ a limit ordinal $< \omega_1\}$ be a ladder system for ω_1. A will be defined so $\chi(x_\alpha) = \chi_\alpha$. Let me talk as though this were already true. If λ is a limit ordinal choose integers k_n so that: $\chi(k_m x_{\eta_\lambda(m)}) \le \chi(k_n x_{\eta_\lambda(n)})$, if $m < n$.

Define $y_{0\lambda} = y_\lambda$ and $y_{n+1\lambda} = \dfrac{y_{n\lambda} - k_n x_{\eta_\lambda(n)}}{n!}$. Also let $\chi(y_{n\lambda}) = \chi(k_n \chi_{\eta_\lambda(n)})$.

Let A be the smallest group which makes the claims about characteristics true and contains $\{x_\alpha | \alpha<\omega_1\} \cup \{y_{n\lambda} | n<\omega$ and λ is a limit ordinal $< \omega_1\}$. For $\nu < \omega_1$ define A_ν as in Theorem 7.

As before A is an $\aleph_1$-separable group which is not completely decomposable. Also for all ν, $A/A(\tau_{\nu+1}) \cong A_{\nu+1}$. So any direct summand the types of whose elements come from a countable set is completely decomposable.

Assuming $V = L$, Theorem 7 can be generalized to all regular non-weakly compact cardinals. The situation with Theorem 8 is more interesting. The generalization to a cardinal κ requires both the combinatorial principle $E(\kappa)$ (the consequence of $V = L$ used in the previous generalizations cf. [E] p. 54) and the existence of a κ-chain of types. If one begins with a model of $V = L$ and forces $MA + \neg CH$ to be true, then both of these principles hold for every regular cardinal $\le 2^{\aleph_0}$ which was not weakly compact in L.

Theorem 9. It is consistent with $2^{\aleph_0} > \aleph_1$ (assuming the consistency of ZFC), that for all regular $\kappa \le 2^{\aleph_0}$ there is a κ-separable group A of cardinality κ which is not completely decomposable. If B is a direct summand of A and the types of elements of B from a set of cardinality $< \kappa$, then B is completely decomposable.

In this paper various $\aleph_1$-separable groups have been constructed. A general theory is still distant. I will conclude this paper with remarks about two open questions.

<u>Question 1</u>. As the direct summand of an $\aleph_1$-separable group $\aleph_1$-separable?

<u>Remarks</u>. If ZFC settles this question for groups of cardinality $\aleph_1$, then it settles this question for all $\aleph_1$-separable groups. Suppose ZFC$\vdash$" every summand of an $\aleph_1$-separable group of cardinality $\aleph_1$ is $\aleph_1$-separable". Further suppose $G = H \oplus K$ and G is $\aleph_1$-separable. I will show H is $\aleph_1$-separable. Take $X \subseteq H$ so that $|X| = \aleph_0$. Choose a countably closed notion of forcing P such that $1 \Vdash |\check{G}| = \aleph_1$. So $1 \Vdash \check{H}$ is $\aleph_1$-separable. Since P adds no countable subsets of H , there is a condition p and a countable subset $H_1 \supseteq X$ so that $p \Vdash \check{H}_1$ is a direct summand of H . Choose $G_1 \supseteq H_1$, so that G_1 is a direct summand of G . Next choose q and G_2 so that $q \Vdash \check{H}_1 \oplus \check{G}_2 = \check{G}_1$. So $H_1 \oplus G_2 = G_1$. Hence H_1 is a direct summand G and so of H .

There are a few cases in which I know the answer to question 1.

<u>Theorem 10</u>. Suppose A is $\aleph_1$-separable and B is a direct summand of A . Then B is $\aleph_1$-separable if:

(1) A is the direct sum of finitely many homogeneous groups;

or (2) the different type of rank 1 summands of A form a finite totally ordered set;

or (3) there is an ω_1-filtration $\{A_\nu | \nu < \omega_1\}$ of A so that $S = \{\nu | A_\nu$ is a direct summand of A$\}$ is stationary.

<u>Proof</u>. (1) The proof is by induction on the number of homogeneous summands. If A is homogeneous, the proof is easy (see Theorem 11 below). Suppose the theorem is true for groups having n homogeneous direct summands and assume A has n + 1 . Let τ be a maximal among the types of rank 1 summands of A . So $B(\tau)$ is a direct summand of $A(\tau)$ which in turn is a direct summand of A . Note that $A(\tau)$ is

$\aleph_1$-separable (by the maximality of τ). So $B(T)$ is $\aleph_1$-separable. Choose B' so that $B = B(\tau) \oplus B'$. Now $B' \cong B/B(\tau) \cong B + A(\tau)/A(\tau)$. This last group is a direct summand of $A/A(\tau)$. By the inductive hypothesis B' is $\aleph_1$-separable. Since B is the direct sum of $\aleph_1$-separable groups, B is $\aleph_1$-separable.

(2) By Theorem 11 every countable subset of B is contained in a countable subgroup which is a direct summand of every countable extension. Suppose C is such a subgroup. Choose $A_1 \supset C$, so that A_1 is a countable direct summand of A. Choose $A_2 \supseteq A_1$ so that A_2 is countable and $B \cap A_2$ is a direct summand of A_2. Since C is a direct summand of $B \cap A_2$, C is a direct summand of A_2. Hence, it is also a direct summand of A_1, A and B.

The proof of (3) is very easy. Suppose $A = B \oplus C$. Then $\{\alpha \in S | A_\alpha = B \cap A_\alpha \oplus C \cap A_\alpha\}$ is stationary. For B to be $\aleph_1$-separable it suffices that this set is unbounded.

__Question 2__. Which subgroups of $\aleph_1$-separable groups are $\aleph_1$-separable?

__Remarks__. A subgroup of a group is called $\aleph_1$-__pure__ if it is a direct summand of every countable extension. Call a group __strongly__ $\aleph_1$-__completely decomposable__ if every countable subset is contained in a completely decomposable $\aleph_1$-pure subgroup. (This notion is analogous to strongly $\aleph_1$-free). A straightforward modification of Shelah's [S] proof that MA +¬CH implies there is a non-free Whitehead group (cf. [E] pp. 68-69) shows, assuming MA + ¬CH, that a group of cardinality $\aleph_1$ is $\aleph_1$-separable iff it is strongly $\aleph_1$-completely decomposable. (This statement is not true without some set theoretic assumption).

__Theorem 11__. (1) If G is a homogeneous strongly $\aleph_1$-completely decomposable group of type τ, then every homogeneous subgroup of type τ is strongly $\aleph_1$-completely decomposable.

(2) If G is strongly $\aleph_1$-completely decomposable and the different types of rank 1 $\aleph_1$-pure subgroups form a finite totally ordered set, then every pure subgroup of G' is strongly $\aleph_1$-completely decomposable.

(3) Assume $MA + \neg CH$. If G is as in (1) and $|G| = \aleph_1$, then every subgroup which is homogeneous of type τ is $\aleph_1$-separable.

(4) Assume $MA + \neg CH$. If G is as in (2) and $|G| = \aleph_1$, then every pure subgroup of G is $\aleph_1$-separable.

Proof. By the comments above (3) and (4) follow from (1) and (2).

(1) Suppose $H \subseteq G$ and H is homogeneous of type τ. Further suppose $X \subseteq H$ and $|X| = \aleph_0$. Choose $G_1 \supseteq X$ a countable $\aleph_1$-pure subgroup of G. Let $H_1 = H \cap G$. Since H_1 is a homogeneous subgroup of G_1 of type τ, H_1 is completely decomposable. Suppose $H_1 \subseteq H_2 \subseteq H$ and $|H_2| = \aleph_0$. Choose $G_2 \geq H_2 \cup G_1$ so that G_2 is countable and homogeneous of type τ. It can be assumed that $H_2 = G_2 \cap H$. Note that H_2/H_1 is homogeneous of type τ. Since $H_2/H_1 \cong H_2 + G_1/G_1 \subseteq G_1/G_2$ and G_2/G_1 is completely decomposable of type τ, H_2/H_1 is completely decomposable. By [F] 86.5, H_1 is a direct summand of H_2.

(2) The proof is by induction on n the number of different types of $\aleph_1$-pure rank 1 groups. For $n = 1$, this case is an instance of (1). Assume the induction hypothesis is true for $n = k$. Suppose H is a pure subgroup of G and the different types of $\aleph_1$-pure rank 1 subgroups are $\tau_0 < \tau_1 < \ldots < \tau_k$. Since $H(\tau_1)$ is a pure subgroup of $G(\tau_1)$, $H(\tau_1)$ is strongly $\aleph_1$-completely decomposable. Also $H/H(\tau_1) \cong H + G(\tau_1)/G(\tau_1) \subseteq G/G(\tau_1)$. So $H/H(\tau_1)$ is strongly $\aleph_1$-completely decomposable. Finally

$$0 \to H(\tau_1) \to H \to H/H(\tau_1) \to 0$$

is a balanced exact sequence. The theorem reduces to proving the following claim.

Claim. If A and C are strongly $\aleph_1$-completely decomposable and $0 \to A \to B \overset{\pi}{\to} C \to 0$ is a balanced exact sequence, then B is strongly $\aleph_1$-completely decomposable.

<u>Proof (of Claim)</u>. Assume that $B = A \times C$, as a set and $f: C \times C \to A$ is a factor set for B (i.e. $(0,c_1) + (0,c_2) = f(c_1,c_2), c_1 + c_2)$ and $f(c,0) = f(0,c) = 0$). Suppose $X \subseteq B$ and X is countable. Choose C_1 a countable $\aleph_1$-pure subgroup of C such that $C_1 \supseteq \{c \in C |$ there is $a \in A$ so that $(a,c) \in X\}$. Choose A_1 a countable $\aleph_1$-pure subgroup of A containing $f(C_1 \times C_1)$ and $\{a \in A |$ there is $c \in C$ so that $(a,c) \in X\}$. Note that $A_1 \times C_1$ is a subgroup of B and $A_1 \times C_1 \supseteq X$. Now I will show A_1 is a direct summand of $A_1 \times C_1$ and hence that $A_1 \times C_1$ is completely decomposable.

Since C_1 is countable and A is strongly $\aleph_1$-completely decomposable, there is completely decomposable subgroup $A' \supseteq A_1$ such that $0 \to A' \to A' \times C_1 \to C_1 \to 0$ is balanced. So A' is a direct summand of $A' \times C$. Since A_1 is a direct summand of A', A_1 is a direct summand of $A_1 \times C_1$.

It remains to show $A_1 \times C_1$ is $\aleph_1$-pure. Consider a countable subgroup of the form $A_2 \times C_2$, where $A_2 \supseteq A_1$, $C_2 \supseteq C_1$ and A_2 is a direct summand of $A_2 \times C_2$ (Any countable subgroup of B is contained in such a subgroup). Choose a homomorphism $g: C_2 \to A_2 \times C_2$ such that πg is the identity on C_2. So $A_1 \times C_1 = A_1 \oplus g(C_1)$ and $A_2 \times C_2 = A_2 \oplus g(C_2)$. Since A_1 (respectively $g(C_1)$) is a direct summand of A_2 (respectively $g(C_2)$), $A_1 \times C_1$ is a direct summand of $A_2 \times C_2$.

One would like to improve Theorem 11 (2) to apply to groups where the different types of $\aleph_1$-pure rank 1 subgroups are inversely well ordered. For completely decomposable and separable groups the analogous result is true ([P1] and [P2]: cf. [F] p. 116).

References

[E] P. Eklof. Set Theoretic Methods in Homological Algebra and Abelian Groups, Les Presses de l'Université de Montréal, Montréal, 1980.

[EM] P. Eklof and A. Mekler. On constructing indecomposable groups in L , J. Algebra 49 (1977), 96-103.

[F] L. Fuchs. Infinite Abelian Groups, Vol. II, Academic Press, New York, 1973.

[G] P. Griffith. Infinite Abelian Groups, Chicago Lecture Notes in Mathematics, Chicago, 1970.

[M] A. Mekler. How to construct almost free groups, Can. J. Math. 32 (1980) 1206-1228.

[P1] L. Procháska. A generalization of a theorem of R. Baer, Comment. Math. Univ. Carolinae 4 (1963), 105-108.

[P2] L. Procházka. Über eine Klasse torsionsfreier abelscher Gruppen, Časopis. Pěst. Mat. 90 (1965), 153-159.

[S] S. Shelah. Infinite abelian groups, Whitehead problem and some constructions, Israel J. Math. 18 (1974), 243-256.

p^{∞}-BASIC SUBGROUPS OF TORSION FREE ABELIAN GROUPS

Ladislav PROCHÁZKA

In this note the notion of the p^{∞}-basic subgroup is introduced, for any torsion free group A and any prime number p. Each p^{∞}-basic subgroup B of A is always $\aleph_1$-free but it need not be free in general. The class $\mathcal{V}_p$ of all torsion free groups containing at least one free p^{∞}-basic subgroup is investigated here. For example it is shown that the class $\mathcal{V}_p$ is closed with respect to direct sums, tensor product and quasi-isomorphism. Some relations concerning $\mathcal{V}_p$ and other classes of torsion free groups are also presented. Finally, the study of this class is used to find necessary and sufficient conditions for a Z_p^*-module (Z_p-module resp.) to be a direct sum of cyclic modules see [11] .

All groups in this paper are supposed to be abelian, all modules are left and unitary. For the terminology and notation we refer to [2]. The letter p is reserved for a prime. Furthermore, the symbol J_p (Z_p resp.) denotes the additive group of the ring Z_p^* of p-adic integers (of the ring Z_p of rational numbers with denominators prime to p resp.) . If A is a torsion free group then the tensor product $J_p \otimes A$ may be considered as a Z_p^*-module, and by the p-rank of A we shall mean the rank $r_p(A)$ of the maximal divisible submodule of $J_p \otimes A$ (see also [3, 4]) . For any subset $M \subseteq A$ the symbol $\langle M \rangle_{*p}^{A}$ represents the p-pure hull of M in A (the smallest p-pure subgroup of A containing M) . If G is any group then $G_{(p)}$ will denote the p-primary component of its maximal torsion subgroup $t(G)$. The relation $x \equiv y \ (H)$ is used to denote that $x - y \in H$, whenever H is a subgroup of G and x , y are elements of G .

Now we recall some notations and notions (see also [5]) which will occur basic for our investigations.

Let A be a torsion free group. A finite sequence $x_1, \dots, x_n$ of elements from A is said to be p-dependent in A, if there are integers $a_1, \dots, a_n$ Z such that

$$a_1x_1 + \dots + a_nx_n \equiv 0 \ (pA)$$

and at least one a_i is prime to p ; in the opposite case the sequence is p-independent in A. If $\alpha_i \in Z_p^*$ (i=1, ... , n) are p-adic integers expressed as sequences $\alpha_i = \{a_i^{(k)}\}_{k=1}^{\infty}$ where $a_i^{(k)} \in Z$ satisfying

$$0 \leqq a_i^{(k)} < p^k \quad , \quad a_i^{(k)} \equiv a_i^{(k+1)} \ (\mathrm{mod}\ p^k)$$

(i=1, ... , n ; k=1,2,...) then we shall write

(1) $$\alpha_1x_1 + \dots + \alpha_nx_n \equiv 0 \quad (p^{\infty}A)$$

whenever the relation

$$a_1^{(k)}x_1 + \dots + a_n^{(k)}x_n \equiv 0 \quad (p^kA)$$

holds for every positive integer k. The sequence $x_1, \dots, x_n$ is called p^{∞}-dependent in A if there is a relation (1) such that at least one α_i is non-zero; in the opposite case the sequence is p^{∞}--independent in A. Any set $M \subseteq A$ is called p-independent (p^{∞}-independent resp.) in A if each its finite subset is so. Every maximal p-independent (p^{∞}-independent resp.) subset of A is a p-basis (p^{∞}--basis resp.) of A (see [5, 2]).

For every torsion free group we have the following sequence of implications:

(2) p-independence $\Longrightarrow$ p^{∞}-independence $\Longrightarrow$ independence .

We begin our investigations by introducing the class $\mathcal{V}_p$.

DEFINITION 1. By $\mathcal{V}_p$ we shall denote the class of all torsion free groups containing at least one p-independent p^{∞}-basis.

It is easily seen that the class $\mathcal{V}_p$ is closed with respect to direct sums (see [5, Lemma 6]). Some further informations concerning the largeness of the class $\mathcal{V}_p$ are contained in the next proposition. In order to present it we recall several notations. By $\mathcal{C}_p$ ($\mathcal{E}_p$

resp.) we denote the class of all torsion free groups A such that the Z_p^*-module $J_p \otimes A$ is completely decomposable (for every torsion group T it is $\mathrm{Ext}(A,T)_{(p)} = 0$ resp.). If $\mathcal{B}$ is the class of all torsion free groups belonging to some Baer class Γ_α (for the definion of Baer classes Γ_α see [8, 6, 2]) then we get:

PROPOSITION 1. i) For every prime p we have the inclusions $\mathcal{B} \subseteq \mathcal{V}_p \subseteq \mathcal{C}_p \subseteq \mathcal{E}_p$. ii) For every torsion free group A and every prime p there is a pure exact sequence

$$0 \longrightarrow A' \longrightarrow A \longrightarrow A'' \longrightarrow 0$$

with $A' \in \mathcal{V}_p$ and $A'' \in \mathcal{V}_p$.

PROOF. i) The inclusions $\mathcal{B} \subseteq \mathcal{V}_p \subseteq \mathcal{C}_p$ are proved in [5, Théoreme 1 and Théoreme 3*] and the inclusion $\mathcal{C}_p \subseteq \mathcal{E}_p$ follows by [6, Proposition 5]. ii) Let A be a torsion free group and take any p-basis Y of A ; hence $\langle Y \rangle$ is a p-basic subgroup of A [see 2, Lemma 32.2] and, consequently, the group $A/\langle Y \rangle$ is p-divisible. If $A' = \langle Y \rangle_*$ is the pure subgroup of A generated by Y then A/A' is torsion free and p-divisible as well. Thus, evidently, the empty set is the unique p-basis and p^∞-basis in A'', whence $A'' \in \mathcal{V}_p$. From the construction of A we deduce that Y is a maximal independent set of A which is p-independent in A . In view of (2) , Y is a p^∞-basis of A' and therefore $A' \in \mathcal{V}_p$. Hence

$$0 \longrightarrow A' \longrightarrow A \longrightarrow A/A' = A'' \longrightarrow 0$$

is a pure exact sequence with the desired properties.

REMARK 1. The class $\mathcal{V}_p$ is not closed with respect to direct products and extensions.

PROOF. Since $Z \in \mathcal{B}$ and $\mathcal{B} \subseteq \mathcal{V}_p$, we have $Z \in \mathcal{V}_p$. As noted in [6, Remark 1] , $Z^{\aleph_0} \notin \mathcal{E}_p$, so that $Z^{\aleph_0} \notin \mathcal{V}_p$. The second assertion follows by applying Proposition 1 to $A = Z^{\aleph_0}$.

The next lemma will occur very useful since it suggest some interesting connections.

LEMMA 1. If Y is an independent set of a torsion free group A then the following assertions are equivalent: i) The set Y is p^∞--independent in A. ii) $r_p(\langle Y\rangle_{*p}^A) = 0$. iii) For each finite subset $Y_o \subseteq Y$ the group $\langle Y_o\rangle_{*p}^A / Y_o$ is finite. iv) For each finite subset $Y_o \subseteq Y$ the group $\langle Y_o\rangle_{*p}^A$ is free.

PROOF. Assume that Y is p^∞-independent in A and put $\langle Y\rangle_{*p}^A = B$. Then Y is a maximal independent set in B which is p^∞-independent in B. By [4, Satz 1] we obtain $r_p(B) = 0$ and hence i) $\Longrightarrow$ ii). For the proof of ii) $\Longrightarrow$ iii) preserve the notation and suppose $r_p(B) = 0$. Then $r_p(S) = 0$ for every subgroup S of B. If $Y_o \subseteq Y$ is finite and $B_o = \langle Y_o\rangle_{*p}^A$, then $B_o \subseteq B$ and hence $0 = r_p(B_o)$. But making use of [3, Theorem 4] we deduce that the p-primary group $B_o/\langle Y_o\rangle$ is finite, which proves ii) $\Longrightarrow$ iii). The implication iii) $\Longrightarrow$ iv) follows from the fact that any extension of a free group by a finite group (by a bounded group) is free as well. Suppose finally that iv) is fulfilled and take any finite subset Y_o of Y. Then $\langle Y_o\rangle_{*p}^A = B_o$ is free and, consequently, $r_p(B_o) = 0$. As Y_o is a maximal independent set in B_o, the group $B_o/\langle Y_o\rangle$ is finite. In view of [4, Lemma 3] the set Y_o is p^∞-independent in B_o and also in A (B_o is p-pure in A). But this means that Y is p^∞-independent in A, whence, iv) $\Longrightarrow$ i). This finishes the proof of lemma.

LEMMA 2. Let Y be an independent set of a torsion free group A. Then Y is a p^∞-basis of A exactly if $r_p(\langle Y\rangle_{*p}^A) = 0$ and for each element $0 \neq z \in A$ with $\langle z\rangle \cap \langle Y\rangle = 0$ there is a finite subset $Y_o \subseteq Y$ such that

$$r_p\left(\langle Y_o \cup \{z\}\rangle_{*p}^A\right) \geq 1 .$$

PROOF. From Lemma 1 it follows that the above mentioned conditions are necessary for Y to be a p^∞-basis of A. Conversely, if Y satisfies the conditions of our lemma then from Lemma 1 we deduce that Y is a p^∞-independent set in A which cannot be properly extended to a p^∞-independent set in A.

In this note it is useful to combine the original definition of the p^∞-independence with the characterizations given in Lemma 1.

LEMMA 3. Let A be a torsion free group and B its p-pure subgroup containing a p^∞-basis Y of A. Then each p^∞-basis of B is a p^∞-basis of A.

PROOF. Take any p^∞-basis Y_o of B and extend Y to a maximal independent set X in the group A; hence $X = Y \cup Y'$ with $Y \cap Y' = \emptyset$. We shall prove that for each $x \in X \setminus Y_o$ the set $Y_o \cup \cup \{x\}$ is not p^∞-independent in A and thus, in view of [4, Lemma 7], it will be shown that Y_o is a p^∞-basis of A. Take now any $x \in X \setminus Y_o$. If $x \in Y \subseteq B$ then, evidently, $Y_o \cup \{x\}$ is p^∞-dependent in B and also in A. So assume $x \in Y'$. Since Y is a p^∞-basis of A and $x \notin Y$, the set $Y \cup \{x\}$ is p^∞-dependent in A; accordingly, there exists a relation

(3) $$\sum_i \alpha_i y_i + \alpha x \equiv 0 \quad (p^\infty A)$$

where $\alpha_i \in Z_p^*$ $(i=1, \ldots, n)$, $0 \neq \alpha \in Z_p^*$ and $\{y_1, \ldots, y_n\}$ is a subset of Y. If all y_i belong to Y_o then (3) represents the desired p^∞-dependence relation. Assume now that $y_i \notin Y_o$ for $i=1, \ldots, r$ and $y_i \in Y_o$ for $r < i \leq n$. If $1 \leq i \leq r$ then $y_i \in \in Y \subseteq B$ but $y_i \notin Y_o$. Therefore the set $Y_o \cup \{y_i\}$ is p^∞-dependent in A_o and also in A, so that we have a relation of the form

(4) $$f_i(Y_o) + \beta_i y_i \equiv 0 \quad (p^\infty A) \qquad (1 \leq i \leq r)$$

where $0 \neq \beta_i \in Z_p^*$ and $f_i(Y_o)$ represents a finite linear combination of elements from Y_o with coefficients in Z_p^*. Each β_i may be written as $\beta_i = p^{k_i} \cdot \beta_i^*$ where β_i^* is invertible in Z_p^*. Without loss of generality we can assume $\beta_i^* = -1$ $(1 \leq i \leq r)$ and $k_1 = k_2 = \ldots = k_r = k$. Therefore (4) may be replaced by

(5) $$f_i(Y_o) - p^k y_i \equiv 0 \quad (p^\infty A) \qquad (1 \leq i \leq r).$$

From (5) and (3) we deduce

$$\sum_{i\leq r} \alpha_i f_i(Y_o) + \sum_{r<i} \alpha_i p^k y_i + p^k \alpha x \equiv 0 \ (p^\infty A)$$

where $p^k \alpha \neq 0$. This means that in this case the set $Y_o \cup \{x\}$ is also p^∞-dependent in A. As explained above this conclude the proof of the fact that Y_o is a p^∞-basis of A.

Now we shall introduce the notion of the p^∞-basic subgroup.

DEFINITION 2. Let A be a torsion free group and p a prime. A subgroup B of A will be called p^∞-basic in A if there is a p^∞-basis Y in A such that $\langle Y\rangle^A_{*p} = B$.

The following proposition gives the posibility to describe the class $\mathcal{V}_p$ in terms of p^∞-basic subgroups.

PROPOSITION 2. A torsion free group A is contained in the class $\mathcal{V}_p$ precisely if there is a p^∞-basic subgroup B in A which is free.

PROOF. Assume $A \in \mathcal{V}_p$ and take a p^∞-basis Y of A which is p-independent in A. From the p-independence of Y in A it follows that $B = \langle Y\rangle^A_{*p} = \bigoplus_{y\in Y}\langle y\rangle$. Therefore B is a p^∞-basic subgroup of A which is free. On the other hand, let there be a p^∞-basic subgroup B in A that is free. Put $B = \bigoplus_{i\in I}\langle x_i\rangle$ and denote $X = \{x_i ; i \in I\}$. Since B is p-pure in A we have $B = \langle X\rangle^A_{*p} = \bigoplus_{i\in I}\langle x_i\rangle$. Hence X is p-independent in A and also in B. Evidently, X is a maximal independent set in B and, making use of (2), we conclude that X is a p^∞-basis of B. By Definition 2, B contains a p^∞-basis Y of A. Hence, in view of Lemma 3, X is a p^∞-basis of A which is p-independent in A. Therefore $A \in \mathcal{V}_p$ and the proof is finished.

LEMMA 4. Let A_o be a subgroup of a torsion free group A such that i) the group A/A_o is torsion and ii) for each finite rank pure subgroup S of A_o the factor group $\langle S\rangle^A_{*p}/S$ is finite. Then $A \in \mathcal{V}_p$ implies $A_o \in \mathcal{V}_p$.

PROOF. Let Y be a p-independent p^{∞}-basis of A ; thus $B = \bigoplus_{y\in Y} \langle y \rangle$ is a p^{∞}-basic subgroup of A. Its subgroup $B_o = B \cap A_o$ is also free and hence $B_o = \bigoplus_{x\in X} \langle x \rangle$ for a free basis X of B_o. As B is p-pure in A, B_o is p-pure in A_o and X is p-independent in A_o. Using Lemma 2 we shall prove that X is a p^{∞}-basis of A_o. First, from the relations

$$B/B_o = B/(B \cap A_o) \cong (B + A_o)/A_o \subseteq A/A_o$$

it follows that the group B/B_o is torsion. Thus if we take $0 \neq z \in A_o$ such that $B_o \cap \langle z \rangle = 0$ then $B \cap \langle z \rangle = 0$ as well. In view of Lemma 2 there exist the elements $y_1, \dots , y_n \in Y$ such that

(6) $$r_p(\langle y_1, \dots , y_n, z\rangle^A_{*p}) \geq 1 .$$

For a suitable $0 \neq k \in Z$ we have $ky_i \in B_o$ $(i=1, \dots , n)$. There is a finite subset $X_o \subseteq X$ satisfying $ky_i \in \langle X_o \rangle$ and hence $y_i \in \langle X_o \rangle^A_*$. Thus we get

$$\langle z, y_1, \dots , y_n \rangle^A_{*p} \subseteq \langle X_o , z\rangle^A_* = S .$$

In view of (6), $1 \leq r_p(S)$. If we put $S_o = \langle X_o , z\rangle^A_*$ then, by ii), the p-component of the group S/S_o is finite. This means (see [3, Theorem 7]) that $r_p(S_o) = r_p(S) \geq 1$. Evidently, the group $\langle X_o , z\rangle^A_{*p}/\langle X_o , z\rangle$ is the p-primary component of the group $S_o/\langle X_o , z\rangle$ and, in view of [3, Theorem 4], it contains $Z(p^{\infty})$ as a subgroup. By the same [3, Theorem 4] we deduce that it holds $r_p(\langle X_o , z\rangle^A_{*p}) \geq 1$. Now Lemma 2 implies that X is a p^{∞}-basis of A_o and hence $A_o \in \mathcal{V}_p$.

The following two assertions are corollaries of the lemma just proved.

LEMMA 5. Let A_o be a p-pure subgroup of a torsion free group A such that the group A/A_o is torsion. Then $A_o \in \mathcal{V}_p$ just if $A \in \mathcal{V}_p$.

PROOF. Suppose first $A_o \in \mathcal{V}_p$ and denote by Y any p^{∞}-basis of A_o which is p-independent in A_o. As A_o is p-pure in A,

Y is p-independent and also p^∞-independent in A. If we extend the set Y to a maximal independent one in A_o, denoted by X, then the hypothesis on A/A_o implies that X is maximal independent in A. Evidently, for every $x \in X \setminus Y$ the set $Y \cup \{x\}$ is p^∞-dependent in A_o and hence in A. By [4, Lemma 7] this means that Y is a p^∞-basis in A. Consequently, $A \in \mathcal{V}_p$. On the other hand, if S is any finite rank pure subgroup of A_o then $\langle S \rangle^A_{*p} = S$, and the converse implication follows by Lemma 4.

LEMMA 6. Let A_o be a subgroup of a torsion free group A such that the group A/A_o is bounded. Then $A_o \in \mathcal{V}_p$ exactly if $A \in \mathcal{V}_p$ or, in other words, the class $\mathcal{V}_p$ is closed with respect to quasi--isomorphism.

PROOF. Evidently, for each finite rank pure subgroup S of A_o the group $\langle S \rangle^A_*/S$ is finite and the implication $A \in \mathcal{V}_p \Longrightarrow A_o \in \mathcal{V}_p$ follows by Lemma 4. Conversely, suppose $A_o \in \mathcal{V}_p$. By the hypothesis, there is a positive integer n with $nA \subseteq A_o$. In view of $nA \cong A$ and the assertion just proved we get the following sequence of implications: $A_o \in \mathcal{V}_p \Longrightarrow nA \in \mathcal{V}_p \Longrightarrow A \in \mathcal{V}_p$. The proof is complete.

Two torsion free groups A_1, A_2 will be called weakly p-quasi--isomorphic if there are subgroups $U_i \subseteq A_i$ $(i=1,2)$ such that A_i/U_i $(i=1,2)$ are torsion groups with bounded p-primary components and $U_1 \cong U_2$. Now we are ready to prove

PROPOSITION 3. i) For a torsion free group A we have $A \in \mathcal{V}_p$ exactly if $Z_p \otimes A \in \mathcal{V}_p$. ii) If $\mathcal{B}_p$ denotes the class of all torsion free groups A such that $Z_p \otimes A \in \mathcal{B}$ then $\mathcal{B} \subseteq \mathcal{B}_p \subseteq \mathcal{V}_p$. iii) The class $\mathcal{V}_p$ is closed with respect to the weak p-quasi-isomorphism.

PROOF. i) From the exact sequence

$$0 \longrightarrow Z \longrightarrow Z_p \longrightarrow Z_p/Z \longrightarrow 0$$

we get the exact sequence

$$0 \longrightarrow Z \otimes A \longrightarrow Z_p \otimes A \longrightarrow (Z_p/Z) \otimes A \longrightarrow 0 .$$

As the group $(Z_p/Z) \otimes A$ is torsion with vanishing p-primary component, we can apply Lemma 5 to the group $Z_p \otimes A$ and its subgroup $Z \otimes A$. Thus $Z \otimes A \cong A \in \mathcal{V}_p$ precisely if $Z_p \otimes A \in \mathcal{V}_p$. ii) The inclusion $\mathcal{B} \subseteq \mathcal{B}_p$ is proved in [6, Proposition 2]. If $A \in \mathcal{B}_p$ then, by the hypothesis, $Z_p \otimes A \in \mathcal{B}$. In view of Proposition 1 we have also $Z_p \otimes A \in \mathcal{V}_p$ and hence, by the part i), $A \in \mathcal{V}_p$; thus $\mathcal{B}_p \subseteq \mathcal{V}_p$. iii) Suppose that the groups A_1, A_2 are weakly p-quasi-isomorphic and $A_1 \in \mathcal{V}_p$. If the subgroups $U_i \subseteq A_i$ are taken as in the definition then in the exact sequence

$$0 \longrightarrow Z_p \otimes U_i \longrightarrow Z_p \otimes A_i \longrightarrow Z_p \otimes (A_i/U_i) \longrightarrow 0$$

the group $Z_p \otimes (A_i/U_i)$ is p-primary and bounded. Since $Z_p \otimes U_1 \cong Z_p \otimes U_2$, the groups $Z_p \otimes A_1$ and $Z_p \otimes A_2$ are quasi-isomorphic. Now, making use of the part i) and Lemma 6, we conclude that $A_1 \in \mathcal{V}_p$ implies $A_2 \in \mathcal{V}_p$.

If T, G, A are groups, T a subgroup of G and A torsion free, then $T \otimes A$ is considered always as a subgroup of $G \otimes A$. Namely, if T is the maximal torsion subgroup of G then $T \otimes A$ is the maximal torsion subgroup of $G \otimes A$, and the relations $G_{(p)} = T_{(p)} = 0$ imply $(G \otimes A)_{(p)} = T_{(p)} \otimes A = 0$.

LEMMA 7. Let A, A_o be two torsion free groups and B a p-pure subgroup of A. Then $B \otimes A_o$ is a p-pure subgroup of $A \otimes A_o$.

PROOF. The exactness of the sequence

$$0 \longrightarrow B \longrightarrow A \longrightarrow A/B \longrightarrow 0$$

implies the exactness of the sequence

$$0 \longrightarrow B \otimes A_o \longrightarrow A \otimes A_o \longrightarrow (A/B) \otimes A_o \longrightarrow 0 .$$

In view of the previous remark, $(A/B)_{(p)} = 0$ yields $[(A/B) \otimes A_o]_{(p)} = 0$.

LEMMA 8. Let A_1, A_2 be two torsion free groups and let X, Y be p^∞-bases of A_1, A_2 respectively. Then $\mathcal{H} = \{x \otimes y ;\ x \otimes y \in A_1 \otimes A_2,\ x \in X,\ y \in Y\}$ is a p^∞-basis of the group $A_1 \otimes A_2$.

PROOF. If one of the sets X, Y is empty then our assertion is trivial. Thus we may suppose $X \neq \emptyset$ and $Y \neq \emptyset$.

a) First we shall show that the set $\mathcal{H}$ is p^∞-independent in $A_1 \otimes A_2$. To this end, take arbitrary finite subsets $\emptyset \neq X_0 = \{x_1, \dots, x_m\} \subseteq X$, $\emptyset \neq Y_0 = \{y_1, \dots, y_n\} \subseteq Y$ and construct $\mathcal{H}_0 = \{x_i \otimes y_j ;\ i=1, \dots, m ;\ j=1, \dots, n\} \subseteq \mathcal{H}$. We shall prove that $\mathcal{H}_0$ is p^∞-independent in $A_1 \otimes A_2$. Put $F_1 = \bigoplus_i \langle x_i \rangle$, $F_2 = \bigoplus_j \langle y_j \rangle$ and $S_i = \langle F_i \rangle_{*p}^{A_i}$ $(i=1, 2)$. In view of Lemma 1, the groups S_i/F_i $(i=1, 2)$ are finite and p-primary. Hence, from the exactness of the sequences

$$
(7)\qquad \begin{array}{l}
0 \longrightarrow F_1 \otimes F_2 \longrightarrow S_1 \otimes F_2 \longrightarrow (S_1/F_1) \otimes F_2 \longrightarrow 0 \\
0 \longrightarrow S_1 \otimes F_2 \longrightarrow S_1 \otimes S_2 \longrightarrow S_1 \otimes (S_2/F_2) \longrightarrow 0
\end{array}
$$

and from the fact that $S_1 \otimes F_2$ and $S_1 \otimes S_2$ are of finite rank it follows that both groups $(S_1/F_1) \otimes F_2$, $S_1 \otimes (S_2/F_2)$ are finite and p-primary. As $(S_1 \otimes S_2)/(F_1 \otimes F_2)$ is p-primary and finite and as $S_1 \otimes S_2$ is p-pure in $A_1 \otimes A_2$, we deduce that $S_1 \otimes S_2$ is just the p-pure hull of $F_1 \otimes F_2$ in $A_1 \otimes A_2$. In view of Lemma 1, $\mathcal{H}_0$ is p^∞-independent in $A_1 \otimes A_2$. Now the p^∞-independence of $\mathcal{H}$ in $A_1 \otimes A_2$ follows from the fact that any finite subset of $\mathcal{H}$ is contained in a subset of the type $\mathcal{H}_0$.

b) In order to prove that $\mathcal{H}$ is a p^∞-basis in $A_1 \otimes A_2$, we extend X to a maximal independent set U in A_1 and Y to a maximal independent set V in A_2. It is not difficult to see that $\mathcal{U} = \{u \otimes v ;\ u \in U,\ v \in V\}$ is a maximal independent set in the group $A_1 \otimes A_2$ containing $\mathcal{H}$. For any element $u \otimes v \in \mathcal{U} \setminus \mathcal{H}$ we prove that the set $\mathcal{H} \cup \{u \otimes v\}$ is p^∞-dependent in $A_1 \otimes A_2$.

Consider first $u \in U \setminus X$ and $v \in Y$. Then $X \cup \{u\}$ is p^∞-dependent in A_1 and hence we get a relation of the form

(8) $$\alpha_1 x_1 + \dots + \alpha_m x_m + \alpha u \equiv 0 \quad (p^\infty A_1) \quad ,$$
where $\alpha_i \in Z_p^*$ $(i=1, \dots, m)$, $0 \neq \alpha \in Z_p^*$ and $X_o = \{x_1, \dots, x_m\}$ is a finite subset of X. From (8) it follows immediately

$$\alpha_1(x_1 \otimes v) + \dots + \alpha_m(x_m \otimes v) + \alpha(u \otimes v) \equiv 0 \quad \left(p^\infty(A_1 \otimes A_2)\right) ,$$

which establishes the p^∞-dependence of $\mathcal{H} \cup \{u \otimes v\}$ in $A_1 \otimes A_2$. Clearly, the case $u \in X$, $v \in V \setminus Y$ is symmetrical. Thus suppose $u \in U \setminus X$ and $v \in V \setminus Y$. Then we have a relation (8) and at the same time a relation

(9) $$\beta_1 y_1 + \dots + \beta_n y_n + \beta v \equiv 0 \left(p^\infty A_2\right)$$
with $\beta_j \in Z_p^*$ $(j=1, \dots, n)$, $0 \neq \beta \in Z_p^*$, $Y_o = \{y_1, \dots, y_n\}$ being a finite subset of Y. From (8) we deduce, for each j

$$\sum_i \alpha_i \beta_j (x_i \otimes y_j) + \alpha \beta_j (u \otimes y_j) \equiv 0 \quad \left(p^\infty(A_1 \otimes A_2)\right),$$

and analogously from (9)

$$\sum_j \alpha \beta_j (u \otimes y_j) + \alpha\beta(u \otimes v) \equiv 0 \quad \left(p^\infty(A_1 \otimes A_2)\right).$$

Finally, we get the relation

$$\sum_{i,j} \alpha_i \beta_j (x_i \otimes y_j) - \alpha\beta(u \otimes v) \equiv 0 \quad \left(p^\infty(A_1 \otimes A_2)\right)$$

establishing the p^∞-dependence of $\mathcal{H} \cup \{u \otimes v\}$ in $A_1 \otimes A_2$. By [4, Lemma 7] we conclude that $\mathcal{H}$ is a p^∞-basis of $A_1 \otimes A_2$.

LEMMA 9. If B_i is a p^∞-basic subgroup of a torsion free group A_i $(i=1,2)$ then $B_1 \otimes B_2$ is a p^∞-basic subgroup in $A_1 \otimes A_2$.

PROOF. Let X_i be a p^∞-basis of A_i with $\langle X_i \rangle_{*p}^{A_i} = B_i$ and let F_i denote the free group generated by X_i $(i=1,2)$. Then the groups B_i/F_i $(i=1,2)$ are p-primary and, consequently, the groups $(B_1/F_1) \otimes F_2$ and $B_1 \otimes (B_2/F_2)$ are p-primary as well. Hence we deduce (compare (7)) that the group $(B_1 \otimes B_2)/(F_1 \otimes F_2)$ is also torsion and p-primary. As $B_1 \otimes B_2$ is p-pure in $B_1 \otimes A_2$ and $B_1 \otimes A_2$ is p-pure in $A_1 \otimes A_2$ (see Lemma 7), we conclude that $B_1 \otimes B_2$ is the p-pure hull of $F_1 \otimes F_2$ in $A_1 \otimes A_2$. Clearly, $\mathcal{H} = \{x \otimes y\,;\ x \in X_1, y \in X_2\}$ is a free basis of the free group $F_1 \otimes F_2$ and hence $B_1 \otimes B_2 = \langle \mathcal{H} \rangle_{*p}^{A_1 \otimes A_2}$. Our assertion follows now from Lemma 8 and the Definition 2.

As a corollary we obtain the following

PROPOSITION 4. The class $\mathcal{V}_p$ is closed with respect to the tensor product.

PROOF. Consider $A_1 , A_2 \in \mathcal{V}_p$. In view of Proposition 2 there are p^∞-basic subgroups B_i in A_i $(i=1, 2)$ which are free. By Lemma 9, $B_1 \otimes B_2$ is a p^∞-basic subgroup in $A_1 \otimes A_2$ which is free and, in view of the same Proposition 2, $A_1 \otimes A_2 \in \mathcal{V}_p$.

The proofs of some next assertions are based on the following author's result:

LEMMA 10. Let A be a torsion free group containing a homogeneous and completely decomposable subgroup B with a complete decomposition $B = \bigoplus_{j \in J} B_j$. Let there exist the elements $0 \neq x_j \in B_j$ $(j \in J)$ such that the set $X = \{x_j \, ; \, j \in J\}$ is p^∞-independent in A . If the group A/B is p-primary and countable then $A \cong B$.

PROOF. The assertion is an immediate consequence of [7, Theorem 5]. (Compare also (9)).

LEMMA 11. Let A be a torsion free group containing a free subgroup F such that A/F is p-primary and countable. If $r_p(A) = 0$ then $A \cong F$.

PROOF. If Y is a free basis of F then $F = \bigoplus_{y \in Y} \langle y \rangle$ and $A = \langle Y \rangle_{*p}^{A}$. By the hypothesis we have $r_p(\langle Y \rangle_{*p}^{A}) = 0$, and Lemma 1 implies that Y is p^∞-independent in A . The relation $A \cong F$ follows now by Lemma 10.

Before stating the following proposition, we remark that in any torsion free group A each p-basis may be expanded to a p^∞-basis (see (2)) and, analogously, each p-basic subgroup is contained in a p^∞-basic one.

PROPOSITION 5. Let A be a torsion free group such that $A \notin \mathcal{V}_p$.

i) If Y_1 is a p-basis and Y_2 a p^∞-basis of A with $Y_1 \subseteq Y_2$ then the set $Y_2 \setminus Y_1$ is uncountable. ii) If B_1 is a p-basic and B_2 a p^∞-basic subgroup of A satisfying $B_1 \subseteq B_2$ then B_2/B_1 is of uncountable torsion free rank.

PROOF. For the proof of i) suppose that $\mathrm{card}(Y_2 \setminus Y_1) \leq \aleph_0$ and put $F_i = \langle Y_i \rangle$ $(i=1, 2)$, $B = \langle Y_2 \rangle_{*p}^A$. As Y_2 is a maximal independent set in B and Y_1 is a p-basis of B (B is p-pure in A), we get by [4, Satz 3]

(10) $$r_p^*(B/F_2) = \mathrm{card}(Y_2 \setminus Y_1) \leq \aleph_0 ;$$

here $r_p^*(B/F_2)$ denotes the rank of the maximal divisible subgroup of the p-primary group B/F_2. Since Y_1 is a p-basis of A, F_1 is p-basic in A and hence A/F_1 is p-divisible. Then B/F_1, as a p-pure subgroup of A/F_1, is p-divisible as well. In view of $F_1 \subseteq F_2$, B/F_2 is an epimorphic image of B/F_1, and as such it is also p-divisible. Thus the inequality (10) means that the group B/F_2 is countable. Since Y_2 is p^∞-independent in A, it is so in B and, by Lemma 10, we have $B \cong F_2$. Thus B is a free p^∞-basic subgroup of A and in view of Proposition 2 we get $A \in \mathcal{V}_p$, a contradiction. Hence, $\aleph_1 \leq \mathrm{card}(Y_2 \setminus Y_1)$.

Now we shall prove ii). According to the choice of the subgroups B_1 and B_2, there are a p-basis Y_1 and a p^∞-basis Y_2 in A such that

$$B_1 = \langle Y_1 \rangle \subseteq B_2 = \langle Y_2 \rangle_{*p}^A .$$

Obviously, the set Y_1 is p-independent in B_2 and by (2) it may be extended to a p^∞-basis Y of B_2. By Lemma 3, Y is already a p^∞-basis of A and, by the part i), $\aleph_1 \leq \mathrm{card}(Y \setminus Y_1)$. As $\langle Y \rangle / B_1$ is a free subgroup of rank $\mathrm{card}(Y \setminus Y_1)$ in B_2/B_1, the proof of ii) is complete.

COROLLARY 1. If a torsion free group A contains a countable p^∞-basis Y then $A \in \mathcal{V}_p$. In particular, each countable torsion free group belongs to $\mathcal{V}_p$ (see [5, Lemma 5] or Proposition 1).

PROOF. Consider the Z_p^*-module $G = J_p \otimes A$. If D is its maximal divisible submodule then there exists a direct module decomposition $G = D \oplus G_1$. From the proof of [4, Satz 1] it follows that the rank of the module G_1 is precisely equal to the cardinality of any p^∞-basis in A. Thus all the p^∞-bases of A have the same cardinality and, in view of the hypothesis, they are countable. If we extend any p-basis Y_1 to a p^∞-basis Y_2 in A we have $\mathrm{card}(Y_2 \setminus Y_1) \leq \aleph_0$, and it suffices to apply Proposition 5.

Recall that a torsion free group A is said to be $\aleph_1$-free if each its countable subgroup is free.

PROPOSITION 6. Let A be a torsion free group. i) Each p^∞-basic subgroup of A is $\aleph_1$-free. ii) The relation $A \in \mathcal{V}_p$ holds if and only if there is a p^∞-basic subgroup B in A such that $B \in \mathcal{B}$.

PROOF. i) Let Y_0 be a countable p^∞-independent set of A and put $F_0 = \langle Y_0 \rangle$, $B_0 = \langle Y_0 \rangle_{*p}^A$. As B_0/F_0 is a countable p-primary group and Y_0 is p^∞-independent in B_0, in view of Lemma 10 we have $B_0 \cong F_0$; therefore B_0 is free. Consider now a p^∞-basic subgroup $B = \langle Y \rangle_{*p}^A$ in A with Y as a p^∞-basis of A and take any countable subgroup B_1 in B. Then there exists a countable subset $Y_0 \subseteq Y$ such that $B_1 \subseteq \langle Y_0 \rangle_{*p}^A = B_0$. The former consideration implies that B_0 is free and hence B_1 is free as well. Thus B is $\aleph_1$-free. ii) If $A \in \mathcal{V}_p$ then, by Proposition 2, there is a p^∞-basic subgroup B in A which is free; consequently, $B \in \mathcal{B}$. On the other hand, let B be a p^∞-basic subgroup of A satisfying $B \in \mathcal{B}$. By the part i), the group B is $\aleph_1$-free. Consider a pure subgroup S in B of finite rank and suppose that T/S is a pure rank one subgroup of B/S. Evidently, T is also pure in B and of finite rank. As T is countable, it is free and therefore $T/S \cong Z$. Thus B/S is a torsion free group which is homogeneous of the type Z.

Now, in view of [8, Theorem 48.2] , the relation $B \in \mathcal{B}$ implies that B is free. By Proposition 2 we get $A \in \mathcal{V}_p$. The proof is complete.

In the following we shall deal with some conditions under which the relation $A \in \mathcal{V}_p$ implies $A \in \mathcal{B}_p$.

LEMMA 12. Let A be a torsion free group which is homogeneous of the type Z_p . Then A is completely decomposable precisely if $0 = r_p(A)$ and $A \in \mathcal{V}_p$.

PROOF. If A is completely decomposable then evidently $r_p(A) = 0$ (see [12, Corollary 2] and [3, Theorem 4]) and $A \in \mathcal{B} \subseteq \mathcal{V}_p$. Conversely, suppose that $r_p(A) = 0$ and $A \in \mathcal{V}_p$. If X is a p-independent p^∞-basis of A then we deduce from [4, Satz 1] that X is already a maximal independent set of A . Clearly, X is a p-basis of A (see (2)) and hence $B = \langle X \rangle$ is a p-basic subgroup of A . Therefore A/B is a torsion group with $(A/B)_{(p)} = 0$. Thus if we consider $Q \otimes B$ as a subgroup of $Q \otimes A$ then necessarily $Q \otimes B = Q \otimes A$. It follows from the exactness of the sequence

$$0 \longrightarrow Z_p \otimes B \longrightarrow Q \otimes B \longrightarrow Q/Z_p \otimes B \longrightarrow 0$$

that the factor group $(Q \otimes B)/(Z_p \otimes B)$ is p-primary. This fact and the inclusions

$$Z_p \otimes B \subseteq Z_p \otimes A \subseteq Q \otimes A = Q \otimes B$$

imply that $(Z_p \otimes A)/(Z_p \otimes B)$ is p-primary as well. But

$$(Z_p \otimes A)/(Z_p \otimes B) \cong Z_p \otimes (A/B) \ , \quad (A/B)_{(p)} = 0 \ ,$$

whence $Z_p \otimes B = Z_p \otimes A$. As A is homogeneous of the type Z_p we have $A \cong Z_p \otimes A$. A complete decomposition of A follows now by the isomorphism $A \cong Z_p \otimes B$ and the freeness of B .

It is obvious that the previous lemma may also be reformulated as follows.

LEMMA 12*. A torsion free Z_p-module A is free precisely if $r_p(A) = 0$ and $A \in \mathcal{V}_p$.

LEMMA 13. For a torsion free group A the group $Z_p \otimes A$ is completely decomposable and homogeneous exactly if either A is p-divisible or $r_p(A) = 0$ and $A \in \mathcal{V}_p$.

PROOF. If $r_p(A) = 0$ and $A \in \mathcal{V}_p$ then evidently $r_p(Z_p \otimes A) = r_p(A) = 0$ and $Z_p \otimes A \in \mathcal{V}_p$ (see Proposition 3). The condition $r_p(Z_p \otimes A) = 0$ implies that $Z_p \otimes A$ has no elements of infinite p-height and therefore $Z_p \otimes A$ is homogeneous of the type Z_p. By Lemma 12, $Z_p \otimes A$ is completely decomposable and homogeneous. The case of p-divisible A is trivial. Suppose conversely that $Z_p \otimes A$ is completely decomposable and homogeneous. In view of Propositions 3,1 we have $A \in \mathcal{V}_p$. If $Z_p \otimes A$ is not divisible then clearly $r_p(A) = r_p(Z_p \otimes A) = 0$. But if $Z_p \otimes A$ is divisible then from the exactness of the sequence

$$0 \longrightarrow Z \otimes A \longrightarrow Z_p \otimes A \longrightarrow (Z_p/Z) \otimes A \longrightarrow 0$$

and from the relation $\left[(Z_p/Z) \otimes A\right]_{(p)} = 0$ we deduce that $Z \otimes A$ is p-pure in $Z_p \otimes A$ and, consequently, $Z \otimes A$ is p-divisible. Now we use the isomorphism $A \cong Z \otimes A$.

LEMMA 14. If $\mathcal{V}_p^o$ denotes the class of all groups $A \in \mathcal{V}_p$ satisfying $r_p(A) = 0$ then $\mathcal{V}_p^o \subseteq \mathcal{B}_p$.

PROOF. If $A \in \mathcal{V}_p^o$ then, in view of Lemma 13, the group $Z_p \otimes A$ is completely decomposable, whence $Z_p \otimes A \in \mathcal{B}$. Therefore, $A \in \mathcal{B}_p$ and we get $\mathcal{V}_p^o \subseteq \mathcal{B}_p$.

LEMMA 15. Let A be a torsion free group. Then $A \in \mathcal{V}_p$ precisely if there is a p^∞-basic subgroup B in A such that the group $Z_p \otimes B$ is completely decomposable.

PROOF. If $A \in \mathcal{V}_p$ then, by Proposition 2, B may be taken free and $Z_p \otimes B$ is completely decomposable. Conversely, suppose that there is a p^∞-basic subgroup B in A with completely decomposable $Z_p \otimes B$; hence $B \in \mathcal{B}_p \subseteq \mathcal{V}_p$ (see Proposition 3). By Definition 1, B contains a p^∞-basis Y_o of B which is p-independent in B and also in A. In view of Lemma 3, Y_o is a p^∞-basis of A

as well, and we conclude $A \in \mathcal{V}_p$.

LEMMA 16. Let A be a torsion free group with $r_p(A) = 0$. Then $A \in \mathcal{V}_p$ precisely if each p^∞-basic subgroup of A contains a free p^∞-basic subgroup of A .

PROOF. If each p^∞-basic subgroup of A contains a free p^∞-basic subgroup of A then $A \in \mathcal{V}_p$ follows by Proposition 2. Conversely, suppose $A \in \mathcal{V}_p$ and consider any p^∞-basic subgroup $B = \langle Y \rangle_{*p}^A$ of A generated by a p^∞-basis Y . As $r_p(A) = 0$, by [4, Satz 1], Y is a maximal independent set in A and therefore A/B is a torsion group. Making use of Lemma 5, we get $B \in \mathcal{V}_p$. Thus there is a p^∞-basis Y_o of B which is p-independent in B . As B is p^∞-basic in A , we deduce from Lemma 3 that Y_o is a p^∞-basis of A which is p-independent in A . Then $B_o = \langle Y_o \rangle_{*p}^A$ is a free p^∞--basic subgroup of A contained in B .

LEMMA 17. Let A be a torsion free group with $r_p(A) = 0$. Then $A \in \mathcal{V}_p$ precisely if for each p^∞-basic subgroup B of A the group $Z_p \otimes B$ is completely decomposable.

PROOF. Assume $A \in \mathcal{V}_p$ and take any p^∞-basic subgroup B of A . As in the proof of Lemma 16, we conclude $B \in \mathcal{V}_p$. Now $r_p(A) = 0$ implies $r_p(B) = 0$, and the complete decomposability of $Z_p \otimes B$ follows by Lemma 13. Conversely, if for each p^∞-basic subgroup B of A the group $Z_p \otimes B$ is completely decomposable then the relation $A \in \mathcal{V}_p$ is a consequence of Lemma 15.

If we summarize the results just obtained we get the following

PROPOSITION 7. If A is a torsion free group with $r_p(A) = 0$ then the following statements are equivalent: i) There is a p^∞-basis in A which is p-independent. ii) There is a Baer class Γ_α such that $Z_p \otimes A \in \Gamma_\alpha$. iii) The group $Z_p \otimes A$ is completely decomposable. iv) There is a p^∞-basic subgroup of A which is free. v) Each p^∞-basic subgroup of A contains a free p^∞-basic subgroup of A .

vi) There is a p^∞-basic subgroup B of A with $Z_p \otimes B$ completely decomposable. vii) For each p^∞-basic subgroup B of A the tensor product $Z_p \otimes B$ is completely decomposable. viii) There is a p^∞-basic subgroup B of A satisfying $B \in \mathcal{V}_p$. ix) Each p^∞-basic subgroup B of A belongs to $\mathcal{V}_p$.

PROOF. The equivalence i) $\Longleftrightarrow$ ii) is proved by Lemma 14, i) $\Longleftrightarrow$ iv) follows by Proposition 2 and i) $\Longleftrightarrow$ v) by Lemma 16. For the proof of i) $\Longleftrightarrow$ vi) see Lemma 15. From Lemma 17 we get i) $\Longleftrightarrow$ vii). Finally the equivalences i) $\Longleftrightarrow$ iii), vi) $\Longleftrightarrow$ viii) and vii) $\Longleftrightarrow$ ix) are consequences of Lemma 13. This completes the proof of the proposition.

Some previous results may be rather generalized.

LEMMA 18. Let A be a torsion free group with a direct decomposition $A = A_o \oplus A_1$ where the group A_o is p-divisible. Then $A \in \mathcal{V}_p$ just if $A_1 \in \mathcal{V}_p$.

PROOF. As A_o is p-divisible, the empty set is the unique p^∞-basis of A_o and therefore $A_o \in \mathcal{V}_p$. Thus if $A_1 \in \mathcal{V}_p$ then also $A \in \mathcal{V}_p$, because $\mathcal{V}_p$ is closed with respect to direct sum. For the proof of the converse consider any system $Y = \{y_i \,;\, i \in I\}$ of elements in A and write $y_i = y_i^{(o)} + y_i^{(1)}$ with $y_i^{(j)} \in A_j$ $(j=0,1\,;\, i \in I)$. If $m_i \in Z$ $(i \in I)$ are integers such that almost all m_i are zero, then the relation

$$\sum m_i y_i \equiv 0 \quad (p^k A) \tag{11}$$

holds if and only if

$$\sum m_i y_i^{(1)} = 0 \quad (p^k A_1) \,. \tag{12}$$

Evidently, (11) $\Longrightarrow$ (12) follows by the fact that A_1 is a summand of A, and the converse is a consequence of the p-divisibility of A_o. Suppose now that $A \in \mathcal{V}_p$ and take for $Y = \{y_i \,;\, i \in I\}$ any p-independent p^∞-basis of A. It follows from the above observation that $Y_1 = \{y_i^{(1)} ; i \in I\}$ is p-independent in A_1. If $y \in A_1$ is any element such that $y \notin Y_1$ then also $y \notin Y$ and therefore the set $Y \cup \{y\}$ is p^∞-dependent in A. But in view of the implication (11) $\Longrightarrow$ (12) we deduce that the set $Y_1 \cup \{y\}$ is p^∞-dependent in A_1. This means that Y_1 is already a p^∞-basis of A_1 and hence the relation $A_1 \in \mathcal{V}_p$.

LEMMA 19. Let A_o be a pure and p-divisible subgroup of a torsion free group A. Then $A \in \mathcal{V}_p$ if and only if $A/A_o \in \mathcal{V}_p$.

PROOF. From the exact sequence

$$0 \longrightarrow A_o \longrightarrow A \longrightarrow A/A_o \longrightarrow 0$$

we obtain another exact sequence

$$(13) \qquad 0 \longrightarrow Z_p \otimes A_o \longrightarrow Z_p \otimes A \longrightarrow Z_p \otimes A/A_o \longrightarrow 0 .$$

Under the hypothesis, the group $Z_p \otimes A_o$ is divisible and therefore the sequence (13) splits. Thus, by Lemma 18, $Z_p \otimes A \in \mathcal{V}_p$ just if $Z_p \otimes (A/A_o) \in \mathcal{V}_p$. Now, making use of Proposition 3, we conclude that $A \in \mathcal{V}_p$ if and only if $A/A_o \in \mathcal{V}_p$.

If A is a torsion free group then we shall denote by $A[p^\infty]$ the set of all elements of infinite p-height in A. Clearly, $A[p^\infty]$ is a pure subgroup in A.

LEMMA 20. If $\mathcal{V}_p^1$ is the class of all groups $A \in \mathcal{V}_p$ satisfying $r_p(A/A[p^\infty]) = 0$ then $\mathcal{V}_p^1 \subseteq \mathcal{B}_p$.

PROOF. If $A \in \mathcal{V}_p^1$ then we have also $A/A[p^\infty] \in \mathcal{V}_p$ by Lemma 19. As $r_p(A/A[p^\infty]) = 0$, we deduce from Lemma 13 that $Z_p \otimes (A/A[p^\infty])$ is completely decomposable. Since the group $Z_p \otimes A[p^\infty]$ is divisible, the exact sequence

$$(14) \qquad 0 \longrightarrow Z_p \otimes A[p^\infty] \longrightarrow Z_p \otimes A \longrightarrow Z_p \otimes (A/A[p^\infty]) \longrightarrow 0$$

splits and therefore the group $Z_p \otimes A$ is completely decomposable. Thus $A \in \mathcal{B}_p$ and we get $\mathcal{V}_p^1 \subseteq \mathcal{B}_p$.

PROPOSITION 8. If A is a torsion free group then the following statements are equivalent. i) $A \in \mathcal{V}_p$ and $r_p(A/A[p^\infty]) = 0$;

ii) $A \in {}_0\mathcal{B}_p$ and $r_p(A/A[p^\infty]) = 0$; iii) the group $Z_p \otimes A$ is completely decomposable; iv) $r_p(A/A[p^\infty]) = 0$ and for each p^∞-basic subgroup B of A the group $Z_p \otimes B$ is completely decomposable; v) $r_p(A/A[p^\infty]) = 0$ and there is a p^∞-basic subgroup B of A with $Z_p \otimes B$ completely decomposable.

PROOF. The equivalence i) $\Longleftrightarrow$ ii) follows by Lemma 20. Suppose now that i) is fulfilled. As $A \in \mathcal{V}_p$, in view of Lemma 19 we get $A/A[p^\infty] \in \mathcal{V}_p$; consequently, Lemma 13 and $r_p(A/A[p^\infty]) = 0$ imply that the group $Z_p \otimes (A/A[p^\infty])$ is completely decomposable. We conclude from the splitting of the sequence (14) and from the divisibility of $Z_p \otimes A[p^\infty]$ that $Z_p \otimes A$ is completely decomposable as well. Thus i) $\Longrightarrow$ iii). For the proof of the implication iii) $\Longrightarrow$ ii) we shall use the following remark: If U is a torsion free group satisfying $U[p^\infty] = 0$ then also $(Z_p \otimes U)[p^\infty] = 0$. In fact, we have the exact sequence

$$0 \longrightarrow Z \otimes U \longrightarrow Z_p \otimes U \longrightarrow (Z_p/Z) \otimes U \longrightarrow 0 \quad .$$

Thus we see that $Z \otimes U$ $(\cong U)$ is a p-pure and essential subgroup in $Z_p \otimes U$. Since $(Z \otimes U)[p^\infty] = 0$ we conclude $(Z_p \otimes U)[p^\infty] = 0$. Now if $Z_p \otimes A$ is completely decomposable then the splitting of (14) implies that $Z_p \otimes (A/A[p^\infty])$ is completely decomposable as well. For $U = A/A[p^\infty]$ we have $U[p^\infty] = 0$ and, in view of the above remark, we have also $(Z_p \otimes U)[p^\infty] = 0$. But this means that the completely decomposable group $Z_p \otimes U$ is homogeneous and reduced. By Lemma 13 we get $0 = r_p(U) = r_p(A/A[p^\infty])$; the relation $A \in {}_0\mathcal{B}_p$ follows immediately from the complete decomposability of $Z_p \otimes A$. Therefore iii) $\Longrightarrow$ ii). If iii) is fulfilled and B denotes any p^∞-basic subgroup of A then B is p-pure in A and hence $Z_p \otimes B$ is pure in $Z_p \otimes A$; by [13] the group $Z_p \otimes B$ is completely decomposable. As we have seen, iii) implies $r_p(A/A[p^\infty]) = 0$, whence iii) $\Longrightarrow$ iv). The implication iv) $\Longrightarrow$ v) is trivial thus assume finally v). From

Lemma 15 we deduce $A \in \mathcal{V}_p$ and we conclude v) $\Longrightarrow$ i). The proof of our proposition is complete.

REMARK 2. If $\mathcal{C}_p^*$ denotes the class of all torsion free groups A with completely decomposable group $Z_p \otimes A$ and $\mathcal{B}_p^1$ the class of all $A \in \mathcal{B}_p$ satisfying $r_p(A/A[p^\infty]) = 0$ then Proposition 8 declares the equality

$$\mathcal{B}_p^1 = \mathcal{V}_p^1 = \mathcal{C}_p^* .$$

Furtermore, if $A \in \mathcal{C}_p^*$ then for any two its p^∞-basic subgroups B_1, B_2 we have $Z_p \otimes B_1 \cong Z_p \otimes B_2$.

LEMMA 21. Let A be a torsion free group such that $A \in \mathcal{V}_p$ and $r_p(A) = n$ finite. Then there is in A a pure subgroup A_o of finite rank satisfying $A/A_o \in \mathcal{V}_p^o$.

PROOF. As $A \in \mathcal{V}_p$, there is in A a p^∞-basis Y which is p-independent in A. If we extend Y to a maximal independent set X of A then, in view of [4, Satz 1], $\operatorname{card}(X \setminus Y) = n$ and hence $X = Y \cup \{x_1, \ldots, x_n\}$. For each i $(1 \leq i \leq n)$, the set $Y \cup \{x_i\}$ is p^∞-dependent in A. Therefore there is a finite subset $Y_o \subseteq Y$ such that for all $i \in \{1, \ldots, n\}$ the sets $Y_o \cup \{x_i\}$ are p^∞-dependent in A. Now we put $A_o = \langle Y_o \cup \{x_1, \ldots, x_n\}\rangle_*$. Evidently, $Y_o \cup \{x_1, \ldots, x_n\}$ is a maximal independent set of A_o and Y_o is p^∞-independent in A_o. Since each $Y_o \cup \{x_i\}$ is p^∞-dependent in A_o, by [4, Lemma 7], Y_o is a p^∞-basis of A_o and hence $r_p(A_o) = n$ (see [4, Satz 1]). If S is any pure subgroup of finite rank in A satisfying $A_o \subseteq S$ then, evidently, $r_p(S) = n$ and, by [3, Theorem 6], $r_p(S/A_o) = 0$. This implies $0 = r_p(A/A_o)$ (see [12, Corollary 2]).

For the proof of the relation $A/A_o \in \mathcal{V}_p$ put $Y_1 = Y \setminus Y_o$ and consider the subgroups U, V, W of A defined as follows: $U = \langle Y_1\rangle^A_{*p}$, $V = \langle Y_o \cup \{x_1, \ldots, x_n\}\rangle^A_{*p}$ and $W = \langle U + V\rangle^A_{*p}$. Since the set Y_1 is p-independent in A, U is a free group with free basis Y_1 . Clearly $U + V = U \oplus V$, and $W/(U \oplus V)$ is a p-primary group. If $a \in W$ and $a \notin U \oplus V$ then for a positive integer k we have $p^k a = u + v$ with $u \in U$ and $v \in V$. Thus with every element $a \in W \setminus (U \oplus V)$ a triplet (k, u, v) is associated. Suppose now that a_1 , a_2 are two elements in $W \setminus (U \oplus V)$ and that (k_i, u_i, v_i) $(i=1,2)$ are their corresponding triplets. If $k_1 = k_2$ and $v_1 = v_2$ then $p^{k_1}(a_1 - a_2) = u_1 - u_2 \in U$, and as U is p-pure in A , we conclude $a_1 - a_2 \in U \subseteq U \oplus V$; therefore in this case we have $a_1 \equiv a_2 \, (U \oplus V)$. Since the set of all couples (k, v) $k \in N$, $v \in V$ is countable we deduce that the group $W/(U \oplus V)$ is also countable. Hence we conclude that the factor group

$$(W/V)/((U \oplus V)/V) \tag{15}$$

is countable and p-primary. By the definition, W is p-pure in A , V is the p-pure closure of the set $Y_o \cup \{x_1, \dots , x_n\}$ in A and so in W , and A_o is the pure closure of the same set in A ; therefore $V = W \cap A_o$. Hence

$$W/V = W/(W \cap A_o) \cong (W + A_o)/A_o \subseteq A/A_o$$

and from $r_p(A/A_o) = 0$ we deduce $r_p(W/V) = 0$. As the group (15) is countable and p-primary we can apply Lemma 11 and we get an isomorphism $W/V \cong (U \oplus V)/V \cong U$. The freeness of U implies a decomposition $W = U_o \oplus V$ with $U_o \cong U$. Since $A_o = \langle V \rangle_*$ we have $U_o + A_o = U_o \oplus A_o$, and also

$$(A/A_o)/((U_o \oplus A_o)/A_o) \cong A/(U_o \oplus A_o) \cong A/(U_o \oplus V)/(U_o \oplus A_o)/(U_o \oplus V).$$

The group W is p-pure in A and contains X , so that the group $A/W = A/(U_o \oplus V)$ is torsion and its p-primary component equals zero. Evidently, the same is true for the group $(A/A_o)/((U_o \oplus A_o)/A_o)$, and we conclude that the free group $(U_o \oplus A_o)/A_o$ is p-pure in A/A_o . Thus $(U_o \oplus A_o)/A_o$ is a p-basic subgroup of A/A_o (see [2, § 32]) and any its free basis is a p-basis of A/A_o . But each such p-basis is a maximal independent set in A/A_o and, consequently, a

p^∞-basis of A/A_o (see (2)). Thus $A/A_o \in \mathcal{V}_p$ and the proof is complete.

Let $\mathcal{A}$ be any non empty class of torsion free groups. We shall define Baer classes $\Gamma_\alpha(\mathcal{A})$ with respect to the class $\mathcal{A}$ as follows. First we put $\Gamma_1(\mathcal{A}) = \mathcal{A}$. Further, if for an ordinal $\alpha > 1$ all classes $\Gamma_\beta(\mathcal{A})$ with $\beta < \alpha$ are defined, then we let a torsion free group A belong to $\Gamma_\alpha(\mathcal{A})$ if $A \notin \Gamma_\beta(\mathcal{A})$ for each $\beta < \alpha$ and there exists a pure subgroup S of finite rank in A such that A/S is a direct sum of groups belonging to classes with indices less then α. We shall denote by $\Gamma(\mathcal{A})$ the union of all the classes $\Gamma_\alpha(\mathcal{A})$.

Below we shall prove a generalization of Lemma 20.

PROPOSITION 9. If $\mathcal{V}_p^2$ denotes the class of all groups $A \in \mathcal{V}_p$ with $r_p(A/A[p^\infty])$ finite then $\Gamma(\mathcal{V}_p^2) \subseteq \mathcal{B}_p$.

PROOF. If $A \in \mathcal{V}_p^2 \subseteq \mathcal{V}_p$ then, by Lemma 19, also $\bar{A} = A/A[p^\infty] \in \mathcal{V}_p$. In view of Lemma 21, there is in $\bar{A}$ a pure subgroup $\bar{A}_o$ of finite rank satisfying $\bar{A}/\bar{A}_o \in \mathcal{V}_p^o$. By Proposition 7, the tensor product $Z_p \otimes (\bar{A}/\bar{A}_o)$ is completely decomposable and, evidently, reduced. Hence $Z_p \otimes (\bar{A}/\bar{A}_o)$, as a Z_p-module, is free. Thus the exact sequence of Z_p-modules

$$(15)\qquad 0 \longrightarrow Z_p \otimes \bar{A}_o \longrightarrow Z_p \otimes \bar{A} \longrightarrow Z_p \otimes (\bar{A}/\bar{A}_o) \longrightarrow 0$$

is splitting. As $Z_p \otimes \bar{A}_o$ is of finite rank and $Z_p \otimes (\bar{A}/\bar{A}_o)$ is completely decomposable, we deduce from (15) that $Z_p \otimes \bar{A} \in \mathcal{B}$. But in view of (14)

$$Z_p \otimes A \cong Z_p \otimes A[p^\infty] \oplus Z_p \otimes \bar{A} ,$$

whence $Z_p \otimes A \in \mathcal{B}$. This means that $A \in \mathcal{B}_p$ and therefore $\mathcal{V}_p^2 \subseteq \mathcal{B}_p$. Making use of the transfinite induction it may be easily shown that for any subclass $\mathcal{A} \subseteq \mathcal{B}_p$ also $\Gamma(\mathcal{A}) \subseteq \mathcal{B}_p$; hence finally $\Gamma(\mathcal{V}_p^2) \subseteq \mathcal{B}_p$.

In the following section we shall apply some methods of this note to the study of Z_p^*-modules (Z_p-modules resp.). In particular, we shall find necessary and sufficient conditions for such a module to be a direct sum of cyclic modules.

LEMMA 22. Let A be a reduced torsion free Z_p^*-module. If $x_1, \ldots, x_n \in A$ and $\alpha_1, \ldots, \alpha_n \in Z_p^*$ then the relation

(16) $$\alpha_1 x_1 + \ldots + \alpha_n x_n \equiv 0 \ (p^\infty A)$$

holds in the group A precisely if

(17) $$\alpha_1 x_1 + \ldots + \alpha_n x_n = 0$$

in the Z_p^*-module A.

PROOF. If $\alpha_i = (a_i^{(k)})_{k=1}^\infty$ is the canonical expression of the p-adic integer α_i then $a_i^{(k)} - \alpha_i \in p^k Z_p^*$ and hence $a_i^{(k)} x_i - \alpha_i x_i \in p^k A$ $(i=1, \ldots, n)$. If (17) holds then for each k we obtain

$$\sum_i a_i^{(k)} x_i = \sum_i (a_i^{(k)} - \alpha_i) x_i \in p^k A$$

and therefore

(18) $$\sum_i a_i^k x_i \equiv 0 \quad (p^k A) \quad .$$

Thus (17) implies (16). Conversely, if (16) is fulfilled then (18) holds for every k. In the p-adic topology of the group A we get

$$\alpha_1 x_1 + \ldots + \alpha_n x_n = \lim_{k \to \infty} (a_1^{(k)} x_1 + \ldots + a_n^{(k)} x_n) = 0 \quad .$$

Hence (16) implies (17).

LEMMA 23. Let A be a reduced torsion free Z_p^*-module. If $A \in \mathcal{V}_p$ then the Z_p^*-module A is free.

PROOF. As $A \in \mathcal{V}_p$, there is in A a p-independent p^∞-basis X. We shall prove that X is a free system of generators for the Z_p^*-module A.

The set X is p^∞-independent in the group A and, in view of Lemma 22, it is independent in the Z_p^*-module A. Thus it remains to prove that X generates A as the Z_p^*-module. Denote by B the subgroup of A generated by X and take any element $y \in A \setminus B$. We shall prove that $y \in \bigoplus_{x \in X} Z_p^* x$. Suppose first that $B \cap \langle y \rangle \neq 0$ and denote by n the smallest natural number with $ny \in B$. As B is p-pure in A , n is prime to p ; therefore $1/n \in Z_p^*$ and we get $y = 1/n(ny) \in \bigoplus_{x \in X} Z_p^* x$. If $B \cap \langle y \rangle = 0$ then the set $X \cup \{y\}$ is independent but p^∞-dependent in the group A . Thus there are a finite subset $\{x_1, \dots, x_n\} \subseteq X$ and non-zero numbers $\alpha_1, \dots, \alpha_n$, $\alpha \in Z_p^*$ such that

(19) $$\alpha_1 x_1 + \dots + \alpha_n x_n - \alpha y \equiv 0 \ (p^\infty A) .$$

Evidently, we can suppose that among the numbers $\alpha_1, \dots, \alpha_n, \alpha$ at least one is prime to p . From the p-independence of X in A we deduce that $p \nmid \alpha$; therefore there exists $\alpha^{-1} \in Z_p^*$. In view of Lemma 22, (19) implies $\alpha y = \alpha_1 x_1 + \dots + \alpha_n x_n$ and hence

$$y = \alpha^{-1} \cdot (\alpha_1 x_1 + \dots + \alpha_n x_n) \in \bigoplus_{x \in X} Z_p^* x .$$

Thus A is generated by X as the Z_p^*-module.

LEMMA 24. If A is a torsion free Z_p^*-module then the following assertions are equivalent. i) The Z_p^*-module A is completely decomposable. ii) $A \in \mathcal{B}$. iii) $A \in \mathcal{V}_p$.

PROOF. If the module A is completely decomposable then the group A is of the form $A = D \oplus (\oplus J_p)$ where D is divisible. Thus evidently $A \in \mathcal{B}$ and hence i) $\Longrightarrow$ ii) . The implication ii) $\Longrightarrow$ iii) follows by Proposition 1. Suppose now that the group A satisfies $A \in \mathcal{V}_p$ and consider a module decomposition $A = D \oplus A_o$ with D divisible and A_o reduced. By Lemma 18, the group A_o satisfies $A_o \in \mathcal{V}_p$ as well and, in view of Lemma 23, the Z_p^*-module A_o is free. Thus we conclude that A is a completely decomposable Z_p^*-module and iii) $\Longrightarrow$ i) is verified.

If G is a Z_p-module or a Z_p^*-module (such modules are studied by L. Kulikov in [11] and are called the generalized primary groups) then $t(G)$ will denote its maximal torsion submodule. The submodule $t(G)$ is in fact a p-primary group (see [11]) . Recall that if a group G is p-primary then there exists the well known Kulikov criterion [2, Theorem 17.1] for G to be a direct sum of cyclic groups. The next proposition concerns the case of p-adic modules (the generalized primary groups).

PROPOSITION 10. 1) A Z_p-module G is a direct sum of cyclic modules precisely if i) t(G) satisfies the Kulikov criterion and ii) $r_p(G/t(G)) = 0$ with $G/t(G) \in \mathcal{V}_p$ (or, equivalently, with $G/t(G) \in \mathcal{B}$). 2) A Z_p^*-module G is a direct sum of cyclic modules precisely if i) t(G) satisfies the Kulikov criterion and ii) $(G/t(G))[p^\infty] = 0$ with $G/t(G) \in \mathcal{V}_p$ (or, equivalently, with $G/t(G) \in \mathcal{B}$).

PROOF. Clearly, G is a direct sum of cyclic modules precisely if t(G) is so and G/t(G) is a free module. Thus in the case 1) it suffices to apply the Kulikov criterion to t(G) and Lemma 12* (or, Proposition 7 and isomorphism $G/t(G) \cong Z_p \otimes (G/t(G))$) to G/t(G). In the case 2) we apply again the Kulikov criterion to t(G) and Lemma 24 to G/t(G) ; the correctness of the condition $(G/t(G))[p^\infty] = 0$ is evident.

We shall conclude this note by stating several quite natural problems concerning the notion of the p^∞-basic subgroup and the class $\mathcal{V}_p$ in general.

1. Is the class $\mathcal{V}_p$ closed with respect to direct summands ?
2. Are any two p^∞-basic subgroups (their tensor product with Z_p resp.) of a torsion free group necessarily isomorphic ?
3. If the group A belongs to $\mathcal{V}_p$, is then each p^∞-basic subgroup B of A free ? Or, is $Z_p \otimes B$ completely decomposable ?
4. What may be said about the equality $\mathcal{B}_p = \mathcal{V}_p$, or analogously, abote the equality $\mathcal{V}_p = \mathcal{C}_p$?

Remark that some partial answers to the last three problems are contained in Proposition 8, Remark 2 and Proposition 9.

REFERENCES

[1] BAER R.: Die Torsionsuntergruppe einer Abelschen Gruppe, Math. Annalen 135 (1958), 219 - 234.

[2] FUCHS L.: Infinite Abelian Groups I, II, Acad. Press 1970, 1973.

[3] PROCHÁZKA L.: O p-range abelevych grupp bez kručenija konečnogo ranga, Czechoslovak Math. J. 12 (1962), 3 - 43.

[4] PROCHÁZKA L.: Bemerkung über den p-Rang torsionsfreier abelscher Gruppen unendlichen Ranges, Czechoslovak Math. J. 13 (1963), 1 - 23.

[5] PROCHÁZKA L.: Sur p-indépendance et p^∞-indépendance en des groupes sans torsion, Symposia Mathematica XXIII (1979), 107 - 120.

[6] PROCHÁZKA L.: Tensor product and quasi-splitting of abelian groups, Comment. Math. Univ. Carolinae 21 (1980), 55 - 69.

[7] PROCHÁZKA L.: Ob odnorodnych abelevych gruppach bez kručenija, Czechoslovak Math. J. 14 (1964), 171 - 202.

[8] FUCHS L.: Abelian Groups, Budapest, 1958.

[9] BICAN L.: Some properties of completely decomposable torsion free abelian groups, Czechoslovak Math. J. 19 (1969), 518 - 533.

[10] SZEKERES G.: Countable abelian groups without torsion, Duke Math. J. 15 (1948), 293 - 306.

[11] KULIKOV L.: Obobščennye primarnye gruppy I, Trudy Moskov. Mat. Obšč. (1952), 247 - 326.

[12] PROCHÁZKA L.: A note on completely decomposable torsion free abelian groups, Comment. Math. Univ. Carolinae 12 (1971), 23 - 31.

[13] PROCHÁZKA L.: A generalization of a theorem of R. Baer, Comment. Math. Univ. Carolinae 4 (1963), 105 - 108.

A New Class of Subgroups of $\prod_{\aleph_0} Z$

by

John Irwin and Tom Snabb

Abstract: Let P be the direct product of a countably infinite number of infinite cyclic groups. We introduce a new class of subgroups of P by considering groups of functions analytic on a disk of radius $\geq \alpha$ whose power series expansions have integer coefficients. The resulting groups A_α, $0 < \alpha \leq 1$, α real, are shown to be monotone subgroups of P and thus a class of $c = 2^{\aleph_0}$ non-isomorphic subgroups of P.

All groups considered in this paper will be additively written abelian groups. We will, for the most part, use the notation and terminology in [1]. In particular, we let P denote the direct product $P = \prod_{n=0}^{\infty} \langle e_n \rangle$ of a countable number of infinite cyclic groups, where $o(e_n) = \infty$. We note that P can be considered to be the product of $\aleph_0$ copies of the integers Z. The elements $\bar{a}$ of P will be written as formal infinite sums $\bar{a} = \sum_{n=0}^{\infty} a_n e_n$ with $a_n \in Z$.

Our intent is to introduce a new class of subgroups of P and to investigate the properties of these subgroups. In order to define our groups, we first consider the power series expansions of analytic functions f(z) of a single complex variable. We let F_α be the set of analytic functions whose power series expansions about 0 have integer coefficients and whose radii of convergence are $\geq \alpha$. The elements f(z) of F_α will be written as power series $f(z) = \sum_{n=0}^{\infty} a_n z^n$ with $a_n \in Z$. Since the sum of two analytic functions, each analytic on a disk of radius $\geq \alpha$, is again analytic on a disk of radius $\geq \alpha$, the F_α are abelian groups. There is a natural mapping $\theta : F_\alpha \to P$ defined by $\theta(\sum_{n=0}^{\infty} a_n z^n) = \sum_{n=0}^{\infty} a_n e_n$ which is an isomorphism of F_α into P. We now use this mapping to define our groups.

Definition. The power series group of radius α, denoted A_α, is defined by $A_\alpha = \theta(F_\alpha)$.

In what follows, we will often rely on properties of analytic functions to prove results concerning our A_α's. We first list a few straightforward facts concerning the A_α's.

(a) The radii α of convergence satisfy $0 < \alpha \leq 1$. This follows since $\alpha > 1$ implies the existence of an integer coefficient power series with an infinite number of nonzero coefficients converging for values of z having magnitude greater than 1, an impossibility.

(b) For $\alpha < \beta$, $A_\beta \leq A_\alpha$. This is immediate since a power series with radius of convergence $\geq \beta$ has radius of convergence $\geq \alpha$, that is, $F_\beta \leq F_\alpha$.

(c) $\cap_\alpha A_\alpha = A_1$ since elements of A_1 correspond to series having radius of convergence 1 and such integer coefficient series always diverge on the boundary.

(d) $A_0 = \cup_\alpha A_\alpha$ is a proper subgroup of P. In fact, $\Sigma_{n=0}^{\infty} n!z^n$ is a series nowhere convergent so $\Sigma_{n=0}^{\infty} n!e_n \notin A_\alpha$ for any α, but $\Sigma_{n=0}^{\infty} n!e_n \in P$.

Our next result concerning the groups A_α shows proper containment in b) above.

<u>Theorem 1</u>. For $\alpha < \beta$, A_β is a proper subgroup of A_α.

Proof. The proof consists of constructing an integer coefficient power series $\Sigma_{n=0}^{\infty} a_n z^n$, having radius of convergence α, which diverges at $z = \alpha$. To do so we let $a_n = [\alpha^{-n}/n]$, where $[\cdot]$ denotes the greatest integer function. Since $a_n = [\alpha^{-n}/n] = \alpha^{-n}/n + \epsilon_n$ $0 \leq \epsilon_n < 1$, and since $\alpha^{-n}/n \to \infty$, we have the radius of convergence $r = \lim_{n\to\infty} |a_n/a_{n+1}| = \alpha$. To see that the series diverges at α, we consider $a_n\alpha^n = [\alpha^{-n}/n]\alpha^n \geq (\alpha^{-n}/n)\alpha^n = 1/n$ and appeal to the comparison test, using the fact that $\Sigma_{n=1}^{\infty} 1/n$ diverges. We have thus shown that F_β is properly contained in F_α, which proves our claim.

Since the above theorem shows that $A_\alpha \neq A_\beta$ for $\alpha \neq \beta$, a natural question to ask is if it is possible that $A_\alpha \approx B_\alpha$ for $\alpha \neq \beta$. To answer this question, we will need the following definition and theorem.

Definition. A subgroup T of P is monotone if the following hold.

(i) If $\Sigma_{k=0}^{\infty} a_k e_k \in T$ and $0 \leqslant b_k \leqslant a_k$, then $\Sigma_{k=0}^{\infty} b_k e_k \in T$.

(ii) If $\Sigma_{k=0}^{\infty} a_k e_k \in T$, then $\Sigma_{k=0}^{\infty} b_k e_k \in T$ for $b_k = \max(1, |a_1|, \ldots, |a_k|)$.

Theorem 2. A_α is a monotone subgroup of P.

Proof. i) For $\Sigma_{k=0}^{\infty} a_k e_k \in A_\alpha$, consider the corresponding power series $\Sigma_{k=0}^{\infty} a_k z^k \in F_\alpha$. If $0 \leqslant b_k \leqslant a_k$ for all k, then since $\Sigma_{k=0}^{\infty} a_k z^k$ has radius of convergence $\geqslant \alpha$, the comparison test implies $\Sigma_{k=0}^{\infty} b_k z^k$ also has radius of convergence $\geqslant \alpha$, so $\Sigma_{k=0}^{\infty} b_k z^k \in F_\alpha$, which implies that $\Sigma_{k=0}^{\infty} b_k e_k \in A_\alpha$.

(ii) For $\Sigma_{k=0}^{\infty} a_k e_k \in A_\alpha$, let $b_k = \max(1, |a_1|, \ldots, |a_k|)$. We consider three cases and will again switch to the power series $\Sigma_{k=0}^{\infty} a_k z^k$ having radius of convergence $r \geqslant \alpha$.

(a) Maximum $|a_k|$ exists.

If there exists k_0 such that $|a_k| \leqslant |a_{k0}|$ for all k, then for all $k \geqslant k_0$, $b_k = a_k$. The radius of convergence for $\Sigma_{k=0}^{\infty} b_k z^k$ is then $1/\lim_k \sup \sqrt[k]{|b_k|} = 1$, so $\Sigma_{k=0}^{\infty} b_k z^k \in F_1 \leqslant F_\alpha$ implies $\Sigma_{k=0}^{\infty} b_k e_k \in A_\alpha$.

(b) The $|a_k|$ are monotone non-decreasing, and no maximum exists. It follows that for sufficiently large k, $b_k = a_k$ implies $\Sigma_{n=0}^{\infty} b_n z^n \in F_\alpha$, so $\Sigma_{n=0}^{\infty} b_n e_n \in A_\alpha$.

(c) The $|a_k|$ are not monotone and no maximum exists. Let $\{c_k\}_{k=0}^{\infty}$ represent the successive maximums of $\{|a_k|\}_{k=0}^{\infty}$. In particular, $c_0 = |a_0|$, $c_1 = |a_{k1}|$ where a_{k1} is the first a_k such that $|a_{k1}| > |a_0|$, and so on. Thus $c_0 < c_1 < c_2 < \cdots$ and $\lim_n \sqrt[n]{c_n} = \lim_n \sup \sqrt[n]{|a_n|} = 1/r$. Now the choice of $b_k = \max(1, |a_1|, \ldots,$

$|a_k|$) implies that for each k, $b_k \in \{c_k\}$. It follows that $\lim_k \sup \sqrt[k]{|b_k|} = \lim_k \sqrt[k]{c_k} = 1/r$, so $\sum_{k=0}^{\infty} b_k z^k \in F_\alpha$ implies $\sum_{k=0}^{\infty} b_k e^k \in A_\alpha$. Since $\{|a_k|\}$ cannot be a strictly monotone decreasing sequence, the above cases are exhaustive.

Having established that our A_α's are monotone subgroups of P, we now only need from Specker [3]:

Theorem. Two monotone subgroups of P are isomorphic if and only if they are equal. Since $A_\alpha \neq A_\beta$ for $\alpha \neq \beta$, it follows that $A_\alpha \not\cong A_\beta$.

Two other results on monotone subgroups which are of interest concerning the structure of our A_α's are the following.

Theorem. (G. Reid, see [1]) Every monotone subgroup of P, different from P, is slender.

Theorem. (Specker [3]) Every monotone subgroup of P, other than the group B of bounded elements, contains a subgroup of power $\aleph_1$ which is not free.

From these theorems we see immediately that the A_α's are slender and not free.

Specker in [3] considered a class of monotone subgroups related to our A_α's. For r real, he defined $F(r)$ by $\sum a_n e_n \in F(r)$ if there exists a constant c such that $|a_n| \leq cn^r$. We make the following observations concerning Specker's groups $F(r)$.

1) $F(0) = B$, the group of bounded sequences.

2) For $r_1 < r_2$, $F(r_1) < F(r_2)$ (proper containment)

3) $F(r) \leq A_1$ for all r.

To see 3), we suppose $\sum_{n=0}^{\infty} a_n e_n \in F(r)$, which implies $|a_n| \leq cn^r$ for all n. Consider the series $\sum_{n=0}^{\infty} a_n z^n$ and $\sum_{n=0}^{\infty} a_n' z^n$, where $a_n' = cn^r$. The radius of convergence of $\sum_{n=0}^{\infty} a_n' z^n$ is given by $\lim_n a_n'/a_{n+1}' = \lim_n cn^r/c(n+1)^r = 1$, and since $a_n < a_n'$, it follows that

$\sum_{n=0}^{\infty} a_n z^n$ has radius of convergence 1. This implies $\sum_{n=0}^{\infty} a_n z^n \in A_1$.

From 3) above, it follows that $\cup_r F(r) \leq A$, which raises the question of proper containment. We answer this question in the following observation.

<u>Observation.</u> $\cup_r F(r)$ is a proper subgroup of A_1.

Proof. Consider the element $\sum a_n e_n$ where $a_n = [n^{\sqrt{n}/\ln(n)}]$. Since there does not exist c and r such that $n^{\sqrt{n}/\ln(n)} < cn^r$ for all n, $\sum_{n=0}^{\infty} a_n e_n$ belongs to no $F(r)$. Since $\sqrt[n]{a_n}$ can be shown to converge to 1, $\sum_{n=0}^{\infty} a_n z^n$ has radius of convergence 1, which implies $\sum_{n=0}^{\infty} a_n e_n \in A_1$.

Fuchs and Salce in [2] have considered monotone subgroups generated by sets of elements of P. They let P^* be the elements $\bar{a} = \sum_{n=0}^{\infty} a_n e_n \in P$ having positive, non-decreasing, and unbounded components, that is, $0 < a_1 \leq a_2 \ldots$ and $\sup_n\{a_n\} = \infty$. They then give the following definition and observation.

<u>Definition.</u> Let $\{\bar{a}^{(i)}\}_{i\in I}$ be a subset of P^*. $\{\bar{a}^{(i)}\}_{i\in I}$ generates a monotone subgroup A if $A = \{\bar{a}\in P \mid |a_k| \leq m(t_k^{(i_1)}+\ldots+t_k^{(i_r)})$ for all k ; $m \in N$; $i_1, \ldots, i_r \in I\}$.

<u>Observation.</u> If A is generated by a set of cardinality $\aleph_0$, then A has a set of generators satisfying $\bar{a}^{(1)} \leq \ldots \leq \bar{a}^{(r)} \leq \ldots$; $a_k^{(r)} = a_k^{(k)}$ for each $k \leq r$.

Concerning our groups A_α, we show in the next theorem that they are not countably generated.

<u>Theorem 3.</u> A_α is not countably generated for any α.

Proof. Suppose A_α has a countable set of generators $\{\bar{a}^{(i)}\}_{i=1}^{\infty}$. Consider the element $\bar{b}$, where $b_k = a_k^{(k)}$. That is, we pick the diagonal elements from the array of coefficients of $\bar{a}^{(i)}$, $\bar{a}^{(2)}$, ... From the definition and the observation above, it follows that $\bar{b}$ does not belong to the monotone group generated by $\{\bar{a}^{(1)}\}_{i=1}^{\infty}$. However, considering the series $\sum_{k=0}^{\infty} b_k z^k$, we see that $\lim_n \sqrt[n]{b_n} \leqslant 1/\alpha$ since $\lim_n \sqrt[n]{a_n^{(k)}} \leqslant 1/\alpha$.. Since the radius of convergence for $\sum_{n=0}^{\infty} b_n z^n$ is $1/\lim_n \sqrt[n]{b_n} \geqslant \alpha$, we have $\bar{b} = \sum_{n=0}^{\infty} b_n e_n \in A_\alpha$. Then A_α cannot have a countable set of generators. We note that the group $F(r)$ can be generated by the single element $\bar{a}$ having $a_k = n^r$.

It is well know that for $f(z) = \sum_{n=0}^{\infty} a_n z^n$ having radius of convergence α that $f'(z) = \sum_{n=1}^{\infty} na_n z^{n-1}$ also has radius of convergence α. This fact allows us to consider d/dz as a map from F_α to F_α and thus leads to the corresponding homomorphism ϕ from A_α to A_α defined by $\phi(\sum_{n=0}^{\infty} a_n e_n) = \sum_{n=1}^{\infty} na_n e_{n-1}$. If we generalize to higher order differentiation for functions, we get composition of ϕ with itself, so we let $A_\alpha^{(n)} = \phi^{(n)}(A_\alpha)$ denote the nth order composition. It follows that $A_\alpha \geqslant \phi(A_\alpha) \geqslant \ldots \geqslant \phi^{(n)}(A_\alpha) \geqslant \ldots$ and that the containment is proper. We note that this is not true in general for monotone subgroups since, for example, $\phi(F(r))$ is not contained in $F(r)$. A problem that arises is to determine $\cap_n A_\alpha^{(n)}$. This is handled in the next theorem.

<u>Theorem 4.</u> $\cap_{n=0}^{\infty} P^{(n)} = 0$.

Proof. Suppose there exists $\bar{a} = \sum_{n=0}^{\infty} a_n e_n \neq \bar{0}$ such that $\bar{a} \in \cap_n P^{(n)}$. Choose n_0 such that $a_n \neq 0$ and $k > a_n$. Now $\bar{a} \in P^{(k)} = \phi^{(k)}(P)$ implies that there exists $\bar{b} = \sum_{n=0}^{\infty} b_n e_n \in P$ such that $\phi^{(k)}(\bar{b}) = \bar{a}$. It follows from the definition of ϕ that

$a_n = (n_0+k)(n_0+k-1) \ldots (n_0+1)b_n$) which implies $n_0 + k$ divides a_n. Since $n_0 + k > a_{n0}$, we have a contradiction.

<u>Corollary.</u> For any α, $\cap_{n=0}^{\infty} A_\alpha^{(n)} = 0$.

Since the group $B = \{\sum_{n=0}^{\infty} b_n e_n \mid \sup_n\{b_n\} < \infty\}$ of bounded elements is a subgroup of each A_α, (in fact it is a subgroup of all monotones) it is of interest to consider $\phi(B)$. In particular, since B is free and $\phi(B) \leq A_\alpha$, we ask whether $\phi(B)$ is also free. The following lemma is the key to answering this question.

Lemma. If $f:Z \to Z$ is a function satisfying $f(n) = f_n \neq 0$ for all n, then the subgroup $C_f = \{\sum_{n=0}^{\infty}(c_n f_n)e_n \mid \sup_n\{c_n\}<\infty\}$ of P is free.

Proof. We show $B \cong C_f$. Define the map $\delta : B \to C_f$ by $\delta : \sum_{n=0}^{\infty} b_n e_n \to \sum_{n=0}^{\infty}(b_n f_n)e_n$. The fact that it is one to one follows from $f_n \neq 0$, and that it is onto from $\sup\{c_n\}<\infty$.

Theorem 5. For each k, $\phi^{(k)}(B)$ is a free group.

Proof. For a fixed k, since for $\bar{b} = \sum_{n=0}^{\infty} b_n e_n \in B$, $\phi^{(k)}(\bar{b}) = \bar{a} = \sum_{n=0}^{\infty} a_n e_n$ has $a_n = ((n+k)!/n!)b_{n+k}$, if we choose f such that $f_n = (n+k)!/n!$, we will have $\phi^{(k)}(B) \cong C_f$.

References

1. Fuchs, L., Infinite Abelian Groups, Vol. 2, Academic Press, New York, 1973.
2. Fuchs, L. and Salce, L., Gruppi monotoni di successioni intere, Seminario Matematico della Universita di Padova, 56 (1977), pp. 147-160.
3. Specker, E., Additive Gruppen von Folgen ganzer Zahlen, Portug. Math., 9 (1950), p. 131-140.

GROUPS OF INTEGER-VALUATED FUNCTIONS*

Rüdiger Göbel, Burkhard Wald and Petra Westphal

0. Introduction. In the following κ will be some infinite cardinal which we will interprete as the set of all ordinals $\alpha < \kappa$. In particular, we will use $\omega = \aleph_0$ for the first infinite cardinal, which may be represented by the set of natural numbers. Let Z denote the group of integers and Z^κ the group of all functions $f : \kappa \to Z$ with addition, defined by components. The value of f at $i \in \kappa$ will be denoted by f(i) and supp(f) $= \{i \in \kappa : f(i) \neq 0\}$ is the support of f. Then Z^ω is the *Baer-Specker group* and $Z^{(\kappa)} = \{f \in Z^\kappa : \text{supp}(f) \text{ finite}\}$ and $Z^{[\kappa]} = \{f \in Z^\kappa : |\text{supp}(f)| \leq \aleph_0\}$ are subgroups of Z^κ.

It is the aim of this paper, to give a detailed analysis of subgroups U of Z^κ which lay between $Z^{(\kappa)}$ and $Z^{[\kappa]}$. This will lead us to a new and simpler proof of the following

THEOREM. *There does not exist a set of groups $\mathcal{H}$ such that $\overline{\mathcal{H}}$, the closure of $\mathcal{H}$ under taking of subgroups, direct sums and extensions is the class of slender groups.*

This result which answers a problem in the book of L. Fuchs [10, Vol.II, p. 184], was proved *differently* in a recent paper by R. Göbel and B. Wald [13]. In order to obtain this results in a natural way, we will generalize or modify methods of S. Sasiada [16], L. Fuchs [9], S. U. Chase [5], S. Balcerzyk [3], R. Nunke [15], E. Specker [17] and others. This leads to many new group constructions which are also useful for other aspects on torsion-free abelian groups of infinite rank.

Notations. $n \mid t$: n devides t

$r^* : Z^\kappa \to Z^\rho$ denotes the stretching monomorphism

$(\sum_{i\in\kappa} x(i)e_i)^{r^*} = \sum_{i\in\kappa} x(i)e_{r(i)}$, where $r : \kappa \to \rho$ is an injective map between the cardinals κ and ρ; c.f. R. Göbel and B. Wald [12, p.216, Definition 5.1].

If $\mathcal{K}$ is a class of groups, then $\overline{\mathcal{K}}$ is the smallest class of groups containing $\mathcal{K}$ and closed under taking subgroups, extensions and arbitrary direct sums.

1. Growth types on κ and Sasiada's Theorem on slender groups.

The group Z^κ carries a natural ring structure, taking multiplication by components. The idempotents are the *characteristic functions* e_M of subsets M of κ defined by $e_M(i) = 1$ if $i \in M$ and $e_M(i) = 0$ if $i \in \kappa \setminus M$. We will write e_j for $e_{\{j\}}$. The set $B = B_\kappa$ of all functions in Z^κ with finitely many values in Z is a ring which is generated by all characteristic functions. Its additive group is free by a well-known result of G. Nöbeling [14] or the more general theorem of G. M. Bergman [4]; also L. Fuchs [10, Vol II, pp. 172-176, Lemma 97.2]. Then Z^κ is naturally a B-module and we are interested in the B-submodules of Z^κ.

A subset U of Z^κ is a B-submodule of Z^κ if and only if U is a subgroup and $u\, e_M \in U$ for all $u \in U$ and all subsets M of κ.

In particular B, $Z^{(\kappa)}$, Z^κ and the subgroups $\langle\langle a\rangle\rangle = \{x \in Z^\kappa : \exists\, k \in \omega\ \forall i \in \kappa\ |x(i)| < k|a(i)|\}$ for all $a \in Z^\kappa$ are B-submodules of Z^κ.

This leads us to the following

DEFINITION 1.1. *A subgroup T of* Z^κ *is a growth type on* κ, *if the following holds:*

(a) $Z^{(\kappa)} \subset T$ *and* T *is pure in* Z^κ.

(b) T *is a* B-*module*.

(c) *If* S *is an infinite subset of* κ, *then there is an element* $t \in T$ *such that for all* $n \in \omega$ *there is an* $s \in S$ *with* $n \nmid t(s) \neq 0$.

By this definition it is clear, that if T is a growth-type and $T \subset U \subset Z^\kappa$, then U is a growth-type as well. Now we want to show that groups defined by growth-types in the sence of E. Specker [17] are growth-types according to (1.1). In order to avoid confusion, we shall refer to Specker's growth-types as *monotone groups*. Let us recall the

DEFINITION 1.2.

(a) *A subgroup* T *of* Z^{κ} *is called minorant-closed if* $x \in Z^{\kappa}$, $t \in T$ *and* $|x(i)| < |t(i)|$ *for all* $i \in \kappa$ *implies that* $x \in T$, *i.e. if* $t \in T$ *then* $\langle\langle t \rangle\rangle$ *is a subgroup of* T.

(b) E. Specker [17], L. Fuchs [8]. *A subgroup* T *of* Z^{ω} *is called monotone subgroup, if the following holds*

(i) T *is minorant-closed*

(ii) *If* $t \in T$, *then the element* $\overline{t}$ *defined by* $\overline{t}(n) = \max_{i=0}^{n} |t(i)|$ *is in* T.

(iii) $e_{\omega} \in T$.

LEMMA 1.3. *Monotone subgroups* $\neq B_{\omega}$ *of* Z^{ω} *are growth-types on* $\aleph_0$.

Proof. Monotone subgroups T are pure subgroups of Z^{ω}. Since $e_{\omega} \in T$, condition (1.2)(b)(i) implies that $Z^{(\omega)} \subset T$. If $t \in T$ and e_M is a characteristic function on ω, then $|(te_M)(i)| = |t(i)| \cdot |e_M(i)| \leq |t(i)|$ for all $i \in \omega$. Therefore $te_M \in T$ by (i) and T is a B-module. Since $B \neq T$, there is $u \in T \setminus B$ and therefore the element $\overline{u}$ is a positive unbounded and monotone sequence of T. Now we choose $t \in Z^{\omega}$ subject to the condition $t(n) = k!$ if $k! \leq u(n) < (k+1)!$. Since $u \in T$ is minorant-closed, t is in T. If S is an infinite subset of ω and $n \in \omega$, choose $s \in S$ such that $u(s) \geq n!$. Therefore $n! \mid t(s)$ and in particular $n \mid t(s)$.

Generalizing J. Łoś (c.f. L. Fuchs [10; Vol. II, p. 159]) and R. Göbel and B. Wald [12; p.207], we will say

DEFINITION 1.4. *Let* $T \subset Z^{\kappa}$ *be a growth-type. Then a group* G *is called* T-*slender if the following holds:*

If $\sigma: T \to G$ *is a homomorphism, then* $e_i^{\sigma} = 0$ *for allmost all* $i \in \kappa$.

The reason for defining growth-type as in (1.1), will now become clear. We want to derive a theorem which generalizes a result of S. Sasiada [16]; c.f. L. Fuchs [10; Vol. II, p. 159, Proposition94.2], c.f.also L. Fuchs [8]. Recall that a group G is called *cotorsion-free* if G contains no cotorsion subgroups different from 0. This class of groups plays a important role in many aspects of group theory; c.f. R. Göbel [11], R. Göbel and B. Wald [12; p. 210, Folgerung 4.2 and p. 213, Satz 4.6] and M. Dugas and R. Göbel [7 ;Main Theorem].

THEOREM 1.5. *Let* $T \subset Z^{\kappa}$ *be a growth-type.*

(a) T-*slender groups are cotorsion-free.*

(b) If $|G| < 2^{\aleph_0}$*, then* G *is* T-*slender if and only if* G *is torsion--free and reduced.*

Proof. (a) Let K be an algebraically compact subgroup of some T-slender group G and fix any $x \in K$. Since $Z^{(\kappa)}$ is freely generated by the set $\{e_i : i \in \kappa\}$, the map $e_i \to x$ for all $i \in \kappa$ can be extended to a homomorphism ϕ from $Z^{(\kappa)}$ to K. The group $Z^{(\kappa)}$ is pure in Z^{κ} and therefore pure in T, and algebraically compact groups are pure-injective; c.f. L. Fuchs [9; Vol. I, p. 160, Theorem 38.1]. Therefore ϕ extends to a homomorphism $\phi: T \to K \subset G$. Since G is slender, we derive $e_i{}^{\phi} = 0$ for almost all $i \in \kappa$. By construction of ϕ we have $e_i{}^{\phi} = x$ and therefore $x = 0$. Hence $K = 0$ and G contains no algebraically compact subgroups different from 0. Since torsion-free cotorsion groups are algebraically compact, we derive that G is cotorsion-free; c.f. L. Fuchs [10; Vol. I, p. 235, Corollary 54.4].

(b) Since T-slender groups are cotorsion-free by (a) and cotorsion-free groups are torsion-free and reduced, we only have to show that

(*) If G is a torsion-free and reduced group, which is not T-slender, i.e. there is a homomorphism $\sigma : T \to G$ with infinite set $S = \{i \in \kappa : e_i{}^{\sigma} \neq 0\}$, then the cardinality of G must be greater or equal to $2^{\aleph_0}$.

Because S is infinite by definition (1.1)(c) there is $t \in T$ such that for all $n \in \omega$ there is $s \in S$ such that $n \mid t(s)$. Since G is torsion--free and reduced, there is a sequence $(k_n)_{n\in\omega}$ in S such that

(**) $0 \neq t(k_n) \mid t(k_{n+1})$ for all $n \in \omega$

(***) $t(k_n)e_{k_n}^{\sigma} \notin t(k_{n+1})G$

Now let $K = \{k_n : n \in \omega\} \subset S$. We want to show that the map

$$K \supset M \to (te_M)^{\sigma} \in G$$

is an injection from the powerset of the infinite set K into G. This proves (*). Consider $M, N \subset K$ with $(te_M)^{\sigma} = (te_N)^{\sigma}$. By induction hypothesis we suppose that for all m less than a given $n \in \omega$ $k_m \in M$ if and only if $k_m \in N$. In terms of characteristic function this is the same as saying that $e_M e_{k_m} = e_N e_{k_n}$ for all $m < n$.

If $L = \{k_m : m > n\}$, we compute

$$te_M - te_N = (te_M - te_N)e_K = (te_M - te_N)(e_{k_0} + \ldots + e_{k_n} + e_L)$$

$$= t(k_n)(e_M e_{k_n} - e_N e_{k_m}) + te_M e_L - te_N e_L$$

By the construction of t and $(k_n)_{n\in\omega}$ for every $m > n$ the k_m-th coordinate of t is devisible by $t(k_{n+1})$. Therefore each coordinate of $te_M e_L = te_{M\cap L}$ respectively $te_N e_L = te_{N\cap L}$ is divisible by $t(k_{n+1})$. Since T is a pure subgroup of Z^{κ} the term $te_M e_L - te_N e_L$ lies in $t(k_{n+1})T$. Now $(te_M - te_N)^{\sigma} = 0$ and $(te_M e_L - te_N e_L)^{\sigma} \in t(k_{n+1})G$ and hence $t(k_n)(e_M e_{k_n} - e_N e_{k_n})^{\sigma} \in t(k_{n+1})G$. By (***) this is only possible if $e_M e_{k_n} - e_N e_{k_n} = 0$ which implies that $k_n \in M$ exactly if $k_n \in N$. Induction proves that $M = N$ and therefore the map between $P(K)$ and G is injective.

2. A weak version of a theorem by L. Fuchs.

It is a well-known fact that direct sums of possibly infinitely many slender groups are slender. This can be seen easily from the nice characterization of slender groups by R. J. Nunke [15]; c.f. L. Fuchs [10; Vol. II, p. 165, Theorem 95.4]. Prior to this L. Fuchs [8] presented an elementary proof of this fact. There is no *nice* characterization of T-slender groups, so we are forced to develop a method of proof similar to Fuchs' proof to show that T-slender is closed under direct sums. This is a consequence of theorem 2.1.

THEOREM 2.1. *Let* T *be a growth-type on* κ, *and* $\{G_i : i \in I\}$ *a family of torsion-free and reduced groups. If* $\sigma : T \to \bigoplus_{i\in I} G_i$ *is a homomorphism, there is a finite subset* $F \subset I$ *such that* $(Z^{(\kappa)})^\sigma \subset \bigoplus_{i\in F} G_i$.

Proof. To ease notations let $\bigoplus_{j\in E} G_j = \bigoplus_E G_j$ and for $i \in E$ let $\pi(i) : \bigoplus_E G_j \to G_i$ be the canonical projection. By way of contradiction we assume that for all finite subsets E of I, there is an element $i \in \kappa$ such that $e_i^{\sigma} \notin \bigoplus_E G_j$. Hence we may choose sequences $(E(n))_{n\in\omega}$ of finite subsets E(n) of I and $(i(n))_{n\in\omega}$ of elements i(n) in κ with the following properties:

(i) $E_o = \emptyset$ and $E(n) \subset E(n+1)$ for all $n \in \omega$

(ii) $e_{i(n)}^{\sigma} \in \bigoplus_{E(n+1)} G_j \setminus \bigoplus_{E(n)} G_j$

(iii) E(n+1) is minimal with (ii).

Let $J = \{i(n) : n \in \omega\}$ and $a \in T$ by (1.1)(c) such that for all $n \in \omega$ there is $j \in J$ with $n \mid a(j) \neq 0$. We will construct inductively a sequence $(r_n)_{n\in\omega}$ of natural numbers such that

(*) $a(i(r_n)) \mid a(i(r_{n+1})) \neq 0$ and $a(i(r_n))e_{i(r_n)}^{\sigma\pi(k)} \notin a(i(r_{n+1})G_k$

for all k $E(r_n + 1)$.

First choose $r_o \in \omega$ such that $a(i(r_o)) \neq 0$. Assume that r_n is already constructed. Since G_k is reduced, for each $k \in E(r_n + 1)$ there is a natural number m(k) such that $a(i(r_n))e_{i(r_n)}^{\sigma\pi(k)} \notin m(k)G_k$. Let $m = \prod_{k\in E(r_n+1)} m(k)\, a(i(r_n))$ and apply (1.1)(c) to obtain $r_{n+1} > r_n$ such that $m \mid a(i(r_{n+1}))$. Hence the sequence $(r_n)_{n\in\omega}$ will satisfy (*).

Let $F = \{i(r_n) ; n \in \omega\}$. Since T is a B-module and $a \in T$, we have $ae_F \in T$. Hence $(ae_F)^{\sigma} \in \bigoplus_I G_i$ and there is a finite subset E of I such that $(ae_F)^\sigma \in \bigoplus_E G_i$. Since $(E(n))_{n\in\omega}$ is a strictly increasing sequence of subsets of I, by (iii) there is a $n \in \omega$ such that $|E(n)| > |E|$ and therefore $E(n) \setminus E \neq \emptyset$. Choose $N \in \omega$ minimal such that $E(r_N+1)\setminus E \neq \emptyset$ and let $k \in E(r_N+1)\setminus E$. Since $(ae_F)^\sigma \in \bigoplus_E G_i$ and $e_{i(r_n)}{}^\sigma \in \bigoplus_{E(r_n+1)} G_i \subset \bigoplus_E G_i$ for all $n > N$, we get by (ii) that $(ae_F)^{\sigma\pi(k)} = 0$ and $e_{i(r_n)}{}^{\sigma\pi(k)} = 0$ for $n < N$. If $L = \{i(r_n) : n > N\}$ we get

$$ae_F = ae_{i(r_0)} + \ldots + ae_{i(r_N)} + ae_L$$

and therefore

$$0 = a(i(r_N))e_{i(r_N)}{}^{\sigma\pi(k)} + (ae_L)^{\sigma\pi(k)}.$$

As in the proof of Theorem 1.5(b) the element ae_L is in $a(i(r_{N+1})T$ and hence $a(i(r_N))e_{i(r_N)}{}^{\sigma\pi(k)} \in a(i(r_{N+1}))G_k$ which contradicts (*).

__3. Strong growth-types on κ and a modification of a Theorem of S. Balcerzyk.__ S. Balcerzyk [3] proved that $Z^\omega / Z^{(\omega)}$ is algebraically compact; c.f. L. Fuchs [10; Vol. I, p. 177, Exercise 7]. This was generalized to a characterization of algebraically compact quotients of filter groups in M. Dugas and R. Göbel [6] and it was shown in R. Göbel and B. Wald [12, p. 211, Theorem 4.3] that quotients of monotone groups *modulo* $Z^{(\omega)}$ are algebraically compact as well. This will be used to generalize Balcerzyk's result as stated in (3.4). For this purpose we shall *restrict* our attention to the remainder of the paper *to the strong growth-types*. Compare Definitions (1.1) and (1.2). We call a map $f : A \to B$ *almost injective* if $0 \in B$ and $f^{-1}b = \{a \in A : f(a) = b\}$ is finite for all $b \in B\setminus\{0\}$.

__DEFINITION 3.1.__ *A subgroup* T *of* Z^κ *is a strong growth-type on* κ *if the following holds*

(i) $Z^{(\kappa)} \subset T$.

(ii) T *is minorant-closed*

(iii) If $a \in T$ *there is* $f \in T$ *such that* $\mathrm{supp}(f) \supset \mathrm{supp}(a)$ *and* $f : \kappa \to Z$ *is almost injective.*

(iv) If $S \subset \kappa$ *and* $|S| = \aleph_0$ *there is* $a \in T$ *such that* $|\mathrm{supp}(a) \cap S| = \aleph_0$.

Remarks. Minorant-closed subgroups of Z^{κ} are B-modules. Hence (3.1)(ii) sharpens (1.1)(ii). Monotone subgroups in Z^{ω} are strong growth-types, c.f. (1,2). Strong growth-types are growth-types.
In order to show this, we only need to prove (1.1)(c) for a strong growth-type T. Let $S \subset \kappa$ be infinite. There is $a \in T$ such that $\mathrm{supp}(a) \subset S$ and $|\mathrm{supp}(a) \cap S| = \aleph_0$ from (3.1)(ii) and (iv). Therefore (3.1)(iii) implies the existence of $f \in T$ such that $\mathrm{supp}(f) \supset \mathrm{supp}(a)$ and f is almost injective. Finally we define

$$b(i) = \begin{cases} n! \text{ if } n! < |f(i)| < (n+1)! \text{ for } i \in \mathrm{supp}(a) \\ 0 \text{ if } i \in \kappa \setminus \mathrm{supp}(a). \end{cases}$$

Therefore b is bounded by a and (3.1)(ii) implies $b \in T$. If $n \in \omega$, there is $i \in \mathrm{supp}(f)$ such that $b(i) = n!$ and (1.1)(c) is shown.

The following properties of strong growth types will be used in this section. For the definition of $\langle\langle y\rangle\rangle$ compare §1.

LEMMA 3.2. *Let* T *be a strong growth-type on* κ.

(a) $Z^{(\kappa)} \subset T \subset Z^{[\kappa]}$

(b) If $t \in T$, *then* $\langle\langle t \rangle\rangle \subset T$.

(c) Let $\mathcal{E}$ *be the set of all almost injective functions* $f \in T$.

Then $T = \sum_{f\in\mathcal{E}} \langle\langle f\rangle\rangle$.

Proof. (a) If $t \in T$ with infinite support $\mathrm{supp}(t)$, by (3.1)(iii) there is $f \in T$ almost injective such that $\mathrm{supp}(t) \subset \mathrm{supp}(f)$. Therefore $\mathrm{supp}(t) \subset \mathrm{supp}(f) = \bigcup_{0\neq z\in Z} f^{-1}(z)$ and $\mathrm{supp}(t)$ is countable since $|f^{-1}(z)| < |Z| = \aleph_0$ for all $z \in Z\setminus\{0\}$.

(b) follows from (3.1)(ii) and this implies $\sum_{f\in\mathcal{E}} \langle\langle f\rangle\rangle \subset T$.

In order to show (c), it remains to prove that $t \in T$ implies

$t \in \sum_{f \in \&} \langle\langle f \rangle\rangle$. By (3.1)(iii) there is $f \in T$ almost injective with countable support $\text{supp}(f) \supset \text{supp}(t)$. If $g \in Z^\kappa$, let $|g| \in Z^\kappa$ be defined by $|g|(i) = |g(i)|$ for all $i \in \kappa$. It is easy to check that $|f| + |t| \in \&$ and obviously t is bounded by $|f| + |t|$. Therefore $t \in \langle\langle |f| + |t| \rangle\rangle$ which implies (c).

The following construction played an important role in many investigations of R. Baer, e.g. [1] and [2]. If π is some group-theoretical property, a group G is called hyper-π if every epimorphic image $\neq 0$ of G contains a normal subgroup $\neq 0$ which has the property π. We will apply this definition in the category of abelian groups to the class of cotorsion groups. We conclude from this definition a

<u>LEMMA 3.3</u>. *For a group* G *are equivalent:*

(1) G *is hypercotorsion.*

(2) Hom(G, F) = 0 *for all cotorsion-free groups* F.

<u>Proof</u>. (1) → (2). Let G be hypercotorsion and $\sigma : G \to F$ some homomorphism into a cotorsion-free group F. If $\sigma \neq 0$, then G^σ contains a non-trivial cotorsion subgroup, which is in F. Therefore $\sigma = 0$.

(2) → (1). If G is not hypercotorsion, there is a cotorsion-free epimorphic image $F \neq 0$ of G. Hence $0 \neq \text{Hom}(G, F)$, which is excluded in (2).

<u>THEOREM 3.4</u>. *If* T *is a strong growth-type on* κ, *then* $T / \dot{Z}^{(\kappa)}$ *is hypercotorsion.*

<u>Proof</u>. (3.4) is by (3.3) equivalent to show

(*) If $\sigma : T \to F$ is some homomorphism into a cotorsion-free group F, such that $e_i^\sigma = 0$ for all $i \in \kappa$, then $\sigma = 0$.

Let f be an almost injective function in T. If supp(f) is finite, then $f^\sigma = 0$ by assumption (*) on σ. Since $f \in Z^{[\kappa]}$ by (3.2)(a) we conclude $|\text{supp}(f)| = \aleph_0$. Hence there is a bijection $r : \omega \to \text{supp}(f)$ such that $f \circ r \in Z^\omega$ is a monotone increasing function. If $r^* : Z^\omega \to Z^\kappa$ is the monomorphism defined in §0, then $(f \circ r)^{r^*} = (\sum_{n\in\omega} f(r(n))e_n)^{r^*}$ $= \sum_{n\in\omega} f(r(n))e_{r(n)} = \sum_{i\in supp(f)} f(i)e_i = f$. Therefore $r^* : \langle\langle f \circ r\rangle\rangle \to \langle\langle f\rangle\rangle$ is an isomorphism of the monotone subgroup $\langle\langle f \circ r\rangle\rangle$ of Z^ω onto the subgroup $\langle\langle f\rangle\rangle$ of T. The map $r^*\sigma : \langle\langle f \circ r\rangle\rangle \to F$ satisfies $e_n^{r^*\sigma} = e_{r(n)}^{\sigma} = 0$ by (*) and therefore $(Z^{(\omega)})^{r^*\sigma} = 0$. Hence $\langle\langle f \circ r\rangle\rangle^{r^*\sigma} \subset F$ is an epimorphic image of $\langle\langle f \circ r\rangle\rangle / Z^{(\omega)}$. However, $\langle\langle f \circ r\rangle\rangle / Z^{(\omega)}$ is algebraically compact by R. Göbel and B. Wald [11; p. 211, Satz 4.3]. Therefore $\langle\langle f \circ r\rangle\rangle^{r^*\sigma}$ is a cotorsion subgroup of the cotorsion-free group F; c.f. L. Fuchs [10; p.234, Proposition 54.1 We conclude that $\langle\langle f \circ r\rangle\rangle^{r^*\sigma} = 0$ and therefore $(f \circ r)^{r^*\sigma} = f^\sigma = 0$. From (3.2)(c) we derive $\sigma = 0$.

Combining (3.4) and (2.1) we obtain the

<u>COROLLARY 3.5</u>. *Let* T *be a strong growth-type on* κ *and* G_i $(i \in I)$ *a family of cotorsion-free groups. For any homomorphism* $\sigma: T \to \bigoplus_I G_i$ *there is a finite subset* F *of* I *such that* $T^\sigma \subset \bigoplus_F G_i$.

<u>Proof</u>. Theorem (2.1) implies that there is a finite subset F of I such that $(Z^{(\kappa)})^\sigma \subset \bigoplus_F G_i$. Let $\pi : \bigoplus_I G_i \to \bigoplus_{I\setminus F} G_i$ be the canonical projection. Then $(Z^{(\kappa)})^{\sigma\pi} = 0$ and therefore $T^{\sigma\pi} = 0$ from (3.4). This is equivalent to $T^\sigma \subset \bigoplus_F G_i$.

From (3.5) we derive easily a generalization of the theorem of L. Fuchs mentioned at the beginning of §2.

<u>COROLLARY 3.6</u>. *Let* T *be a strong growth-type on* κ. *Then direct sums of* T-*slender groups are* T-*slender.*

The following result generalizes a theorem of E. Specker [17] on monotone subgroups to strong growth-types.

COROLLARY 3.7. *Let* T *be a strong growth-type on* κ.

(i) If $\sigma : T \to G$ *is a homomorphism of* T *into a* T*-slender group* G, *there are a finite subset* $F \subset I$ *and elements* $g_i \in G$ *for all* $i \in F$ *such that* $x^\sigma = \sum_{i\in F} x(i)g_i$ *for all* $x \in T$.

(ii) Hom(T, Z) *is a free group of rank* κ.

Proof. (i) Let $F = \{i \in \kappa : e_i{}^\sigma \neq 0\}$. Since G is T-slender, the set F is finite. If $\sigma^* : T \to G$ is the homomorphism defined by $x^{\sigma^*} = x^\sigma - (xe_F)^\sigma$ for all $x \in T$, then $e_i{}^{\sigma^*} = e_i{}^\sigma - (e_ie_F)^\sigma = 0$ for all $i \in \kappa$. Since G is T-slender, from (1.5)(b) it follows that G is cotorsion-free and therefore $\sigma^* = 0$ from (3.3) and (3.4). From the definition of σ^* we derive $x^\sigma = (xe_F)^\sigma$ for all $x \in T$ which implies (i).
The integers Z are T-slender by (1.5)(b). Therefore (ii) follows from (i).

4. Definition of κ-slender groups. The following is the central tool of the paper.

DEFINITION 4.1. *Let* κ *be some cardinal. Then a group* G *is called* κ*-slender if for all strong growth-types* T *on* κ *and all homomorphisms* $\sigma : T \to G$ *the set* $\{i \in \kappa : e_i{}^\sigma \neq 0\}$ *has a cardinality less than* κ.

The main result of this section will be that slender groups are κ-slender for sufficiently large κ. We will first collect the closure properties of κ-slenderness.

LEMMA 4.2. *(a)* κ*-slender groups are cotorsion-free.*

(b) If T *is a strong growth-type on* κ, *then* T *is not* κ*-slender.*

(c) The class of κ*-slender groups is closed under taking subgroups extensions and arbitrary direct sums.*

Proof. (a) Let κ be an algebraically compact subgroup of the

κ-slender group G and let T be any strong growth-type on κ. If $x \in K$, let $e_i^{\sigma} = x$ for all $e_i \in T$. Since $\bigoplus_{i\in\kappa} \langle e_i \rangle = Z^{(\kappa)}$ is a free and pure subgroup of T and K is algebraically compact, σ can be extended to an homomorphism σ^* from T to G such that $e_i^{\sigma^*} = x$ for all $i \in \kappa$. Therefore $|\{i \in \kappa : e_i^{\sigma} = x\}| = \kappa$ and $x = 0$ since G is κ-slender. Hence $K = 0$ and G is cotorsion-free, cf. R. Göbel and B. Wald [12, p.210, Folgerung 4.2]

(b) The identity map $1 : T \to T$ implies (b).

(c) Subgroups of κ-slender groups are κ-slender by definition. Therefore let $U \subset G$ such that U and $G/_U$ are κ-slender. In order to show that G is κ-slender, let $\sigma : T \to G$ be a homomorphism from the strong growth-type T into G and τ the canonical projection from G onto $G/_U$. Since $G/_U$ is κ-slender, the set $\{i \in \kappa : e_i^{\sigma\tau} \neq 0\}$ has cardinality less than κ and therefore its complement $S = \{i \in \kappa : e_i^{\sigma\tau} = 0\}$ is of cardinality κ. Since $|S| = \kappa$, the set $T_S = \{xe_S : x \in T\}$ is again a strong growth-type on κ. We will leave it to the reader to check the conditions (3.1) for T_S identifying S and κ. Let $\sigma^* = \sigma|_{T_S}$ be the restriction of σ. Since $G/_U$ is κ-slender, we derive from (4.2)(a) that $G/_U$ is cotorsion-free. If $i \in S$ then $e_i^{\sigma^*\tau} = 0$ and therefore $\sigma^*\tau = 0$ by (3.3) and (3.4). We conclude $(T_S)^{\sigma^*} \subset U$. Since U is κ-slender and T_S is a growth-type on κ, the set $\{i \in S : e_i^{\sigma} \neq 0\}$ has cardinality less than κ as well. Therefore,

$\{i \in \kappa : e_i^{\sigma} \neq 0\} \subset \{i \in \kappa : e_i^{\sigma\tau} \neq 0\} \cup \{i \in S : e_i^{\sigma} \neq 0\}$

has cardinality less than κ and G is κ-slender.

Let $G = \bigoplus_I G_i$ $(i \in I)$ be a direct sum of κ-slender groups. In order to show that G is κ-slender, let $\sigma : T \to G$ be a homomorphism from the strong growth-type T on κ into G. By (3.5) there is a finite subset F of I such that $T^{\sigma} \subset \bigoplus_F G_i$. Since extensions of κ-slender groups by κ-slender groups are κ-slender, we conclude by complete induction that $\bigoplus_F G_i$ is κ-slender. Therefore σ is a homomorphism from T into a κ-slender group, hence $|\{i \in \kappa : e_i^{\sigma} \neq 0\}| < \kappa$ and G is κ-slender.

The following result was shown by B. Wald [18, p. 45, Satz 12] for monotone groups. This was generalized to strong growth-types by P. Westphal [20] . The result and its proof are of special interest, since they lead to new methods for investigating slender groups.

<u>THEOREM 4.4</u>. *If* $\sigma : T \to G$ *is a homomorphism from a strong growth-type* T *on* κ *into a cotorsion-free group* G *, then the set* $\{i \in \kappa : e_i^\sigma = x\}$ *is finite for all* $x \in G \setminus \{0\}$.

<u>Proof</u>. Assume that $S = \{i \in \kappa : e_i^\sigma = x\}$ is infinite for some $x \in G$. We want to show that $x = 0$. The basic idea is, to construct a homomorphism $\xi : \hat{Z} \to G$ from the Z-adic completion $\hat{Z}$ of Z into G with $1^\xi = x$. Since G is cotorsion-free and epimorphic images of $\hat{Z}$ are cotorsion, ξ must vanish and therefore $x = 0$.

If $k \in \hat{Z}$, there is a Z-adic zero-sequence $(k(n))_{n\in\omega}$ of integers such that $k = \sum_{n\in\omega} k(n)$ in the Z-adic topology on $\hat{Z}$. Because there is an almost injective map in T which has infinite support and the support is a subset of S, we can find an injective map $r : \omega \to S$, such that $\sum_{n\in\omega} k(n)e_{r(n)} \in T$. We define $k^\xi = (\sum_{n\in\omega} k(n)e_{r(n)})^\sigma$. Since ξ will automatically become a homomorphism, it remains to show that ξ is well defined. Let $k = \sum_{n\in\omega} q(n)$ be another Z-adic representation of $k \in \hat{Z}$ and $s : \omega \to \mathrm{supp}(f)$ be an injective map such that $\sum_{n\in\omega} q(n)e_{s(n)} \in T$. We will show that $d = (\sum_{n\in\omega} k(n)e_{r(n)})^\sigma - (\sum_{n\in\omega} q(n)e_{s(n)})^\sigma$ is zero. We have for all $N \in \omega$ that

$$d = \sum_{n=0}^{N} (k(n) - q(n))x + (\sum_{n=N+1}^{\infty} k(n)e_{r(n)})^\sigma - (\sum_{n=N+1}^{\infty} q(n)e_{s(n)})^\sigma .$$

The first sum converges to 0 as N goes to infinity. Since $(k(n))_{n\in\omega}$ and $(q(n))_{n\in\omega}$ are Z-adic zerosequences, this shows that each sum is divisible by a given natural number if N is sufficiently large. Therefore $d \in \bigcap_{m\in\omega} mG = G^1$. Since G is cotorsion-free, $G^1 = 0$ and therefore $d = 0$.

To compute 1^{ξ}, let $k(0) = 1$, $k(n) = 0$ $(n > 0)$ and let $r:\omega \rightarrow S$ be an arbitrary injective map. Then $1 = \sum_{n\in\omega} k(n)$, $\sum_{n\in\omega} k(n)e_{r(n)} = e_{r(0)} \in T$ and hence $1^{\xi} = e_{r(0)}{}^{\sigma} = x$.

<u>THEOREM 4.5</u>. *If* G *is cotorsion-free and* $|G| < \kappa$, *then* G *is* κ*-slender.*

<u>Proof</u>. Let $\sigma : T \rightarrow G$ be some homomorphism from a strong growth-type T on κ into a cotorsion-free group G. Since the set $(\phi^{-1}, x) = \{i \in I : e_i{}^{\phi} = x\}$ is finite for all $x \in G \setminus \{0\}$ by (4.4), we conclude $|\{i \in I : e_i{}^{\phi} \neq 0\}| = |\bigcup_{0\neq x\in G} (\phi^{-1}, x)| = |G|\aleph_0 < \kappa$. Therefore G is κ-slender.

Since slender groups are cotorsion-free, (4.5) is a theorem on slender groups as well. We obtain the

<u>COROLLARY 4.6</u>. *Let* $\mathcal{K}$ *be a class of cotorsion-free groups of cardinality less than* κ. *Then all groups in* $\overline{\mathcal{K}}$ *are* κ*-slender.*

<u>Proof</u>. Apply (4.2)(c) and (4.5).

<u>5. A modification of a Theorem of E. Specker</u>. L. Fuchs [9, Theorem 1] showed that all monotone subgroups of Z^{ω} except Z^{ω} itself are slender groups. This result cannot be generalized immediately to strong growth--types on κ, because Z^{ω} will be isomorphic to some proper subgroup for many growth-types on κ. However, we can single out slender growth--types by the following

<u>THEOREM 5.1</u>. *For a strong growth-type* T *on* κ *are equivalent*

(1) T *is slender*

(2) For each countable infinite subset S *of* κ *there is an element* $s \in Z^{\kappa} \setminus T$ *such that* $\mathrm{supp}(s) \subset S$.

Proof. (1) ⇒ (2) There is a bijection $r: \omega \to S$. Since T is slender; T cannot contain a subgroup isomorphic to Z^{ω}. Therefore $(Z^{\omega})^{r*} \subset Z^{\kappa}$ but $(Z^{\omega})^{r*} \not\subset T$. There is an element $s \in Z^{\kappa} \setminus T$ such that $s \in (Z^{\omega})^{r*}$. Hence $\mathrm{supp}(s) \subset r(\omega) = S$ and (2) is shown.

(2) ⇒ (1) If $r : \omega \to I$ is an injective map, we will show that $(Z^{\omega})^{r*} \not\subset T$. Since T is minorant-closed, a characterization of slender groups by B. Wald [18] applies, showing that T is slender. In order to show $(Z^{\omega})^{r*} \not\subset T$, take any $s \in Z^{\kappa} \setminus T$ with $\mathrm{supp}(s) \subset r(\omega)$ according to (2). Then $u = \sum_{n\in\omega} s(r(n))e_n \in Z^{\omega}$ is mapped onto

$$u^{r*} = \sum_{n\in\omega} s(r(n))e_{r(n)} = \sum_{i\in\kappa} s(i)e_i = s,$$ which shows $s \in (Z^{\omega})^{r*} \setminus T$.

Finally we will show the existence of strong growth-type on κ which are slender. This can be done in a simpler way than in R. Göbel and B. Wald [13]. Because of the new notion of κ-slender, we can restrict the construction to the basic step used in [13]. As in [13], we have by the maximum principle of set theory the following

LEMMA 5.2. *Let κ be a fixed infinite cardinal. Then there is a system $\mathbb{D}$ of subsets of κ with the following properties:*

(1) $\cup \mathbb{D} = \kappa$

(2) If $D \in \mathbb{D}$, then $|D| = \aleph_0$

(3) If $D_1, D_2 \in \mathbb{D}$ and $|D_1 \cap D_2| = \aleph_0$ then $D_1 = D_2$

(We say that the elements in $\mathbb{D}$ are almost disjoint)

(4) If X is an infinite subset of κ, there is an element $D \in \mathbb{D}$ such that $|D \cap X| = \aleph_0$

(We say that $\mathbb{D}$ is maximal).

Now we associate with any $D \in \mathbb{D}$ an almost injective element D^* in Z^{κ} with $\mathrm{supp}(D^*) = D$. Since $|D| = \aleph_0$, this can be arranged easily, taking any bijection between D and ω. We will consider the group $T = \sum_{D\in\mathbb{D}} \langle\langle D^*\rangle\rangle$ c.f. §1.

THEOREM 5.3. *The group* $T = \sum_{D \in \mathbb{D}} \langle\langle D^* \rangle\rangle$ *is slender and is a strong growth-type on* κ.

Proof. First we show, that T satisfies Definition 3.1:

If $i \in \kappa$, there is $D \in \mathbb{D}$ with $i \in D$ by (5.2)(1). Therefore $e_i \in \langle\langle D^* \rangle\rangle \subset T$ and we get $Z^{(\kappa)} \subset T$. Since $\langle\langle D^* \rangle\rangle$ is minorant-closed for every $D \in \mathbb{D}$ the group T is minorant-closed as well. If $a \in T$, there are $D_1, \ldots, D_n \in \mathbb{D}$ such that $a \in \sum_{i=1}^{n} \langle\langle D_i^* \rangle\rangle$, and therefore $\mathrm{supp}(a) \subset \bigcup_{i=1}^{n} D_i$. If we define $|D_i^*| \in Z^{\kappa}$ by $|D_i^*|(k) = |D_i^*(k)|$ for all $k \in \kappa$, then $f = \sum_{i=1}^{n} |D_i^*| \in T$ is almost injective and $\mathrm{supp}(a) \subset \bigcup_{i=1}^{n} D_i = \mathrm{supp}(f)$. Let $S \subset \kappa$ and $|S| = \aleph_0$, then there is $D \in \mathbb{D}$ such that $|S \cap D| = \aleph_0$, by the maximality (5.2)(4) of $\mathbb{D}$. Therefore $D^* \in T$ has the required property and (3.1) is shown. In order to prove that T is slender, we will now show (2) of Theorem 5.1. Let S be a countable infinite subset of κ. By the maximality of $\mathbb{D}$ there is an $D \in \mathbb{D}$ such that $|S \cap D| = \aleph_0$. We want to show that $s = D^{*2} e_S$ is the element required in (2). Obviously $\mathrm{supp}(s) \subset S$ and we assume that $s \in T$. There are $D_1, \ldots, D_n \in \mathbb{D}$ such that $D = D_1$ and $s \in \sum_{i=1}^{n} \langle\langle D_i^* \rangle\rangle$. Since $\mathrm{supp}(s) \subset D$ and the elements in $\mathbb{D}$ are almost disjoint, we conclude $|\mathrm{supp}(s) \cap D_i| < \infty$ for all $D_i \neq D$ and therefore $s \in \langle\langle D^* \rangle\rangle$ by definition of "$\langle\langle\ \rangle\rangle$", §1. Therefore $D^*(i)^2 = s(i) < k|D^*(i)|$ for all $i \in D \cap S$ and some fixed $k \in \omega$. This is impossible since D is unbounded on $D \cap S$. Hence T is slender by (5.1).

From (5.3), (4.5) and (4.2) we obtain easily the corollary, which answers a problem in L. Fuchs [9, Vol. II , p. 184, Problem 78 b].

COROLLARY 5.4. *If* $\mathcal{K}$ *is a set of cotorsion-free groups, there is a slender and strong growth-type* T, *which is not in* $\overline{\mathcal{K}}$.

Recall: $\overline{\mathcal{K}}$ is the closure of $\mathcal{K}$ under taking subgroups, extensions and direct sums.

Proof. Since $\mathcal{K}$ is a set, there is a cardinal κ such that $|G| < \kappa$ for all G $\in \mathcal{H}$. Theorem 4.5 implies that all groups in $\mathcal{H}$ are κ - slender and from (4.2)(a) we derive that all groups in $\mathcal{H}$ are κ - slender as well. In (5.3) we constructed a slender growth-type T on κ. However, T is not κ - slender by (4.2)(b) and therefore T $\notin \mathcal{H}$.

6. References.

1. R. Baer: The hypercenter of a group, Acta Mathematica 89 (1953), 165-208
2. R. Baer: Gruppentheoretische Eigenschaften, Math. Annalen 149 (1963), 181-210
3. S. Balcerzyk: On factor groups of some subgroups of a complete direct sum of infinite cyclic groups, Bull. Acad. Polon. Sci. 7 (1959), 141-142
4. G. Bergman: Boolean rings of projection maps, Journ. London Math.Soc. (2) 4 (1972), 593-598
5. S. U. Chase: On direct sums and products of modules, Pacif. J. Math. 12 (1962), 847- 854
6. M. Dugas and R. Göbel: Algebraisch kompakte Faktorgruppen, Journal für die reine und angewandte Mathematik 307/308 (1979), 341-352
7. M. Dugas and R. Göbel: Every cotorsion-free ring is an endomorphism ring, to appear in Proceedings London Math. Soc.
8. L. Fuchs: "Infinite Abelian Groups" Publ. House of the Hungar. Acad. Sciences, Budapest 1958
9. L. Fuchs: Note on certain subgroups of products of infinite cyclic groups, Comment. Math. Univ. St. Pauli 19 (1970), 51-54
10. L. Fuchs: "Infinite Abelian Groups" Vol. I (1970), Vol. II (1973), Academic Press, New York
11. R. Göbel: On stout and slender groups, Journ. Algebra 53 (1975), 39-55
12. R. Göbel and B. Wald: Wachstumstypen und schlanke Gruppen, Symp. Math. Vol. XXIII (1979), 201-239

13. R. Göbel and B. Wald: Lösung eines Problems of L. Fuchs, to appear in Journal of Algebra

14. G. Nöbeling: Verallgemeinerung eines Satzes von Herrn E. Specker, Inventiones Math. 6 (1968), 41-55

15. R. Nunke: Slender groups, Acta Sci. Math. Szeged 23 (1962), 67-73

16. S. Sasiada: Proof that every countable and reduced torsion-free abelian group is slender, Bull. Acad. Polon. Sci. 7 (1959), 143-144

17. E. Specker: Additive Gruppen von Folgen ganzer Zahlen, Portugaliae Math. 9 (1950), 131-140

18. B. Wald: A note on slender groups, Arch. Math. 31 (1978), 432-434

19. B. Wald: Schlankheitsgrade kotorsionsfreier Gruppen, Dissertation Essen, 1979

20. P. Westphal: Gewisse Untergruppen der Gruppe Z^I, Diplomarbeit Essen, 1981

* Financial support for this paper was furnished by the Ministerium für Wissenschaft und Forschung des Landes Nordrhein-Westfalen under the title Überabzählbare abelsche Gruppen.

ITERATED DIRECT SUMS AND PRODUCTS OF MODULES*

M. Dugas and B. Zimmermann-Huisgen

1. INTRODUCTION

In his 1967 Lecture Notes on abelian groups of integer-valued functions [5] Reid gave a comprehensive account of previous efforts to cope with two major problems of this area. Whereas the first - to determine the Whitehead groups - has received a great deal of attention since 1967, the second has hardly been attacked at all in the meantime. It asks for a characterization of the $\mathbb{Z}$-duals of abelian groups (i.e. the groups of the form $\mathrm{Hom}_{\mathbb{Z}}(A,\mathbb{Z})$ for some abelian group A). For the following reason, the latter were baptized kernel groups by Reid: If L is a compact, connected abelian group, the kernel of the evaluation map $\mathrm{Hom}_c(\mathbb{R},L) \to L$, which assigns $\varphi(1)$ to each continuous $\mathbb{Z}$-homomorphism φ from $\mathbb{R}$ to L, equals $\mathrm{Hom}_{\mathbb{Z}}(\hat{L},\mathbb{Z})$, where $\hat{L}$ denotes the (discrete) Pontrjagin dual of L. Another motivation for an investigation of kernel groups is their occurrence as homotopy groups: L being still compact and connected, $\pi_1(L)$ is again isomorphic to $\mathrm{Hom}_{\mathbb{Z}}(\hat{L},\mathbb{Z})$.

The list of known properties of kernel groups is short. They are locally free groups (i.e. every pure subgroup of finite rank is a direct summand) occurring as kernels of homomorphisms between direct products of copies of $\mathbb{Z}$ [4, Thm. 8]. The fact that the class of kernel groups is closed under direct products is trivial, the observation (due to Łoś, see [2, Thm. 94.4]) that it is also closed under direct sums over non-measurable index sets is sharp, however. As a consequence, all of the groups $\mathbb{Z}$, $\mathbb{Z}^{\mathbb{N}}$, $(\mathbb{Z}^{\mathbb{N}})^{(\mathbb{N})}$, $((\mathbb{Z}^{\mathbb{N}})^{(\mathbb{N})})^{\mathbb{N}}$,... obtained by starting with $\mathbb{Z}$ and alternatingly forming direct products and direct sums, are kernel groups. Answering a question of Fuchs [2, Problem 76] and Reid [5,p.28], the second author has shown in [7] that this construction process does not become stationary, i.e. the groups we started listing above are pairwise non-isomorphic.

The purpose of the present paper is to pursue Reid's initiative to clas-

*One of the authors (M. Dugas) was supported by the Ministerium für Wissenschaft und Forschung des Landes Nordrhein-Westfalen under the title Überabzählbare abelsche Gruppen

sify the groups which are obtained by closing $\{\mathbb{Z}\}$ under direct products and direct sums; we call them $\mathbb{Z}$-kernels for short. (If we exclude the existence of measurable cardinals, e.g. by adopting the axiom "V = L", all $\mathbb{Z}$-kernels are kernel groups. On the other hand, it is still not known whether the class of $\mathbb{Z}$-kernels comprises all kernel groups.) Each $\mathbb{Z}$-kernel is labeled with an ordinal number according to the minimal number of sums and products involved in its construction. We show that each ordinal number μ actually occurs as the type of a $\mathbb{Z}$-kernel (even a little better, see Corollary 5). Our candidates for $\mathbb{Z}$-kernels belonging to μ are the "natural" ones, obtained by iterating the alternate formation of infinite products and sums μ times. It is not at all obvious, however, that these groups are not of a type smaller than μ, i.e. that there is no "shorter way" to reach them.

Since group theory does not enter the core of our arguments, we establish the main results for modules (with certain restrictions) over an arbitrary commutative ring (Section 3). Our principal tool, a descendant of Chase's theorem [1, Thm. 1.2], does not even require commutativity of the ring; it will be provided in Section 2. The concluding section contains the solution to a related problem, posed by Irwin and O'Neill [3, Problem (1)], which roughly reads as follows: Characterize the direct products M of abelian p-groups with $M \cong M^{(\mathbb{N})}$.

2. EXTENSION OF A THEOREM OF CHASE ON DIRECT-SUM DECOMPOSITIONS OF DIRECT PRODUCTS

Throughout this section, R denotes an arbitrary associative ring with identity, and R-module means left R-module.

Roughly speaking, the result of Chase under discussion [1, Thm. 1.2] says that each representation of a *countable* direct product $\prod_{n\in\mathbb{N}} U_n$ of R-modules U_n as a direct sum $\bigoplus_{j\in J} V_j$ of modules V_j resembles a finite sum decomposition in the following sense: for each descending chain $r_1R \supset r_2R \supset r_3R \supset \cdots$ of principal right ideals of R there are a natural number n_o and a finite subset J' of J such that the restricted infinite product $\prod_{n\geq n_o} r_{n_o} U_n$ is contained in $\bigoplus_{j\in J'} V_j + \bigcap_{n\in\mathbb{N}} (r_n \bigoplus_{j\in J} V_j)$.

It is immediate what a generalization of this theorem to direct products over arbitrary index sets should look like. But it appeared inaccessible to proof as well as to refutation. The reason lies in set theory: It turns out that the statement extends precisely to those products which run over index sets of non-measurable cardinality. (The existence of non-measurable cardinal numbers is excluded by the axiom $V = L$ for instance.) We start with the positive assertion, setting out from a rather general position in order to reveal the affinity between our result and Baire's category theorem.

Let P be a subfunctor of the forgetful functor R-Mod $\to \mathbb{Z}$-Mod (i.e., given ${}_R X \in$ R-Mod, PX is a subgroup of the underlying abelian group X, and $f(PX) \subset PY$ for $f \in \mathrm{Hom}_R(X,Y)$). Moreover, suppose that M is an R-module equipped with a linear topology. Call the functor P compatible with the topology of M if PM is a closed subgroup of M. Prototypes of this situation are the following:

1. The topology of M is induced by 0-neighborhoods of the form $\mathfrak{a}M$, $\mathfrak{a} \in \mathfrak{A}$, where $\mathfrak{A}$ is a set of ideals of R, and P is given by $X \mapsto \mathfrak{a}X$ for some $\mathfrak{a} \in \mathfrak{A}$.
2. $M = \prod_{n\in\mathbb{N}} U_n$ for some family of R-modules $(U_n)_{n\in\mathbb{N}}$, with the linear topology determined by the basis of 0-neighborhoods $\prod_{i\geq n} U_i$, $n \in \mathbb{N}$. Compatible with this topology are all subfunctors of the forgetful functor R-Mod $\to \mathbb{Z}$-Mod which commute with direct products. Among these are e.g. the functors $X \mapsto \mathfrak{a}X$ for some finitely generated right ideal $\mathfrak{a}$ of R and annihilators $X \mapsto \mathrm{Ann}_S(X) = \{x \in X\colon Sx = 0\}$, $S \subset R$. Moreover, this class of functors is closed under arbitrary intersections and finite sums.

<u>Lemma 1</u>. Suppose that M is an R-module which is complete and Hausdorff in a linear topology with a countable basis of 0-neighborhoods. Let $f\colon M \to \bigoplus_{j\in J} V_j$ be a homomorphism from M into a direct sum of R-modules V_j. Then, given any descending chain $P_1 \supset P_2 \supset P_3 \supset \cdots$ of functors R-Mod $\to \mathbb{Z}$-Mod which are compatible with the topology of M, there exist a natural number n_0, a neighborhood U of 0 and a finite subset J' of J such that

$$f(P_{n_0}U) \subset \bigoplus_{j\in J'} V_j + \bigcap_{n\in\mathbb{N}} (P_n \bigoplus_{j\in J} V_j) .$$

($P_1 \supset P_2$ means $P_1X \supset P_2X$ for each R-module X.)

Proof. Our argument was inspired by that of Chase in [1]. Let $U_1 \supset U_2 \supset U_3 \supset \cdots$ be a basis of neighborhoods of 0 consisting of submodules of M.

Assume that the assertion is false. Then, by a straightforward induction, one can find an increasing sequence $n_1 < n_2 < n_3 < \cdots$ of natural numbers, elements m_k of $P_{n_k} U_{n_k}$ respectively and elements $j_k \in J$ for $k \in \mathbb{N}$ such that

$$q_{j_k} f(m_k) \notin P_{n_{k+1}} V_{j_k} \quad \text{and} \quad q_{j_l} f(m_k) = 0 \quad \text{for} \quad l > k,$$

where $q_j \colon \bigoplus_{h \in J} V_h \to V_j$ denotes the canonical projection. Let m be the limit of $\left(\sum_{l=1}^{k} m_l\right)_{k \in \mathbb{N}}$. By the compatibility of the functors P_n, we have $m - \sum_{l=1}^{k} m_l \in P_{n_{k+1}} M$ and therefore

$$q_{j_k} f(m) = q_{j_k} f(m_k) + q_{j_k} f\left(m - \sum_{l=1}^{k} m_l\right) \neq 0$$

for all k (the first term on the right-hand side is not contained in $P_{n_{k+1}} V_{j_k}$ by construction, whereas the second is). But since $f(m)$ has only a finite number of non-zero components, the assumption is unsustainable. ∎

By specializing the lemma to Example 2 above and considering the functors $X \mapsto rX$, $r \in R$, Chase's theorem is retrieved. But the lemma also includes the following well-known fact which was derived from Baire's category theorem by Warfield [6, Cor. on p.254]: If M is a complete Hausdorff abelian group (in the $\mathbb{Z}$-topology) and if $f \colon M \to \bigoplus_{j \in J} V_j$ is a homomorphism from M to a Hausdorff abelian group $\bigoplus_{j \in J} V_j$, then there exists a finite subset J' of J such that, for some $n > 0$, $f(nM)$ is contained in $\bigoplus_{j \in J'} V_j$. A similar result for torsion-complete p-groups, due to Hill resp. Enochs (see [2, 71.1 and 71.2]), arises as a special case of Lemma 1 as well: Retain the setting just described with the exception that M is now a torsion-complete p-group. Then, for some finite subset J' of J and some $n \in \mathbb{N}$, we have the inclusion $f((p^n M)[p]) \subset \bigoplus_{j \in J'} V_j$, where $X[p] = \{x \in X \colon px = 0\}$ for an abelian group X. (Use the functors $P_n \colon \mathbb{Z}\text{-Mod} \to \mathbb{Z}\text{-Mod}$, $X \mapsto (p^n X)[p]$ in Lemma 1.)

Recall that a cardinal number is called measurable if a set I of this cardinality carries a $\{0,1\}$-valued, countably additive measure μ whose σ-algebra is the power set of I and which takes the prescribed values $\mu(I) = 1$, $\mu(\{x\}) = 0$ for all x.

Theorem 2. Start with two families $(U_i)_{i\in I}$ and $(V_j)_{j\in J}$ of R-modules and a homomorphism $f: \prod_{i\in I} U_i \to \bigoplus_{j\in J} V_j$. Moreover, suppose that $|I|$ is non-measurable. Then, for each descending chain $\mathfrak{a}_1 \supset \mathfrak{a}_2 \supset \mathfrak{a}_3 \supset \cdots$ of finitely generated right ideals of R, there exist a natural number n_o and finite subsets I' of I, J' of J such that

$$f\Big(\mathfrak{a}_{n_o} \prod_{i\in I\setminus I'} U_i\Big) \subset \bigoplus_{j\in J'} V_j + \bigcap_{n\in \mathbb{N}} \Big(\mathfrak{a}_n \bigoplus_{j\in J} V_j\Big)$$

Proof. By Lemma 1, the theorem holds for $|I| = \aleph_o$. For the general case, let $\mathfrak{F}$ be the set of those subsets T of I for which the restriction $f: \prod_{i\in T} U_i \to \bigoplus_{j\in J} V_j$ satisfies our claim, i.e.

$$\mathfrak{F} = \Big\{T\subset I : \text{ there are } n_o \in \mathbb{N} \text{ and finite subsets } T'\subset T, J'\subset J \text{ with } f\Big(\mathfrak{a}_{n_o} \prod_{T\setminus T'} U_i\Big) \subset \bigoplus_{j\in J'} V_j + \bigcap_{n\in \mathbb{N}} \Big(\mathfrak{a}_n \bigoplus_{j\in J} V_j\Big)\Big\}$$

It follows immediately from the definition that $\mathfrak{F}$ contains all finite subsets of I and is closed under subsets and finite unions. As a consequence of the following assertion, $\mathfrak{F}$ is even closed under countable unions. Given an arbitrary family $(Y_n)_{n\in \mathbb{N}}$ of pairwise disjoint subsets of I there is a natural number n_o with $\bigcup_{n\geq n_o} Y_n \in \mathfrak{F}$. In fact, apply the countable version of the theorem (which is already available) to the restriction $f: \prod_{n\in \mathbb{N}} \Big(\prod_{i\in Y_n} U_i\Big) \to \bigoplus_{j\in J} V_j$ in order to obtain $n_o \in \mathbb{N}$ and a finite subset J' of J such that
$f\Big(\mathfrak{a}_{n_o} \prod_{n\geq n_o}\Big(\prod_{i\in Y_n} U_i\Big)\Big) \subset \bigoplus_{j\in J'} V_j + \bigcap_{n\in \mathbb{N}} \Big(\mathfrak{a}_n \bigoplus_{j\in J} V_j\Big)$. The latter inclusion shows that $\bigcup_{n\geq n_o} Y_n$ belongs to $\mathfrak{F}$. In particular, we have $Y_n \in \mathfrak{F}$ for $n \geq n_o$.

Our claim amounts to $I \in \mathfrak{F}$ or, equivalently, to the coincidence of $\mathfrak{F}$ with the power set of I. Assume the contrary. Then, using the information just obtained, we can find a subset X of I, outside $\mathfrak{F}$, with the property that for any partition $X = X_1 \uplus X_2$ at least one of the sets X_1, X_2 belongs to $\mathfrak{F}$. Now define a $\{0,1\}$-valued function μ on the power set of I via

$$\mu(Y) = \begin{cases} 1 & \text{if } Y\cap X \notin \mathfrak{F} \\ 0 & \text{otherwise} \end{cases}$$

Clearly, we have $\mu(I) = 1$, $\mu(\{x\}) = 0$ for $x \in I$. It is a matter of routine to check that μ is countably additive. But this means that I is measurable. ∎

The theorem can definitely *not* be extended to products over measurable index sets. We demonstrate this by an example based on an idea of Łoś (see [2, Vol.II, p.161]).

Example 3. Let I be a set of measurable cardinality, μ a countably additive probability measure on the power set of I with $\mu(\{x\}) = 0$ for all $x \in I$. Choose a ring R and a countable R-module A which admits a strictly descending chain of subgroups $\mathfrak{a}_1 A \supset \mathfrak{a}_2 A \supset \mathfrak{a}_3 A \supset \cdots$, where $\mathfrak{a}_1 \supset \mathfrak{a}_2 \supset \mathfrak{a}_3 \supset \cdots$ are finitely generated right ideals (e.g. $R = A = \mathbb{Z}$, $\mathfrak{a}_n = (n!)$). Set $U_i = A^{(\mathbb{N})}$ for $i \in I$, $J = \mathbb{N}$ and $V_j = A$ for $j \in J$. Then define an R-homomorphism $f\colon \prod_{i \in I} U_i \to \bigoplus_{j \in J} V_j = A^{(\mathbb{N})}$ as follows: $A^{(\mathbb{N})}$ being countable, each element $(u_i)_{i \in I}$ of $\prod_{i \in I} U_i$ gives rise to a countable partition of I, namely $I = \bigcup_{a \in A^{(\mathbb{N})}} I_a$, where $I_a = \{i \in I\colon u_i = a\}$. On precisely one of these sets μ attains the value 1, on I_{a_0} say. That the definition $f((u_i)) = a_0$ yields a homomorphism, is ensured by the properties of μ. That the conclusion of the preceding theorem is violated, is obvious. ∎

3. MAIN RESULT

Since our techniques are by no means group-theoretic in nature, we continue to look at modules over a ring R which, from now on, is assumed to be *commutative*. In order to avoid clumsy statements, we exclude the existence of measurable cardinals in the sequel. The reader who is not willing to follow us there can maintain the results of this section by imposing the restriction of non-measurability to all of the occurring ordinals and product-index sets.

Let A be an R-module. In accordance with the definition of $\mathbb{Z}$-kernels, we introduce A-kernels and their types:

Definition: An R-module M is an A-kernel of type 0 if $M \cong A^n$ for some natural number n. Given an ordinal number $\mu \geq 1$, call M an A-kernel of type μ, if M is a direct product or a direct sum of A-kernels of types less than μ, whereas M itself is not of a type smaller than μ. Specify further by saying that M is of type (μ,P) or (μ,S) depending on whether M is a direct product or a direct sum of A-kernels of types smaller than μ .

Since formation of direct products is associative, each A-kernel of type (μ,P), $\mu \geq 1$, is a direct product of A-kernels of types (σ,S) with $\sigma < \mu$ and of A-kernels of type 0 (the analogous statement is true for sums). The index set over which the direct product extends must be infinite in this situation, for a direct *sum* of A-kernels of types (σ_i,S) is an A-kernel of type (σ,S) with $\sigma \leq \sup \sigma_i$. In contrast to this, even formation of finite direct sums of A-kernels may increase the type beyond the maximum of the types of the factors if no precautions are taken: the direct sum $\mathbb{Z}^{\mathbb{N}} \oplus \mathbb{Z}^{(\mathbb{N})}$ of the $\mathbb{Z}$-kernels $\mathbb{Z}^{\mathbb{N}}$ and $\mathbb{Z}^{(\mathbb{N})}$ of type (1,P) resp. (1,S) is of type 2. This example shows in addition that the types (σ,S) and (σ,P) do not exclude each other: in fact, $\mathbb{Z}^{\mathbb{N}} \oplus \mathbb{Z}^{(\mathbb{N})}$ is simultaneously of type (2,P) and (2,S). Of course, it cannot be expected, not even over $R = \mathbb{Z}$, that for arbitrary choice of A all ordinal numbers occur as types of A-kernels. Take $A = \mathbb{Q}$ for instance: each $\mathbb{Q}$-kernel has a type ≤ 1 . (In Cor. 6 we will see that, among the abelian groups of finite rank, the divisible and the bounded groups play an outsider role with respect to their classes of kernels.)

If, given some A , there is any hope for the existence of A-kernels of arbitrary type (μ,P) and (μ,S) , it is an immediate guess that the simplest candidates should look as follows: As is well-known, each ordinal number μ can be written uniquely as a sum $\mu = \alpha + m$, where α is a limit ordinal and $0 \leq m < \omega$. We will call μ odd resp. even according to m being odd resp. even.

Fix an arbitrary infinite set I_μ for each successor ordinal μ (e.g. $I_\mu = \mathbb{N}$) and define

$$\mathcal{P}_0(A) = A$$

and for $\mu \geq 1$

$$\mathcal{P}_\mu(A) = \begin{cases} (\mathcal{P}_{\mu-1}(A))^{I_\mu} & \text{if } \mu \text{ is an odd successor ordinal} \\ (\mathcal{P}_{\mu-1}(A))^{(I_\mu)} & \text{if } \mu \text{ is an even successor ordinal} \\ \bigoplus_{\tau<\mu} \mathcal{P}_\tau(A) & \text{if } \mu \text{ is a limit ordinal} \end{cases}$$

Similarly, but with the sequence of sums and products reversed, set

$\mathcal{Q}_0(A) = A$

and for $\mu \geq 1$

$$\mathcal{Q}_\mu(A) = \begin{cases} (\mathcal{Q}_{\mu-1}(A))^{(I_\mu)} & \text{if } \mu \text{ is an odd successor ordinal} \\ (\mathcal{Q}_{\mu-1}(A))^{I_\mu} & \text{if } \mu \text{ is an even successor ordinal} \\ \prod_{\tau<\mu} \mathcal{Q}_\tau(A) & \text{if } \mu \text{ is a limit ordinal} \end{cases}$$

Theorem 4. Suppose that A is an R-module of finite Goldie dimension and that there exists a family $(\mathfrak{t}_n)_{n\in\mathbb{N}}$ of finitely generated ideals $\mathfrak{t}_n$ of R such that $A\mathfrak{t}_1\cdot\ldots\cdot\mathfrak{t}_n \neq 0$ for all n, whereas $\bigcap_{n\in\mathbb{N}} (A\mathfrak{t}_1\cdot\ldots\cdot\mathfrak{t}_n) = 0$.

Then, for each ordinal number $\mu > 0$, $\mathcal{P}_\mu(A)$ is neither isomorphic to a direct summand of an A-kernel of type $\nu < \mu$ nor isomorphic to a direct summand of an A-kernel of type (μ,P) if μ is even resp. (μ,S) if μ is odd.

The analogous statement for $\mathcal{Q}_\mu(A)$ is true.

Proof. We will only prove the assertion concerning $\mathcal{P}_\mu(A)$. The finite product $\mathfrak{t}_1\cdot\ldots\cdot\mathfrak{t}_n$ will be abbreviated by $\mathfrak{a}_n$.

Assume that our claim is false and let μ be the smallest ordinal number such that, for some R-module A and some family $(\mathfrak{t}_n)_{n\in\mathbb{N}}$ of finitely generated ideals of R satisfying the above hypotheses, the conclusion of the theorem fails. This means that $\mathcal{P}_\mu(A)$ is isomorphic to a direct summand of an A-kernel M of type $\nu \leq \mu$, where the equality $\nu = \mu$ is coupled with the additional restriction that M is of type (ν,P) if ν is even, of type (ν,S) if ν is odd.

First suppose that μ is even. Then $\mathcal{P}_\mu(A) = \bigoplus_{j\in J} V_j$ with $(V_j)_{j\in J} = (\mathcal{P}_{\mu-1}(A))_{j\in I_\mu}$ or $(V_j)_{j\in J} = (\mathcal{P}_\tau(A))_{\tau<\mu}$ according to μ being a successor or a limit ordinal. Since $\mu > 0$, $\dim(A) < \infty$ implies $\nu > 0$. If M were not of type (ν,P), we would necessarily have $\nu < \mu$, and consequently we could find $j \in J$ such that $V_j \cong \mathcal{P}_\tau(A)$ for some odd ordinal τ with $\nu \leq \tau < \mu$. But since each V_j is a fortiori isomorphic to a direct summand of M, this would contradict the minimal choice of

μ . Therefore we may assume that M is of type (ν,P) , i.e. $M = \prod_{i\in I} U_i$ where $\text{type}(U_i) = 0$ or $\text{type}(U_i) = (\alpha_i,S)$ with $\alpha_i < \nu$.

Our assumption yields a commutative diagram:

$$\begin{array}{lcl} \bigoplus_{j\in J} V_j & \overset{\iota}{\longrightarrow} & \\ \Vert & & \prod_{i\in I} U_i \\ \bigoplus_{j\in J} V_j & \overset{f}{\longleftarrow} & \end{array}$$

By Theorem 1 we can choose a natural number n_o and finite subsets I' of I resp. J' of J such that

$$f\Big(\prod_{I\setminus I'} \mathfrak{a}_{n_o} U_i\Big) \subset \bigoplus_{J'} V_j + \bigcap_{n\in \mathbb{N}} \Big(\mathfrak{a}_n \bigoplus_J V_j\Big) .$$

Since both procedures - multiplication of an R-module X with $\mathfrak{a}_n$ and passing to the intersection $\bigcap_{n\in\mathbb{N}} \mathfrak{a}_n X$ - commute with the formation of direct products and direct sums, we derive $\bigcap_{n\in\mathbb{N}} \Big(\mathfrak{a}_n \bigoplus_J V_j\Big) = 0$ from the hypothesis. Consequently, $pf\Big(\prod_{I\setminus I'} \mathfrak{a}_{n_o} U_i\Big) = 0$, where p denotes the projection $\bigoplus_J V_j \to \bigoplus_{J\setminus J'} V_j$ along $\bigoplus_{J'} V_j$. If we write q resp. q' for the canonical projections $\prod_I \mathfrak{a}_{n_o} U_i \to \prod_{I\setminus I'} \mathfrak{a}_{n_o} U_i$ resp. $\prod_I \mathfrak{a}_{n_o} U_i \to \prod_{I'} \mathfrak{a}_{n_o} U_i$, we infer that the identity map on $\bigoplus_{J\setminus J'} \mathfrak{a}_{n_o} V_j$ is equal to the appropriate restriction of $pfq'\iota = pf(q+q')\iota = pf\iota$. All that is left to do, is to interpret the resulting commutative diagram:

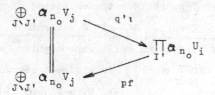

$$\begin{array}{lcl} \bigoplus_{J\setminus J'} \mathfrak{a}_{n_o} V_j & \overset{q'\iota}{\longrightarrow} & \\ \Vert & & \prod_{I'} \mathfrak{a}_{n_o} U_i \\ \bigoplus_{J\setminus J'} \mathfrak{a}_{n_o} V_j & \overset{pf}{\longleftarrow} & \end{array}$$

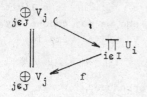

First observe that the type σ of the $(\mathfrak{a}_{n_o} A)$-kernel $\prod_{i\in I'} \mathfrak{a}_{n_o} U_i$ does not exceed the maximum of the types of the A-kernels U_i, $i \in I'$, and hence is strictly below ν , because $\text{type}(U_i) = (\alpha_i,S)$ with $\alpha_i < \nu$ for all i . The case $\sigma = 0$ is ruled out by the fact that the Goldie dimension of $\mathfrak{a}_{n_o} A$ is finite, whereas $J \setminus J'$ is infinite and $\mathfrak{a}_{n_o} V_j \neq 0$ for all j.

We can find $j \in J \setminus J'$ such that $V_j = \mathcal{P}_\tau(A)$ for some ordinal τ with $\sigma \leq \tau < \mu$: If μ is a successor ordinal, we have $V_j = \mathcal{P}_{\mu-1}(A)$ for all j, and from $\mu \geq \nu > \sigma$ we conclude $\mu-1 \geq \sigma$; if μ is a limit ordinal, then J is the set of ordinals $< \mu$, and we may choose any $\tau \in J \setminus J'$ which is larger than σ. - Clearly, $V_j = \mathcal{P}_\tau(A)$ implies $\mathfrak{a}_{n_o} V_j \cong \mathcal{P}_\tau(\mathfrak{a}_{n_o} A)$.

Next observe that the hypotheses for A are reproduced for $\mathfrak{a}_{n_o} A$: Use the family $(\mathfrak{t}_{n_o+k})_{k \in \mathbb{N}}$ instead of $(\mathfrak{t}_k)_{k \in \mathbb{N}}$. On the other hand, the above commutative diagram shows that $\mathcal{P}_\tau(\mathfrak{a}_{n_o} A)$ is isomorphic to a direct summand of $\prod_{i \in I'} \mathfrak{a}_{n_o} U_i$, which, as we will see, contradicts the minimal choice of μ: This is clear if $\tau > \sigma$, i.e. if μ is either a limit ordinal or $\mu-1 > \sigma$. Suppose that μ is a successor ordinal and $\mu-1 = \sigma$. In view of $\mu \geq \nu > \sigma$, we infer $\mu = \nu$ and $\sigma = \nu-1$. But $\mu-1$ is odd, and the type of the $(\mathfrak{a}_{n_o} A)$-kernel $\prod_{I'} \mathfrak{a}_{n_o} U_i$ is necessarily $(\mu-1,S)$ (if U is an A-kernel, then the type of the $(\mathfrak{a}_{n_o} A)$-kernel $\mathfrak{a}_{n_o} U$ does not exceed the type of U, so $\prod_{I'} \mathfrak{a}_{n_o} U_i$ is a direct sum of $(\mathfrak{a}_{n_o} A)$-kernels of types $< \mu-1$). We have thus excluded the case that μ is even.

Now let μ be odd. Then μ is a successor ordinal and $\mathcal{P}_\mu(A) = \prod_{i \in I} U_i$ with $(U_i)_{i \in I} = (\mathcal{P}_{\mu-1}(A))_{I_\mu}$. An argument analogous to the one carried out in the first case shows that we may assume type$(M) = (\nu,S)$ with $\nu > 0$, i.e. $M = \bigoplus_{j \in J} V_j$ with type$(V_j) = 0$ or type$(V_j) = (\alpha_j,P)$, where $\alpha_j < \nu$. By assumption, there is a split monomorphism $f\colon \prod_{i \in I} U_i \to \bigoplus_{j \in J} V_j$. Once more, Theorem 1 guarantees the existence of $n_o \in \mathbb{N}$ and finite subsets I' of I resp. J' of J such that $f\left(\prod_{I \setminus I'} \mathfrak{a}_{n_o} U_i\right) \subset \bigoplus_{J'} \mathfrak{a}_{n_o} V_j$. Now f induces a split monomorphism $\prod_I \mathfrak{a}_{n_o} U_i \to \bigoplus_J \mathfrak{a}_{n_o} V_j$, and therefore $f\left(\prod_{I \setminus I'} \mathfrak{a}_{n_o} U_i\right)$ is a direct summand of $\bigoplus_{J'} \mathfrak{a}_{n_o} V_j$. In particular, for each $i \in I \setminus I'$, $\mathfrak{a}_{n_o} U_i \cong \mathcal{P}_{\mu-1}\left(\mathfrak{a}_{n_o} A\right)$ is isomorphic to a direct summand of the module $\bigoplus_{J'} \mathfrak{a}_{n_o} V_j$ which, in turn, is an $(\mathfrak{a}_{n_o} A)$-kernel whose type σ is located strictly between 0 and ν; as above, the latter type can be specified further to be $(\mu-1,P)$ in case μ is a successor ordinal and $\sigma = \mu-1$. Again we have reached a contradiction to the choice of μ. ∎

Theorem 4 fits exactly our original problem concerning kernel groups, namely it shows that the groups $\mathcal{P}_\mu(\mathbb{Z})$, $\mathcal{Q}_\mu(\mathbb{Z})$ represent all possible types (choose $\mathfrak{r}_n = (n)$). Calling a module bounded if its annihilator ideal is non-zero, we obtain more generally:

Corollary 5. Let R be a Dedeking domain and A an R-module of finite rank. Then the following statements are equivalent:

(1) A is not a direct sum of a divisible and a bounded module.

(2) For each ordinal number μ, the A-kernel $\mathcal{P}_\mu(A)$ is of type (μ,S) if μ is even and of type (μ,P) otherwise.

(3) For each ordinal number μ, the A-kernel $\mathcal{Q}_\mu(A)$ is of type (μ,P) if μ is even and of type (μ,S) otherwise.

Moreover, if (1) is satisfied, none of the modules $\mathcal{P}_\mu(A)$, $\mathcal{Q}_\mu(A)$ is simultaneously of type (μ,P) and (μ,S). In particular, $\mathcal{P}_\mu(A)$, $\mathcal{Q}_\mu(A)$, μ,ν running through all ordinals, are pairwise non-isomorphic.

Proof. (1) $\Rightarrow$ (2). By construction, $\text{type}(\mathcal{P}_\mu(A)) \le \mu$. The hypothesis yields a family $(\mathfrak{r}_n)_{n\in\mathbb{N}}$ of ideals in R such that $\mathfrak{r}_1\cdot\ldots\cdot\mathfrak{r}_n A \not\subset \mathfrak{r}_1\cdot\ldots\cdot\mathfrak{r}_{n+1} A$ for $n\in\mathbb{N}$. Set $\bar{A} = A/\bigcap_{n\in\mathbb{N}}(\mathfrak{r}_1\cdot\ldots\cdot\mathfrak{r}_n A)$ and observe that, given an A-kernel M, the factor $M/\bigcap_{n\in\mathbb{N}}(\mathfrak{r}_1\cdot\ldots\cdot\mathfrak{r}_n M)$ is an $\bar{A}$-kernel whose type does not exceed the type of M. It therefore suffices to verify condition (2) for $\mathcal{P}_\mu(\bar{A})$ instead of $\mathcal{P}_\mu(A)$, i.e. we may assume $\bigcap_{n\in\mathbb{N}}(\mathfrak{r}_1\cdot\ldots\cdot\mathfrak{r}_n A) = 0$. But then the hypotheses of Theorem 4 are satisfied, and the claim follows.

(2) $\Rightarrow$ (1). Assuming that A is a direct sum of a divisible and a bounded module, we have $\text{type}(\mathcal{P}_\mu(A)) \le 1$ for all μ.

The equivalence of (1) and (3), as well as the supplementary statement, can be similarly derived from Theorem 4. ■

There are, of course, large classes of modules of infinite Goldie dimension for which the corresponding modules $\mathcal{P}_\mu$, $\mathcal{Q}_\mu$ are also precisely of type μ. In fact, inspection of the proof shows that in Theorem 4 the hypothesis about the Goldie dimension can be weakened to the following condition: If $A_n = \mathfrak{r}_1\cdot\ldots\cdot\mathfrak{r}_n A$, none of the modules $(A_n)^m$ contains a direct summand isomorphic to $(A_n)^{(\mathbb{N})}$ or $(A_n)^{\mathbb{N}}$, $n,m\in\mathbb{N}$ (i.e. we only have to take care of the out-set of our transfinite induction). Specializing to $R=\mathbb{Z}$, this yields

Corollary 6. Let A be an abelian group and denote by $f(p,\sigma)$ the σ-th Ulm-Kaplansky invariant of A with respect to a prime p (i.e. $f(p,\sigma)$ is the rank of the p-socle of the factor group $p^\sigma A/p^{\sigma+1}A$).

If either
(a) $0 < f(p,\sigma) < \aleph_0$ for infinitely many pairs (p,σ) or
(b) $0 < f(p,\sigma) < \aleph_0$ for some prime p and some ordinal $\sigma \geq \omega$,

then the A-kernels $\mathcal{P}_\mu(A)$, $\mathcal{Q}_\mu(A)$ are of type μ for all choices of the infinite sets I_μ (stronger: the statements (2), (3) of Corollary 5 hold).
On the other hand, if A is a countable primary group, this implication is reversible.

Proof. Suppose that (a) or (b) is satisfied: In view of the preceding remark, all we have to show is that, for each pair n,m of natural numbers, the group $(nA)^m$ does not contain a direct summand isomorphic to $(nA)^{(\mathbb{N})}$ or to $(nA)^{\mathbb{N}}$. But since the Ulm-Kaplansky invariants behave additive with respect to direct sums and nA has at least one finite non-zero invariant by hypothesis, both of the forbidden situations are ruled out.

Conversely, let A be a countable p-group such that only a finite number $f(p,n_1),\ldots,f(p,n_k)$, $n_i < \omega$, of Ulm-Kaplansky-invariants are finite and non-zero. By a theorem of Zippin (see [2,Thm.76.2]), there is a countable p-group B whose Ulm-Kaplansky invariants are those of A, except the ones belonging to $n_1,\ldots,n_k$ which are replaced by 0. The invariants of $C = B \oplus \bigoplus_{i=1}^{k} \left(\mathbb{Z}/(p^{n_i})\right)^{f(p,n_i)}$ and A are then identical hence $A \cong C$ by Ulm's theorem (see [2,Thm.77.3]). Another application of Ulm's result shows $B \cong B^{(\mathbb{N})}$. Choosing $I_1 = I_2 = \mathbb{N}$ and writing $C = B \oplus D$, we infer that $\mathcal{Q}_2(A) = (B^{(\mathbb{N})})^{\mathbb{N}} \oplus (D^{(\mathbb{N})})^{\mathbb{N}} \cong B^{\mathbb{N}} \oplus D^{\mathbb{N}} \cong \mathcal{P}_1(A)$, in particular $\mathcal{Q}_2(A)$ is of type ≤ 1. ∎

Let A be an R-module satisfying the hypothesis of Theorem 4 (e.g. $R = A = \mathbb{Z}$). Then, as is guaranteed by the theorem, none of the modules $\mathcal{P}_\mu(A)$, $\mathcal{Q}_\nu(A)$, where μ,ν run through all ordinals, is isomorphic to any other. It is therefore tempting to believe that, if one starts with A and alternatingly forms infinite direct products and direct sums, one can end up with isomorphic objects only if the well-ordered sequences of sum and product functors are the same (we forget about the cardinalities of the index sets for the moment). This is true if the process is finite, it may fail, however, if the procedure passes beyond a limit ordinal.

<u>Example 7</u>. Let A be an arbitrary R-module and define
$B_o = A,\ B_1 = A^{\mathbb{N}},\ B_2 = (A^{\mathbb{N}})^{(\mathbb{N})}, \ldots$
$C_o = A,\ C_1 = A^{(\mathbb{N})},\ C_2 = (A^{(\mathbb{N})})^{\mathbb{N}}, \ldots$
(i.e. $B_n = \mathcal{P}_n(A)$, $C_n = \mathcal{Q}_n(A)$ with $I_n = \mathbb{N}$ for all n).
We will show that $B = \bigoplus_{n<\omega} B_n$ and $C = \bigoplus_{n<\omega} C_n$ are isomorphic.
Observe that for each $n \geq 1$ we have $B_n \cong B_n \oplus C_{n-1}$ and $C_n \cong C_n \oplus B_{n-1}$. For $n \in \mathbb{N}$ decompose B_n in the form $B_n = B_{n1} \oplus B_{n2}$ with $B_{n1} \cong B_n$, $B_{n2} \cong C_{n-1}$ and analogously $C_n = C_{n1} \oplus C_{n2}$ with $C_{n1} \cong C_n$, $C_{n2} \cong B_{n-1}$. Setting $C_o = C_{o1}$, we infer
$B_n \cong C_{n+1,2} \oplus C_{n-1,1}$ for all n and conclude
$B = \bigoplus_{n<\omega} B_n \cong B_o \oplus \bigoplus_{1\leq n<\omega} (C_{n+1,2} \oplus C_{n-1,1}) \cong \bigoplus_{n<\omega} C_n$ (the isomorphism $B_o \cong C_{1,2}$ takes care of the two outsider terms).

Continuing this game of exchanging summands, one obtains
$\bigoplus_{\tau<\mu} \mathcal{P}_\tau(A) \cong \bigoplus_{\tau<\mu} \mathcal{Q}_\tau(A)$ for $\mu = \omega, 2\omega, 3\omega, \ldots$ ($I_\nu = \mathbb{N}$ for all ν). In essence: if μ is the number of iteration steps performed and if α is the largest limit ordinal below μ, the final result of the construction hinges on the alternative "product or sum" in the α-th step.

4. SOLUTION TO A PROBLEM OF IRWIN AND O'NEILL

In [3, Thm. 4.28] Irwin and O'Neill have shown that, given an abelian p-group M, the countable product $M^{\mathbb{N}}$ is an infinite direct sum of pairwise isomorphic groups if and only if M is bounded modulo its largest divisible subgroup. In response to question (1) of [3, p.544] we show that no substantially different phenomenon occurs when $M^{\mathbb{N}}$ is replaced by a direct product $\prod_{i\in I} U_i$ of arbitrary abelian torsion groups U_i, where I is some index set of non-measurable cardinality (we will not stress the non-measurability in the sequel, see beginning of Section 2). The problem is related to the one treated in the previous section, since the existence of a group V and an infinite set J such that $M \cong V^{(J)}$ is obviously equivalent to $M \cong M^{(J)}$.

Of course, it means no loss of information to concentrate on the case where all of the groups U_i are reduced. For an arbitrary abelian group X let $\bar{X}$ denote the corresponding Hausdorff group $X/\bigcap_{n\in\mathbb{N}} nX$.

Lemma 8. Let $(U_i)_{i\in I}$ be a family of reduced torsion groups. Then, for each direct-sum decomposition $\bigoplus_{j\in J} V_j$ of the direct product $\prod_{i\in I} U_i$, almost all of the groups $\bar{V}_j$ are torsion groups.

Proof. In view of $\prod_{i\in I} \bar{U}_i \cong \bigoplus_{j\in J} \bar{V}_j$ we may assume $\bigcap_{n\in\mathbb{N}} nV_j = 0$ for all j. If we set $P_n X = n!X$, an application of Theorem 2 yields a natural number n_o and finite subsets I' resp. J' of I resp. J such that $(n_o!)\prod_{i\in I\setminus I'} U_i$ is contained in $(n_o!)\bigoplus_{j\in J'} V_j$. This implies that the homomorphism $p:(n_o!)\bigoplus_{j\in J\setminus J'} V_j \to (n_o!)\prod_{i\in I'} U_i$, induced by the canonical projection $\prod_{i\in I} U_i \to \prod_{i\in I'} U_i$, is a monomorphism. Thus, the V_j's, $j\in J\setminus J'$, are torsion groups, and our claim is established. ■

Theorem 9. Given a family $(U_i)_{i\in I}$ of reduced torsion groups, the following conditions are necessary and sufficient for $\prod_{i\in I} U_i$ to be an infinite direct sum of pairwise isomorphic groups:

There is a finite subset I' of I such that $\prod_{i\in I\setminus I'} U_i$ is isomorphic to a direct sum $C_1^{(\Gamma_1)} \oplus \ldots \oplus C_m^{(\Gamma_m)}$ where $C_1,\ldots,C_m$ are non-isomorphic cyclic groups of prime power orders and $\Gamma_1,\ldots,\Gamma_m$ non-empty sets. The remaining finite product $\prod_{i\in I'} U_i$ is of the form $C_1^{(\Delta_1)} \oplus \ldots \oplus C_m^{(\Delta_m)} \oplus \bigoplus_{k\in\mathbb{N}} Q_k$, where the Q_k's are pairwise isomorphic groups and where Δ_1 is an infinite set if Γ_1 is finite, arbitrary if Γ_1 is infinite.

Proof. The sufficiency is obvious. Necessity: From $\prod_{i\in I} U_i \cong V^{(J)}$ for some group V and some infinite set J we deduce via Lemma 8 that $\bar{V}$ is a torsion group. This, in turn, implies that $\prod_{i\in I} \bar{U}_i$ is torsion, i.e. there exists a finite subset I' of I and a natural number n_o with $n_o \prod_{i\in I\setminus I'} \bar{U}_i = 0$. But the equality $n_o \prod_{i\in I\setminus I'} U_i = \bigcap_{n\in\mathbb{N}} n\left(\prod_{i\in I\setminus I'} U_i\right)$ means that $n_o \prod_{i\in I\setminus I'} U_i$ is divisible and hence zero by hypothesis. We are thus authorized to write $\prod_{i\in I\setminus I'} U_i = C_1^{(\Gamma_1)} \oplus \ldots \oplus C_m^{(\Gamma_m)}$, where

$C_1, \dots, C_m$ are groups of different prime power orders, $\Gamma_1 \neq \emptyset$.

The rest of the proof is a simple task in counting. Set $A = \prod_{i \in I'} U_i$, $B = \prod_{i \in I \setminus I'} U_i$ and decompose A in the form $A = A' \oplus A''$, where $A' \cong C_1^{(\Delta_1)} \oplus \dots \oplus C_m^{(\Delta_m)}$ is a maximal direct summand of A which is a direct sum of cyclic groups of type C_l, $1 \leq l \leq m$. According to the same pattern, write $V = V' \oplus V''$. In view of

$$(A' \oplus B) \oplus A'' = \prod_{i \in I} U_i = V'^{(J)} \oplus V''^{(J)},$$

we know by [2, Thms. 33.2 and 35.2] that

$$A' \oplus B \cong V'^{(J)} \quad \text{and} \quad A'' \cong V''^{(J)}.$$

First we check that Δ_1 is infinite if Γ_1 is finite: Since C_1 appears as a direct summand of B and hence is isomorphic to a direct summand of V', we have $|\Delta_1| \geq |J|$. Moreover, by suitably bundling the summands V'', we can write $V''^{(J)}$ in the form $\bigoplus_{k \in \mathbb{N}} Q_k$ with pairwise isomorphic groups Q_k. ■

Remark. It would be of interest to decide whether non-measurability of $|I|$ is really required for the validity of Theorem 9. The exceptional epimorphisms from "measurable direct products" to direct sums, which we have exhibited in Example 3, are far from being injective.

References.

1. S.U. Chase, On direct sums and products of modules, Pac. J. Math. 12 (1962), 847-854.
2. L. Fuchs, Infinite abelian groups, Vol. I, II, Academic Press, 1973.
3. J.M. Irwin and J.D. O'Neill, On direct products of abelian groups, Can. J. Math. 22 (1970), 525-544.
4. R.J. Nunke, On direct products of infinite cyclic groups, Proc. Amer. Math. Soc. 13 (1962), 66-71.
5. G.A. Reid, Almost free abelian groups, Lecture Notes, Tulane University, 1966/67.
6. R.B. Warfield, Jr., An isomorphic refinement theorem for abelian groups. Pac. J. Math. 34 (1970), 237-255.
7. B. Zimmermann-Huisgen, On Fuchs' problem 76, J. reine angew. Math. 309 (1979), 86-91.

Note added after finishing this paper: S.V. Rychkow has kindly informed us that A.V. Ivanov has recently obtained a result similar to our Theorem 2.

A general theory of slender groups and Fuchs-44-groups

R. Göbel, S.V. Richkov and B. Wald

§1 Introduction. Laszlo Fuchs asked in his book "Infinite abelian groups" [F2, p.251, Problem 44] the following question:

PROBLEM 44. *Investigate groups A with the following property: If A is contained in a direct sum of reduced groups, then there is an integer $n > 0$ such that nA is contained in the direct sum of a finite number of them.*

This problem was investigated already in 1962 by S.U. Chase [C] and later on by A.V. Ivanov, cf. A.B. Иванов [И1 and И2]. Following A.B. Иванов [И1], groups with the property described in Problem 44 will be called Fuchs-44-groups. A.B. Иванов[И1] obtained the following interesting characterization of Fuchs-44-groups.

THEOREM 1.1 (A.B. Иванов [И1]; *Engl. Transl.: p.462, Proposition 1.1, and p.463 Theorem 1.3): The following conditions on a group G are equivalent:*

(1) *G is a Fuchs-44-group*

(2) *For any homomorphism* $\sigma: G \to \sum_{i\in I} A_i$ *there are an integer* $n > 0$ *and a finite subset* $J \subseteq I$ *such that* $\sigma(nG) \subseteq \sum_{i\in I} A_i$

(3) *If M is an epimorphic image of G and M is a direct sum of cyclic groups, then M is bounded.*

(4) *If M is an epimorphic image of G and M is torsion, then M is a direct sum of a divisible group and a bounded group.*

In this paper we will elaborate a duality which apparently links the properties of Fuchs-44-groups and a class of generalized slender groups. This leads to a method to detect theorems for one or the other class of groups if one of the companions is known.

In the following let $\aleph_m$ be the first measurable cardinal (if it exists). Now we dualize condition (2) of (1.1) to obtain the following

DEFINITION 1.2: *A group G is called almost-slender, if the following holds:*

If I is a set of cardinality $< \aleph_m$ *and* $A_i\,(i\in I)$ *is a family of abelian groups, then for any homomorphism* $\sigma: \prod_{i\in I} A_i \to G$ *there are a natural number* $n > 0$ *and a finite subset* $J \subseteq I$ *such that*
$n(\prod_{i\in I\setminus J} A_i) = 0$.

Apparently (1.2) generalizes the notion of slender groups due to J. Łoś, cf. L. Fuchs [F1, p.169]. The Definition 1.2 will be generalized in (2.1*). However, there are still <u>different</u> extensions of slender groups, which have been used for <u>other</u> investigations on abelian and non-abelian groups; cf. R. Göbel [G] and R. Göbel and B. Wald [GW]. Definition 2.1 of $\mathcal{K}$-slender groups can be obtained dualizing the notion of Fuchs-44-groups relatively to a class $\mathcal{K}$ of groups (= $\mathcal{K}$-44-groups), cf. (2.1). Slender groups are closed under taking direct sums; this is a non-trivial result, first proved by L. Fuchs [F1, p.172, Theorem 47.4] which of course follows easily from Nunke's characterization [N1] of slender groups. The dual result on Fuchs-44-groups says, that non-measurable products of Fuchs-44-groups are Fuchs -44-groups. This result is due to A.B. Иванов [И2], where he applies a generalization of a theorem of S.U. Chase and J. Łoś, cf. (2.4). We will extend Ivanov's result to $\mathcal{K}$-44-groups (2.5) and prove its dual result on $\mathcal{K}$-slender groups (2.8). The final section is devoted to almost-slender groups, which are called generalized slender groups in C.B. Рычков [P1, p.301].

§2 Fuchs-44-groups and almost-slender groups with respect to a class $\mathcal{K}$ of groups

Let $\mathcal{K}, \mathcal{L}$ be classes of abelian groups. Let $\prod_{i\in I} G_i = \prod_I G_i$ and $\sum_{i\in I} G_i = \sum_I G_i$ be the direct product respectively the direct sum of groups $G_i (i\in I)$. If $I' \subseteq I$ assume $\prod_{I'} G_i \subseteq \prod_I G_i$ and $\sum_{I'} G_i \subseteq \prod_I G_i$ to be canonical. A.B. Иванов [И1] introduced the following

DEFINITION 2.1: *A group* G *is called Fuchs-44-group relative to* $\mathcal{K}$ *if for any groups* $A_i \in \mathcal{K}$ *and any homomorphism* $\sigma : G \to \sum_I A_i$ *there are an integer* $n > 0$ *and a finite subset* $J \subseteq I$ *such that* $n\sigma(G) \subseteq \sum_J A_i$.

We will call "Fuchs-44-groups relative to $\mathcal{K}$" for short $\mathcal{K}$-44-groups. Dualizing (2.1) we obtain the

DEFINITION 2.1*: *A group* G *is called almost-slender relative to* $\mathcal{K}$ *(in short* $\mathcal{K}$*-slender) if for any groups* $A_i \in \mathcal{K}$ *where* $i \in I$ *with* $|I| < \aleph_m$ *and any homomorphism* $\sigma : \prod_I A_i \to G$ *there are an integer* $n > 0$ *and a finite subset* $J \subseteq I$ *such that* $n\sigma(\prod_{I\setminus J} A_i) = 0$.

Let $S\mathcal{K}(Q\mathcal{K}, E\mathcal{K}, \Pi\mathcal{K}, \Sigma\mathcal{K})$ denote (as usual) the smallest class of abelian groups containing $\mathcal{K}$ and closed under taking subgroups (quotients, extensions, direct products of less than $\aleph_m$ groups, direct sums). Let $\mathcal{K} \circ \mathcal{K}$ be the class of all extensions of $\mathcal{K}$-groups by $\mathcal{K}$-groups. Then the two classes defined by (2.1) respectively (2.1) have the following obvious closure properties:

PROPOSITION 2.2: *Let $\mathcal{K}_{44}$ be the class of $\mathcal{K}$-44-groups.*
(a) $Q(\mathcal{K}_{44}) = \mathcal{K}_{44}$, $E(\mathcal{K}_{44}) = \mathcal{K}_{44}$
(b) *If* $\mathcal{K} \leq \mathcal{E}$, *then* $\mathcal{E}_{44} \leq \mathcal{K}_{44}$.

PROPOSITION 2.2*: *Let $\mathcal{K}_s$ be the class of $\mathcal{K}$-slender groups.*
(a) $S\mathcal{K}_s = \mathcal{K}_s$, $E\mathcal{K}_s = \mathcal{K}_s$
(b) *If* $\mathcal{K} \leq \mathcal{E}$, *then* $\mathcal{E}_s \leq \mathcal{K}_s$.

PROPOSITION 2.3:
(a) *A group is $\mathcal{K}$-44 iff it is $S\mathcal{K}$-44.*
(b) *A group is $\mathcal{K}$-44 iff it is $\mathcal{K}\circ\mathcal{K}$ -44.*
(c) *A group is $\mathcal{K}$-44 iff it is $\Sigma\mathcal{K}$-44.*

PROPOSITION 2.3*:
(a) *A group is $\mathcal{K}$-slender iff it is $Q\mathcal{K}$-slender*
(b) *A group is $\mathcal{K}$-slender iff it is $\mathcal{K}\circ\mathcal{K}$-slender*
(c) *A group is $\mathcal{K}$-slender iff it is $\Pi\mathcal{K}$-slender.*

The proof for(a) and (c) is trivial, and (2.3*)(b) follows by the next argument: Let $A_i, C_i \in \mathcal{K}$ for all $i \in I$.
If $0 \to A_i \to B_i \to C_i \to 0$ is short exact, then

$$0 \to \Pi A_i \to \Pi B_i \to \Pi C_i \to 0 \quad \text{is a}$$
$$\sigma: \Pi B_i \to G, \quad \tau$$

short exact sequence (using only canonical maps). There are $n \in \mathbb{N}$ and $J \subseteq I$ finite such that $n\sigma(\prod_{I\setminus J} A_i) = 0$. Hence $n\sigma$ induces a map τ from $\prod_{I\setminus J} C_i$ into G. There are $m \in \mathbb{N}$ and $X \subseteq I \setminus J$ finite such that $m\tau(\prod_{I\setminus Y} C_i) = 0$ for $Y = X \cup J$. Then $n\circ m\sigma(\prod_{I\setminus Y} B_i) = 0$. □
The proof of (2.3)(b) is similar.
Remark: In (2.3)(b) and (2.3*)(b) the class $\mathcal{K} \circ \mathcal{K}$ can not be replaced by $E\mathcal{K}$ as follows from the example $\mathcal{K} = \{Z_p\}$. Observe that $\prod_{n\in\mathbb{N}} Z_{p^n}$ is $\mathcal{K}$-slender but not $E\mathcal{K}$-slender. In order to investigate $\mathcal{K}$-44-groups and $\mathcal{K}$-slender groups it is useful to apply the following theorem.
This Theorem 2.4 is a result due to A.B. Иванов [И2] which generalizes S.U. Chase [C, p.846, Theorem 2.4] and a well known theorem of J. Łoś, cf. L. Fuchs [F1, p.170, Theorem 47.2], [F2, vol II, p.161, Theorem 94.4]. Independently Theorem 2.4 has been shown recently by M. Dugas and B. Zimmermann-Huisgen [DZ, Theorem 2].
THEOREM 2.4: (A.B. Иванов [И2, p.96]): *Let J and I be sets such that $|I| < \aleph_m$. Let A_i $(i\in I)$ be a family of arbitrary abelian groups and let G_j $(j\in J)$ be a set of reduced groups. Then for every homomorphism $\varphi : \prod_I A_i \to \bigoplus_J G_j$ there are a natural number $n > 0$ and finite subsets*

$I' \subseteq I$ *and* $J' \subseteq J$ *such that* $\varphi(n \prod_{I \setminus I'} A_i) \subseteq \sum_{J'} G_j$.

From this follows immediately the

COROLLARY 2.5: *Let* $\mathcal{K}$ *be a class of reduced groups. Then non-measurable products of* $\mathcal{K}$*-44-groups are again* $\mathcal{K}$*-44-groups.*

In the case $\mathcal{K} = \{Q\}$ the $\mathcal{K}$-44-groups are the torsion groups. This shows that the word "reduced" in (2.5) cannot be removed. Dual to (2.5) we derive from (2.4) the following

COROLLARY 2.5*: *Direct sums of reduced* $\mathcal{K}$*-slender groups are again* $\mathcal{K}$*-slender.*

Hence it remains the question, whether we can drop the word "reduced" in this case(2.5*). This is possible as shown in the next section, cf. (2.8). First we remark

LEMMA 2.6 : *For a group* A *and a prime* p *are equivalent:*

(1) $\mathrm{Hom}(A, Z(p^\infty))$ *is bounded*

(2) A *is torsion and the primary component* A_p *is bounded.*

Proof: (2) → (1) is trivial.

(1) → (2): Let A be unbounded or A not torsion and $n > 0$. Then we can find an element a of infinite order or of p-power-order $>n$ in A. Hence there is a homomorphism $\sigma : \langle a \rangle \to Z(p^\infty)$ such that $n\sigma \neq 0$. This extends to $\sigma_n \in \mathrm{Hom}(A, Z(p^\infty))$ with $n\sigma_n \neq 0$. Condition (1) implies that A is torsion and A_p is bounded.

THEOREM 2.7: *For a class* $\mathcal{K}$ *are equivalent:*

(1) *Every abelian group is* $\mathcal{K}$*-slender.*

(2) *There is a natural number* $n > 0$ *such that* $nA = 0$ *for all* $A \in \mathcal{K}$

(3) *There is a non-reduced* $\mathcal{K}$*-slender group.*

(4) *Either* Q *or one of the Prüfer groups* $Z(p^\infty)$ *is* $\mathcal{K}$*-slender.*

Proof. (2) → (1) → (3) is trivial and (3) → (4) follows from (2.2*). It remains to show (4) → (2): Let D be the $\mathcal{K}$-slender divisible group which exists after (4) and assume that (2) is false. There are $A_n \in \mathcal{K}$ such that $nA_n \neq 0$ for all $n \in \mathbb{N}$. Decompose $\mathbb{N}$ into infinitely many infinite subsets $S_n (n \in \mathbb{N})$. Then $B_n = \prod_{S_n} A_i$ is not torsion and $\mathrm{Hom}(B_n, D)$ is unbounded by (2.6). There are $\varphi_n \in \mathrm{Hom}(B_n, D)$ such that $n\varphi_n \neq 0$. Let $\varphi = \bigoplus_{\mathbb{N}} \varphi_n : \bigoplus_{\mathbb{N}} B_n \to D$ and $\hat{\varphi}$ its extension acting on $\prod_{\mathbb{N}} B_n$. This is possible, since D is injective. Because D is $\mathcal{K}$-slender and $\prod_{\mathbb{N}} B_n = \prod_{\mathbb{N}} A_i$, there are a finite subset $E \subset \mathbb{N}$ and $k \in \mathbb{N}$ such that $k\hat{\varphi}(\prod_{\mathbb{N} \setminus E} A_i) = 0$. In particular there is an $n > 0$ such that $k | n$ and $k\varphi_n = 0$. This contradicts $n\varphi_n \neq 0$ and $\mathcal{K}$ is simultaneously bounded .

<u>*COROLLARY 2.8:*</u> *Direct sums of $\mathcal{K}$-slender groups are $\mathcal{K}$-slender.*

<u>Proof.</u> Apply (2.7) and (2.5*).

Similar to the well-known Theorem of J. Łoś we have the following more general result.

Here it is useful to have the following

<u>*DEFINITION 2.9:*</u> *Call a group G <u>countable $\mathcal{K}$-slender</u> if the following condition on G is satisfied. If $A_n \in \mathcal{K}$ for all $n \in \mathbb{N}$ and $\sigma: \prod_{\mathbb{N}} A_n \to G$ there are a finite subset $E \subset \mathbb{N}$ and a natural $n > 0$ such that $n\sigma(\prod_{\mathbb{N}\setminus E} A_n) = 0$.*

<u>*THEOREM 2.10:*</u> *If $\Pi\mathcal{K} = \mathcal{K}$, then a group G is $\mathcal{K}$-slender if and only if it is countable $\mathcal{K}$-slender.*

<u>Proof.</u> Assume that G is countable $\mathcal{K}$-slender and let $\varphi : \prod_I A_i \to G$ be a homomorphism such that $A_i \in \mathcal{K}$ for all $i \in I$ and $|I| < \aleph_m$. Consider the ideal

$$\Phi = \{J \subseteq I, \exists k \in \mathbb{N}, \exists E \subset J \text{ finite such that } k\varphi(\prod_{J\setminus F} A_i) = 0\}$$

First we show that

(*) Φ is closed under taking countable unions.

Complete induction shows that it is sufficient to show that a countable <u>disjoint</u> union $\dot{\bigcup}_{n\in\mathbb{N}} J_n$ belongs to Φ if $J_n \in \Phi$ for all $n \in \mathbb{N}$. Let $B_n = \prod_{J_n} A_i$ then $B_n \in \mathcal{K}$ by assumption $\Pi\mathcal{K} = \mathcal{K}$. Since $\varphi : \prod_{\mathbb{N}} B_n \to G$, condition (*) implies $k\cdot\varphi(\prod_{\mathbb{N}\setminus E} B_n) = 0$ for some $k \in \mathbb{N}$ and some finite subset $E \subset \mathbb{N}$.

Since $J_n \in \Phi$ for all $n \in E$ there are $k_n \in \mathbb{N}$ and finite subsets $E_n \subseteq J_n$ such that $k_n\varphi(\prod_{J_n\setminus E_n} A_i) = 0$. Let $F = \bigcup_{n\in E} E_n$ and $m = k \cdot \prod_{n\in E} k_n$. Then $m\,\varphi(\prod_{J\setminus F} A_i) = 0$ and therefore $J \in \Phi$ and (*) is shown. Following the pattern of the proof of the mentioned theorem by J. Łoś, we can show that $B(I)/\Phi$ is finite, cf. L. Fuchs [F2, Vol.2 p.161]. Since $|I| < \aleph_m$, we derive $B(I) = \Phi$ and in particular $I \in \Phi$, which proves the theorem.

<u>§3 $\mathbb{Z}$-slender groups</u> Here we will concentrate on the class of $\mathcal{K}$-slender groups for $\mathcal{K} = \{\mathbb{Z}\}$. It will turn out that this is the class of almost -slender groups, mentioned in §1. We want to show that a group is $\mathbb{Z}$-slender iff it is countable $\mathbb{Z}$-slender. Let $e_i \in \mathbb{Z}^{\mathbb{N}}$ be defined by the Kronecker-symbol $e_i = (\delta_{ij})_{j\in\mathbb{N}}$ and let $\mathfrak{U}$ be the universal class of all abelian groups. We will prove the

THEOREM 3.1: For a group G *are equivalent:*

(1) G *is* $\mathbb{Z}$*-slender.*

(2) G *is countable* $\mathbb{Z}$*-slender.*

(3) *If* $\varphi : \mathbb{Z}^{\mathbb{N}} \to G$ *there is a* $k \in \mathbb{N}$ *such that* $k\varphi(e_i) = 0$ *for almost all* $i \in \mathbb{N}$.

(4) G *is countable* $\mathfrak{U}$*-slender.*

(5) G *is* $\mathfrak{U}$*-slender, compare Def. 1.2.*

(6) G *is* $\mathcal{K}$*-slender for an arbitrary class* $\mathcal{K}$ *containing a group* A *with* $A^* = \mathrm{Hom}(A,\mathbb{Z}) \neq 0$.

Proof: (1) → (2) → (3) is trivial.

(3) → (4). Let $\varphi : \prod_{\mathbb{N}} A_n \to G$ and assume that for all $k \in \mathbb{N}$ and all finite $E \subset \mathbb{N}$ so that $k\circ\varphi(\prod_{\mathbb{N}\setminus E} A_n) \neq 0$. Hence there is $a_k \in \prod_{\mathbb{N}\setminus E} A_n$ such that $k\circ\varphi(a_k) \neq 0$ where $E = \{1,\dots,k-1\}$. Therefore the infinite sum $\sum_{k\in\mathbb{N}} x(k)a_k$ is well-defined and let

$\psi: \mathbb{Z}^{\mathbb{N}} \to \prod A_n$ be defined by $\psi\left(\sum_{k\in\mathbb{N}} x(k)e_k\right) = \sum_{k\in\mathbb{N}} x(k)a_k$. Hence

$k(\varphi\circ\psi)(e_k) = k\varphi(a_k) \neq 0$ for all $k \in \mathbb{N}$. Since $\varphi\circ\psi: \mathbb{Z}^{\mathbb{N}} \to G$, this contradicts (3).

(4) → (5). Since $\Pi\mathfrak{U} = \mathfrak{U}$, Theorem 2.10 implies (5).

(5) → (6) is trivial.

(6) → (1). Since $A^* \neq 0$, the infinite cyclic group $\mathbb{Z}$ is in $Q\mathcal{K}$ and (2.3*)(a) implies (1).

Using Nunke's characterization [N, p.67] of slender groups, we derive the following characterization of $\mathbb{Z}$-slender groups.

THEOREM 3.2: For a group G *are equivalent*

(1) G *is* $\mathbb{Z}$*-slender.*

(2) (a) $\mathbb{Z}^{\mathbb{N}}$ *is not a subgroup of* G.
 (b) *If* $C \subseteq G$ *and* C *is cotorsion, then* C *is bounded.*

Proof. Apply R.J. Nunke [N, p.67, Theorem] and the fact that torsion cotorsion groups are direct sums of a divisible and a bounded group, cf. L. Fuchs [F2, vol.I, p.235, Corollary 54.4]. Therefore reduced torsion groups which are epimorphic images of the Baer-Specker group $\mathbb{Z}^{\mathbb{N}}$ are bounded. Hence (2) implies (1). The assertion (1) → (2) is trivial, if we apply some well-known facts on cotorsion groups, cf. [F1, §54]. It is very interesting to remark, that (2) → (1) follows already by a different argument given by Reinhold Baer [B, p.231] in 1958.

Similar to a result of E.C. Zeeman (cf.[F2, p.162, Corollary 94.6]) we have the

COROLLARY 3.3: If H *is* $\mathbb{Z}$*-slender, then*

$$\mathrm{Hom}(\prod_I G_i, H) \cong \sum_I \mathrm{Hom}(G_i, H) + U$$

for some torsion group U.

Proof. This follows from (3.1)(5).

COROLLARY 3.4: All groups $\prod_I G_i / \sum_I G_i$ *for* $|I| < \aleph_m$ *are Fuchs-44-groups.*

Proof. Let $\sigma : P = \prod_I G_i / \sum_I G_i \to T$ be a homomorphism into a reduced torsion group T. Then T is $\mathbb{Z}$-slender by (3.2). Let π be the canonical homomorphism from $\prod_I G_i$ onto P. Then $k\sigma\pi(\prod_{I \setminus E} G_i) = 0$ for some $k \in \mathbb{N}$ and some finite subset $E \subseteq I$ by (3.1)(5). Therefore $k \circ \sigma = 0$ and $\sigma(P)$ is bounded. Finally Theorem 1.1(4) implies, that P is a Fuchs-44-group.

§4 References

B R. Baer, Die Torsionsuntergruppe einer abelschen Gruppe, Mathematische Annalen 135(1979), 219-234

C S.U. Chase, On direct sums and products of modules, Pacific Journal of Math. 12(1962), 847-854

DZ M. Dugas and B. Zimmermann-Huisgen, Iterated direct Sums and Products of Modules, this volume

F1 L. Fuchs, Abelian groups, Publ. House of the Acad. Sci, Budapest 1958

F2 L. Fuchs, Infinite abelian groups, Academic Press, New York, Vol.1(1970), Vol.2(1973); R.uss. transl.: Л.Фукс, Бесконечные абелевы группы, том 1, Москва 1974, том 2, Москва 1977

G R. Göbel, On stout and slender groups, Journal of Algebra 35 (1975), 39-45

GW R. Göbel and B. Wald, Wachstumstypen und schlanke Gruppen, Sympos. Math 23(1979), 201-239

И1 А.В.Иванов, Об одной проблеме абелевых групп, Математический сборник, том 105 (147) вып. 4, (1978) 525 - 542
Engl. transl. A.V.Ivanov, A problem on abelian groups, Math. USSR Sbornik 34 (1978) 461 - 474

И2 А.В.Иванов, Прямые и полные прямые суммы абелевых групп, Вестник Московский университет, Сер. 1, Математика, Механика, Том 6, (1979) стр. 96

N R.J.Nunke, Slender groups, Acta Sci. Math., Szeged 23 (1962) 67 - 73

P1 С.В.Рычков, О прямых произведениях абелевых групп, Доклады Академий наук СССР, вып. 2, стр. 301 - 302 (1978)

P2 С.В.Рычков, О фактор-группе $\Pi A_\alpha / \Sigma A_\alpha$, Вестник Московский университет, Сер. 1, Математика, Механика том 6 (1979) 96

HOMOMORPHISMS BETWEEN CARTESIAN POWERS OF AN ABELIAN GROUP

Martin Huber[1] and R. B. Warfield, Jr.[2]

0. Introduction. In this paper we study the kernels and cokernels of homomorphisms between Cartesian powers $A^X \to A^Y$, where A is a torsion-free reduced Abelian group of finite rank, and X,Y are sets of non-measurable cardinality. From a general duality for such powers we deduce an exact sequence giving information about the cokernel of such a homomorphism. We obtain particularly strong results for groups A whose endomorphism ring is hereditary.

In Section 1 we first establish a duality between direct summands of powers A^X and projective left End(A)-modules where A is a torsion-free reduced group of finite rank (more generally, slender) and the cardinality of X is non-measurable (a standing hypothesis which will not be mentioned again in this introduction). This duality is then applied to investigate the cokernel C of a homomorphism $f: A^X \to A^Y$. It is shown that C fits into a four term exact sequence

$$\mathrm{Ext}^1_E(M,A) \rightarrowtail C \to C^{**} \twoheadrightarrow \mathrm{Ext}^2_E(M,A)$$

where $C^{**} = \mathrm{Hom}_E(\mathrm{Hom}_{\mathbb{Z}}(C,A),A)$, $E = \mathrm{End}(A)$, and M is a suitable left E-module, (Theorem 1.3). In the second section we show that if E is the endomorphism ring of a torsion-free reduced group of finite rank, and if E is left or right semihereditary, then E is left and right hereditary and left and right Noetherian. This gives us a great deal of in-

---

1. This work was done while the first author was a fellow of the "Schweizerische Nationalfonds"
2. Research of the second author supported in part by a grant from the (U.S.) National Science Foundation.

formation about groups with such endomorphism rings, (Theorem 2.4). The third section consists of results on extensions of E-modules, including some conditions guaranteeing that the Ext-terms in the above sequence are finite. The main technical result of this section is a variant of Stein's theorem which we use in the following section. The final section contains a number of results on Cartesian powers of torsion-free groups A of finite rank with hereditary endomorphism rings. Two samples are the following: (i) if C is the cokernel of $f: A^X \to A^Y$ then $C \cong P \oplus \mathrm{Ext}_E^1(M,A)$ for a suitable E-module M, where P is a product of groups of finite rank, each quasi-isomorphic to a summand of A; (ii) if $f: A^X \twoheadrightarrow A^Y$ is an epimorphism and Y is countable, then f splits. (If we add the set theoretic hypothesis that V = L, then this remains valid without the countability condition on Y.)

This work has its origins in Nunke's work [N2,N3] on products of infinite cyclic groups. Some of his ideas were carried further by Dugas and Göbel in [DG1] for products of infinite cyclic groups, by Huber in [H] for powers of a group of rank one, and again by Dugas and Göbel in [DG2] for modules and groups satisfying certain reflexivity conditions (satisfied, e.g., for powers of a group of finite rank whose endomorphism ring is a subring of the rationals). The purpose of this paper is to extend this study, and also to proselytize a certain point of view - that decomposition properties of torsion-free groups of finite rank can profitably be studied by looking at properties of their endomorphism rings and modules over these rings. We should point out that in general one has little knowledge about the behavior of arbitrary homomorphic images of powers or infinite products, with the exception of Nunke's work in [N3].

Most of our notation is standard. For a group A and a set X, we let A^X denote the product of copies of A indexed by X, and we let $A^{(X)}$ denote the corresponding direct sum. We use the arrows $\rightarrowtail$ and $\twoheadrightarrow$ for monomorphisms and epimorphisms, respectively.

1. Duality with respect to a slender group. In this section we establish a duality with respect to a slender group and prove the existence of the four term exact sequence mentioned in the introduction.

Let us first recall the definition of a slender group. To this end we denote by $\iota_n: \mathbb{Z} \to \mathbb{Z}^{\mathbb{N}}$ the injection of the nth factor of the power $\mathbb{Z}^{\mathbb{N}}$, $n \in \mathbb{N}$. An Abelian group A is termed slender if for every homomorphism $f: \mathbb{Z}^{\mathbb{N}} \to A$ the restriction to the nth factor, $f\iota_n: \mathbb{Z} \to A$, is zero for almost all n. Nunke [N3] proved that the slender groups are precisely those torsion-free groups which contain no copy of $\mathbb{Q}$, $\mathbb{Z}^{\mathbb{N}}$ or $\hat{\mathbb{Z}}_p$ (the p-adic integers) for any prime p. Thus, in particular, every torsion-free reduced group of finite rank is slender. If A is any Abelian group we let its endomorphism ring, End(A), act on the left. Thus for any group B, $\mathrm{Hom}_{\mathbb{Z}}(B,A)$ carries a natural left End(A)-module structure.

The following is a particular case of a deep result of Łoś (cf. [F, Theorem 94.4]).

Theorem 1.1. Let A be a slender group, let E = End(A) and let X be a set of non-measurable cardinality. Then $\mathrm{Hom}_{\mathbb{Z}}(A^X,A)$ is a free left E-module, and the set of coordinate projections $\{\pi_x \in \mathrm{Hom}_{\mathbb{Z}}(A^X,A) \mid x \in X\}$ is a basis.

We recall the definition and some basic facts about measurable cardinals. A cardinal κ is said to be measurable if it admits a countably additive measure ν such that ν assumes only values 0 and 1, and $\nu(\kappa) = 1$ and $\nu(\{i\}) = 0$ for all $i \in \kappa$. If there exist measurable cardinals at all, then there is a smallest one, say μ, and all larger cardinals are measurable. If μ exists then it is strongly inaccessible i.e., μ is a regular limit cardinal such that $\rho < \mu$ implies $2^\rho < \mu$. By a well-known result of Scott there are no measurable cardinals if one assumes Gödel's Axiom of Constructibility, V = L.

Let Ab denote the category of Abelian groups, A a fixed slender group with endomorphism ring E and ${}_E\underline{\mathrm{Mod}}$ the category of left E-modules. Let $H: \underline{\mathrm{Ab}} \to {}_E\underline{\mathrm{Mod}}$ be the contravariant functor $\mathrm{Hom}_{\mathbb{Z}}(-,A)$ and $H_E: {}_E\underline{\mathrm{Mod}} \to \underline{\mathrm{Ab}}$ the contravariant functor $\mathrm{Hom}_E(-,A)$. There are natural transformations

$$\sigma: 1_{\underline{\mathrm{Ab}}} \to H_E H \quad \text{and} \quad \tau: 1_{{}_E\underline{\mathrm{Mod}}} \to H H_E$$

where σ_B associates to each $b \in B$ the E-homomorphism taking $f \in \mathrm{Hom}_{\mathbb{Z}}(B,A)$ to $f(b)$, and τ_M is defined similarly. If there is no danger of confusion we shall simply write B^* for $H(B)$, f^* for $H(f)$, as well as M^* for $H_E(M)$ and φ^* for $H_E(\varphi)$. Instead of σ_B we shall sometimes write $\sigma[B]$. By $|X|$ we denote the cardinality of the set X.

Theorem 1.2. Let A be a slender group. Let $\underline{D}$ denote the category of direct summands of powers A^X with $|X|$ non-measurable, and let $\underline{P}$ be the category of projective left End(A)-modules admitting a set of generators of non-measurable cardinality. The functor H is a duality between the categories $\underline{D}$ and $\underline{P}$ with inverse H_E.

The essence of this theorem was known in the case that A is a torsion-free reduced group of rank one (cf. e.g. [Ch, Thèorème 7]).

Proof. It is clear that H_E sends projective left E-modules to direct summands of powers A^X, and by Theorem 1.1 H sends objects of $\underline{D}$ to objects of $\underline{P}$. So it remains only to prove that for $B \in \underline{D}$ (resp. for $M \in \underline{P}$) the map $\sigma_B: B \to B^{**}$ (resp. $\tau_M: M \to M^{**}$) is an isomorphism. We verify this for σ_B; the verification for τ_M is similar. Since both H and H_E preserve direct sums, it suffices to show that for non-measurable $|X|$, $\sigma[A^X]$ is an isomorphism. Of course, $\sigma[A^X]$ is injective. Let $\varphi: (A^X)^* \to A$ be any E-homomorphism and let $a = (\varphi(\pi_x)) \in A^X$ where $\pi_x: A^X \to A$ is the xth coordinate projection. Now if $f \in \mathrm{Hom}_{\mathbb{Z}}(A^X,A)$ then by Theorem 1.1 there are $f_x \in E$, $x \in X$, with $f_x = 0$ for almost all x and $f = \Sigma_{x\in X} f_x \pi_x$.

Therefore

$$\varphi(f) = \Sigma_{x \in X} f_x \varphi(\pi_x) = \Sigma_{x \in X} f_x \pi_x(a) = f(a).$$

Hence $\sigma[A^X]$ is surjective, and our proof is complete.

The following theorem is the principal result of this section.

<u>Theorem 1.3.</u> Let A be a slender group and let E = End(A). Let C be the cokernel of a homomorphism $f: A^X \to A^Y$ where $|X|$, $|Y|$ are non-measurable. Then the natural map $\sigma_C: C \to C^{**}$ fits into an exact sequence

$$\mathrm{Ext}^1_E(M,A) \rightarrowtail C \xrightarrow{\sigma_C} C^{**} \twoheadrightarrow \mathrm{Ext}^2_E(M,A)$$

where M is the cokernel of $f^*: H(A^Y) \to H(A^X)$. If, in particular, E is left hereditary then σ_C is a splitting epimorphism.

This theorem is an easy consequence of a result of Auslander's [Au, Prop. 5.8], once one knows Theorem 1.2. Nevertheless, we have chosen to include a complete (but elementary) proof because this is fundamental for all what follows.

<u>Proof of Theorem 1.3.</u> By hypothesis there is an exact sequence

$$A^X \xrightarrow{f} A^Y \twoheadrightarrow C.$$

By dualizing we obtain two exact sequences

(i) $C^* \rightarrowtail E^{(Y)} \twoheadrightarrow L$

and

(ii) $L \rightarrowtail E^{(X)} \twoheadrightarrow M$

where $L = \mathrm{Im}(f^*: (A^Y)^* \to (A^X)^*)$ and $M = \mathrm{Coker}\, f^*$. Dualizing sequence (i) again yields the bottom row of the following commutative diagram with exact rows:

$$\text{(iii)}\qquad \begin{array}{ccccccc} & & A^X & \xrightarrow{f} & A^Y & \twoheadrightarrow & C & & \\ & & & & \downarrow{\scriptstyle \sigma[A^Y]} & & \downarrow{\scriptstyle \sigma_C} & & \\ L^* & \rightarrowtail & & & (A^Y)^{**} & \longrightarrow & C^{**} & \twoheadrightarrow & \mathrm{Ext}^1_E(L,A). \end{array}$$

Since (by Theorem 1.2) $\sigma[A^Y]$ is an isomorphism we conclude that $\mathrm{Coker}\, \sigma_C \cong \mathrm{Ext}^1_E(L,A)$, and it follows from (ii) that $\mathrm{Ext}^1_E(L,A) \cong \mathrm{Ext}^2_E(M,A)$. To establish the four term exact sequence it thus remains to identify $\mathrm{Ker}\, \sigma_C$. Let $\theta: A^X \to L^*$ be the unique homomorphism making the left hand square of (iii) commute. By the "snake lemma" of Homological Algebra we have $\mathrm{Ker}\, \sigma_C \cong \mathrm{Coker}\, \theta$. Furthermore, by dualizing sequence (ii) we obtain the exact bottom row of the diagram

$$\begin{array}{ccccc} A^X & & & & \\ \downarrow{\scriptstyle \sigma[A^X]} & \searrow^{\theta} & & & \\ (A^X)^{**} & \longrightarrow & L^* & \twoheadrightarrow & \mathrm{Ext}^1_E(M,A) \end{array},$$

where the triangle commutes by uniqueness of θ. We infer that $\mathrm{Coker}\, \theta \cong \mathrm{Ext}^1_E(M,A)$, and hence $\mathrm{Ker}\, \sigma_C \cong \mathrm{Ext}^1_E(M,A)$, as desired.

Suppose now that E is left hereditary. In this case by (ii) L is a projective left E-module. Therefore sequence (i) and its dual sequence split. Hence we see from diagram (iii) that σ_C is a splitting epimorphism. This completes our proof.

The above theorem applies in particular if A is a torsion-free reduced group of finite rank. It is this case which seems to us of particular interest and on which we focus our attention in the remaining sections of the paper.

2. Modules over torsion-free rings. In this section we consider modules over rings which as Abelian groups are torsion-free and have finite rank. These rings have been studied extensively by Beaumont and Pierce [BP]. Following their terminology we call such a ring a torsion-free ring of finite rank. If A is a torsion-free group of finite rank then of course End(A) is a torsion-free ring of finite rank. The converse of this is less obvious but nevertheless true in the reduced case: Corner proved that if R is a torsion-free ring of rank n which as a group is reduced, then there is a torsion-free reduced group A of rank $\leq 2n$ such that End(A) $\cong$ R (cf. [F, Theorem 101.2]). There is an even more precise result, due to Zassenhaus [Z], in the case that R is free of finite rank (as an Abelian group). Then R is the endomorphism ring of a torsion-free reduced group of the same rank.

Lemma 2.1. Let E be a torsion-free ring of finite rank, let $A = E \otimes \mathbb{Q}$, let N be the radical of the Artinian ring A and $P = N \cap E$. Then (i) P is the prime radical of E; (ii) $(E/P) \otimes \mathbb{Q}$ is a semisimple $\mathbb{Q}$-algebra; and (iii) as an Abelian group, E/P is isomorphic to a subgroup of E.

Proof. All of this is contained in [BP, Theorem 1.4]. Part (iii) depends on the Beaumont-Pierce generalization of Wedderburn's Principal theorem. Parts (i) and (ii), however, are elementary consequences of the following remarks: P is nilpotent and therefore contained in the prime radical of E. If I is a nilpotent ideal of E then $I \otimes \mathbb{Q}$ is a nilpotent ideal of A, from which it follows that P is precisely the prime radical and E/P is semiprime. Finally, we easily identify A/N with $(E/P) \otimes \mathbb{Q}$.

We recall that a ring is semiprime if its prime radical is zero, or, equivalently, if it has no nilpotent ideals. The above lemma implies that a torsion-free ring E of finite rank which is semiprime has the property that $E \otimes \mathbb{Q}$ is a semisimple algebra. The next result contains stronger properties of these rings.

Theorem 2.2. Let E be a torsion-free ring of finite rank which is semiprime. Then (i) If I is an essential left ideal of E, then E/I is finite, and more generally if F is a finitely generated free E-module and F' an essential submodule then F/F' is finite; (ii) E is left and right Noetherian; and (iii) if M is a finitely generated left E-module and t(M) is the $\mathbb{Z}$-torsion subgroup of M, then t(M) is finite and M/t(M) is a submodule of a finitely generated free E-module.

(J. Reid informed us that statement (ii) is folklore.)

Proof. If I is a left ideal of E, then $I \otimes \mathbb{Q}$ is a left ideal of $E \otimes \mathbb{Q}$, and therefore a summand (since $E \otimes \mathbb{Q}$ is semisimple Artinian). If I is essential, then it follows immediately that $I \otimes \mathbb{Q} = E \otimes \mathbb{Q}$. It follows that I is essential in E as a subgroup - i.e. that for $x \in E$, there is a positive integer n with $nx \in I$. In particular, if 1 is the identity of E, then $n1 \in I$ and therefore $nE \leq I$ for some positive integer n. It follows that E/I is a homomorphic image of E/nE and is therefore finite. (We remark that for any torsion-free group A of finite rank and any positive integer n, A/nA is finite. To see this, note that the rank r of a torsion-free group can be described as the smallest integer r with the property that every finitely generated subgroup can be generated by r elements. This property is inherited by homomorphic images, so every finitely generated subgroup of A/nA can be generated by r elements, and thus has order at most n^r. This clearly implies that A/nA is finite, of order at most n^r.) The apparently stronger statement about free modules is clearly equivalent to what we have just proved.

For statement (ii), we remark first that if I is any left ideal, there is a left ideal J such that $I \cap J = 0$ and $I \oplus J$ is essential, and it suffices to show that $I \oplus J$ is finitely generated. From our previous remarks, we know that for some positive integer n, $n \in I \oplus J$ (where we identify n with n1), $(I \oplus J)/nE$ is finite, so $I \oplus J$ is generated by n and a finite number of additional elements.

We now pass to statement (iii). If M is a finitely generated left E-module, then t(M) is a fully invariant subgroup, and hence a submodule. Since E is Noetherian, t(M) is finitely generated. If $t(M) \cong F/F'$, where F is a finitely generated free module, then F' must be essential in F as a group, and therefore as a module. It follows from (i) that t(M) is finite. For the rest of (iii) we may assume that t(M) = 0 and show that M must be a submodule of a free module. Since $E \otimes \mathbb{Q}$ is semisimple Artinian, the $(E \otimes \mathbb{Q})$-module $M \otimes \mathbb{Q}$ must be projective, and thus there are monomorphisms $M \rightarrowtail M \otimes \mathbb{Q} \rightarrowtail E^n \otimes \mathbb{Q}$ for some positive integer n. If we regard E^n as an E-submodule of the E-module $E^n \otimes \mathbb{Q}$, then it suffices to show that every finitely generated E-submodule of $E^n \otimes \mathbb{Q}$ is isomorphic to a submodule of E^n. If A is the submodule and X is a set of generators of A, then since X is finite, there is a single positive integer k such that $kx \in E^n$ for each $x \in X$. Hence $A \cong kA \leq E^n$. This completes the proof of the theorem.

<u>Theorem 2.3.</u> Let E be a torsion-free ring of finite rank which is reduced (as an Abelian group) and left or right semihereditary. Then E is left and right hereditary and semiprime. Further, $E = E_1 \times E_2 \times \ldots \times E_n$ for suitable ideals E_i, such that $E_i \otimes \mathbb{Q}$ is simple Artinian and E_i is hereditary, Noetherian and prime. Finally, if P is any projective left E-module, then $P = \oplus_{x \in X} P_x$ where each P_x is a left ideal of E such that $P_x \otimes \mathbb{Q}$ is a simple module over the $\mathbb{Q}$-algebra $A = E \otimes \mathbb{Q}$.

<u>Proof.</u> Let N be the prime radical of E. If we can show that N = 0, then it will follow from Theorem 2.2 that E is left and right Noetherian, and hence by Small's theorem [S] that E is left and right hereditary. Suppose then that E is any torsion-free ring of finite rank, reduced as a group, with prime radical N, and suppose $N \neq 0$. We will find a finitely generated right ideal in E which is not projective,

thus showing that $N = 0$ if E is right semihereditary. Suppose $N^{m+1} = 0$, $N^m \neq 0$, $x \in N^m$, and $x \neq 0$. Since E is reduced as a group, there is an integer $k \neq 0$ such that $x \notin kE$. Let $I = kE + xE$. We claim that I is not projective. If I were projective then I/IN would be projective as an E/N-module and hence torsion-free as a group (by Lemma 2.1). However, $IN = kN$, so $x \notin IN$ but $kx \in IN$. Hence I/IN is not torsion-free, and I is not projective. This completes the proof of the first statement of the theorem.

The second statement is a standard remark about semihereditary rings which have a semisimple Artinian classical right quotient ring (cf. [Le]). In our case $E \otimes \mathbb{Q}$ $[E_i \otimes \mathbb{Q}]$ is a classical two-sided quotient ring of E [respectively E_i].

Finally, we consider a projective E-module P. It is well known that P is a direct sum of left ideals of E ([CE], p. 13). What we must show is that if I is an indecomposable left ideal, then $I \otimes \mathbb{Q}$ is a simple A-module. We suppose, then, that $I \otimes \mathbb{Q} = L \oplus M$, where L and M are A-modules, and let π be the projection of $L \oplus M$ onto M. Then $\pi(I)$ is torsion-free as a group, and hence a submodule of a free module (Theorem 2.2), and therefore projective. Assuming $M \neq 0$, it follows that $\pi(I) \cong I$, so (since I is essential in $L \oplus M$) it follows that $L = 0$. This shows that $I \otimes \mathbb{Q}$ is indecomposable, and therefore (since $A = E \otimes \mathbb{Q}$ is semisimple) $I \otimes \mathbb{Q}$ is a simple A-module.

<u>Theorem 2.4.</u> Let A be a reduced torsion-free group of finite rank and $E = \mathrm{End}(A)$, and assume that E is left or right semihereditary. Then A is a direct sum of fully invariant subgroups $A = A_1 \oplus \ldots \oplus A_n$, such that for each i, $\mathrm{End}(A_i)$ is an hereditary Noetherian prime ring. Let B_i be an indecomposable summand of A_i and let B be an object in $\underline{D}$, the category of direct summands of powers A^X with $|X|$ non-measurable. Then B is a direct product of indecomposable groups, there are up to isomorphism only a finite number of indecomposable groups in $\underline{D}$, and each

of these is quasi-isomorphic to B_i for some i, $1 \leq i \leq n$.

Proof. The first statement is a direct consequence of Theorem 2.3. For the second, we apply the duality of section 1 and see that every indecomposable summand of B is of the form P^* for some indecomposable projective E-module P, where $P^* = \mathrm{Hom}_E(P,A)$. We infer from Theorem 2.3 that P is finitely generated, and hence P is a direct summand of a finitely generated free E-module. It follows that P^* is a direct summand of a finite direct sum of copies of A. The finiteness statement now follows from Arnold's finiteness theorem [A, Theorem 3]. Using the product decomposition $E = E_1 \times \ldots \times E_n$ we can identify $A_i = E_i^*$ and $B_i = P_i^*$ where P_i is an indecomposable summand of E_i. Suppose that P is an indecomposable projective E-module. Then P is an E_i-module for some i, $1 \leq i \leq n$, and by Theorem 2.3 $P \otimes \mathbb{Q}$ is a simple $(E \otimes \mathbb{Q})$-module and thus a simple $(E_i \otimes \mathbb{Q})$-module. Since there is up to isomorphism only one simple module over a simple Artinian ring, it follows that $P \otimes \mathbb{Q} \cong P_i \otimes \mathbb{Q}$. So there are positive integers j and m such that $jP \leq P_i$ and $mP_i \leq P$. If $\iota: P \rightarrowtail P_i$ and $\mu: P_i \rightarrowtail P$ denote the corresponding embeddings, the composition $\mu\iota: P \to P$ is just multiplication by mj. Thus the induced map $(\mu\iota)^*: P^* \to P^*$ is likewise multiplication by mj, from which we infer that $\mu^*: P^* \to P_i^*$ is a monomorphism with finite cokernel. Hence P^* is quasi-isomorphic to $B_i = P_i^*$ as stated.

Example 2.5. An indecomposable torsion-free group A of finite rank such that End(A) is hereditary, and such that there are indecomposable groups B and C with $A \oplus A \cong B \oplus C$ but A is not a summand of a power of B. (In particular, A and B are not nearly isomorphic in the sense of Lady [L], so that in Theorem 2.4 we cannot improve "quasi-isomorphic" to say "nearly isomorphic".) In our example we will have $A = pA$ for all primes p, $p \neq 3$. We let $H = \mathbb{Z} \oplus \mathbb{Z}i \oplus \mathbb{Z}j \oplus \mathbb{Z}(1+i+j+k)/2$, the Hurwitz ring of integral quaternions, which is well known (e.g. [R, p. 229]) to be a left and right principal ideal domain. Let $H_{(3)}$ be its localization at the prime 3 and let $\varphi: H_{(3)} \to H_{(3)}/3H_{(3)}$ denote the natural map. Since $H_{(3)}/3H_{(3)}$ is a homomorphic image of the group ring of the quaternion group, the complete reducibility of the representations of this group over the field $\mathbb{Z}/3\mathbb{Z}$ shows that $H_{(3)}/3H_{(3)}$ is semi-simple, and since it is non-commutative and of dimension 4 over $\mathbb{Z}/3\mathbb{Z}$, it follows immediately that $H_{(3)}/3H_{(3)} \cong M_2(\mathbb{Z}/3\mathbb{Z})$. Let

$$E = \varphi^{-1}\left(\begin{bmatrix} \mathbb{Z}/3\mathbb{Z} & \mathbb{Z}/3\mathbb{Z} \\ 0 & \mathbb{Z}/3\mathbb{Z} \end{bmatrix}\right), \quad P = \varphi^{-1}\left(\begin{bmatrix} \mathbb{Z}/3\mathbb{Z} & \mathbb{Z}/3\mathbb{Z} \\ 0 & 0 \end{bmatrix}\right), \text{ and}$$

$$Q = \varphi^{-1}\left(\begin{bmatrix} 0 & \mathbb{Z}/3\mathbb{Z} \\ 0 & \mathbb{Z}/3\mathbb{Z} \end{bmatrix}\right).$$ Since E has no zero divisors, for every $x \in E$, $x \neq 0$, we have $Ex \cong E$, and therefore Ex has finite index in E (by Theorem 2.2). It follows that every left ideal of E is essential, of finite index in E, and indecomposable as a left module. Furthermore, E is hereditary (either by a complete localization, or because it is an idealizer [Ro], or by Jacobinski's theorem as in [R, Theorem 40.10]). We infer that P is an indecomposable projective E-module. We will show that $P^2 = P$, that E is not a summand of a direct sum of copies of P, but that P is a summand of $E \oplus E$.

We first note that $H_{(3)} = H_{(3)}P$. Since P is a right ideal in $H_{(3)}$, we have $P^2 = PH_{(3)}P = PH_{(3)} = P$. We next note that E/P is a simple E-module, but E/P is not a homomorphic image of a direct sum of copies of P, since if F is such a direct sum, then $PF = F$. Therefore, E cannot be a (left module) summand of a direct sum of copies of P.

It remains to show that P is a summand of $E \oplus E$. We let J be the Jacobson radical of E. We note that the Jacobson radical of $H_{(3)}$ is $3H_{(3)}$, from which it follows that $3H_{(3)} \leq J$. Since (as before) $H_{(3)}P = H_{(3)}$, it follows that $(3H_{(3)})P = 3H_{(3)}$, so $JP \geq 3H_{(3)}$. If T is the simple module E/Q, then it is routine to check that $P/3H_{(3)} \cong T \oplus T$. Therefore, $JP = 3H_{(3)}$ and $P/JP \cong T \oplus T$. In particular, it follows that P/JP is isomorphic to a summand of $(E \oplus E)/J(E \oplus E)$. Recall that if M and

N are finitely generated projective modules over any ring with Jacobson radical J, then M is a summand of N if and only if M/JM is a summand of N/JN. In our case, we conclude that P is a summand of $E \oplus E$, so we can write $E \oplus E \cong P \oplus L$, for some projective module L. (The reader who is interested may check that $L \cong Q$ and $J = P \cap Q$, but we will not need these facts.)

We now let A be a torsion-free group of finite rank with endomorphism ring E (which exists by Corner's theorem). Using the duality of section 1, we have $E^* \cong A$, and we let $B = P^*$, $C = L^*$. Then we obtain $A \oplus A \cong B \oplus C$, as desired, and yet A is not a summand of a power of B.

3. <u>Extensions of modules over endomorphism rings.</u> The four term exact sequence of Theorem 1.3 motivates the study of the groups $\mathrm{Ext}_E^k(M,A)$. We first consider the case where M is a finitely generated E-module.

<u>Theorem 3.1.</u> Let E be a torsion-free ring of finite rank which is semiprime, let M be a finitely generated left E-module and A a left E-module which as an Abelian group is torsion-free and of finite rank. Then for every positive integer k, $\mathrm{Ext}_E^k(M,A)$ is a finite Abelian group.

<u>Proof.</u> As before we let t(M) denote the torsion subgroup of M, and consider the short exact sequence $t(M) \rightarrowtail M \twoheadrightarrow M/t(M)$. From the corresponding long exact sequence of Ext, it follows that it will suffice to prove the result when $M = t(M)$ and when $t(M) = 0$. If $M = t(M)$, then M is finite (by Theorem 2.2) so there is an integer n, $n > 0$, such that $nM = 0$. It follows that $n\mathrm{Ext}_E^k(M,A) = 0$ for all k. Now we recall that the groups $\mathrm{Ext}_E^k(M,A)$ are computed by taking a projective resolution of M (all terms of which are torsion-free groups of finite rank, since M is finitely generated and E is Noetherian), taking the sequence of homomorphism groups of these modules into A (the results again being

torsion-free of finite rank) and then taking homology. The resulting Ext groups are therefore homomorphic images of torsion-free groups of finite rank, and the fact that they are bounded shows that they are finite.

We now pass to the case in which $t(M) = 0$, so by Theorem 2.2 M is a submodule of a finitely generated free module. Thus we can find a module N so that $M \oplus N$ is an essential submodule of a finitely generated free module F. Without loss of generality, we may therefore assume that M is essential in a free module F, and hence for some integer n, $nF \leq M$ by Theorem 2.2. We therefore look at the sequence $nF \rightarrowtail M \twoheadrightarrow M/nF$ in which the third term is finite. It follows from the long exact sequence of Ext that for every $k \geq 1$, $\mathrm{Ext}_E^k(M,A)$ is a homomorphic image of $\mathrm{Ext}_E^k(M/nF,A)$, which we have already shown is finite. This proves Theorem 3.1.

Corollary 3.2. Let A be a torsion-free reduced group of finite rank such that End(A) is semiprime, let $f: A^X \to A^Y$ be a homomorphism such that X is a finite set and $|Y|$ is non-measurable, and let $C = \mathrm{Coker}\, f$. Then the natural map $\sigma_C: C \to C^{**}$ has finite kernel and cokernel. In particular, t(C) is finite.

Example 3.3. The statement of this corollary may fail if End(A) is not semiprime, and if E is not semiprime then the conclusion of Theorem 3.1 may be false. We give an example of a torsion-free group of rank 2, whose endomorphism ring is not semiprime, and a homomorphism $f: A \to A$ such that if C is the cokernel of f, then the kernel and cokernel of the natural map $\sigma_C: C \to C^{**}$ are both isomorphic to $\mathbb{Z}(2^\infty)$. This is not only an example relevant to Corollary 3.2, but also (because of Theorem 1.3) to Theorem 3.1.

Let J be the subgroup of the additive group $\mathbb{Q}$ of rationals generated by the numbers $1/p$ for all primes p. Let $B = 2^{-\infty}J$ (the 2-divisible hull of J). According to [W2], $\mathrm{Ext}_{\mathbb{Z}}(J,B)$ is a torsion-free divisible group of uncountable rank. We choose a non-split extension $B \rightarrowtail A \twoheadrightarrow J$. The group A so constructed is the group we want.

We first note that A is strongly indecomposable (i.e., every subgroup of finite index is indecomposable). If A' is a subgroup of finite index, then $A' \cap B$ is a rank-one fully invariant subgroup of A' and hence must be a summand if A' is decomposable. If $A' = (A' \cap B) \oplus L$ then the natural map $A' \to J$ takes L isomorphically onto a subgroup of finite index in J. This contradicts the fact that the original extension is not quasi-split (since the element we chose in $\mathrm{Ext}_{\mathbb{Z}}(J,B)$ was not of finite order).

We next note that End(A) consists of all sums $n+f$, where n is an integer and $f\colon A \to A$ is an endomorphism with $B \leq \mathrm{Ker}\, f$ and $\mathrm{Im}\, f \leq B$. To see this, note first that B is a fully invariant subgroup and $\mathrm{End}(J) = \mathbb{Z}$ so we see that for any endomorphism g, there is an integer n such that $\mathrm{Im}(g-n) \leq B$. We need only to show that $(g-n)(B) = 0$. If not, then for some positive integer m, $(g-n)(B) = mB$. We let $A' = (g-n)^{-1}(mB)$, and note that A' is a subgroup of finite index in A. We then consider the homomorphism $\frac{1}{m}(g-n)\colon A' \to B$ and we note that this is surjective and its restriction to B is an automorphism. Composing this with the inverse of this automorphism, we obtain a projection of A' onto B, which is impossible since A' is indecomposable (as we showed in the previous paragraph). It follows that $(g-n)(B) = 0$ and we have established our statement about the endomorphism ring of A.

We now choose an endomorphism f of A such that $f(B) = 0$, $\mathrm{Im}\, f \leq B$ and $C = \mathrm{Coker}\, f \cong \mathbb{Z}(2^\infty) \oplus J$. We note that

$$C^* \cong \mathrm{Hom}_{\mathbb{Z}}(J, 2^{-\infty}J) \cong 2^{-\infty}\mathbb{Z}.$$

(Here we use the fact that A is strongly indecomposable, and hence every subgroup isomorphic to J is contained in B. We refer to [W1,

Prop. 2] for the computation of $\mathrm{Hom}_{\mathbb{Z}}(G,H)$ for arbitrary torsion-free groups G and H of rank one.)

If $E = \mathrm{End}(A)$, then our above structure result on E implies that every $\mathbb{Z}$-homomorphism: $C^* \to A$ is an E-homomorphism. Hence we can compute

$$C^{**} = \mathrm{Hom}_E(C^*,A) \cong \mathrm{Hom}_{\mathbb{Z}}(2^{-\infty}\mathbb{Z},B) \cong 2^{-\infty}J.$$

We therefore see that $C \cong \mathbb{Z}(2^\infty) \oplus J$ and $C^{**} \cong 2^{-\infty}J$, and since $\sigma_C \neq 0$, it is immediate that the kernel of σ_C is $\mathbb{Z}(2^\infty)$ while the cokernel is of the form $\mathbb{Z}(2^\infty) \oplus F$ where F is finite. In this case it is easy to compute that $F = 0$.

The rest of this section is devoted to the proof of the following result. The particular case $A = \mathbb{Z}$ is known as Stein's Theorem.

<u>Theorem 3.4.</u> Let A be a reduced torsion-free Abelian group of finite rank such that $E = \mathrm{End}(A)$ is left or right semihereditary. If M is a left E-module which as an Abelian group has countable rank, then $\mathrm{Ext}^1_E(M,A) = 0$ implies that M is projective.

The proof of this theorem is based on several lemmas. By Theorem 2.3 E is left and right hereditary, semiprime, and left and right Noetherian. We shall frequently use these facts without reference. If $k \in \mathbb{Z}$ and M is any E-module, we let $M[k] = \{m \in M \mid km = 0\}$, a fully invariant subgroup of M and hence an E-submodule. If M is torsion-free as an Abelian group and if L is a subgroup of M, we let $P_{\mathbb{Z}}(L)$ denote the pure closure of L in M; we note that if L is an E-submodule of M then so is $P_{\mathbb{Z}}(L)$.

<u>Lemma 3.5.</u> Let E be a torsion-free ring of finite rank which is reduced (as an Abelian group), and let S be a simple left E-module. Then (i) S is finite, (ii) there is a simple left E-module T such that $\mathrm{Ext}^1_E(S,T) \neq 0$, and (iii) there is a simple left E-module T' such that

$\mathrm{Ext}^1_E(T',S) \neq 0$.

Proof. If N is the prime radical of E, then Lemma 2.1 implies that E/N is torsion-free and reduced, so since every simple module is annihilated by N, we may assume without loss of generality that N = 0 and that E is semiprime. In this case, Theorem 2.2 implies that every finitely generated E-module which is torsion-free as a group is a submodule of a free module, and hence reduced. It follows that S cannot be torsion-free and divisible. We infer that for some prime p, pS = 0, so S is finite (again using Theorem 2.2).

If S = E/M, then we show that $\mathrm{Ext}^1_E(S, M/p^2E) \neq 0$ where p is the prime such that pS = 0. Since M/p^2E is finite, it will follow that for some simple composition factor T of M/p^2E, $\mathrm{Ext}^1_E(S,T) \neq 0$. We consider the sequence

$$M/p^2E \rightarrowtail E/p^2E \twoheadrightarrow S.$$

If this sequence splits, then E/p^2E has a summand annihilated by p. This, however, is impossible since for any torsion-free group B, B/p^2B is a direct sum of copies of $\mathbb{Z}/p^2\mathbb{Z}$.

For (iii), note that $S \cong p^2E/p^2M$. We therefore have a short exact sequence

$$S \rightarrowtail M/p^2M \twoheadrightarrow M/p^2E.$$

Just as before, we conclude that this sequence cannot split and therefore that for some simple module T', $\mathrm{Ext}^1_E(T',S) \neq 0$.

Lemma 3.6. Let A and E be as in Theorem 3.4. For any left E-module M, if $\mathrm{Ext}^1_E(M,A) = 0$ then M is torsion-free as a group.

Proof. Since E is left hereditary, it suffices to prove that $\mathrm{Ext}^1_E(M,A) \neq 0$ for every E-module M which is torsion as a group. By Theorem 2.2 (iii) M contains a simple E-module, say S. Thus we need only to show that $\mathrm{Ext}^1_E(S,A) \neq 0$. Now by the previous lemma there is a simple E-module T such that $\mathrm{Ext}^1_E(S,T) \neq 0$. Since T is finite, there is a prime p such that $pT = 0$, so if I is the annihilator ideal of T then $I \geq pE$. Therefore E/I is a finite primitive ring, and thus simple Artinian. It follows that every E/I-module is isomorphic to a direct sum of copies of T. Now by Theorem 1 of [AL] we have $A \neq IA$, so A/IA is a non-trivial E/I-module, from which we infer that A has a homomorphic image isomorphic to T. Therefore there is an epimorphism $\mathrm{Ext}^1_E(S,A) \twoheadrightarrow \mathrm{Ext}^1_E(S,T)$, hence as desired, $\mathrm{Ext}^1_E(S,A) \neq 0$.

Lemma 3.7. ("Pontryagin's Criterion"). Let E be a torsion-free ring of finite rank which is left or right semihereditary and reduced as a group. Let M be a left E-module which as a group has countable rank. If every E-submodule of M which as a group has finite rank is projective, then so is M.

Proof. Let M be represented as the union of an ascending chain of finitely generated E-submodules M_n, $n \in \mathbb{N}$. Let $\overline{M}_n = P_{\mathbb{Z}}(M_n)$. (We note that M is torsion-free as a group, so the definition of $\overline{M}_n$ makes sense.) Since for each n, M_n as a group has finite rank so does $\overline{M}_n$, and thus by hypothesis $\overline{M}_n$ is a projective E-module. Therefore $\overline{M}_n \cong \oplus_{x\in X} P_x$, where for each x, P_x is a finitely generated left ideal of E. Since $\overline{M}_n$ has finite rank, the index set X must be finite, hence $\overline{M}_n$ is finitely generated. We infer that $\overline{M}_{n+1}/\overline{M}_n$ is a finitely generated E-module and, since it is torsion-free as a group, Theorem 2.2 implies that $\overline{M}_{n+1}/\overline{M}_n$ is projective. Since this is true for each $n \in \mathbb{N}$, we conclude that M is a projective E-module, as required.

Lemma 3.8. Let R be any ring, let A be a left R-module and let $\{M_n\}$, $n \in \mathbb{N}$, be a direct system of left R-modules. Then the natural map

$$\theta:\ \mathrm{Ext}^1_R(\varinjlim M_n, A) \longrightarrow \varprojlim \mathrm{Ext}^1_R(M_n, A)$$

is an epimorphism.

Proof. By examination we see that the proof given in [N1] (Lemma 6.1) for R a Dedekind domain applies in general.

Proof of Theorem 3.4. By Lemmas 3.6 and 3.7 it suffices to consider the case that M as a group is torsion-free and of finite rank. Let F be a maximal free subgroup of M and let M_o denote the E-submodule generated by F which of course is finitely generated. It suffices to prove that $T = M/M_o$ is finitely generated, because then M is finitely generated and hence projective by Theorem 2.2. Since T is a torsion Abelian group we have $T = \oplus_p T_p$ where p ranges over the rational primes and T_p is the p-primary part, an E-submodule of T. Suppose that T is not finitely generated. Then either $T_p \neq 0$ for infinitely many primes p or T_p contains an element of infinite height for at least one p. In the first case, by Lemma 3.6 we have $\mathrm{Ext}^1_E(T_p, A) \neq 0$ for infinitely many p's, hence $\mathrm{Ext}^1_E(T,A)$ is uncountable. Considering the exact sequence

$$\mathrm{Hom}_E(M_o, A) \longrightarrow \mathrm{Ext}^1_E(T,A) \longrightarrow \mathrm{Ext}^1_E(M,A)$$

we see that $\mathrm{Ext}^1_E(M,A) \neq 0$ (in fact, uncountable), since $\mathrm{Hom}_E(M_o,A)$ is countable. Now suppose that for some prime p, T_p contains an element of infinite height. We note that T_p is the union of the ascending chain of E-submodules

$$T[p] \leq T[p^2] \leq \ldots \leq T[p^n] \leq \ldots$$

which in this case is strictly increasing. We consider the induced inverse system $\{\mathrm{Ext}^1_E(T[p^n],A)\}$, $n \in \mathbb{N}$, which consists of epimorphisms

$$\theta_n: \mathrm{Ext}^1_E(T[p^{n+1}],A) \twoheadrightarrow \mathrm{Ext}^1_E(T[p^n],A)$$

with $\mathrm{Ker}\ \theta_n \cong \mathrm{Ext}^1_E(T[p^{n+1}]/T[p^n],A)$. Since by Lemma 3.6 these kernels are non-zero, we conclude that $\varprojlim \mathrm{Ext}^1_E(T[p^n],A)$ is uncountable, and hence by Lemma 3.8 so is $\mathrm{Ext}^1_E(T_p,A)$. Thus $\mathrm{Ext}^1_E(T,A)$ is uncountable and as in the first case we show that $\mathrm{Ext}^1_E(M,A) \neq 0$, a contradiction. We infer that T is finitely generated, thus completing our proof.

Remark. Theorem 3.4 is still valid for uncountable M if we assume V = L (or the weaker hypothesis that every stationary subset of any regular uncountable cardinal is non-small). It is indeed not hard to verify that the methods developed by Shelah in order to solve Whitehead's Problem can also be applied in our more general situation. (See [E] for an excellent exposition of Shelah's work on Whitehead's Problem and for the set-theoretic terminology used here.)

4. Cartesian powers of a group with hereditary endomorphism ring.

In this final section we return to the study of homomorphisms $f: A^X \to A^Y$ where A is now a torsion-free reduced group of finite rank such that End(A) is (left and right) hereditary. We note that the theorems of Corner and Zassenhaus provide non-trivial examples of such groups A, for, e.g., every maximal order in a finite-dimensional semisimple $\mathbb{Q}$-algebra is hereditary (cf. [R, Theorem 21.4]).

The following theorem contains the main result of [H]. As before we let $\underline{D}$ denote the category of direct summands of Cartesian powers A^X where X is a set of non-measurable cardinality.

<u>Theorem 4.1.</u> Let A be a torsion-free reduced group of finite rank such that E = End(A) is left or right semihereditary. For a pair (C,K) of Abelian groups the following conditions are equivalent:

(1) There is a homomorphism $f: A^X \to A^Y$ with $|X|, |Y|$ non-measurable such that $C \cong \mathrm{Coker}\, f$ and $K \cong \mathrm{Ker}\, f$.

(2) There is a left E-module M and there are sets X_k, $1 \le k \le n$, all of non-measurable cardinality, such that

$$C \cong \mathrm{Ext}^1_E(M,A) \oplus C_1^{X_1} \times \ldots \times C_n^{X_n} \quad \text{and} \quad K \cong \mathrm{Hom}_E(M,A),$$

where $C_1,\ldots,C_n$ are the finitely many indecomposable groups in the category <u>D</u>.

<u>Proof.</u> (1) ⇒ (2). Let $M = \mathrm{Coker}(f^*: E^{(Y)} \to E^{(X)})$. Since E is in fact hereditary, Theorem 1.3 implies that $C \cong \mathrm{Ext}^1_E(M,A) \oplus C^{**}$, and by duality we have $K \cong \mathrm{Hom}_E(M,A)$. Now C^* is the kernel of f^* and hence projective. Therefore C^{**} is in <u>D</u> and thus of the form $C_1^{X_1} \times \ldots \times C_n^{X_n}$ by Theorem 2.4. This proves the first implication.

(2) ⇒ (1). Let $P_1 \overset{h}{\rightarrowtail} P_o \twoheadrightarrow M$ be a projective presentation of M such that $P_o = E^{(S)}$ for some S of non-measurable cardinality. Let P_2 be the dual of $C_1^{X_1} \times \ldots \times C_n^{X_n}$, a projective E-module (of non-measurable cardinality), and let $P = P_1 \oplus P_2$. Thus we obtain an exact sequence

$$P_2 \rightarrowtail P \xrightarrow{g'} E^{(S)} \twoheadrightarrow M$$

where $g' = h\pi$, π denoting the projection $P \twoheadrightarrow P_1$. Let now Y be a set of cardinality $|P|$ and let X be the disjoint union of S and Y. Then P is a direct summand of $E^{(Y)}$, and so is $P^{(\mathbb{N})}$, since $E^{(Y)}$ is isomorphic to an infinite direct sum of copies of itself. We conclude that $E^{(Y)} \oplus P \cong E^{(Y)}$. Therefore, defining $g: E^{(Y)} \oplus P \to E^{(X)}$ to be g' on P and the identity on $E^{(Y)}$, we obtain an exact sequence

$$P_2 \rightarrowtail E^{(Y)} \xrightarrow{g} E^{(X)} \twoheadrightarrow M.$$

If we let $f = g^*: A^X \to A^Y$ it is easy to verify, applying Theorem 1.3, that Coker f and Ker f are of the required form. This completes our proof.

Remarks. (1) In the following particular cases of Theorem 4.1 we can say a little more: If X is a countable set, then M can be chosen countable; if Y is countable, then M can be chosen $M = F \oplus M'$ where F is a free E-module and M' is countable. Since the obvious choice is $M = \mathrm{Coker}(f^*: E^{(Y)} \to E^{(X)})$ the first case is clear while the second statement holds because Im f^* is contained in a countable direct summand of $E^{(X)}$ with a free complement.

(2) If A is of rank 1 and M is a countable E-module then by [HW] $\mathrm{Ext}_E^1(M,A)$ admits a compact topology. This remains valid, more generally, in the case that M is countable and A is indecomposable, strongly homogeneous, and $\dim_{\mathbb{Z}/p\mathbb{Z}} A/pA \leq 1$ for every prime p. Note that for such A, E is a principal ideal domain and A is a torsion-free E-module of rank 1 (cf. [M]). Therefore, in this case, if $f: A^X \to A^Y$ is a homomorphism with X or Y countable then $C = \mathrm{Coker}\ f$ is of the form $C \cong T \oplus A^W$ where T admits a compact topology (cf. [H, Cor. 1]).

We now concentrate on the case where f is an epimorphism and discuss this situation in some detail.

Theorem 4.2. Let A be a torsion-free reduced group of finite rank such that $E = \mathrm{End}(A)$ is left or right semihereditary. Let $f: A^X \to A^Y$ be an epimorphism where $|X|$ is non-measurable and Y is countable. Then $K = \mathrm{Ker}\ f$ is a direct summand of A^X, and $K \cong \Pi_{k=1}^n C_k^{X_k}$ where $C_1,\dots,C_n$ are the finitely many indecomposable groups in $\underline{D}$.

<u>Proof.</u> Applying the four term exact sequence of Theorem 1.3 we obtain $\mathrm{Ext}^1_E(M,A) = 0$ with $M = \mathrm{Coker}(f^*: E^{(Y)} \to E^{(X)})$, where we note that f^* is a monomorphism. Now since Y is countable, by Remark (1) above $M = F \oplus M'$ where F is free and M' is countable. Then we have $\mathrm{Ext}^1_E(M',A)=0$, so by Theorem 3.4 M is projective. Hence f^* splits and so does f. Therefore K is a direct summand of A^X. So by Theorem 2.4 it is of the form $K \cong \Pi_{k=1}^n C_k^{X_k}$.

<u>Remark.</u> The same proof works for arbitrary Y (of non-measurable cardinality) if for any left E-module M, $\mathrm{Ext}^1_E(M,A) = 0$ implies M is projective. By the remark at the end of Section 3 the latter holds if V = L (or if every stationary subset of any regular uncountable cardinal is non-small). Note that in L, X and Y may be arbitrary sets.

If A is of rank 1 (or if E is any principal ideal domain) the group K in Theorem 4.2 is a power of A. In the general case this is still true under various conditions on X and Y.

<u>Corollary 4.3.</u> Let A,X,Y,f and K be as in Theorem 4.2. If either (i) X is uncountable or (ii) Y is finite and $|X| \geq |Y| + 2$, then $K \cong A^W$ for some set W of non-measurable cardinality.

<u>Proof.</u> By Theorem 4.2 we have $A^X \cong K \oplus A^Y$ where by hypothesis Y is countable. Suppose first that X is uncountable. By duality we obtain $E^{(X)} \cong K^* \oplus E^{(Y)}$ where $K^* = \mathrm{Hom}_{\mathbb{Z}}(K,A)$, a projective E-module. Clearly, in the terminology of [B] K^* is uniformly $|X|$-big. Thus by Theorem 2.2 of [B], $K^* \cong E^{(X)}$ and hence $K \cong A^X$. Now suppose that Y is finite, say $|Y| = n$, and $|X| \geq n + 2$. Choose a subset X' of X such that $A^X \cong A^{X'} \oplus A^{n+2}$; so we have $A^{X'} \oplus A^{n+2} \cong K \oplus A^n$. Now we infer from Theorem 5.6 of [W3] that $A^{X'} \oplus A^{n+1} \cong K \oplus A^{n-1}$, and iterating this argument we finally obtain $K \cong A^{X'} \oplus A^2$. This completes our proof.

If X,Y are both countably infinite sets, by the shifting argument used in the proof of 4.1 we see that K may be the dual of any countable projective E-module. It remains to consider the case where X and Y are both finite and $|X| = |Y| + 1$. In this case the structure of K depends on the simple components of the semisimple $\mathbb{Q}$-algebra $E \otimes \mathbb{Q}$. We refer to [R, p. 293] for the definition and properties of a totally definite quaternion algebra.

Corollary 4.4. (1) Let A and E be as in Theorem 4.2 and suppose that no simple component of $E \otimes \mathbb{Q}$ is a totally definite quaternion algebra (over its center). If $f: A^{n+1} \twoheadrightarrow A^n$ is an epimorphism, then Ker f is a direct summand of A^{n+1} isomorphic to A.

(2) There exists a torsion-free reduced group A of rank 16 with End(A) hereditary and an epimorphism $f: A^2 \twoheadrightarrow A$ such that Ker $f \not\cong A$.

Proof. (1) By the same argument as in the proof of Corollary 4.3 (ii) we obtain $A^2 \cong K \oplus A$ with K = Ker f. Hence by [A, Theorem 4] we have $K \cong A$.

(2) According to [Sw, Remark 4] there is a maximal R-order Λ in a division algebra D of $\mathbb{Q}$-dimension 16 (where R is the ring of integers of the center of D) such that $\Lambda^2 \cong \Lambda \oplus P$ for some non-principal left ideal P. As a maximal R-order, Λ is hereditary. Furthermore, Λ is $\mathbb{Z}$-free and thus, by Zassenhaus's theorem, Λ = End(A) for some torsion-free reduced group A of rank 16. Hence by duality there is a group $B \not\cong A$ such that $A \oplus B \cong A^2$.

In this section we have mainly been occupied with epimorphisms $f: A^X \twoheadrightarrow A^Y$. We conclude with a result concerning monomorphisms between powers of A.

Corollary 4.5. Let A be a torsion-free reduced group of finite rank such that End(A) is hereditary. Let X be a finite set, and let Y be a set of non-measurable cardinality. Suppose that $f: A^X \rightarrowtail A^Y$ is a monomorphism such that Im f is a pure subgroup of A^Y. Then Im f is a direct summand of A^Y.

<u>Proof.</u> By Theorem 1.3 we have $C = \text{Coker } f \cong \text{Ext}^1_E(M,A) \oplus C^{**}$ where M is the cokernel of $f^*: E^{(Y)} \to E^{(X)}$; thus M is finitely generated. Since by hypothesis Im f is pure in A^Y, we infer from Theorem 3.1 that $\text{Ext}^1_E(M,A)=0$. Hence by Theorem 3.4 M is projective. On the other hand, we have $M^* \cong \text{Ker } f = 0$, hence M itself must be zero. Therefore f^* is a splitting epimorphism, and hence f splits as well.

REFERENCES

[A] Arnold, D.: Genera and direct sum decompositions of torsion free modules, pp. 197 - 218 in: <u>Abelian Group Theory</u>, Lecture Notes in Math. 616, Springer-Verlag 1977.

[AL] Arnold, D. and Lady, E.: Endomorphism rings and direct sums of torsion-free Abelian groups, Trans. Amer. Math. Soc. 211 (1975), 225 - 237.

[Au] Auslander, M.: Coherent functors, pp. 189 - 231 in: Proceedings of a Conference on Categorical Algebra (La Jolla 1965), Springer-Verlag 1966.

[B] Bass, H.: Big projective modules are free, Illinois J. Math. 7 (1963), 24 - 31.

[BP] Beaumont, R. and Pierce, R.: Torsion-free rings, Illinois J. Math. 5 (1961), 61 - 98.

[CE] Cartan, H. and Eilenberg, S.: <u>Homological Algebra</u>, Princeton University Press, Princeton 1956.

[Ch] Charles, B.: Méthodes topologiques en théorie des groupes Abéliens, pp. 29 - 42 in: Proceedings of a Colloquium on Abelian Groups (Tihany 1963), Akadémiai Kiadó, Budapest 1964.

[DG1] Dugas, M. and Göbel, R.: Die Struktur kartesischer Produkte ganzer Zahlen modulo kartesische Produkte ganzer Zahlen, Math. Z. 168 (1979), 15 - 21.

[DG2] Dugas, M. and Göbel, R.: Quotients of reflexive modules, Fund. Math. (to appear)

[E] Eklof, P.: <u>Set-theoretic Methods in Homological Algebra and Abelian Groups</u>, Les Presses de l'Université de Montréal, Montréal 1980.

[F] Fuchs, L.: <u>Infinite Abelian Groups</u>, Vol. II, Academic Press, New York 1973.

[H] Huber, M.: On Cartesian powers of a rational group, Math. Z. 169 (1979), 253 - 259.

[HW] Huber, M. and Warfield, R.: On the torsion subgroup of Ext(A,G), Arch. Math. (Basel) 32 (1979), 5 - 9.

[L] Lady, E.: Nearly isomorphic torsion free Abelian groups, J. Algebra 35 (1975), 235 - 238.

[Le] Levy, L.: Torsion-free and divisible modules over non-integral domains, Can. J. Math. 15 (1963), 132 - 151.

[M] Murley, C.: The classification of certain classes of torsion-free Abelian groups, Pacific J. Math. 40 (1972), 647 - 665.

[N1] Nunke, R.: Modules of extensions over Dedekind rings, Illinois J. Math. 3 (1959), 222 - 241.

[N2] Nunke, R.: On direct products of infinite cyclic groups, Proc. Amer. Math. Soc. 13 (1962), 66 - 71.

[N3] Nunke, R.: Slender groups, Acta Sci. Math. Szeged 23 (1962), 67 - 73.

[R] Reiner, I.: Maximal Orders, Academic Press, London 1975.

[Ro] Robson, J.: Idealizers and hereditary Noetherian prime rings, J. Algebra 22 (1972), 45 - 81.

[S] Small, L.: Semihereditary rings, Bull. Amer. Math. Soc. 73 (1967), 656 - 658.

[Sw] Swan, R.: Projective modules over group rings and maximal orders, Ann. Math. 76 (1962), 55 - 61.

[W1] Warfield, R.: Homomorphisms and duality of torsion-free groups, Math. Z. 107 (1968), 189 - 200.

[W2] Warfield, R.: Extensions of torsion-free Abelian groups of finite rank, Arch. Math. (Basel) 23 (1972), 145 - 150.

[W3] Warfield, R.: Cancellation of modules and groups and stable range of endomorphism rings, Pacific J. Math. (to appear)

[Z] Zassenhaus, H.: Orders as endomorphism rings of modules of the same rank, J. London Math. Soc. 42 (1967), 180 - 182.

ALGEBRAIC COMPACTNESS OF FILTER QUOTIENTS

Berthold Franzen[1]

The remarkable result of S. Balcerzyk [1] that $\mathbb{Z}^{\mathbb{N}}/\mathbb{Z}^{(\mathbb{N})}$ is an algebraically compact group, was generalized by A. Hulanicki [8] to arbitrary quotients of the form $\prod_{n\in\mathbb{N}} G_n / \bigoplus_{n\in\mathbb{N}} G_n$. This theorem was set into the more general context of filter quotients by L. Fuchs [4]: Let φ be a filter on a set I (i.e. a system of subsets of I closed under forming finite intersections and supersets). For a family $(G_i)_{i\in I}$ of groups their φ-direct sum is denoted by $\Sigma_\varphi(G_i) = \{g \in \prod_{i\in I} G_i \mid z(g) := \{i\in I \mid g(i) = 0\} \in \varphi\}$. Fuchs proved the algebraic compactness of groups of the form $\Sigma_{\varphi^*}(G_i)/\Sigma_\varphi(G_i)$, where the filter arising from φ by adding countable intersections is denoted by φ^*.

On the other hand O. Gerstner [6] observed that $\mathbb{Z}^I/\mathbb{Z}^{(I)}$ is not algebraically compact, provided the index set I is uncountable. G. de Marco [9] generalized this to a theorem on filters: If for two filters $\varphi \subset \psi$ satisfying $\varphi^* \subsetneq \psi$ the filter quotient $\Sigma_\psi(G)/\Sigma_\varphi(G)$ is algebraically compact, then G is as well algebraically compact. M. Dugas and R. Göbel [2] gave a complete characterization of those filters φ,ψ such that the corresponding filter quotient is algebraically compact. Therefore they introduced the notion of purity to filters: φ is pure in ψ, if for every descending chain $(B_n)_{n\in\mathbb{N}}$ in ψ there are elements $(A_n)_{n\in\mathbb{N}}$ from φ and $X \in \psi$ such that $X \cap B_n = X \cap A_n$ for all n. Their results are the following:

(a) If the group G is not algebraically compact, then $\Sigma_\psi(G)/\Sigma_\varphi(G)$ is algebraically compact iff φ is pure in ψ and $\psi \subset \varphi^*$.

---

[1]Financial support for this paper was furnished by the Ministerium für Wissenschaft und Forschung des Landes Nordrhein-Westfalen under the title Überabzählbare abelsche Gruppen.

(b) If G is algebraically compact with unbounded reduced part, then $\Sigma_\psi(G)/\Sigma_\varphi(G)$ is algebraically compact iff φ is pure in ψ.

Their proofs are essentially based on the fact that algebraically compact groups are exactly those groups being complete in the $\mathbb{Z}$-adic topology and having divisible first Ulm subgroup. In the first section we give proofs for modules making no use of this characterization. Thus statements (a) and (b) can be generalized to the module case replacing "algebraically compact" by "$\aleph_0$-compact" and "with unbounded reduced part" by "not Σ-pure-injective". A module M is $\aleph_0$-compact, if every finitely solvable system of countable many equations possesses a global solution. M is Σ-pure-injective if every direct sum of arbitrarly many copies of M is pure-injective, i.e. algebraically compact.

In the second section we investigate the structure of filter quotients as regards abelian groups inspired by results of K. Golema and A. Hulanicki [7], who determined the structure of $\prod_{n\in\mathbb{N}} G_n / \bigoplus_{n\in\mathbb{N}} G_n$. U. Felgner [3] generalized this by proving their results for groups of the form $\prod_{i\in I} G_i / \Sigma_{\varphi_\kappa}(G_i)$, where $\kappa = \text{card } I$ and φ_κ is the filter of all subsets of I, whose complements have cardinality less than κ, provided κ has cofinality ω. In particular, groups of this form are compact. We give a characterization of those filters φ, such that $\Sigma_\psi(G_i)/\Sigma_\varphi(G_i)$ is always compact.

1. $\aleph_0$-compactness of modules .

For $m \in \prod_{i\in I} M_i$ its zero set is always denoted by $z(m) = \{i\in I \mid m(i)=0\}$.

<u>(1.1). Theorem:</u> For a family $(M_i)_{i\in I}$ of R-modules and filters $\varphi \subset \psi$ on I hold:

(a) $A = \Sigma_\psi(M_i)/\Sigma_\varphi(M_i)$ is $\aleph_0$-compact, if φ is pure in ψ and $\psi \subset \varphi^*$.

(b) If all M_i are $\aleph_0$-compact and φ is pure in ψ, then again $A = \Sigma_\psi(M_i)/\Sigma_\varphi(M_i)$ is $\aleph_0$-compact.

<u>Remark:</u> Note that φ is always pure in φ^* and therefore by (a) the corresponding filter quotient is always $\aleph_0$-compact.

<u>Proof:</u> (a) We have to establish a solution in A for a system of equations

$$(1) \qquad \sum_{k\in K} r_{jk} x_k = \bar{a}_j \qquad (\bar{a}_j \in A,\ j \in \mathbb{N})$$

under the hypothesis that every finite subsystem of (1) admits a solution in A. This gives us solutions $\bar{m}(k,n) \in A$ $(k \in K)$ for the finite subsystem $(1)_n$ of the first n equations of (1). We choose representatives $m(k,n) \in \bar{m}(k,n)$ and $a_j \in \bar{a}_j$ with $z(m(k,n)) \in \psi$ and $z(a_j) \in \psi$. The $m(k,n)$ can be choosen in such a way that $m(k,n) = 0$ for $k \notin K_n = \{k \in K \mid r_{jk} \neq 0 \text{ for some } j \leq n\}$. This implies for all $n \in \mathbb{N}$ $\bigcap_{k\in K} z(m(k,n)) = \bigcap_{k\in K_n} z(m(k,n))$. Clearly K_n is a finite set, therefore all $B_n = \bigcap_{j=1}^{n} \bigcap_{k\in K} z(m(k,n))$ belong to ψ. By definition $(B_n)_{n\in\mathbb{N}}$ is a descending chain in ψ, which, by purity of φ in ψ, assures the existence of some $X \in \psi$ and $C_n \in \varphi$ $(n \in \mathbb{N})$ such that

$$(2) \qquad X \cap B_n = X \cap C_n$$

holds for all n. The C_n $(n \in \mathbb{N})$ can be choosen to be descending.

Since the system $(1)_n$ is solved by $\bar{m}(k,n)$ $(k \in K)$, the sets $z(\sum_{k\in K} r_{jk}\, m(k,n) - a_j)$ belong to φ for $j = 1,\dots,n$ and therefore a descending chain $(A_n)_{n\in\mathbb{N}}$ in φ is defined by

$$A_n = \bigcap_{l=1}^{n} \bigcap_{j=1}^{l} z(\sum_{k\in K} r_{jk}\, m(k,l) - a_j).$$

Moreover $\bigcap_{j\in\mathbb{N}} z(a_j) \in \psi^* \subset \varphi^*$, since ψ is contained in φ^* by assumption. Hence there is a descending chain $(E_j)_{j\in\mathbb{N}}$ in φ such that $\bigcap_{j\in\mathbb{N}} z(a_j) = \bigcap_{j\in\mathbb{N}} E_j$.

In order to give a solution of (1) we define by $W_n = A_n \cap C_n \cap E_n$ a descending chain $(W_n)_{n\in\mathbb{N}}$ in φ and hence the setting

$$\bar{m}_k(i) = \begin{cases} m(k,n)(i) & \text{if } i \in W_n \setminus W_{n+1} \text{ for some } n \\ 0 & \text{else} \end{cases}$$

is well defined. Now $\bar{m}_k$ belongs to A, since $X \subset z(m_k)$. For $i \in X$ either $m_k(i) = 0$ or there is some n such that $m_k(i) = m(k,n)(i)$. In the second case (2) implies $i \in X \cap W_n \subset X \cap C_n = X \cap B_n \subset z(m(k,n)$, hence in both cases i belongs to $z(m_k)$. It remains to show that $m_k (k \in K)$ actually solve the system (1). But this is an immediate consequence of $W_j \subset z(\sum_{k \in K} r_{jk} m_k - a_j)$. Suppose $i \in W_j$. Then either $i \in \bigcap_{n \in \mathbb{N}} W_n$ or $i \in W_l \setminus W_{l+1}$ for some $l \geq j$. In the first case we have $m_k(i) = 0$ for all $k \in K$ as well $a_j(i) = 0$ for all $j \in \mathbb{N}$. Hence $i \in z(\sum_{k \in K} r_{jk} m_k - a_j)$. In the second case we have $m_k(i) = m(k,l)(i)$ and with $i \in W_l \subset A_l \subset z(\sum_{k \in K} r_{jk} m(k,l) - a_j)$ the claim is proved.

(b) Let (1) as in (a) be a finitely solvable system. Then again there are solutions $\bar{m}(k,n) (k \in K)$ of the finite subsystems $(1)_n$. Again A_n defined as in (a), belong to φ and form a descending chain. For $i \in Y = \bigcap_{n \in \mathbb{N}} A_n$ the system

$$(3)_i \qquad \sum_{k \in K} r_{jk} x_k = a_j(i) \quad (j \in \mathbb{N})$$

is finitely solvable in M_i. By assumption all M_i are $\aleph_0$-compact. This assures the existence of certain $m_{k,i} \in M_i$ $(k \in K)$ satisfying $(3)_i$ for $i \in Y$. By $B_n = \bigcap_{j=1}^{n} z(a_j) \cap A_n \in \psi$ we get a descending chain in ψ. Because φ is pure in ψ there are $X \in \psi$ and $C_n \in \varphi$ such that $X \cap B_n = X \cap C_n$ for all n. Now we define

$$m_k(i) = \begin{cases} m(k,n)(i) & \text{if } i \in (A_n \setminus A_{n+1}) \cap X^c \text{ for some } n \\ m_{k,i} & \text{if } i \in Y \cap X^c \\ 0 & \text{else} \end{cases}$$

Thus $X \subset z(m_k)$ holds and hence $\bar{m}_k \in A$. In order to show that $\bar{m}_k$ $(k \in K)$ satisfy (1), we have to prove $A_j \cap C_j \subset T = z(\sum_{k \in K} r_{jk} m_k - a_j)$ for all $j \in \mathbb{N}$. Therefore assume $i \in A_j \cap C_j$. We consider the following cases:

(i) $i \in X$: Then $X \cap C_j = X \cap B_j \subset z(a_j)$ implies $a_j(i) = 0 = m_k(i)$ for all k and therefore $i \in T$.

(ii) $i \in (A_n \setminus A_{n+1}) \cap X^c$ for some $n \geq j$: Then $m_k(i) = m(k,n)(i)$ and $A_n \subset z(\sum_{k\in K} r_{jk} m(k,n) - a_j)$ imply $i \in T$.

(ij) $i \in Y \cap X^c$: $i \in T$ follows immediately by the choice of $m_{k,i}$.

All cases are treated and the solvability of (1) in A is proved.

(1.2). Lemma : Suppose $\varphi \subset \psi$ to be filters on an index set I and M a not $\aleph_0$-compact module. Then $\aleph_0$-compactness of $A = \Sigma_\psi(M) / \Sigma_\varphi(M)$ implies $\psi \subset \varphi^*$.

Proof : By assumption there is a finitely solvable system

$$(1) \qquad \sum_{k\in K} r_{jk} x_k = m_j \qquad (m_j \in M;\ j \in \mathbb{N})$$

without a global solution in M. Let $W \in \psi$. We have to show $W \in \varphi^*$. We define $a_j(i) = m_j$ if $i \in W^c$ and $a_j(i) = 0$ if $i \in W$ for all $j \in \mathbb{N}$. $W \in \psi$ implies $\bar{a}_j \in A$. Clearly the system

$$(2) \qquad \sum_{k\in K} r_{jk} x_k = \bar{a}_j \qquad (j \in \mathbb{N})$$

is finitely solvable in A which gives solutions $\bar{m}_k \in A$ $(k \in K)$ satisfying (2) by the $\aleph_0$-compactness of A. Choosing representatives $m_k \in \bar{m}_k$ we obtain $W_j = z(\sum_{k\in K} r_{jk} m_k - a_j) \in \varphi$. The existence of an $i \in \bigcap_{j\in\mathbb{N}} W_j \setminus W$ gives a global solution of (1) due to the validity of the equations $\sum_{k\in K} r_{jk} m_k(i) = a_j(i) = m_j (j \in \mathbb{N})$. This would be a contradiction to our assumption and hence $W \in \varphi^*$ holds.

(1.3). Definition (W. Zimmermann) : Let M be a R-module, $A = (r_{jk})_{j\in J, k\in K}$ a row and column finite matrix with coefficients $r_{jk} \in R$, $l \in K$ and $K' = K \setminus \{l\}$. Then the subgroup $[A,l]M$ of M consists of those elements $x_l \in M$ for which there exists a $(x_k)_{k\in K'} \in M^{K'}$ such that $\sum_{k\in K} r_{jk} x_k = 0$ holds for all $j \in J$, called a matrix subgroup of M (relative to $[A,l]$).

(1.4). Proposition (W. Zimmermann) : For a module M is equivalent:

(a) M fulfils the minimum condition for matrix subgroups.

(b) M is Σ-pure-injective.

(c) M is Σ_φ-pure-injective, i.e. $\Sigma_\varphi(M)$ is pure-injective for all filters φ on an arbitrary index set I.

Proof : The equivalence of (a) and (b) was observed by W. Zimmermann [10], Folgerung 3.4. Assuming (a) $\Sigma_\varphi(M)$ as well fulfils the minimum condition on matrix subgroups by the following lemma (1.5)(a) and is hence pure-injective. (b) follows from (c) by choosing the filter of cofinite subsets.

(1.5). Lemma : Let M be a R-module.

(a) For a matrix $(r_{jk})_{j\in J,k\in K}$ over $R(J,K$ finite) and $l \in K$ we have:

(i) ${[A,l]M+N}/{N} \subset [A,l]({M}/{N})$ for every submodule N. Equality holds, if N is a pure submodule.

(ii) $[A,l](\Sigma_\varphi(M)) = \Sigma_\varphi([A,l]M)$ for all filters φ, i.e. Σ_φ commutes with $[A,l]$.

(b) Let $\varphi \subset \psi$ be filters on I, $A^s = (r^s_{jk})_{j\in J_s,k\in K_s}$ row and column finite matrices and $l_s \in K_s$ $(s = 1,2)$. Then for $\bar{M} = {\Sigma_\psi(M)}/{\Sigma_\varphi(M)}$ holds: $[A^1,l_1]M \subset [A^2,l_2]M$ implies $[A^1,l_1]\bar{M} \subset [A^2,l_2]\bar{M}$.

Proof : (a)(i) For $x_l \in [A,l]M$ there exist $x_k \in M (k \in K')$ such that $\sum_{k\in K} r_{jk}x_k = 0$ holds for all $j \in J$. Modding by N we obtain $x_l + N \in [A,l]({M}/{N})$. Suppose now N is a pure submodule and $x_l + N \in [A,l]({M}/{N})$. Then there are $x_k + N \in {M}/{N}$ $(k \in K')$ such that $\sum_{k\in K} r_{jk}x_k = n_j$ for certain $n_j \in N (j \in J)$. Since N is pure in M and J is finite, we get elements $z_k \in N (k \in K)$ satisfying the above system.

By setting $y_k = x_k - z_k$ we obtain $y_1 \in [A,1]M$ and hence $x_1 + N = y_1 + N \in {}^{[A,1]M + N}/_{N}$.

To show (ii) let $x_1 \in [A,1](\Sigma_\varphi(M))$. Then $x_1(i) \in [A,1]M$ for all $i \in I$ and hence $x_1 \in \Sigma_\varphi([A,1]M)$. The other inclusion follows in the same way.

(b) Clearly $\Sigma_\varphi(M)$ is a pure submodule of $\Sigma_\psi(M)$. Hence by (a)(i) and (ii) $[A^s,1_s]M = {}^{\Sigma_\psi([A^s,1_s]M) + \Sigma_\varphi(M)}/_{\Sigma_\varphi(M)}$ for $s = 1,2$. The assertion is now obvious.

<u>(1.6). Proposition :</u> Let M be a $\aleph_0$-compact module. Then for any countable family $(M_j)_{j\in J}$ of matrix subgroups and elements $m_j \in M (j \in J)$ the finite intersection property of the system $(m_j + M_j)_{j\in J}$ implies that the total intersection is nonempty.

<u>Proof:</u> W. Zimmermann [10], Satz 2.1, (2) ⇒ (3), proved this for algebraically compact M and arbitrary index set J. His proof holds for this case almost verbatim.

<u>(1.7). Lemma:</u> Let M be a module which does not satisfy the minimum condition for matrix subgroups, $\varphi \subset \psi$ filters on an index set I. If $A = {}^{\Sigma_\psi(M)}/_{\Sigma_\varphi(M)}$ is $\aleph_0$-compact, then φ is pure in ψ.

<u>Proof:</u> By assumption there is a properly descending chain of matrix subgroups $M_n = [A^n,1_n]M$ $(n \in \mathbb{N})$ of M. Pick $m_n \in M_n \setminus M_{n+1}$ for every n. Further let $(B_n)_{n\in\mathbb{N}}$ be a descending chain in ψ with $B_1 = I$. Setting

$$a_n(i) = \begin{cases} m_s & \text{if } i \in B_s \setminus B_{s+1} \text{ for some } s \leq n \\ 0 & \text{else} \end{cases}$$

and $\bar{a}_n = a_n + \Sigma_\varphi(M)$ we obtain $\bar{a}_n \in A$ because of $B_{n+1} = z(a_n)$. By $U_n = [A^n,1_n]A$ we will show that $(\bar{a}_n + U_n)_{n\in\mathbb{N}}$ is a descending chain. By construction we have $a_n(i) - a_{n-1}(i) \in M_n$ for all $i \in I$ and

$a_n - a_{n-1} \in \Sigma_\psi(M_n) = [A^n, 1_n](\Sigma_\psi(M))$ by lemma (1.5)(a)(i). Part (a)(ii) of the same lemma gives $\bar{a}_n - \bar{a}_{n-1} \in U_n$ and therefore $\bar{a}_n + U_n = \bar{a}_{n-1} + U_n$. Part (b) of this lemma and $M_n \subset M_{n-1}$ implies $U_n \subset U_{n-1}$. Consequently the system $(\bar{a}_n + U_n)_{n\in\mathbb{N}}$ has the finite intersection property ensuring the existence of an element $x + \Sigma_\varphi(M) \in \bigcap_{n\in\mathbb{N}} \bar{a}_n + U_n$ by proposition (1.6), since A is $\aleph_0$-compact. Hence we get elements $\bar{u}_n = u_n + \Sigma_\varphi(M) \in U_n$ satisfying $x - a_n - u_n \in \Sigma_\varphi(M)$. By $\bar{u}_n \in U_n$ the set $C_n = \{i \in I \mid u_n(i) \in M_n\}$ belongs to φ as well as $A_n' = z(x - a_n - u_n) \cap C_n$. Now we get $z(x) \cap A_n' \subset z(x) \cap B_n$ for all n, since $a_n(i) = -u_n(i) \in M_n$ holds for $i \in z(x) \cap A_n'$, what is possible only if $i \in B_n$. Setting $X = z(x)$ and $A_n = A_n' \cup B_n \in \varphi$ the equality $X \cap B_n = X \cap A_n$ holds for all n. Hence φ is pure in ψ.

<u>(1.8). Theorem :</u> Let M be a module and $\varphi \subset \psi$ filters on an index set I.

(a) If M is not $\aleph_0$-compact, the following is equivalent:

(i) $\Sigma_\psi(M)/\Sigma_\varphi(M)$ is $\aleph_0$-compact.

(ii) φ is pure in ψ and $\psi \subset \varphi^*$.

(b) If M is $\aleph_0$-compact, but not Σ-pure-injective, the following are equivalent:

(i) $\Sigma_\psi(M)/\Sigma_\varphi(M)$ is $\aleph_0$-compact.

(ii) φ is pure in ψ.

<u>Proof:</u> (a) Suppose (i). Since M is not $\aleph_0$-compact, it is not Σ-pure-injective and does not satisfy the minimum condition for matrix subgroups by proposition (1.4). Lemma (1.7) gives the purity of φ in ψ. Lemma (1.2) delivers the other filter condition of (ii). The opposite direction is a consequence of theorem (1.1)(a).

(b) (i) implies (ii) by lemma (1.7) and proposition (1.4). The other direction is contained in theorem (1.1)(b).

Remark: Note that for a Σ-pure-injective module M every filter quotient is even algebraically compact, since Σ-pure-injectivity and Σ_φ-pure-injectivity are the same by proposition (1.4) and $\Sigma_\varphi(M)$ is a pure submodule of $\Sigma_\psi(M)$.

2. Compactness of filter quotients.

In this section we restrict to abelian groups. The structure of the algebraically compact groups is well-known, cf. L. Fuchs [5],(23.1), (40.1) and (40.2). An algebraically compact group A is of the form

$$A = \prod_p \widehat{\left(\bigoplus_{n\in\mathbb{N}} Z(p^n)^{(\alpha_{p,n})} \oplus J_p^{(\beta_p)} \right)} \oplus \bigoplus_p Z(p^\infty)^{(\gamma_p)} \oplus \mathbb{Q}^{(\delta)}$$

where $\widehat{\ }$ means p-adic completion. These cardinal number invariants determine A up to isomorphism. We characterize those filters φ, ψ on an index set I of cardinality κ, for which the corresponding filter quotient is always algebraically compact and the cardinal number invariants ore either 0 or 2^κ. Further they fulfil some inequalities such that the filter quotients admit a compact topology.

For a filter φ on an index set I there is a natural equivalence relation on P(I) : Two subsets X,Y of I are φ-equivalent, if the complement of their symmetric difference $(X \Delta Y)^c$ belongs to φ. According to this we define the index $[\psi:\varphi]$ of φ in another filter ψ to be the number of different representatives in ψ. Further for $T \subset P(I)$ let $\langle T\rangle$ always denote the smallest filter on I containing T.

Definition : Let $\varphi \subset \psi$ be filters on I, card I = κ. Then ψ is big over φ if for every descending chain $(X_n)_{n\in\mathbb{N}}$ in P(I) satisfying $\psi \not\subset \langle \varphi \cup \{X_n\}\rangle$ for all n, we have $[\psi:\langle \varphi \cup \{X_n \mid n \in \mathbb{N}\}\rangle] = 2^\kappa$.

Example : P(I) is big over φ_κ. This remains true for filters obtained from φ_κ by adding any countable descending chain of subsets of I of cardinality κ. Hence there are for every cardinal filters satisfying the conditions in (a) of theorem (2.2).

An immediate consequence of the definition is that filters $\varphi \subset \psi$, ψ being big over φ, satisfy $[\langle X \rangle : \varphi] = 2^\kappa$ for every $X \notin \varphi$.

<u>(2.1) Lemma :</u> Let ψ be big over φ on an index set I of cardinality κ, $(G_i)_{i\in I}$ a family of groups and $A = \Sigma_\psi(G_i) / \Sigma_\varphi(G_i)$. Then

(a) card $p^{n-1}A[p] / p^n A[p]$ is 1 or not less than 2^κ.

(b) Suppose $\psi \subset \varphi^*$ and $A \neq A[p^n] + pA$ for all n, then card $A / t(A) + pA \geq 2^\kappa$.

(c) If $D = \bigcap_{n\in\mathbb{N}} nA$ is the divisible part of A, then its cardinal number invariants satisfy :

(i) $\gamma_p = 0$ or $\gamma_p \geq 2^\kappa$.

(ii) $\delta \geq 2^\kappa$, provided A is not reduced and $\psi \subset \varphi^*$.

<u>Proof:</u> (a) Assume there is an element $\bar{a} = a + \Sigma_\varphi(G_i) \in p^{n-1} A[p] \setminus p^n A[p]$. Setting

$$b(i) = \begin{cases} a(i) & \text{if } a(i) \in p^n G_i[p] \\ 0 & \text{else} \end{cases}$$

we obtain $\bar{b} \in p^{n-1}A[p] \setminus p^n A[p]$ and especially $z(b) \in \psi \setminus \varphi$. Since ψ is big over φ there is a system $T \subset \langle \{z(b)\} \rangle$ of different representatives relative to φ-equivalence with card $T = 2^\kappa$. Defining

$$a_V(i) = \begin{cases} b(i) & \text{if } i \in V^c \\ 0 & \text{else} \end{cases}$$

for $V \in T$ we have $z(b) \subset V = z(a_V)$ and hence $\bar{a}_V \in A$. A simple calculation shows $z(a_V - a_W) = (V \Delta W)^c$ for all $V, W \in T$. Supposing $a_V \equiv a_W \bmod p^n A[p]$ we get an element $M \in \varphi$ such that $a_V(i) - a_W(i) \in p^n G_i[p]$ for all $i \in M$. By construction of b this is only possible if $a_V(i) - a_W(i) = 0$ and therefore $M \subset (V \Delta W)^c$ holds. Consequently $(V \Delta W)^c \in \varphi$ implies $V = W$, since V, W were from T. We obtain card $p^{n-1}A[p] / p^n A[p] \geq$ card $\{\bar{a}_V \mid V \in T\} = 2^\kappa$.

(b) $X_n = \{i \in I \mid G_i \neq G_i[p^n] + pG_i\}$ is a descending chain in $P(I)$. Now $\psi \subset \langle \varphi \cup \{X_l\}\rangle$ for some l implies $A = A[p^l] + pA$, because for every $\bar{a} \in A$ there is some $M \in \varphi$ such that $z(a) \supset M \cap X_l$, which clearly implies $M \subset \{i \in I \mid a(i) \in G_i[p^l] + pG_i\}$ and therefore $\bar{a} \in A[p^l] + pA$. Since ψ is big over φ we get by assumption a system $T \subset \psi$ of size 2^κ, whose elements are pairwise different under $\langle \varphi \cup \{X_n \mid n \in \mathbb{N}\}\rangle$-equivalence. Since $T \subset \psi \subset \varphi^*$ holds we obtain a descending chain $(M_n^{(V)})_{n\in\mathbb{N}}$ in φ satisfying $M_1^{(V)} = I$ and $\bigcap_{n\in\mathbb{N}} M_n^{(V)} = V$ for every $V \in T$. Picking some $g_{i,n} \in G_i \setminus G_i[p^n] + pG_i$ for every $i \in X_n$ we set for $V \in T$

$$a_V(i) = \begin{cases} g_{i,n} & \text{if } i \in (X_n \setminus X_{n+1}) \setminus V \text{ or } i \in \bigcap_{l\in\mathbb{N}} X_l \cap (M_n^{(V)} \setminus M_{n+1}^{(V)}) \\ 0 & \text{else} \end{cases} .$$

Clearly $z(a_V) = V \cup X_1^c \in \psi$ and thus $\bar{a}_V \in A$. We have to show that the $\bar{a}_V (V \in T)$ are pairwise different mod $t(A) + pA$. So let us assume $\bar{a}_V - \bar{a}_W \in A[p^n] + pA$ for some n, which gives a set $L \in \varphi$ such that $a_V(i) - a_W(i) \in G_i[p^n] + pG_i$ for all $i \in L$. Now $L \cap (V \triangle W) \cap X_n = \emptyset$, since $i \in (V \triangle W) \cap X_n$ implies $a_V(i) - a_W(i) = g_{i,l} \in G_i[p^n] + pG_i$ for some $l \geq n$. Hence we have $(V \triangle W)^c \cup X_n^c \in \varphi$ and further $(V \triangle W)^c \in \langle \varphi \cup \{X_m \mid m \in \mathbb{N}\}\rangle$ implying $V = W$. This ensures $\operatorname{card} A/_{t(A) + pA} \geq \operatorname{card}\{\bar{a}_V \mid V \in T\} = \operatorname{card} T = 2^\kappa$.

(c) (i) Assume that there is some $0 \neq \bar{a} \in D$ with $p\bar{a} = 0$. Since ψ is big over φ and $z(a) \in \psi \setminus \varphi$, there is a system $T \subset \langle\{z(a)\}\rangle$ of 2^κ pairwise not φ-equivalent sets. For $V \in T$ define

$$a_V(i) = \begin{cases} a(i) & \text{if } i \in V^c \\ 0 & \text{if } i \in V \end{cases}$$

As before we get $z(a_V) = V$, $z(a_V - a_W) = (V \triangle W)^c$. Clearly $p\bar{a}_V = 0$ and $\bar{a}_V \in D$ hold. Hence there 2^κ different elements of order p in D by which $\gamma_p \geq 2^\kappa$ follows.

(c) (ii) There is an element $0 \neq \bar{a} \in D$, since A is not reduced. Then $L_n = \{i \in I \mid (n!)^2 \mid a(i)\}$ $(n \in \mathbb{N})$ is a descending chain in φ satisfying $L_1 = I$ and $z(a) \subset \bigcap_{n\in\mathbb{N}} L_n$. For $z(a) \in \psi \subset \varphi^*$ there is a descending chain $(K_n)_{n\in\mathbb{N}}$ in φ satisfying $K_1 = I$ and $\bigcap_{n\in\mathbb{N}} K_n = z(a)$.

Setting $M_n = K_n \cap L_n$ the descending chain $(M_n)_{n\in\mathbb{N}}$ fulfils the same requirements. Pick for $i \in M_n$ some $g_{i,n} \in G_i$ given by $a(i) = (n!)^2 g_{i,n}$. Again let $T \subset \langle \{z(a)\} \rangle$ be a system of 2^κ different representatives under φ-equivalence and set for $V \in T$

$$a_V(i) = \begin{cases} n!g_{i,n} & \text{if } i \in V^c \cap (M_n \setminus M_{n+1}) \\ 0 & \text{else} \end{cases}$$

Then $z(a_V) = V$, since $n!g_{i,n} = 0$ implies $i \in z(a) \subset V$. Clearly $a_V(i)$ is divisible by $n!$ for $i \in M_n$ and therefore $\bar{a}_V \in D$. Again we have $z(a_V - a_W) = (V \Delta W)^c$ for $V,W \in T$ so that the $\bar{a}_V (V \in T)$ are pairwise different. It remains to show that they are torsion free. Suppose $n\bar{a}_V = 0$ for some n and $V \in T$. Thus we get $M_n \cap z(na_V) \subset V$. For, if $i \in M_n \cap z(na_V)$, we have either $i \in V$ or $i \in M_l \setminus M_{l+1}$ for some $l \geq n$, since $V \supset z(a) = \bigcap_{j\in\mathbb{N}} M_j$. But $0 = na_V(i) = nl!g_{i,l}$ gives $0 = (l!)^2 g_{i,l} = a(i)$, which shows that the second case cannot occur. Hence V belongs to φ and $\bar{a}_V = 0$. Therefore there are 2^κ different torsion free elements in D, by which $\delta \geq 2^\kappa$ follows.

<u>(2.2). Theorem :</u> Let $\varphi \subset \psi$ be filters on a set I of cardinality κ. Then the following statements are equivalent :

(a) $\psi \subset \varphi^*$, φ is pure in ψ and ψ is big over φ.

(b) For every family $(G_i)_{i\in I}$ of nonzero groups with card $G_i \leq 2^\kappa$ the filter quotient $A = \Sigma_\psi(G_i) / \Sigma_\varphi(G_i)$ is algebraically compact and its invariants $\alpha_{p,n}$, β_p, γ_p and δ are either zero or 2^κ and satisfy for all primes p :

  (i) $\beta_p \geq \limsup_{n\to\infty} \alpha_{p,n}$

  (ii) $\delta \geq \gamma_p$

(c) (b) holds for all families $(G_i)_{i\in I}$ with $G_i \in \{\mathbb{Z}, Z(2^n)(n \in \mathbb{N})\}$.

<u>Remark :</u> Restricting condition (b) to those families of groups where all the groups are isomorphic, the filter condition "ψ big over φ" can be replaced by "$[\psi : \varphi] = 2^\kappa$". Also in this case it suffices to

claim (b) only for $\mathbb{Z}$. The proofs are slightly different though they are based on the same ideas.

<u>Proof: (a) ⇒ (b) :</u> A is algebraically compact by theorem (1.1)(a). Since card $G_i \leq 2^\kappa$ for all i, also card $A \leq 2^\kappa$ and therefore all occuring invariants are at most 2^κ. In [7] was shown that $\alpha_{p,n} = \dim p^{n-1}A[p]/p^nA[p]$ and $\beta_p = \dim A/t(A) + pA$ (dim denotes the dimension as $Z(p)$-vectorspaces). By lemma (2.1)(a) $\alpha_{p,n}$ is either zero or 2^κ. If $A = A[p^n] + pA$ holds for some n, then $A = t(A) + pA$ gives $\beta_p = 0$. In the other case lemma (2.1)(b) implies $\beta_p = 2^\kappa$. In part (c) of this lemma we have learnt that γ_p and δ are either zero or 2^κ and the inequality (ii) holds. If for some prime p there are infinitely many n with $\alpha_{p,n} \neq 0$, then is $A \neq A[p^n] + pA$ for all n and therefore by (b) of the preceeding lemma $\beta_p = 2^\kappa$ holds.

<u>(b) ⇒ (c) :</u> Trivial

<u>(c) ⇒ (a) :</u> $\psi \subset \varphi^*$ and the purity of φ in ψ are a consequence of theorem (1.8)(a). In order to show ψ big over φ let $(X_n)_{n\in\mathbb{N}}$ be a descending chain in $P(I)$. We may assume $X_1 = I$. Setting

$$G_i = \begin{cases} Z(2^{n+1}) & \text{if } i \in X_n \setminus X_{n+1} \quad \text{for some } n \in \mathbb{N} \\ \mathbb{Z} & \text{else} \end{cases}$$

We obtain $X_n = \{i \in I \mid G_i \neq G_i[2^n] + 2G_i\}$. In case $\beta_2 = 0$ there is some l satisfying $\alpha_{2,n} = 0$ for all $n \geq l$ and thus $A = A[2^l] + 2A$. We show $\psi \subset \langle \varphi \cup \{X_1\}\rangle$. For $V \in \psi$ define $a_V(i) = 1$ if $i \in V^c$ and 0 if $i \in V$. (Here 1 means $1 \in \mathbb{Z}(2^n)$ or $1 \in \mathbb{Z}$.) By $a_V \in A[2^l] + 2A$ there is a $M \in \varphi$ satisfying $a_V(i) \in G_i[2^l] + 2G_i$ for all $i \in M$. This clearly gives $M \subset V \cup X_l^c$ and therefore V belongs to $\langle \varphi \cup \{X_l\}\rangle$. So let us assume $\beta_2 = 2^\kappa$. Hence there exist 2^κ many different elements $\bar{a}_\xi$ $(\xi < 2^\kappa)$ in A satisfying $\mathbb{Z}\bar{a}_\xi \cap \mathbb{Z}a_\eta = 0$ and $\bar{a}_\xi \not\equiv \bar{a}_\eta$ mod $t(A) + 2A$ for $\xi \neq \eta$. Easily we can pick representatives $a_\xi \in \bar{a}_\xi$ with $a_\xi(i) \in \{0,1\}$ for every i. This ensures $z(2^na_\xi) = z(a_\xi) \cup X_n^c$ for all n. Assume

$(z(a_\xi) \Delta za_\eta))^c \in \langle\varphi \cup \{X_n\}\rangle$ for some n. Then by

$\varphi \ni X_n^c \cup (z(a_\xi) \Delta z(a_\eta))^c = ((z(a_\xi) \cup X_n^c) \Delta (z(a_\eta) \cup X_n^c))^c =$

$= (z(2^n a_\xi) \Delta z(2^n a_\eta))^c = z(2^n a_\xi - 2^n a_\eta)$ we obtain $2^n \bar{a}_\xi = 2^n \bar{a}_\eta$ and hence $\bar{a}_\xi \equiv \bar{a}_\eta \bmod t(A) + 2A$. Thus $\xi = \eta$ and we have $[\psi : \langle\varphi \cup \{X_n \mid n \in \mathbb{N}\}\rangle] \geq \text{card } \{z(a_\xi) \mid \xi < 2^\kappa\} = 2^\kappa$. This proves that ψ is big over φ.

This is part of a thesis written at the University of Essen under the supervision of Priv.-Doz. Dr. M. Dugas and Prof. Dr. R. Göbel.

References

1. S. Balcerzyk, On factor groups of some subgroups of the complete direct sum of infinite cyclic groups, Bull. Acad. Polon. Sci. 7(1959), 141-142.
2. M. Dugas and R. Göbel, Algebraisch kompakte Faktorgruppen, J. reine angew. Math. 307/308, 341-352 (1979).
3. U. Felgner, Reduced Products of Abelian Groups, unpublished.
4. L. Fuchs, Note on factor groups in complete direct sums, Bull. Acad. Polon. Sci. 11(1963), 39-40.
5. L. Fuchs, Infinite Abelian Groups I, New York 1970.
6. O. Gerstner, Algebraische Kompaktheit bei Faktorgruppen von Gruppen ganzzahliger Abbildungen, Manuscripta math. 11(1974), 104-109.
7. K. Golema and A. Hulanicki, The structure of the Factor Groups of the Unrestricted Sum by the Restricted Sum of Abelian Groups II, Fundamenta Mathem. 53(1964), 177-185.
8. A. Hulanicki, The Structure of the Factor Group of the Unrestricted Sum by the Restricted Sum of Abelian Groups, Bull. Acad. Polon. Sci. 10(1962), 77-80.
9. G. de Marco, On the Algebraic Compactness of some Quotients of Product Groups, Rend. Sem. Math. Univ. Padova 53(1975), 329-333.
10. W. Zimmermann, Rein injektive direkte Summen von Moduln, Communications in Algebra 5(10) (1977), 1083-1117.

COTORSION MODULES OVER NOETHERIAN HEREDITARY RINGS

Ray Mines*

The purpose of this paper is to develop a theory of cotorsion modules over a Noetherian hereditary ring. In order to define a cotorsion theory one must first have a torsion theory. Such a theory is given by a pair $t = (\mathcal{T},\mathcal{F})$ of R-modules where $\mathcal{T}$ is the class of torsion modules and $\mathcal{F}$ is the class of torsion-free modules [3]. If N is an R-module then the torsion submodule of N is denoted by tN. Most of the results obtained hold for any hereditary stable torsion theory. However to be more definite and because some of the results are stronger the Goldie torsion theory is used throughout. Because of notational convenience the ring is assumed to be left nonsingular; that is 0 is the only element of R whose left annihilator is an essential left-ideal.

A theory of cotorsion modules was also worked out in [1] using different methods. Our development, when applied to the ring of integers gives a description of cotorsion abelian groups which does not depend on homological algebra and can thus be thought of as a natural extension of the theory of algebraic compact groups obtained in [2].

The Goldie torsion theory is developed in section 1. In section 2 the equivalences of various possible definitions of a cotorsion theory and the properties of cotorsion modules are studied. In the third section it is shown that there are plenty of cotorsion modules for the Goldie theory over a Noetherian hereditary ring. Finally, in the last section, a description of the cotorsion hull, in terms of the Ext functor, is obtained.

Throughout, unless explicitly stated otherwise, all modules are left unital R-modules.

---

*This research was partially supported by the Ministerium für Wissenschaft und Forschung des Landes Nordrhein-Westfalen under the title Überabzählbare abelsche Gruppen.

1. Divisibility. Let R be a left nonsingular ring with identity, and let Q be the injective envelope of R. The Goldie torsion theory is the torsion theory cogenerated by Q [3,p.17]. That is an R-module M is torsion if $\mathrm{Hom}(M, Q) = 0$. If M is an R-module, then the torsion submodule tM of M is $tM = \{x \in M : fx = 0 \text{ for all } f \in \mathrm{Hom}(M, Q)\}$. If N is an essential submodule of M then M/N is a torsion module in the Goldie theory. In particular $K = Q/R$ is a torsion module.

A torsion theory t is hereditary if t is a left exact functor. That is if $N \subseteq M$ are R-modules, then $tN = N \cap tM$. The Goldie theory is hereditary [3,p.9]. A torsion theory t is stable if the injective envelope of a torsion module is a torsion module. The Goldie theory is stable. To see this let M be a torsion module with $E(M)$ the injective envelope. As M is essential in $E(M)$, the quotient module $E(M)/M$ is torsion. But extensions of torsion modules are again torsion. If t is a stable torsion theory and M is an injective R-module, then tM is injective. Thus each injective can be written $M = tM \oplus N$ where N is torsion-free injective.

A module M is Q-reduced if $\mathrm{Hom}(Q, M) = 0$. Let M be an R-module. The submodule generated by all homomorphic images of Q in M will be denoted by DM and is the Q-divisible submodule of M. That is the functor D is the socle generated by Q.

Lemma 1.1. Let M be an R-module and let $x \in DM$. Then there exists a homomorphism $g : Q \to M$ so that $g(1) = x$. If M is torsion-free then g is unique.

Proof. As $x \in DM$ there exists a finite set of homomorphisms $f_i : Q \to M$ and elements $q_i \in Q$ such that $x = \Sigma f_i(q_i)$. Define $f : R \to \sum_i Q_i$, with $Q_i \cong Q$, by $f(1) = \Sigma q_i$. As the finite direct sum ΣQ_i is injective there exists a map $\overline{f} : Q \to \Sigma Q_i$ so that $\overline{f}(1) = \Sigma q_i$. Composing $\overline{f}$ with the map Σf_i gives the desired map. The uniqueness follows as Q/R is torsion. □

Lemma 1.2. Let $N \subseteq M$ be R-modules such that M/N is torsion-free. Then $N \cap DM = DN$.

Proof. Clearly $DN \subset N \cap DM$. So let $x \in N \cap DM$. Then there exists $f : Q \to M$ so that $f(1) = x$. As M/N is torsion-free, Q/R is torsion, and $fR \subseteq N$, it follows that $f(Q) \subseteq N$. That is $x \in DN$. □

<u>2. Cotorsion Modules.</u> In the category of abelian groups the Goldie torsion theory gives the usual torsion theory. Here Q is the group of rational numbers, and a Q-divisible group is a divisible group. If N is a reduced abelian group then each of the following three conditions are equivalent and a group N which satisfies them is called <u>cotorsion</u>.

C1. If $N \subseteq M$, then $D(M/N) = (N + DM)/N$.

C2. If $N \subseteq M$ and $M/N \cong Q$, then N is a summand of M.

C3. If $N \subseteq M$ and M/N is torsion-free then N is a summand of M.

In the category of R-modules with the torsion theory cogenerated by Q, any module satisfying C3 also satisfies C2 as Q is a torsion-free module.

<u>Theorem 2.1.</u> If R is a left nonsingular ring, and t is the Goldie torsion theory then a Q-reduced module N satisfies C1 if, and only if, it satisfies C2.

<u>Proof.</u> Let N satisfy C1 and assume that $N \subseteq M$ with $M/N \cong Q$. Then by C1 $M/N = D(M/N) = (N + DM)/N$. Lemma 1.2 shows that $N \cap DM = DN = 0$, as M/N is torsion-free.
So $M \cong N \oplus DM$, and $DM \cong Q$.

Let N satisfy C2 and let $N \subseteq M$. Assume, without loss of generality, that $M/N \neq 0$ and is Q-divisible. Let $\bar{x} \in M/N$, $\bar{x} \neq 0$. By Lemma 1.1 there exists a homomorphism $f : Q \to M/N$ so that $f(1) = \bar{x}$. Form the pullback diagramm

$$\begin{array}{ccccccccc} 0 & \longrightarrow & N & \longrightarrow & X & \longrightarrow & Q & \longrightarrow & 0 \\ & & \| & & \downarrow g & & \downarrow f & & \\ 0 & \longrightarrow & N & \longrightarrow & M & \longrightarrow & N/M & \longrightarrow & 0 \end{array}$$

As N satisfies C2, the module $X \cong N \oplus Q$ and thus $g(1) \in DM$ is a representative of $\bar{x}$, showing $\bar{x} \in (N + DM)/N$. □

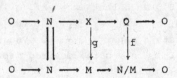

Theorem 2.2. Let R be a left nonsingular hereditary ring and let t be the Goldie torsion theory. Then a Q-reduced R-module N satisfies C1 if, and only if, N satisfies C3 .

Proof. As already remarked, if N satisfies C3 then it satisfies C2 and hence C1. So suppose that N satisfies C1. Let $N \subseteq M$ with M/N torsion-free. Let $E(M)$ be the injective envelope of M. Then $M/N \subseteq E(M)/N$ and the latter module is injective as R is left hereditary. Therefore, the injective envelope $E(M/N)$ of M/N is a submodule of $E(M)/N$. Let $M' \subseteq E(M)$ be such that $M'/N = E(M/N)$. Since M/N is torsion-free and t is a stable torsion theory it follows that $E(M/N)$ is torsion-free; $E(M/N)$ is also Q-divisible. So $E(M/N) = M'/N = (N + DM')/N$ by condition C1 . But M'/N torsion-free implies, by Lemma 1.2, that $N \cap DM' = DN = 0$. Thus $M' = N \oplus DM'$. But $M = M \cap M' = N \oplus (M \cap DM')$, and so N satisfies C3. □

A Q-reduced module satisfying C1, and hence C2, is a *cotorsion module*. We shall now give some properties of cotorsion modules.

Lemma 2.3. Let N be cotorsion and let $A \subseteq B$ with $A/B = Q$. Then any homomorphism from A to N can be extended uniquely to a homomorphism from B to N. If N satisfies C3, then the same result holds if A/B is torsion-free.

Proof. Let $f : A \to G$. Then using the pushout we obtain

$$\begin{array}{ccccccccc} 0 & \longrightarrow & A & \longrightarrow & B & \longrightarrow & Q & \longrightarrow & 0 \\ & & \downarrow f & & \downarrow & & \| & & \\ 0 & \longrightarrow & N & \longrightarrow & X & \longrightarrow & Q & \longrightarrow & 0 \end{array}$$

By C2 the bottom now splits. Thus there is an extension of f to a homomorphism $\bar{f} : B \to N$ which is unique as Q is Q-divisible and N is Q-reduced. A similar proof works if A/B is torsion-free and N satisfies C3. □

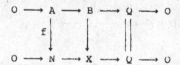

<u>Theorem 2.4.</u> If $\{N_i : i \in I\}$ is a family of cotorsion modules, then ΠN_i is cotorsion.

<u>Proof.</u> Let $\Pi N_i \subseteq M$ with $M/\Pi N_i \cong Q$. By Lemma 2.3 the map $f_j : \Pi N_i \to N_j$ extends to a unique map $f_j' : M \to N_j$. The homomorphism $\Pi f_j : M \to \Pi N_j$ is a splitting map. □

<u>Theorem 2.5.</u> Let M be cotorsion and $N \subseteq M$.

i) If M/N is Q-reduced, then N is cotorsion.

ii) If N is cotorsion then M/N is Q-reduced.

iii) If R is hereditary and M/N is Q-reduced, then it is cotorsion.

<u>Proof.</u> i) Let $N \subseteq A$ with $A/N \cong Q$. By Lemma 2.3 there exists a unique map $f : A \to M$ extending the embedding of N into M. As M/N is Q-reduced and A/N is Q-divisible it follows that $f(A) \subseteq N$. So f is a splitting map. Thus N is cotorsion by C2.

ii) As N is cotorsion and M is Q-reduced it follows by C1 that M/N is Q-reduced.

iii) Let $M/N \subseteq A$ with cokernel isomorphic to Q. If R is hereditary then there exists an R-module $A' \supseteq M$ so that the following diagram is a pushout

$$\begin{array}{ccccccccc} 0 & \longrightarrow & M & \longrightarrow & A' & \longrightarrow & Q & \longrightarrow & 0 \\ & & \downarrow & & \downarrow & & \| & & \\ 0 & \longrightarrow & M/N & \longrightarrow & A & \longrightarrow & Q & \longrightarrow & 0 \end{array}$$

As M is cotorsion the top now splits by C2. Thus the bottom now splits and so by C2 it follows that M/N is cotorsion. □

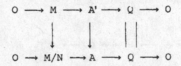

<u>3. The Existence of Cotorsion Modules.</u> In order to show that there exist lots of cotorsion modules the ring R will be assumed to be left nonsingular, Noetherian, and hereditary. The module Q will continue to represent the injective envelope of R and t the Goldie torsion theory. In this set up each Q-divisible module is injective, as the injectives are closed under direct sums and homomorphic images and each Q-divisible is an image of a direct sum of Q's.

Let N be a Q-reduced module. We shall show that there exists a functor c and a natural transformation $N \subseteq cN$ such that cN/N is torsion-free divisible and cN is cotorsion. To do this write $Q = \Sigma Q_i$ a finite direct sum of irreducible injectives. Choose a set of representatives of the equivalence classes of all extensions of N by each of the Q_i. Form the direct sum of these extensions and amalgamate the subgroup N. That is form the pushout diagram relative to the codiagonal map ∇ :

$$\begin{array}{ccccccccc} 0 & \longrightarrow & \Sigma\Sigma_i N & \longrightarrow & \Sigma\Sigma_i M & \longrightarrow & \Sigma\Sigma_i Q_i & \longrightarrow & 0 \\ & & \nabla \downarrow & & \downarrow & & \| & & \\ 0 & \longrightarrow & N & \longrightarrow & K & \xrightarrow{g} & \Sigma\Sigma_i Q_i & \longrightarrow & 0 \end{array}$$

As $\Sigma\Sigma_i Q_i$ is torsion-free, it follows that $N \cap DK = DN = 0$. Thus DK is isomorphic to a submodule of $\Sigma\Sigma_i Q_i$ under the map g. Let $cN = K/DK$ and let $D = \Sigma\Sigma_i Q_i/DK$. As $g(DK)$ is Q-divisible it is injective and so D is again torsion-free Q-divisible. As $N \cap DK = 0$ it follows that N is embedded as a submodule of cN, and $cN/N \cong D$. If N is not Q-reduced, then let $rN = N/DN$, and define $cN = c(rN)$. Then the sequence $0 \to DN \to N \to cN$ is exact.

<u>Theorem 3.1.</u> There exists an idempotent functor $c : R\text{-Mod} \to R\text{-Mod}$ and a natural transformation $rN \subseteq cN$ such that

i) cN is cotorsion.

ii) cN/rN is torsion-free Q-divisible.

iii) If M is Q-reduced, and $rN \subseteq M$ with M/rN torsion-free Q-divisible, then there exists a unique monomorphism of M into N fixing rN pointwise.

iv) If N is cotorsion (hence reduced) then $N = cN$.

<u>Proof.</u> We may assume that N is reduced. Let cN be as described above, and suppose that $cN \subseteq M$ with $M/cN \cong Q$. Form the pushout diagramm

$$\begin{array}{ccccccccc} & & N & = & N & & & & \\ & & \downarrow & & \downarrow & & & & \\ 0 & \longrightarrow & cN & \longrightarrow & M & \longrightarrow & Q & \longrightarrow & 0 \\ & & \downarrow & & \downarrow & & \| & & \\ 0 & \longrightarrow & cN/N & \rightarrow & M/N & \longrightarrow & Q & \longrightarrow & 0 \end{array}$$

The bottom now splits as cN/N is Q-divisible, and so M/N is torsion-free Q-divisible. If we prove iii) then the middle now will also split showing cN satisfies C2. To this end suppose $N \subseteq M$ with M/N torsion-free Q-divisible and $DM = 0$. Since M/N is torsion-free injective it follows that $M/N \cong \Sigma Q_j$ where each Q_j is an irreducible summand of Q. Let $M_j \subseteq M$ be such that $M_j/N \cong Q_j$. By the description of cN there exists for each j a unique $f_j : M_j \to cN$ such that f_j is the identity on N. Then Σf_j defines a unique homomorphism f of M into cN. Since cN/N is torsion-free, M/N is Q-divisible, and $N \cap \ker f = 0$, it follows that $\ker f = 0$. This shows iii) and hence i). Statement ii) follows from the description of c. To show iv) note that if N is cotorsion then every extension of N by Q splits and so $cN = N$. Thus, c is idempotent.

It remains to show that c is a functor. Again we may assume that $DN = 0$ and let $f : N \to M$. Then we have a homomorphism $N \to M \to cM$ which we will also call f. As cN/N is torsion-free and cM is cotorsion, and satisfies C3, it follows by Lemma 2.3 that there exists a unique map $cf : cN \to cM$ extending f. □

A cotorsion module is <u>adjusted</u> if it has no torsion-free summands.

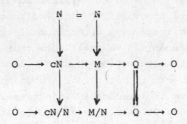

Theorem 3.2. If N is a Q-reduced torsion module, then cN is adjusted cotorsion.

Proof. As cN/N is torsion-free Q-divisible, this is clear. □

Theorem 3.3. Let N be cotorsion and $aN \subseteq N$ be such that $aN/tN = D(N/tN)$. Then aN is adjusted cotorsion and $N = aN \oplus M$ where M is torsion-free cotrosion. Moreover, $aN \cong ctN$.

Proof. By Theorem 2.5 it follows that aN and N/aN are cotorsion. As N/aN is torsion-free ($D(N/tN)$ is a summand of N/tN) it follows that $N \cong aN \oplus N/aN$. To show that aN is adjusted it is sufficient to show that $aN \cong ctN$, by Theorem 3.2. As c is a functor and $cN = N$ we may assume that $ctN \subseteq N$. By Theorem 2.5 ii) N/ctN is Q-reduced. As ctN/N is Q-divisible it follows that $ctN = aN$. □

4. Cotorsion modules and Ext. Let R be a Noetherian, left nonsingular, hereditary ring. Then the injective envelope Q of R is a left nonsingular ring [3]. Let $K = Q/R$. Then the sequence $0 \to R \to Q \to K \to 0$ is an exact sequence of R-bimodules. Let N be a left R-module and consider the exact sequence

$$0 \to \mathrm{Hom}_R(K,N) \to \mathrm{Hom}_R(Q,N) \to \mathrm{Hom}_R(R,N) \to \mathrm{Ext}_R(K,N) \to \mathrm{Ext}_R(Q,N) \to \\ \to \mathrm{Ext}_R(R,N) \to 0.$$

Using the right module structure on R, Q and K this becomes an exact sequence of left R-modules. Moreover $\mathrm{Hom}_R(R,N)$ is naturally isomorphic to N as a left module.

If N is cotorsion, then $\mathrm{Ext}_R(Q,N) = 0$ by C2 and $\mathrm{Hom}_R(Q,N) = 0$ as N is Q-reduced. So the sequence becomes

$$0 \to N = \mathrm{Hom}_R(R,N) \to \mathrm{Ext}_R(K,N) \to \mathrm{Ext}_R(Q,N) = 0$$

Therefore, $N \cong \mathrm{Ext}_R(K,N)$. Next notice that if N is Q-reduced, the sequence $0 \to N \to cN \to D \to 0$ is exact with D torsion-free Q-divisible. Therefore, $\mathrm{Hom}_R(K,D) = 0$ and $\mathrm{Ext}_R(K,D) = 0$. Which implies that $\mathrm{Ext}_R(K,N) \cong \mathrm{Ext}_R(K,cN) \cong cN$. Thus the functor c can also be described in terms of the functor $\mathrm{Ext}_R(K,\_)$. So, for a left nonsingular Noetherian hereditary ring, the theory of cotorsion modules associated with the Goldie torsion theory could also be developed using the Ext functor.

REFERENCES

[1] Fuchs, L., Cotorsion modules over Noetherian hereditary rings, Houston J. Math., 3(1977) 33-46.

[2] Legg, M.W. and Walker, E.A., An algebraic treatment of algebraically compact groups, Rocky Mountain J. Math., 5(1975) 291-299.

[3] Stenström, Bo, Rings and Modules of Quotients, Springer Lecture Notes No.237(1971)

BASIC CONCEPTS OF FUNCTORIAL TOPOLOGIES

Adolf Mader

I. INTRODUCTION AND PRELIMINARIES. In his 1925 paper [32] Pruefer enlarged a given group ("group" means abelian group) G by adding "ideal elements" in the hope that the new group $\hat{G}$ would be a direct product of groups of "rank one". In today's terminology $\hat{G}$ is the completion of G when G is given the "Pruefer topology", i.e. the linear topology in which a subgroup U is open if and only if G/U is a group with minimum condition. Pruefer showed that his "ideal groups" are linearly compact (terminology due to Lefschetz [19]) and proceeded to investigate such groups. The questions he raised led to a series of papers, Pietrkowski [31], Krull [17], Leptin [20], [21], Schoeneborn [33], culminating in a duality theorem for linearly compact groups and modules (Leptin [20], [21], Schoeneborn [33], Kaplansky [15], MacDonald [22], Fleischer [9]). The p-adic and $\mathbb{Z}$-adic topologies and the associated topological terminology proved inescapable and have been used routinely. A generalization of the p-adic topology is the p^{λ}-topology (also called λ-topology and natural topology). Here a group A is topologized via the neighborhood basis $\{p^{\alpha}A: \alpha<\lambda\}$ of 0 where λ is a limit ordinal. Major contributions to this area are contained in Kulikov [18], Harrison [11], [12], Mines [28], [29], Waller [36], Megibben [23], Cutler [6], and Salce [34]. In 1964 Charles [4] abstracted from the above examples the concept of a "functorial topology".

1.1 DEFINITION. A functorial topology on the category $\mathcal{A}$ of all abelian groups is a functor T on $\mathcal{A}$ to the category of topological abelian groups such that TA is the group A with a topology T_A, and for every homomorphism f we have Tf=f. □

The last condition means that every homomorphism between groups so topologized is continuous. Charles pointed out how an abundance of such topologies can be constructed. Fuchs [10] generalized Charles' method somewhat and Boyer-Mader [3] noted that this amounted to fixing the class of discrete groups and giving all other groups the least topology required by the continuity of all homomorphisms. Functorial topologies constructed in this fashion are called minimal and they are in bijective correspondence with classes of groups satisfying certain closure properties. Thus studying groups with a minimal functorial topology amounts to doing group theory with respect to a distinguished class of groups, namely the class of discrete groups. Not all functorial topologies are minimal (see Boyer-Mader [3]) but most of the commonly used topologies are, and it is now possible to deal with basic properties and main questions in reasonable generality. This will be done in this paper. In addition we survey some earlier results. The content is roughly the following.

Section I contains the introduction as well as basic conventions and notation. In Section II we summarize the results of Boyer-Mader [3] on the classification of functorial topologies, give examples, and show that the functorial topologies generated by so-called topological families of radicals are included among the minimal functorial topologies. In Section III we develop the concepts which arise naturally for any minimal functorial topology. There are no deep results but it is interesting to see how properties of minimal functorial topologies are reflected in the associated discrete classes as closure properties for example. Section IV contains a number of results on the category $\mathcal{A}_{lt}$ of abelian groups with a linear topology and the completion functor $C:\mathcal{A}_{lt} \to \mathcal{A}_{lt}$. There are only a few new results in this section but it should be useful to have some of the basic facts collected and proved. They lead to deeper questions which are indicated and references to the relevant literature are given. The final Section V contains a completability criterion, i.e. a criterion for deciding whether the completion topology of the completion of a group TA, T a minimal functorial topology, is the functorial topology. The result is essentially the criterion of Mines-Oxford [30] stated in greater generality with a self-contained proof. Some applications are given.

In this paper we consider exclusively <u>linear</u> <u>topologies</u>, i.e. the topology $\mathcal{T}$ on the group A is such that there exists a neighborhood basis $\mathcal{U}$ at 0 consisting of subgroups. Such a neighborhood basis will be called a <u>linear</u> <u>basis</u> of the topological group. Adopting the notation of Koethe [16] we write $A[\mathcal{T}]$ for the group A with topology $\mathcal{T}$. We also write $A[\mathcal{U}]$ if $\mathcal{U}$ is a linear basis of the topological group A. If T is a functorial topology then T_A denotes the topology of TA, so $TA = A[T_A]$. Let $\mathcal{A}_{lt}$ denote the category of abelian groups with a linear topology. Concepts in $\mathcal{A}_{lt}$ are distinguished from those in $\mathcal{A}$ by stating explicitly the topologies involved. Thus $A[\mathcal{T}] \cong A[\mathcal{S}]$, $\oplus_i A_i[\mathcal{T}_i]$, $\Pi_i A_i[\mathcal{T}_i]$ denote isomorphism, direct sum, and product in $\mathcal{A}_{lt}$. We write functors on the left, maps on the right. For a group A, let A_t denote its torsion part and $A_{tf}=A/A_t$. For the rest we use the standard notation of Fuchs [10].

II. FUNCTORIAL TOPOLOGIES- CLASSIFICATION. In this section we essentially survey the results of Boyer-Mader [3]. We first note that there is no gain of generality by considering functorial topologies defined only on a full subcategory $\mathcal{B}$ of $\mathcal{A}$, such as the category of torsion groups. If T is defined on $\mathcal{B}$, and A $\in \mathcal{A}$, let $\mathcal{U}(A)$ be the family of all finite intersections of groups Uf^{-1} where f $\in$ Hom (A,B), B $\in \mathcal{B}$, and U is an open subgroup of TB. Then $\bar{T}A=A[\mathcal{U}(A)]$, $\bar{T}(f)=f$, is easily checked to define a functorial topology defined on $\mathcal{A}$ which extends T. We call $\bar{T}$ the "canonical extension" of T, and we have:

2.1 THEOREM. If $\mathcal{B}$ is a full subcategory of $\mathcal{A}$ and T is a functorial topology on $\mathcal{B}$ then T is the restriction of the functorial topology $\bar{T}$ on $\mathcal{A}$. □

We associate with a functorial topology T its class of discrete groups $\mathcal{C}(T)=\{D:TD \text{ is discrete}\}$. This class has certain closure properties which follow easily from the continuity of all homomorphisms, and there is a related refinement of 2.1.

2.2 THEOREM. a) If T is a functorial topology then $\mathcal{C}(T)$ is closed under imbedded groups and finite direct sums.

b) Let $\mathcal{B}$ be a full subcategory of $\mathcal{A}$ which is closed under imbedded groups and finite direct sums. If T is a functorial topology on $\mathcal{B}$ and $\bar{T}$ its canonical extension then $\mathcal{C}(\bar{T})=\mathcal{C}(T)$. □

2.3 DEFINITION. A class of groups $\mathcal{C}$ closed under imbedded groups and finite direct sums is called a <u>discrete class</u>. If $\mathcal{C}$ is, in addition, closed under epimorphic images it is called an <u>ideal discrete</u> class. □

We may now start with any discrete class $\mathcal{C}$ and let T be the discrete functorial topology on $\mathcal{C}$ i.e. TD is discrete for every D $\in \mathcal{C}$, and then construct the functorial topology $\bar{T}$ on $\mathcal{A}$ with $\mathcal{C}(\bar{T})=\mathcal{C}$. This is the construction of Charles [4] and Fuchs [10]. Not all functorial topologies are obtained in this fashion. This motivates the following definition.

2.4 DEFINITION. Let T be a functorial topology, $\mathcal{C}=\mathcal{C}(T)$ its discrete class, and for every group A let $\mathcal{U}_A=\{U\leq A:A/U \in \mathcal{C}\}$. If $TA=A[\mathcal{U}_A]$ for all A then T is called a <u>minimal functorial topology</u>. Furthermore, we call T <u>ideal</u> if every epimorphism f:A→B is open as a map f: TA→TB. □

The following theorem reduces questions about minimal functorial topologies to questions about a distinguished class of groups. It will become clear that this is a real advantage.

2.5 THEOREM. The map $T\to\mathcal{C}(T)$ is a bijective correspondence between minimal functorial topologies and discrete classes as well as between ideal functorial topologies and ideal discrete classes. The inverse of the correspondence assigns to the class C the canonical extension $\bar{T}$ of the discrete functorial topology T on C. □

An example of a functorial topology which is not minimal is furnished by the so-called "large subgroup topology" (see [3,2.8]). This is the only example we know but it is very easy to give examples of minimal functorial topologies via discrete classes.

2.6 EXAMPLES. The following are ideal discrete classes.

a) All bounded p-groups (p-adic topology).

b) All bounded groups ($\mathbb{Z}$-adic topology).

c) All finitely co-generated groups (Prüfer topology).

d) All finite groups (finite index or co-$\aleph_0$-topology).

e) All groups of cardinality $< \aleph$ where $\aleph$ is an infinite cardinal (co-$\aleph$-topology).

The following are discrete classes which are not ideal.

f) All p-groups of length $<\lambda$ where λ is a limit ordinal (p^λ-topology).

g) All direct sums of cyclic groups ($\oplus_C$-topology).

h) All torsion free groups of finite rank. □

In fact, every class of groups X is contained in a smallest discrete class C which consists of all groups imbedded in finite direct sums of groups of X, and in a smallest ideal discrete class C' which consists of all epimorphic images of the groups of C.

It seems hopeless to classify discrete classes because of their abundance. However, it might be possible to require certain features of interest regarding the associated functorial topologies, and to survey the resulting classes. We will do this by restricting attention to ideal discrete classes. Our results are based on ideas of Balcerzyk [2] who classified Serre classes of groups these being ideal discrete classes closed under extensions. For a given discrete class C consider the following subclasses:

C_{cb}: the class of all torsion groups of C whose primary components are bounded (cb-groups).

C_{dt}: the class of all divisible torsion groups of C (dt-groups).

C_t : the class of all torsion groups of C (t-groups).

C_{tf}: the class of all torsion free groups of C (tf-groups).

It is clear that C_{tf} is a discrete class, and that C_{cb} and C_t are themselves ideal discrete classes while C_{dt} is closed under finite direct sums and epimorphic images. The groups of C_{cb} and C_{dt} are determined up to isomorphism by cardinal invariants. If A is a cb-group let $i_A(p,n)$ be the cardinality of cyclic summands of order p^n in an indecomposable direct decomposition of A. Following Balcerzyk the possible families $\{i_A : A \in C_{cb}\}$ are characterized, and the correspondence $C_{cb} \to \{i_A : A \in C_{cb}\}$ is clearly injective ([3, 3.1]). Similarly, the classes C_{dt} are characterized in terms of the invariants $r_A(p) = \dim A[p]$ for $A \in C_{dt}$ ([3, 3.6]). The class C_t is determined by its subclasses C_{cb} and C_{dt} as follows.

(2.7) $A \in C_t$ if and only if A is imbedded in a group $B \oplus D$ with $B \in C_{cb}$ and $D \in C_{dt}$. □

This leads to a satisfactory description of the possible classes C_t: they can all be obtained via (2.7) from possible classes C_{cb} and C_{dt} satisfying the compatibility condition that any cb-group imbedded in a group of C_{dt} lies in C_{cb} ([3, 3.9]).

Finally, we consider the classes C_{tf}. If C_{tf} contains all tf-groups then $C = A$. If not, either C_{tf} is the class of all tf-groups of cardinality $< \aleph$, where $\aleph$ is an uncountable cardinal, or else C_{tf} contains finite rank groups only. The classes C_{tf} of the latter kind can be characterized in terms of the set of types of its rank one elements ([3, 3.13]). It remains to construct the class C from its subclasses C_t and C_{tf}. We have a useful reduction theorem.

(2.8) $A \in C$ if and only if A is imbedded in a group $D \oplus M$ with $D \in C_{dt}$, $M \in C$ and $M_t \in C_{cb}$. □

If $\aleph$ is an uncountable cardinal and $C_{tf} = \{A : A \text{ is torsion free and } |A| < \aleph\}$ it can be shown that $A \in C$ if and only if A is imbedded in a group $T \oplus B$ with $T \in C_t$ and $|B| < \aleph$. It follows that the ideal discrete classes C with C_{tf} of this kind are characterized by $\aleph$, its subclass C_t and the compatibility condition that C_t contain all torsion groups of cardinality $< \aleph$ ([3, 3.17]).

The remaining case of those ideal discrete classes which contain torsion free groups of finite rank only has not been settled. Let C be the smallest ideal discrete class containing all rank one torsion free groups of type $\leq [\chi]$ where $\chi(p) = 1$ for all p. Then $C_t = C_{cb}$ and there exists a non-split mixed group M with $M_t \in C_{cb}$ and $M_{tf} \in C_{tf}$ which does not belong to C. If C' is the smallest ideal discrete class containing C and M then $C'_{tf} = C_{tf}$ but C'_t is strictly larger than C_t. Problems of independent interest and relevant to the classification of ideal discrete classes are: finding the epimorphic images of finite rank torsion free groups and the study of extensions of cb-groups by finite rank torsion free groups.

Frequently functorial topologies are obtained from radicals defined on A. Motivated by Mines-Oxford [30, p. 190] we give a general definition.

2.9 DEFINITION. If $\mathcal{R}$ is a family of subfunctors of the identity on $\mathcal{A}$ which is directed downwards under the partial order of subfunctors of 1 then $\mathcal{R}$ is called a <u>topological family of subfunctors of 1</u>. For a group A the family $\mathcal{U}(\mathcal{R})=\{RA:R\in\mathcal{R}\}$ defines a topology $\mathcal{T}(\mathcal{R})$ on A which will be called the <u>$\mathcal{R}$-topology on</u> A. □

2.10 PROPOSITION. Let $\mathcal{R}$ be a topological family of subfunctors of 1.

a) If $T(\mathcal{R})A=A[\mathcal{U}(\mathcal{R})]$ and $T(\mathcal{R})f=f$ then $T(\mathcal{R})$ is a functorial topology on $\mathcal{A}$.

b) If $C=\{A:T(\mathcal{R})A \text{ is discrete}\}=\{A:RA=0 \text{ for some } R\in\mathcal{R}\}$ then C is a discrete class closed under arbitrary powers, i.e. if $A\in C$ then $A^{\aleph}\in C$ for any cardinal $\aleph$.

c) If $\mathcal{R}$ is a topological family of radicals, i.e. if $R(A/RA)=0$ for all $A\in\mathcal{A}$ and all $R\in\mathcal{R}$ then $T(\mathcal{R})$ is a minimal functorial topology with discrete class C.

PROOF. a) If $f:A\to B$ and $R\in\mathcal{R}$ then $RA<(RB)f^{-1}$.

b) Suppose $f:A\to B$ is a monomorphism and $RB=0$ for some $R\in\mathcal{R}$. Then $(RA)f<RB=0$ so $RA=0$. Hence C is closed under imbedded groups. Let A_i, $i\in I$, be a family of groups in C and let $R_i\in\mathcal{R}$ be such that $R_iA_i=0$. Suppose there exists $S\in\mathcal{R}$ such that $S\le R_i$ for all $i\in I$. Then $S(\Pi_{i\in I}A_i)\le\Pi_{i\in I}SA_i\le\Pi_{i\in I}R_iA_i=0$. So $\Pi_{i\in I}A_i\in C$. A functor S exists if I is finite showing that C is closed under finite direct sums. If $A_i=A$ for all $i\in I$, and $R_i=R$ then $S=R$ satisfies the above inequality and C is closed under powers.

c) Let $A\in\mathcal{A}$ be given. If $R\in\mathcal{R}$ then $R(A/RA)=0$ so RA is open in $T(C)A$ where $T(C)$ is the minimal functorial topology corresponding to C. Conversely, suppose $U\le A$ such that $R(A/U)=0$ for $R\in\mathcal{R}$. Then $RA<U$ so U is open in $T(\mathcal{R})A$. □

The minimal functorial topologies which arise from topological families of radicals are in fact characterized by 2.10 b).

2.11 PROPOSITION. The minimal functorial topology T equals $T(\mathcal{R})$ for some topological family of radicals if and only if $C(T)$ is closed under arbitrary powers.

PROOF. The "only if" part of the proposition is 2.10 b). Thus assume T is given and $C(T)$ is closed under arbitrary powers. For each $D\in C(T)$ and $A\in\mathcal{A}$ define $R_D(A)=\cap\{\text{Ker } f:f\in \text{Hom}(A,D)\}$. By Charles [5, 2.7-2.10] R_D is a radical, and $R_D(A)=0$ if and only if A is imbedded in a power of D. Let $\mathcal{R}=\{R_D:D\in C(T)\}$. If $D,E\in C(T)$ then $D\oplus E\in C(T)$ and $R_{D\oplus E}\le R_D\wedge R_E$ hence $\mathcal{R}$ is a topological family of radicals. Clearly $A\in C(T)$ if and only if $RA=0$ for some $R\in\mathcal{R}$. Hence, by 2.10 c) and 2.5, $T=T(\mathcal{R})$. □

It is easily seen that a minimal functorial topology $T(R)$ is ideal only in very special cases.

2.12 PROPOSITION. If $T=T(R)$ is an ideal functorial topology then there exist integers $0\le n_1\le n_2\le n_3\le \ldots$ such that for each group A the set $\{n_1A, n_2A, \ldots\}$ is a linear basis of TA.

PROOF. Let $\{n_i\mathbb{Z}\}$ be a linear basis of $T\mathbb{Z}$ with $0\le n_1\le n_2\le \ldots$. Let $F=\oplus\mathbb{Z}_i$ be a free group, $\mathbb{Z}_i\cong\mathbb{Z}$. By hypothesis TF has a linear basis of radical subgroups, so in particular fully invariant subgroups which are known to be of the form nF. Using the continuity of the insertions $\mathbb{Z}_i\to F$ and of the projections $F\to\mathbb{Z}_i$ it follows that $\{n_iF\}$ is a linear basis of TF. Now let A be any group, F a free group and $f:F\to A$ an epimorphism. By hypothesis f is continuous and open, hence $\{n_iA=(n_iF)f\}$ is a linear basis for TA. □

III. <u>BASIC PROPERTIES</u>. Throughout this section T will be a minimal functorial topology on A, $C=C(T)$ will denote the associated discrete class.

There is a second class of groups associated with T which is significant. This is the class $C^{\perp}$ consisting of all groups A for which TA is indiscrete.

(3.1) a) $C^{\perp}=\{A:\mathrm{Hom}(A,D)=0 \text{ for all } D\in C\}$.

b) $C^{\perp}$ is closed under finite direct sums, epimorphic images and extensions. □

The following clarifies the connection between dense and indiscrete.

(3.2) a) If B is dense in TA then $T(A/B)$ is indiscrete.

b) If T is ideal then B is dense in TA if and only if $T(A/B)$ is indiscrete.

PROOF. a) Let $f\in\mathrm{Hom}(A/B,D)$ where $D\in C$. Let U be the kernel of the composite map $A\overset{g}{\to}A/B\overset{f}{\to}D$. Then U is open in TA, and since B is dense in A we have $U+B=A$. But $B<U$ so $g=0$ and $f=0$ also.

b) If T is ideal then the quotient topology on each A/B coincides with the functorial topology. Hence if $T(A/B)$ is indiscrete then B is dense in A. □

Two simple invariants of C determine the class of indiscrete groups to a large extent.

3.3 DEFINITION. For a discrete class C let $\pi_d=\{p:Z(p^{\infty})\in C\}$ and let $\pi_t=\{p:Z(p)\in C\}$. □

For a set of primes π, a group A is a π-<u>group</u> if it is a torsion group with $A[p]=0$ for $p\notin\pi$, and π' is the complement of π in the set of all primes. A group is π-<u>divisible</u> if it is p-divisible for all $p\in\pi$.

The following immediate observations hold for the minimal functorial topology T.

(3.4) a) If TA is indiscrete than A is π_t-divisible.

b) Suppose $\pi_d \neq 0$. Then TA is indiscrete if and only if A is a π_t-divisible π_d'-group.

c) If $\pi_d = \emptyset$ then TA is an indiscrete torsion group if and only if A is a π_t-divisible torsion group.

d) If $\mathbb{Q} \in C$ then TA is indiscrete if and only if A is a π_t-divisible π_d'-group. □

In the ideal case there is a complete answer.

3.5 PROPOSITION. Let T be an ideal functorial topology. Then TA is indiscrete if and only if A is a π_t-divisible π_d'-group. □

It follows that some groups TA have no proper dense subgroups, not even proper subgroups with indiscrete quotients. This is of interest in the search for "basic subgroups" as will be seen below. It appears that in some situations the concept "subgroup with indiscrete quotient" should replace the more restricted concept "dense". Evidence is provided in the following observation.

(3.6) If B is a subgroup of A such that T(A/B) is indiscrete, C[T] is a Hausdorff topological group, and $f: TB \to C[T]$ is continuous then f has at most one extension to A.

PROOF. $0 \to \mathrm{Hom}(A/B,C) \to \mathrm{Hom}(A,C) \to \mathrm{Hom}(B,C)$ is exact. If $g \in \mathrm{Hom}(A/B,C)$ then $\{0\}g^{-1}$ is closed in T(A/B) so $\{0\}g^{-1} = A/B$, i.e. $g=0$. □

We next characterize the Hausdorff groups.

3.7 DEFINITION. For each prime number p let $e(p) = \sup\{n : Z(p^n) \in C\}$ where $e(p) = \omega$ if C contains all cyclic p-groups. □

(3.8) Let $\bar{0}$ denote the closure of 0 in TA. Then $\bar{0} < [\cap_p p^{e(p)} A]$, and if $\pi_d \neq \emptyset$ then $\bar{0} < \oplus\{A_p : p \notin \pi_d\}$.

PROOF. Recall that $\bar{0}$ is the intersection of all open subgroups of TA. Let $x \in \bar{0}$ and consider $x + p^{e(p)}A$. If $e(p) = 0$ there is nothing to prove. If $e(p) \neq 0$, then $Z(p) \in C$ and if $x + p^{e(p)}A \neq 0$ there is a homomorphism $f: A \to Z(p)$ with $xf \neq 0$. This is a contradiction since $x \notin \mathrm{Ker} f$ which is open in TA. Now suppose $\pi_d \neq \emptyset$. Then $Z(p^\infty) \in C$ for some p. If x has infinite order or a non-zero p-component then there is a map $f: A \to Z(p^\infty)$ with $xf \neq 0$, and we have a contradiction as before. □

Again, there is a complete answer in the ideal case.

3.9 PROPOSITION. Let T be an ideal functorial topology. If $\pi_d = \emptyset$ then $\bar{0} = \cap_p p^{e(p)} A$. If $\pi_d \neq \emptyset$ then $\bar{0} = [\cap_p p^{e(p)} A] \cap [\oplus\{A_p : p \in \pi_d\}]$. Since TA is Hausdorff if and only if $\bar{0} = 0$ this characterizes the Hausdorff groups.

PROOF. It was shown in 3.8 that $\bar{0}$ is contained in the right hand sides of the equalities. It remains to show that the right hand sides are contained in every open subgroup of TA, or, equivalently, that the right hand sides are mapped to 0 by any map $f:A\to D$, $D\in C$. Suppose first that $\pi_d\neq\emptyset$, $x=\Sigma\{x_p : p\notin\pi_d\}$, $x_p\in A_p$, and $x\in\cap_p p^{e(p)}A$. Then $x_p\in p^{e(p)}A$, and if f is as above, then $x_p f\in p^{e(p)}D_p=0$. So $xf=0$. Suppose $\pi_d=0$ and $x\in\cap p^{e(p)}A$. If $f:A\to D$, $D\in C$, and if xf is torsion then $xf=0$ as above. If xf is torsion free then $\mathbb{Z}\in C$ and $e(p)=\omega$ for all p. But C_{tf} contains no p-divisible groups since $\pi_d=\emptyset$, so $xf+D_t\in(D/D_t)^1=0$ and $xf=0$ since xf is torsion free. □

A familiar complication in dealing with functorial topologies is the fact that subgroup topologies and quotient topologies need not be the functorial topology. We adopt terminology of Harrison [11].

3.10 DEFINITION. A subgroup B of A is called T-concordant if the subspace topology of B as a subset of TA coincides with the functorial topology of B. Dually, a quotient group C of A is called T-coconcordant if the quotient topology induced by TA on C is the functorial topology of C. □

An interesting property arises when a subgroup is both dense and concordant.

3.11 PROPOSITION. If B is a concordant and dense subgroup of A then every $D\in C$ is injective with respect to the exact sequence $0\to B\to A\to A/B\to 0$.

PROOF. Let $D\in C$ and let $f\in Hom(B,D)$. Since B is concordant in A there is an open subgroup U of A such that $U\cap B<Ker f$. Now $B/U\cap B\cong B+U/U=A/U$, and f induces a map $\bar{f}:B/U\cap B\to D$. The natural map $A\to A/U\to B/U\cap B \xrightarrow{\bar{f}} D$ extends f. □

If T is the $\mathbb{Z}$-adic topology then C(T) is the class of all bounded groups and it is easily seen that every bounded group is injective with respect to a given short exact sequence if and only if the sequence is pure exact. This motivates the following definition.

3.12 DEFINITION. The subgroup B of A is T-pure in A if every $D\in C(T)$ is injective with respect to the exact sequence $0\to B\to A\to B/A\to 0$. □

It is easily seen that B is T-pure in A if and only if for every $U\leq A$ with $A/U\in C$ it follows that A/U is a direct summand of B/U. The latter is the definition of "C-copure" in C. Walker [35]. We quote some of the results of [35].

3.13 THEOREM ([35, 2.1]). Let $C\leq B\leq A$. Then the following hold.

(a) If B is a summand of A then B is T-pure in A.

(b) If C is T-pure in A then C is T-pure in B.

(c) If B is T-pure in A then B/C is T-pure in A/C.

(d) If C is T-pure in B and B is T-pure in A then C is T-pure in A.

(e) If C is T-pure in A and B/C is T-pure in A/C then B is T-pure in A. □

An exact sequence $0\to A\to B\to C\to 0$ is T-pure exact if A is T-pure in B. If a short exact sequence is T-pure exact then each of its equivalent short exact sequences is T-pure exact also. Let $\mathrm{Pext}_T(A,B)$ be the set of equivalence classes of T-pure exact sequences $0\to B\to G\to A\to 0$.

3.14 THEOREM ([35, 2.2]). Pext_T is a subfunctor of Ext. If $0\to A\to B\to C\to 0$ is T-pure exact and X is any group then the following sequences are exact: $0\to \mathrm{Hom}(X,A)\to \mathrm{Hom}(X,B)\to \mathrm{Hom}(X,C)\to \mathrm{Pext}_T(X,A)\to \mathrm{Pext}_T(X,B)\to \mathrm{Pext}_T(X,C)\to 0$ and $0\to \mathrm{Hom}(C,X)\to \mathrm{Hom}(B,X)\to \mathrm{Hom}(A,X)\to \mathrm{Pext}_T(C,X)\to \mathrm{Pext}_T(B,X)\to \mathrm{Pext}_T(A,X)\to 0$. □

A group I is T-pure injective if $\mathrm{Pext}_T(X,I)=0$ for all groups X.

3.15 THEOREM ([35, 2.4]). For any group A there is a monomorphism $f:A\to I$ such that I is T-pure injective and A is T-pure in I. It is possible to choose $I=I_1\oplus I_2$ with I_1 injective and I_2 a product of discrete groups. If TA is Hausdorff then I may be chosen to be a product of discrete groups. □

Finally there is a characterization of the T-pure injectives.

3.16 THEOREM ([35, 2.6]). A group I is T-pure injective if and only if $I=I_1\oplus I_2$ with I_1 injective and I_2 a summand of a direct product of T-discrete groups. □

If T is the $\mathbb{Z}$-adic topology then the completions are T-pure injective. It is an interesting question whether this is true for other functorial topologies. A first step is 4.12 below.

Basic subgroups are a useful tool in dealing with the p-adic topology. We propose the following definition which generalizes the concept of p-basic subgroup.

3.17 DEFINITION. A subgroup B of A is a T-basic subgroup if

(a) B is T-concordant in A.

(b) T(A/B) is indiscrete.

(c) B has a simpler structure than A. □

Alternatives to conditions (a) and (b) are

(a') B is T-pure in A.

(b') B is T-dense in A.

Condition (a') is more restrictive than (a), and (b') is more restrictive than (b). It follows from (a) and (b) that every homomorphism $B\to C$ extends uniquely to a homomorphism $A\to \hat{C}$ where $\hat{C}$ is the completion of C. Conditions (a') and (b') imply the equality $\hat{B}=\hat{A}$ of the completions of B and A. If T is ideal then (a) and (b) together are equivalent to (a') and (b').

Condition (c) may have to be specified in each concrete case. In order that discrete groups have basic subgroups basics should at least be allowed to be direct sums of discrete groups. But this is not enough even for the p-adic topology. If B is allowed to be a direct sum of discrete groups and a free group then p-basic subgroups are T-basic subgroups if T is the p-adic topology or if C(T) is the class of finite p-groups. Except for the above comments we know nothing about the existence and use of T-basic subgroups.

We conclude the discussion of basic subgroups with an example showing that (a) and (b) together do not imply (a').

3.18 EXAMPLE. Let T be the minimal functorial topology whose discrete class C consists of all free groups. Then $T(\mathbb{Z}/n\mathbb{Z})$ is indiscrete, $n\mathbb{Z}$ is concordant in $\mathbb{Z}$ since both groups are discrete but the map $n\mathbb{Z}\to\mathbb{Z}: n\mapsto 1$ cannot be lifted to a map $\mathbb{Z}\to\mathbb{Z}$. □

Categorical properties of functorial topologies are best viewed within the encompassing category A_{lt} of groups with linear topology. A detailed discussion of A_{lt} is contained in Janvier [13]. We list the following properties.

3.19 PROPOSITION. a) A_{lt} is an additive category.

b) A_{lt} has arbitrary products. Products carry the product topology.

c) A_{lt} has arbitrary direct sums. We have $A[\mathcal{U}]=\oplus_i A_i[\mathcal{U}_i]$ if $A=\oplus_i A_i$ and $\mathcal{U}$ consists of all groups $\oplus U_i$ with $U_i\in\mathcal{U}_i$.

d) A_{lt} has kernels and cokernels. If $f:A[\mathcal{T}]\to B[\mathcal{T}]$ is a morphism then Kerf is the algebraic kernel of f with the subspace topology and Cokerf is the algebraic cokernel with the quotient topology. Thus Imf is Af with the subspace topology.

e) A sequence $A[\mathcal{T}_1]\xrightarrow{f}B[\mathcal{T}_2]\xrightarrow{g}C[\mathcal{T}_3]$ is exact in A_{lt} if and only if it is algebraically exact and both f and g are continuous and open maps onto their ranges, the latter carrying the subspace topologies. □

For functorial topologies we can now show:

3.20 PROPOSITION. Let T be a minimal functorial topology and let C=C(T). Then the following hold.

a) $0\to TA\to TB$ is exact in A_{lt} if and only if A is concordant in B. This is the case for each exact sequence $0\to A\to B$ if and only if C is closed under injective hulls.

b) $TB\to TC\to 0$ is exact in A_{lt} if and only if C is coconcordant with B. This is the case for each exact sequence $B\to C\to 0$ if and only if C is closed under epimorphic images.

c) The sequence $0\to TU\to TA\to T(A/U)\to 0$ is exact for all A and all $U\leq A$ with $A/U\in C$ if and only if C is closed under extensions.

PROOF. a) The first statement is clear. Suppose C is closed under injective hulls. Let $f:A\to D$ be given with $D\in C$. Without loss of generality D is injective and f extends to $\bar{f}:B\to D$. Now $\mathrm{Ker}f=A\cap\mathrm{Ker}\bar{f}$ showing that A is concordant in B. Conversely, let $D\in C$ and H its injective hull. By hypothesis D is concordant in H, hence there is an open subgroup U of H with $D\cap U=0$. Since D is essential in H, it follows that $U=0$ so TH is discrete.

b) The first part is trivial, the second part was noted in II.

c) The exactness at $T(A/U)$ is trivial. Suppose C is closed under extensions, and suppose $V\leq U$ with $U/V\in C$. Then $0\to U/V\to A/V\to A/U\to 0$ is exact, so $A/V\in C$ and U is concordant in A. Conversely, suppose $0\to D\to E\to F\to 0$ is exact with $D\in C$ and $F\in C$. By hypothesis $0\to TD\to TE\to TF\to 0$ is exact. There exists an open subgroup U of E such that $0=U\cap D$ since D is discrete. But D is open in E so $\{0\}$ is open in E and TE is discrete. □

Another property worth noting is the following.

(3.21) For any two groups A and B and any minimal functorial topology we have $T(A\oplus B)=TA\oplus TB$. Given a cardinal $\aleph$, for any family of groups $\{A_i:i\in I\}$ with $|I|\leq\aleph$ we have $T(\oplus_i A_i)=\oplus_i TA_i$ if and only if $C(T)$ is closed under direct sums with $\leq\aleph$ summands.

PROOF. If TA_i is discrete and $T(\oplus_i A_i)=\oplus_i TA_i$ then this last group is clearly discrete. Now suppose $\{A_i:i\in I\}$ with $|I|\leq\aleph$ is given. Let $U_i\leq A_i$ such that $A_i/U_i\in C$. Then $\oplus A_i/\oplus U_i\cong\oplus A_i/U_i\in C$ hence the functorial topology on $\oplus A_i$ is finer than the sum topology. On the other hand if $f:\oplus_i A_i\to D$, $D\in C$, is given then $\mathrm{Ker}f\supset\oplus_i\mathrm{Ker}(f|A_i)$ so the sum topology is finer than the functorial topology. □

One may ask when each group TA has a totally ordered linear basis. This question has been answered for ideal functorial topologies by Mader-Mines [27].

3.22 THEOREM. If every group TA has a totally ordered linear basis then $C(T)$ is closed under powers. Hence $T=T(R)$ for some topological family of radicals R and if T is ideal it is as described in 2.12. □

IV. <u>COMPLETION</u>. It is well-known that any Hausdorff group $A[T]$ has a completion $\hat{A}$. If $A[T]$ is not Hausdorff there still is a "Hausdorff completion" which is the completion of $(A/\bar{0})[S]$ where $\bar{0}$ is the closure of $\{0\}$ in $A[T]$ and S is the quotient topology. The completion process is, in fact, functorial, and we let $C:A_{lt}\to A_{lt}$ denote the completion functor. Some basic properties of the completion functor are the following.

4.1 PROPOSITION. a) C is an additive functor.

b) If $0\to A_1[T_1]\to A_2[T_2]\to A_3[T_3]\to 0$ is exact in A_{1t} then $0\to CA_1[T_1]\to CA_2[T_2]\to CA_3[T_3]$ is exact in A_{1t}.

c) $C(\oplus_i A_i[T_i])=\oplus_i CA_i[T_i]$.

d) $C(\Pi_i A_i[T_i])=\Pi_i CA_i[T_i]$. □

Given $A[\mathcal{U}]$ and $U\in\mathcal{U}$, consider the exact sequence $0\to U[U\cap\mathcal{U}]\to A[\mathcal{U}]\to(A/U)[\text{discrete}]\to 0$. Applying the completion functor we get

4.5 PROPOSITION. Let $A[\mathcal{U}]\in A_{1t}$ and $U\in\mathcal{U}$. Then $0\to CU\to CA\to A/U\to 0$ is exact in A_{1t}. The family $\{CU: U\in\mathcal{U}\}$ is a linear basis for CA, and if V is an open subgroup of CA then $V=C(A\cap V)$. Also, CU is the topological closure of U in CA. □

This simple result is surprisingly useful. It implies that A is complete if U is complete. Hence, if a group contains a complete subgroup $B[\mathcal{U}]$ then $A[\mathcal{U}]$ is complete. This observation is the basis of the solution of Problem 8 in Fuchs [10] by DeMarco-Orsatti [7]. If T is a minimal functorial topology and $A/U\in C(T)$ then 4.5 implies that the functorial topology on the underlying group of CA is finer than the topology of CA.

It is well-known (Fuchs [10; 13]) and commonly used that the Hausdorff completion of a group $A[\mathcal{U}]$ can be represented as an inverse limit $\hat{A}=\varprojlim\{A/U: U\in\mathcal{U}\}$. Let us introduce some relevant maps.

(4.6) Let $p_U: A\to A/U$ be the natural epimorphism. □

(4.7) For $V\le U\le A$, let $p_{UV}: A/V\to A/U$ be the unique homomorphism with $p_V p_{UV}=p_U$. □

Now $\hat{A}$ can be identified with the subgroup of $\Pi\{A/U: U\in\mathcal{U}\}$ consisting of all elements $[a_U+U]$ such that for $V\le U$, $(a_V+V)p_{UV}=a_U+U$ or equivalently $a_V-a_U\in U$. There are projections

(4.8) $\pi_U: \hat{A}\to A/U$: $[a_Z+Z]\pi_U=a_U+U$ and $\{\mathrm{Ker}\,\pi_U: U\in\mathcal{U}\}$ is a linear basis for the topology of $\hat{A}$. Furthermore, $\mathrm{Ker}\,\pi_U=\{[a_Z+Z]: a_Z\in U\}=CU$. The map

(4.9) $\eta: A\to\hat{A}$: $a\eta=[a+Z]$

has kernel $\bar{0}$, hence is injective if and only if $A[\mathcal{U}]$ is Hausdorff. We will frequently pretend that A is a subgroup of $\hat{A}$ although this is true only for $A/\bar{0}$.

A subfamily $\mathcal{B}$ of $\mathcal{U}$ will be called <u>directed</u> if for any $U,V\in\mathcal{B}$ there is $W\in\mathcal{B}$ such that $W\subset U\cap V$. If $\mathcal{B}$ is such a family then $A[\mathcal{B}]$ is again a topological group and we have an exact sequence

(4.10) $0\to\cap\{CV: V\in\mathcal{B}\}\to CA[\mathcal{U}]\to CA[\mathcal{B}]$

where $CA[\mathcal{U}]\to CA[\mathcal{B}]$: $[a_Z+Z]_{Z\in\mathcal{U}}\to[a_Z+Z]_{Z\in\mathcal{B}}$ is the projection. It is an interesting question which conditions will force this projection to be either injective or surjective. An easy result is the following.

4.11 PROPOSITION. Let $A[\mathcal{U}]$ be given and suppose $\mathcal{B}$ is a directed subfamily of $\mathcal{U}$ such that each $U\in\mathcal{U}$ is closed in $A[\mathcal{B}]$. Then

$CA[\mathcal{U}] \leq CA[\mathcal{B}]$.

PROOF. Suppose $[a_Z+Z] \in CA[\mathcal{U}]$ and $[a_V+V]_{V\in\mathcal{B}}=0$. Then $a_V \in V$ for all $V\in\mathcal{B}$. Consider a_U with $U\in\mathcal{U}$. Since U is closed in $A[\mathcal{B}]$ we have $U=\cap\{U+V: V\in\mathcal{B}\}$. For $V\in\mathcal{B}$ choose $W\in\mathcal{U}$ such that $W\subset U\cap V$. Then $a_U=(a_U-a_W)+(a_W-a_V)+a_V \in U+V+V$ so $a_U \in \cap\{U+V: V\in\mathcal{B}\}=U$, i.e. $a_U+U=0$. □

As an example it follows that the $\mathbb{Z}$-adic completion of a group is imbedded in its completion in the finite index topology.

If T is a minimal functorial topology we may view C(T) as a category of "models" and completions as objects constructed via inverse limits from these models. This point of view is developed in Applegate-Tierney [1]; see also V for more details. It is natural to ask which properties of the models are inherited by the completions. In particular, discrete groups are T-pure injective, and the question arises whether completions are T-pure injective. We have one result in this direction.

4.12 PROPOSITION. If T is a minimal functorial topology and $\hat{A}=CTA$ is T-pure in B with $T(B/\hat{A})$ indiscrete then $\hat{A}$ is a direct summand of B.

PROOF. Let $\mathcal{U}=\{U\leq A: A/U\in C(T)\}$ and let $g:\hat{A}\to B$ be the insertion. For each $U\in\mathcal{U}$ there is $f_U:B\to A/U$ such that $gf_U=\pi_U$. For $U\supset V$ we have $g(f_V p_{UV}-f_U)=\pi_V p_{UV}-\pi_U=0$ so $f_V p_{UV}-f_U$ factors through $B/\hat{A}$. But $T(B/\hat{A})$ is indiscrete so $f_V p_{UV}=f_U$. Hence there is a unique map $f:B\to\hat{A}$ such that $f\pi_U=f_U$. Since $gf\pi_U=gf_U=1\pi_U$ for each $U\in\mathcal{U}$ it follows that $gf=1$ and $\hat{A}$ is a summand of B. □

In the case of the $\mathbb{Z}$-adic topology this result is the first step in showing that $\mathbb{Z}$-adic completions are algebraically compact.

If T is such that every discrete group is torsion then the discrete groups have a primary decomposition. We will show that a group CTA has a "primary decomposition" provided that each discrete group A/U has only finitely many non-zero primary components. We begin with greater generality. Let F be a subfunctor of 1 on A and $A[\mathcal{U}]$ a topological group. For $U\in\mathcal{U}$ let $U_F/U=F(A/U)$. If $V\in\mathcal{U}$ and $V\leq U$ then the function p_{UV}: $A/V\to A/U$ maps $F(A/V)$ into $F(A/U)$ hence $V_F\subset U_F$. Thus $\mathcal{U}_F=\{U_F: U\in\mathcal{U}\}$ is a directed family of open subgroups of $A[\mathcal{U}]$ and by (4.10) $0\to\cap\{CU_F: U\in\mathcal{U}\}\to CA[\mathcal{U}]\to CA[\mathcal{U}_F]$ is exact in A_{1t}. Further, let G be another subfunctor of 1 such that for all $U\in\mathcal{U}$, $A/U=U_F/U\oplus U_G/U$. If $U\supset V$ and $a+V=(a_F+V)+(a_G+V)$ with $a_F\in V_F$ and $a_G\in V_G$ then $a+U=(a_F+U)+(a_G+U)$ with $a_F\in U_F$ and $a_G\in U_G$ since F and G are subfunctors of the identity. Hence $\{A/U\}=\{A/U_F\}\oplus\{A/U_G\}$ is valid in the category of inverse systems over $\mathcal{U}$. Thus we have:

4.13 PROPOSITION. If each A/U, $U\in\mathcal{U}$, has a decomposition $A/U=F(A/U)\oplus G(A/U)$ where F and G are subfunctors of the identity then $CA[\mathcal{U}]=CA[\mathcal{U}_F]\oplus CA[\mathcal{U}_G]$ in A_{1t}. □

We now consider the case of infinite decompositions.

4.14 PROPOSITION. Suppose that $\{F_i: i\in I\}$ is a family of subfunctors of the identity and that $A[\mathcal{U}]$ is a topological group. Then $CA[\mathcal{U}]\cong\Pi\{CA[\mathcal{U}_{F(i)}]: i\in I\}$ if $A/U\to\Pi\{A/U_{F(i)}: i\in I\}: a+U\to[a+U_{F(i)}]$ is an isomorphism for each $U\in\mathcal{U}$.

PROOF. This follows immediately from the fact that inverse limits commute with products. □

We now apply the last proposition to primary decompositions.

4.15 PROPOSITION. Let $A[\mathcal{U}]$ be a topological group such that each A/U, $U\in\mathcal{U}$, is a direct sum of finitely many primary groups. If U_p/U is the direct complement of the p-primary part of A/U and $\mathcal{U}_p=\{U_p: U\in\mathcal{U}\}$ then $CA[\mathcal{U}]\cong\Pi_p CA[\mathcal{U}_p]$. Furthermore, each $CA[\mathcal{U}_p]$ is a module over the ring J_p of p-adic integers.

PROOF. The map $A/U\to\Pi_p A/U_p: a+U\to[a+U_p]$ is an isomorphism since the product is by hypothesis finite. Since A/U_p is isomorphic to the p-primary part of A/U it is a J_p-module and so is $CA[\mathcal{U}_p]$. □

The preceeding proposition fails if there is no linear basis such that A/U, $U\in\mathcal{U}$, has only finitely many non-zero primary components. For example, if A is a discrete group with infinitely many non-zero primary components then it is not isomorphic to the product of its p-primary parts.

Proposition 4.15 can be used to determine the completion of $\mathbb{Z}$ in any linear topology.

4.16 PROPOSITION. Let $\mathcal{T}$ be a linear topology on $\mathbb{Z}$ which is not indiscrete. Let $\mathcal{U}=\{n_i\mathbb{Z}\}$ be a linear basis for $\mathbb{Z}[\mathcal{T}]$, and let $e(p)=\sup\{k: p^k|n_i \text{ for some } i\}$ $(\le\omega)$. Then $C\mathbb{Z}[\mathcal{T}]\cong\Pi_p C\mathbb{Z}[\mathcal{U}_p]$ and $C\mathbb{Z}[\mathcal{U}_p]$ is the topological group of p-adic integers if $e(p)=\omega$ and $C\mathbb{Z}[\mathcal{U}_p]$ is the discrete group $\mathbb{Z}/p^{e(p)}\mathbb{Z}$ if $e(p)\neq\omega$. □

If $0\to A_1[\mathcal{T}_1]\to A_2[\mathcal{T}_2]\xrightarrow{f} A_3[\mathcal{T}_3]\to 0$ is exact in A_{1t} and if $\mathcal{U}$ is a linear basis for $A_2[\mathcal{T}_2]$ then $0\to\{A_1/A_1\cap U\}\to\{A_2/U\}\to\{A_3/Uf\}\to 0$ is exact in the category of inverse systems over $\mathcal{U}$ and using properties of inverse systems (see Jensen [14]) we obtain an exact sequence in A $0\to CA_1[\mathcal{T}_1]\to CA_2[\mathcal{T}_2]\to CA_3[\mathcal{T}_3]\to C^{(1)}A_1[\mathcal{T}_1]\to\ldots\to C^{(n)}A_1[\mathcal{T}_1]\to C^{(n)}A_2[\mathcal{T}_2]\to C^{(n)}A_3[\mathcal{T}_3]\to\ldots$ where the "derived completion functors" $C^{(n)}$ do not depend on the choice of $\mathcal{U}$. This creates a dimension concept for a single topological group as well as for a functorial topology and a discrete class. Results on this topic are contained in Mader [24] and Mader-Mines [26].

Finally, we mention that a new method for computing completions was developed in Mader [25] using dual pairs. This technique allows the determination of completions of groups with the finite index topology (done previously by Zelger [37]), the Prüfer topology and we conjecture that it will yield a proof showing that every group is complete in the co-$\aleph$-topology for $\aleph > \aleph_0$.

V. COMPLETABILITY. Let T be a functorial topology on abelian groups and let C be the completion functor. The completion CTA of the group TA carries a certain topology, called the completion topology. The functorial topology on CTA is always finer than the completion topology (see 4.5), and the question arises whether the completion topology coincides with the functorial topology on the underlying group of CTA. Following Mines-Oxford [30] we call A completable if this is the case and we call T and C(T) completable if each A is completable. It is well-known that the $\mathbb{Z}$-adic and p-adic topologies are completable; for $\mathrm{cof}\lambda = \omega$ the p^{λ}-topology is completable (Kulikov [18]) but it is not necessarily completable if $\mathrm{cof}\lambda \neq \omega$ (Mines [28], Megibben [23, p. 109]); the finite index topology (Zelger [37]), and the Prüfer topology (Mader [24], also 5.12) are not completable. We will state and prove the compatibility criterion essentially due to Mines-Oxford [30, Theorem 1.2], and give several applications and examples.

In the following T denotes a fixed minimal functorial topology, and C=C(T). To simplify the notation let LA denote the completion of TA as an abstract group. Then L is a functor on A to A. Associated with L are two natural transformations.

(5.1) $\eta: \mathrm{Id} \to L$ is uniquely determined by the commutativity of the diagrams

$$\begin{array}{ccc} A & \xrightarrow{\eta_A} & LA \\ & {\scriptstyle p_U}\searrow \quad \swarrow{\scriptstyle \pi_U} & \\ & A/U & \end{array} \qquad (A \in A,\ A/U \in C)$$

where p_U is the natural epimorphism and π_U is the projection. Since CTA is the completion of TA the map $\eta_A: TA \to CTA$ is continuous. □

(5.2) $\mu: L^2 \to L$ is uniquely determined by the commutativity of the diagrams

$$\begin{array}{ccc} L^2A & \xrightarrow{\mu_A} & LA \\ {\scriptstyle \pi_{LU}}\big\downarrow & & \big\downarrow{\scriptstyle \pi_U} \\ LA/LU & \xrightarrow{\sigma_U} & A/U \end{array} \qquad (A \in A,\ A/U \in C)$$

where π_U is the projection, LU and σ_U are as in 4.5, and π_{LU} is the projection. It follows directly from these diagrams that $\mu_A: CTLA \to CTA$ is continuous. □

It is easy to check that η and μ are natural transformations. In fact, (L, μ, η) is a triple in A. This was shown by Applegate-Tierney [1] in much greater generality, and we simply state what this means.

5.3 PROPOSITION. (L, η, μ) is a triple in A i.e. $\eta: \mathrm{Id} \to L$ and $\mu: L^2 \to L$ are natural transformations such that the diagrams

$$\text{(5.4)} \qquad \begin{array}{ccc} L & \xrightarrow{L\eta} & L^2 \\ & \searrow & \swarrow \mu \\ & L & \end{array}$$

$$\text{(5.5)} \qquad \begin{array}{ccc} L & \xrightarrow{\eta L} & L^2 \\ & \searrow & \swarrow \mu \\ & L & \end{array}$$

and

$$\text{(5.6)} \qquad \begin{array}{ccc} L^3 & \xrightarrow{\mu L} & L^2 \\ L\mu \downarrow & & \downarrow \mu \\ L^2 & \xrightarrow{\mu} & L \end{array}$$

commute. η is the <u>unit</u> and μ the <u>multiplication</u> of the triple. □

Immediate consequences of (5.4) and (5.5) are the following direct decompositions.

5.7 PROPOSITION. For any abelian group A, $L^2A=\text{Im}(L\eta_A)\oplus\text{Ker}(\mu_A)$ $=\text{Im}(\eta_{LA})\oplus\text{Ker}(\mu_A)$, so $L^2A\cong LA\oplus\text{Ker}(\mu_A)$. □

It is opportune to note at this point that

(5.8) $L\eta_A:TLA\to CTLA$ is continuous.

PROOF. $\eta_A:TA\to TLA$ is continuous since T is a functorial topology hence $L\eta_A:CTA\to CTLA$ is continuous. Since the topology of TLA is finer than that of CTA it follows that $L\eta_A:TLA\to CTLA$ is continuous also. □

The maps η_{LA} and $L\eta_A$ are compared in the next proposition.

5.9 PROPOSITION. For any group A, $\text{Ker}(\eta_{LA}-L\eta_A)$ is the topological closure of $A\eta_A$ in TLA.

PROOF. The map $\eta_{LA}-L\eta_A:TLA\to CTLA$ is continuous by 5.1 and 5.8. Since CTLA is Hausdorff $\text{Ker}(\eta_{LA}-L\eta_A)$ is closed in TLA. Since η is a natural transformation the diagram

$$\text{(5.10)} \qquad \begin{array}{ccc} A & \xrightarrow{\eta_A} & LA \\ \eta_A \downarrow & & \downarrow \eta_{LA} \\ LA & \xrightarrow{L\eta_A} & L^2A \end{array}$$

commutes. Hence $\text{Ker}(\eta_{LA}-L\eta_A)$ contains $A\eta_A$ and its closure as well. Conversely, suppose $x\in\text{Ker}(\eta_{LA}-L\eta_A)$ i.e. $x\eta_{LA}=xL\eta_A$. Let $U\le LA$ be such that $LA/U\in C$. It suffices to show that $x\in A\eta_A+U$ since $\cap\{A\eta_A+U: LA/U\in C\}$ is the closure of $A\eta_A$ in TLA. Let $V=U\eta_A^{-1}$. Then V is open in A. Let ξ be the map which makes the diagram

$$\begin{array}{ccc} A & \xrightarrow{\eta_A} & LA \\ p_V \downarrow & & \downarrow p_U \\ A/V & \xrightarrow{\xi} & LA/U \end{array}$$

commutative. By definition of L the diagram

$$\begin{array}{ccc} LA & \xrightarrow{L\eta_A} & L^2A \\ \pi_V \downarrow & & \downarrow \pi_U \\ A/V & \xrightarrow{\xi} & LA/U \end{array}$$

is commutative. Hence $xp_U = x\eta_{LA}\pi_U = xL\eta_A\pi_U = x\pi_V\xi = ap_V\xi = a\eta_A p_U$ for some $a \in A$. Thus $x = a\eta_A + U$ as desired. □

We are now in a position to prove the completability criterion.

5.10 THEOREM. For any group A the following are equivalent.

(1) TA is completable.

(2) $\eta_{LA}: LA \to L^2A$ is bijective.

(3) $\mu_A: L^2A \to LA$ is bijective.

(4) $\eta_{LA} = L\eta_A$.

(5) $A\eta_A$ is dense in TLA.

(6) $T(LA/A\eta_A)$ is indiscrete.

(7) For every $U \leq LA$ with $LA/U \in C$, U is the closure of $U \cap A\eta_A$ in CTA.

PROOF. (1)⇒(2): By hypothesis TLA=CTA and the latter is complete and Hausdorff, so η_{LA} is bijective.

(2)⇒(3): (5.5).

(3)⇒(4): By (5.4) and (5.5), $L\eta_A \circ \mu_A = \eta_{LA} \circ \mu_A$.

(4)⇒(5): By 5.9 the closure of $A\eta_A$ is $\mathrm{Ker}(\eta_{LA} - L\eta_A) = LA$.

(5)⇒(6): (3.2) a).

(6)⇒(7): If $LA/U \in C$ then $A \cap U\eta_A^{-1} = V$ is open in A and we obtain the following commutative diagram with exact rows.

$$\begin{array}{ccccc} 0 \to V & \subset & A & \xrightarrow{p_V} & A/V \to 0 \\ \alpha \downarrow & & \eta_A \downarrow & & \downarrow \beta \\ 0 \to U & \subset & LA & \xrightarrow{p_U} & LA/U \to 0 \end{array}$$

where α, β are the maps induced by η_A. The projection $\pi_V: LA \to A/V$ satisfies $\eta_A\pi_V = p_V$. Thus $\eta_A(p_U - \pi_V\beta) = 0$. Hence $p_U - \pi_V\beta$ induces a map $LA/A\eta_A \to LA/U \in C$ which must be 0 by hypothesis. So $p_U = \pi_V\beta$, and the closure of $V\eta_A = U \cap A\eta_A$ is $\mathrm{Ker}\,\pi_V = \mathrm{Ker}\,\pi_V\beta = \mathrm{Ker}\,p_U = U$.

(7)⇒(1): If $LA/U \in C$ then U=the closure of $U \cap A\eta_A$ in CTA hence is open in CTA. □

As we have seen earlier a group is indiscrete in a minimal functorial topology if and only if it has certain divisibility properties. This allows for the following criterion.

5.11 PROPOSITION. Let T be a minimal functorial topology with discrete class C which contains reduced torsion groups only. Let $\pi_t(C) = \{p: Z(p) \in C\}$. Then TA is completable if and only if LA/A is p-divisible for all $p \in \pi_t(C)$.

PROOF. 5.10(6) and 3.5. □

It is easy to derive from 5.11 the well-known facts that the p-adic, $\mathbb{Z}$-adic, and for $\mathrm{cof}\lambda=\omega$ the p^{λ}-topology are completable. Negative results are obtained when C contains divisible groups. The difficulty here is that information is needed on the completion of at least one suitable group. If C contains no torsion free groups then $\mathbb{Z}$ can be used as a test case.

5.12 PROPOSITION. Let T be a minimal functorial topology with discrete class C which contains torsion groups only. If C contains a divisible group then T is not completable.

PROOF. Let $e(p)=\sup\{n: Z(p^n)\in C\}$ $(\leq\omega)$. By hypothesis C contains some group $Z(p^{\infty})$, so $e(p)=\omega$ for this prime p. By 4.16 $CT\mathbb{Z}$ has J_p as an epimorphic image. Since $J_p/\mathbb{Z}$ is not torsion there is a nontrivial homomorphism $J_p/\mathbb{Z}\to Z(p^{\infty})$. So $\mathbb{Z}$ is not completable by 5.10(6). □

An interesting example of a functorial topology T which is not completable although each group TA is metrizable is due to Mines-Oxford [30].

5.13 EXAMPLE. Let $C=\{D: D_t \text{ is bounded}\}$, and let T be the corresponding minimal functorial topology. Then T is not completable.

PROOF. For any group TA its torsion subgroup A_t is open since C contains all torsion free groups. Hence we have by 3.20 and 4.5 that $A+LA_t=LA$ and $A\cap LA_t=A_t$, thus $LA/A\cong LA_t/A_t$. However, LA_t is just the $\mathbb{Z}$-adic completion of A_t and it is well-known that LA_t/A_t is divisible but not torsion unless A_t is bounded. Hence no group with unbounded torsion subgroup is completable by 5.10(6). □

In the positive direction Mines-Oxford [30, 2.3] showed:

5.14 PROPOSITION. Let R be a radical which commutes with countable products. If $C=\{D: R^nD=0 \text{ for some } n\}$ then C is completable. □

D'Este [8] showed that every p-group is completable in the $\oplus_C$-topology.

REFERENCES

[1] H. Applegate-M. Tierney; Categories with models, Lecture Notes in Mathematics 80(1969), 156-244.

[2] S. Balcerzyk; On classes of abelian groups, Fund. Math. 51(1962), 149-178; (Corrections) 56(1964), 199-202.

[3] D. Boyer-A. Mader; Functorial topologies on abelian groups, Rocky Mountain J. Math, to appear.

[4] B. Charles; Methodes topologiques en theorie des groupes abeliens, Proceedings of the Colloquium on Abelian Groups, Budapest 1964, 29-42.

[5] B. Charles; Sous-groupes fonctoriels et topologies, Etudes Sur Les Groupes Abeliens, Paris 1968, 75-92.

[6] D. O. Cutler; Completions of topological abelian p-groups, Acta Math. Acad. Sci. Hung. 22(1972), 331-335.

[7] G. DeMarco-A. Orsatti; Complete linear topologies on abelian groups, Symp. Math. XIII(1974), 153-161.

[8] G. D'Este; The ω_C-topology on abelian p-groups, Ann. Scuola Normale Sup. Pisa, Serie 4, 7(1980), 241-256.

[9] I. Fleischer; Ueber Dualitaet in lineartopologischen Moduln, Math. Z. 72(1960), 439-445.

[10] L. Fuchs; Infinite Abelian Groups I, II, New York and London, 1970 and 1973.

[11] D. K. Harrison; Infinite abelian groups and homological methods, Ann. of Math. 69(1959), 366-391.

[12] D. K. Harrison; On the structure of Ext, Topics in Abelian Groups, Chicago 1963, 195-209.

[13] M. Janvier; Topologies lineaires sur des groupes abeliens, Universite des Sciences et Techniques du Languedoc, U. E. R. Mathematiques, 1970-1971.

[14] C. U. Jensen; Les foncteurs derives de $\varprojlim$ et leurs applications en theorie des modules, Lecture Notes in Mathematics 254(1972).

[15] I. Kaplansky; Dual modules over a valuation ring I, Proc. Amer. Math. Soc. 4(1953), 213-219.

[16] G. Koethe; Topological Vector Spaces I, New York 1969.

[17] W. Krull; Ueber separable, insbesondere kompakte separable Gruppen, J. Reine u. Angew. Math. 184(1942), 19-48.

[18] L. Ya. Kulikov; Generalized primary groups [Russian], Trudy Moskov. Mat. Obsc. 1(1952), 247-326; 2(1953), 85-167.

[19] S. Lefschetz; Algebraic Topology, Colloq. Lect. Amer. Math. Soc. 27(1942).

[20] H. Leptin; Ueber eine Klasse linearkompakter abelscher Gruppen I,II, Abh. Math. Sem. Univ. Hamburg 19(1954), 23-40; 19(1955), 221-243.

[21] H. Leptin; Linearkompakte Moduln und Ringe, I, II, Math. Z. 62 (1955), 241-267; 66(1957), 289-327.

[22] I. G. MacDonald; Duality over complete local rings, Topology I(1962), 213-235.

[23] Ch. Megibben; On p^{α}-high injectives, Math. Z. 122(1971), 104-110.

[24] A. Mader; Exact sequences of completions of abelian groups with linear topology, Houston Math. J., to appear.

[25] A. Mader; Duality theory and completions of linearly topologized modules, preprint.

[26] A. Mader-R. Mines; Completions of linearly topologized vector spaces, J. Algebra, to appear.

[27] A. Mader-R. Mines; Functorial topologies with totally ordered neighborhood bases, Archiv d. Math. (Basel) 34(1980), 272-275.

[28] R. Mines; A family of functors defined on generalized primary groups, Pac. J. Math. 26(1968), 349-360.

[29] R. Mines; Torsion and cotorsion completions, Etudes sur les groupes abeliens, Paris 1968, 301-303.

[30] R. Mines- E. Oxford; Model induced triples and completions of abelian groups, Symposia Mathematica 23(1979), 189-199.

[31] S. Pietrkowski; Theorie der unendlichen Abelschen Gruppen, Math. Annalen 104(1931), 535-569.

[32] H. Pruefer; Theorie der abelschen Gruppen II, Math. Z. 22(1925), 222-249.

[33] H. Schoeneborn; Ueber gewisse Topologien in Abelschen Gruppen I, II, Math. Z. 59(1954), 455-473; 60(1954), 17-30.

[34] L. Salce; The λ-inductive topology on abelian p-groups, Rend. Sem. Mat. Univ. Padova 59(1978), 167-177.

[35] C. Walker; Relative homological algebra and abelian groups, Ill. J. Math. 10(1966), 186-209.

[36] J. Waller; Generalized torsion complete groups, Etudes sur les groupes abeliens, Paris 1968, 345-356.

[37] A. Zelger; Sul completamento di un gruppo abeliano nella topologia dei sottogruppi di indice finito, Rend. Sem. Mat. Univ. Padova 52 (1974), 59-69.

AUTOMORPHISM GROUPS OF LOCALLY COMPACT ABELIAN GROUPS

Peter Plaumann

If G is a locally compact group its group of continuous automorphisms has a strong tendency not to be locally compact in its natural topology. In this note we study this phenomenon for the topological analoga of abelian torsion groups, i.e. for the class $\mathfrak{K}$ of locally compact, totally disconnected abelian groups, where every element lies in a compact subgroup. If $G \in \mathfrak{K}$ is discrete, the (local) compactness of AutG forces G to have artinian primary components. By Pontrjagin duality this settles the question for compact $G \in \mathfrak{K}$ too. To obtain results for a general $G \in \mathfrak{K}$ we need the stronger hypothesis that every factor of G has a locally compact group of automorphisms. We show that this condition is equivalent to the fact that the p-adic components of G are minimax groups. For arbitrary locally compact groups G the local compactness of AutG will be discussed in [P].

1 Groups of automorphisms. For topological groups G and H we denote by Hom(G,H) the set of all continuous homomorphisms from G to H. For $C \subseteq G$, $\mathcal{U} \subseteq H$ and $\eta \in \mathrm{Hom}(G,H)$ put

$$\Phi(C,\mathcal{U};\eta) = \{\varphi \in \mathrm{Hom}(G,H) \mid \forall c \in C \;\; c^{\varphi} c^{-\eta} \in \mathcal{U}\}.$$

If $G = H$ put

$$\Omega'(C,\mathcal{U}) = \Phi(C,\mathcal{U};1) \cap \mathrm{Aut}G$$

$$\Omega(C,\mathcal{U}) = \Omega'(C,\mathcal{U}) \cap \Omega'(C,\mathcal{U})^{-1}.$$

The following is well known (of [B], Chap. IV; [HR], 23.34, 26.5; [GLM]):

(1.1) *Let G and H be locally compact groups, let $\mathfrak{U}$ be a neighbourhood basis at the identity in H and let $\mathcal{L}$ be a family of compact subsets of G such that every compact subset of G is contained in an element of $\mathcal{L}$.*

(a) *The family $\Phi(\mathcal{L},\mathfrak{U};\mathrm{Hom}(G,H))$ is a basis for the compact-open topology on $\mathrm{Hom}(G,H)$.*

(b) *If H is abelian, then $\Phi(\mathcal{L},\mathfrak{U};0)$ is a basis at 0 for a group topology of $\mathrm{Hom}(G,H)$.*

(c) *If $G = H$, then $\Omega(\mathcal{L},\mathfrak{U})$ is a basis at 1 for a group topology on $\mathrm{Aut}G$.*

For a subgroup Γ of $\mathrm{Aut}G$ and subsets $C,U \subseteq G$ we put

$$\Omega_\Gamma(C,U) = \Omega(C,U) \cap \Gamma .$$

A direct computation shows

(1.2) *Let C be a subset and let U be a subgroup of the topological group G and let Γ be a subgroup of $\mathrm{Aut}G$. Then:*

(a) *If $C^\Gamma = C$ or $U^\Gamma = U$, then $\Omega_\Gamma(C,U)$ is a subgroup of Γ.*

(b) *If $C^\Gamma = C$ and $U^\Gamma = U$, then $\Omega_\Gamma(C,U)$ is a normal subgroup of Γ.*

A factor H/K of a topological group G is a quotient of a closed subgroup H of G. The factor H/K is called Γ-admissible for a subgroup Γ of $\mathrm{Aut}G$ if $H^\Gamma = H$ and $K = K^\Gamma$. Neglecting first the topologies it is clear that Γ induces on H/K an automorphism group $\tilde{\Gamma} = \Gamma/\Omega_\Gamma(H,K)$. On $\tilde{\Gamma}$ we now have two topologies: τ_q, the quotient topology induced from Γ, and τ_c, the toplology of $\tilde{\Gamma}$ as a group of automorphisms of H/K. In general we only know that τ_c is coarser than τ_q. There are however two important special cases where the two topologies coincide:

(1.3) _Let_ G _be a locally compact group and let_ Γ _be a subgroup of_ _AutG. Assume that for a_ Γ_-admissible factor_ H/K _of_ G _one of the following conditions holds._

a) τ_q _is a compact toplogy,_

b) $K = 1$ _and_ H _is a direct factor of_ G .

Then τ_q _and_ τ_c _coincide._

For a family $\{(G_\alpha, H_\alpha) \mid \alpha \in A\}$ of topological groups G_α with open normal subgroups H_α the local direct product is formed as in [B]. Algebraically it is the group of all elements (g_α) of the cartesian product $\prod_{\alpha \in A} G_\alpha$ for which almost all components lie in H_α. Its topology is given by the product topology on the normal subgroup $\prod_{\alpha \in A} H_\alpha$. As in [B], p. 64, Théorème 1 one shows

(1.4) _For_ $L = \bigwedge_\alpha (G_\alpha, H_\alpha)$ _the natural embedding_ $\mu : \prod_\alpha \mathrm{Aut}G_\alpha \longrightarrow \mathrm{Aut}L$ _is a continuous and open monomorphism._

The key to prove compactness of a group of automorphisms is the Ascoli-Theorem of Grosser and Moskowitz ([GM], Theorem (0.1)).

(1.5) _Let_ G _be a locally compact group and let_ Γ _be a subgroup of_ _AutG. The following two conditions are necessary and sufficient for the compactness of_ $\mathrm{cl}_{\mathrm{AutG}}(\Gamma)$ _, i.e. the closure of_ Γ _in_ AutG:

$[SIN]_\Gamma$: G _has a basis at_ 1 _consisting of_ Γ_-invariant sets,_

$[FC]^-_\Gamma$: _For all_ $g \in G$ _the orbits_ g^Γ _are relatively compact in_ G .

Denote by $\mathcal{R}$ the class of all locally compact, totally disconnected abelian groups, in which every element lies in a compact subgroup. In this class the following propositions are valid:

a) G has a basis at 1 consisting of compact and open subgroups,

b) Every compact subset of G lies in a compact subgroup.

Furthermore $\mathcal{R}$ is self-dual: the character group XG of a group $G \in \mathcal{R}$ is again contained in $\mathcal{R}$ ([HR], 24.18). Noting the fact that AutG and AutXG are isomorphic as topological groups we can formulate the following version of (1.5) for our purpose:

(1.6) *For* $G \in \mathcal{R}$ *and* $\Gamma \subseteq \mathrm{Aut}G$ *the following are equivalent:*

(1) $\mathrm{cl}_{\mathrm{Aut}G}(\Gamma)$ *is compact*

(2) (S): *The compact and open* Γ*-invariant subgroup of* G *form a basis at* 1

(F): *Every element of* G *lies in a compact* Γ*-invariant subgroup*

(3) $(\hat{S})$: *Every compact subgroup of* G *is contained in a compact and open* Γ*-invariant subgroup of* G

$(\hat{F})$: *The intersection of all open* Γ*-invariant subgroups of* G *is trivial.*

If an abelian group G has a direct decomposition $G = A \times B$ and if Γ is the subgroup of all automorphisms of G fixing A, consider in Γ the centralizer Σ of the series $1 \subseteq A \subseteq G$. We clearly have $\Sigma \cong \mathrm{Hom}(B,A)$ and $\Gamma/\Sigma \cong (\mathrm{Aut}A) \times (\mathrm{Aut}B)$. For locally compact groups the topologies behave correctly. Thus we obtain the following modest but useful tool:

(1.7) *Let* G *be a locally compact abelian group with a direct decomposition* $G = A \times B$, *where* A *is a characteristic subgroup of* G. *Then* AutG *is [locally] compact if and only if* AutA, AutB *and* Hom(B,A) *are [locally] compact.*

(1.8) *Let* L *be a locally, compact group with a compact open normal subgroup* $H \neq L$ *and let* A *be an infinite set. If* $L_\alpha = L$

<u>and</u> $H_\alpha = H$ <u>for all</u> $\alpha \in A$ <u>and if</u> $G = \underset{\alpha\in A}{\Lambda}(L_\alpha, H_\alpha)$ <u>then</u> AutG <u>is not locally compact</u>.

<u>Proof</u> (adapted from [HR], 26.18 (j)): Assume AutG locally compact. Then there are a compact subset C of G and a neighbourhood U of 1 in G such that $\Omega(C, U)$ is compact. Since C is compact there is a finite subset A_1 of A with

$$C \subseteq \prod_{\alpha\in A_1} L_\alpha \times \prod_{\alpha\in A_2} H_\alpha = F ,$$

where $A_2 = A \setminus A_1$. Without loss of generality we can assume that

$$E = \prod_{\alpha\in A_1} \{1\} \times \prod_{\alpha\in A_2} H_\alpha \subseteq U .$$

Then the inclusion $\Omega(F,E) \subseteq \Omega(C, U)$ implies that $\Omega(F,E)$ has compact closure in AutG . We now chose a fixed element $x_o \in L \setminus H$ and define elements $x^{(\alpha)} \in G$ for all $\alpha \in A_2$ by

$$x^{(\alpha)}_\beta = \begin{cases} x_o & \alpha = \beta \\ 1 & \alpha \neq \beta \end{cases} .$$

Then all the elements of the set $S = \{x^{(\alpha)} \mid \alpha \in A_2\}$ are conjugate under $\Omega(F,E)$. Hence S is relatively compact, and this implies that

$$S \cdot (\prod_{\alpha\in A} H_\alpha) \Big/ (\prod_{\alpha\in A} H_\alpha)$$ is finite, q. e.a.

<u>2 Discrete and compact groups.</u> From now on all groups are abelian and we use the additive notation.

(2.1) <u>Theorem:</u> <u>For a discrete abelian p-group G the following are equivalent:</u>

(1) AutG <u>is compact,</u>

(2) <u>Every finite subset of</u> G <u>is contained in a finite characteristic subgroup of</u> G ,

(3) G <u>is artinian</u> .

Proof: By the Ascoli-Theorem (2) is a consequence of (1). Suppose that (2) holds and assume first G reduced. By [F], 27.3 there is a cyclic subgroup $A = gp\{a\}$ with $G = A \oplus C$. For $B = \{c \in C \mid o(c) \leq o(a)\}$ and every $b \in B$ let α_b be the automorphism of G defined by

$$a^{\alpha_b} = a + b \quad , \quad c^{\alpha_b} = c \quad \text{for all } c \in C .$$

If G was not artinian then B would be infinite. Hence $a^{\mathrm{Aut}G}$ would be infinite which is impossible. For the general case we observe that a divisible group satisfying (2) has finite rank and that (2) as well as (3) hold in G iff they hold respectively in the maximal divisible subgroup D of G and the factor group G/D. Hence (2) implies (3). If finally G is artinian then $G \cong \mathbb{Z}(p^\infty)^r \times E$ with a finite subgroup E and a natural number r. The groups $\mathrm{Aut}E$ and $\mathrm{Aut}\ \mathbb{Z}(p^\infty)^r \cong GL(r, \Delta_p)$ are compact and $\mathrm{Hom}(E, \mathbb{Z}(p^\infty)^r)$ is finite. Using (1.7) we conclude that $\mathrm{Aut}G$ is compact.

(2.2) **Corollary:** The automorphism group of a discrete abelian p-group G without elements of infinite height is locally compact if and only if it is compact.

Proof: Assume $\mathrm{Aut}G$ locally compact. Then $\Omega(E,\{o\})$ is compact for a suitable finite subset E of G. We may take for E a subgroup and observing [F], 65.1 we may assume that $G = E \oplus C$ holds. By (1.3) we know that $\mathrm{Aut}C$ is isomorphic to subgroup of the compact group $\Omega(E,\{o\})$. Hence $\mathrm{Aut}C$ itself is compact and we can apply (2.1).

By duality we can convert (2.1) and (2.2) in statements about compact, totally disconnected abelian groups, which are projective limits of finite p-groups or equivalently which admit a continuous operation of the ring Δ_p of p-adic integers. We remark that such a group G has a dense torsion subgroup if and only if its discrete character group XG has no elements of infinite height (compare KIEFER's contribution to these Lecture Notes).

(2.1)* Theorem: For a compact abelian pro-p-group the following propositions are equivalent:

(a) $\mathrm{Aut}G$ is compact,

(b) Every open subgroup of G contains an open characteristic subgroup of G,

(c) G is noetherian (w.r.t. closed subgroups).

(2.2)* Corollary: The automorphism group of a compact abelian pro-p-group with a dense torsion subgroup is locally compact if and only if it is compact.

To conclude this section we remark that using (1.4) it is now very easy to characterize the compactness of $\mathrm{Aut}G$ for a discrete abelian torsiongroup G or a compact, totally disconnected abelian group G in general. We do not bother to formulate these theorems explicitely.

3 Minimax groups. We turn our attention now to arbitrary locally compact abelian groups which admit a Δ_p-operation and call these groups (p)-groups. As new phenomen on we have to care about groups like $\mathbb{H}_p$, the additive group of the quotient field of Δ_p. The class of abelian (p)-groups is obviously self-dual, in particular $X\mathbb{H}_p \cong \mathbb{H}_p$.

(3.1) Lemma: For a locally compact abelian (p)-group G the following are equivalent:

(a) G has finite rank as a Δ_p-module,

(b) $G \cong \Delta_p{}^n \oplus \mathbb{H}_p{}^m \oplus T$ with a discrete artinian p-group T and natural numbers m and n.

(c) G is a minimax group (w.r.t. closed subgroups).

Proof: By [B], p. 30, Proposition 7, we know that (b) is a consequence of (a). If G has the structure indicated under (b) then there is a subgroup $U \cong \Delta_p^{m+n}$ such that $G/U \cong \mathbb{Z}(p^\infty)^m \oplus T$, i.e. (c) holds. Finally we remark that an artinian (p)-group is discrete and dually a noetherian (p)-group is compact. If G has a noetherian subgroup N with artinian factor group G/N then XN is artinian too. Hence both N and G/N have finite Δ_p-rank.

(3.2) Lemma: If the locally compact abelian (p)-group G is a minimax group then there is a compact and open subgroup C of G for which $\Omega(C,C)$ is compact.

Proof: By (3.1) and (1.3) we may assume that G is isomorphic to $\Delta_p{}^n \oplus \mathbb{H}_p{}^m$. Take any open subgroup $C_1 \cong \Delta_p{}^m$ of $\mathbb{H}_p{}^m$ and put $C = \Delta_p{}^n \oplus C_1$. Then every (continuous) automorphism of C extends uniquely to a (continuous) automorphism of G. Hence $\Omega = \Omega(C,C)$ and AutC coincide neglecting the topology first. Since AutC is compact by (2.1)* and C is open in G we obtain from (1.5) that G has the property $[SIN]_\Omega$. If g is any element of G then $(g+C)^\Omega \subseteq (g + C)^{\mathrm{Aut}(G/C)}$ is finite by (2.1). Therefore $\bar{g}^\Omega + C$ is compact, because C is compact, i.e. G has property $[FC]_\Omega^-$. Hence Ω is relatively compact by (1.5). As $\Omega(C,C)$ is obviously closed in AutG our assertion holds.

(3.3) Theorem: For a locally compact abelian group the following propositions are equivalent:

(a) AutF is locally compact for every factor F of G,

(b) G is a minimax group.

Proof: Assume (a). If A is a discrete factor of G by (1.8) with $L = \mathbb{Z}(p)$, $H = o$ every Ulm factor of A is artinian. Then according to [F], 33.2 we know that A itself is artinian. Dually every compact factor of G is noetherian. Hence G is a minimax group. Conversely by choosing C for any factor F of G as in Lemma (3.2) it is clear that $\Omega(C,C)$ is a compact and open subgroup of $\mathrm{Aut}F$. Hence $\mathrm{Aut}F$ is locally compact.

(3.4) **Corollary:** *For a locally compact abelian (p)-group the following propositions are equivalent:*

(a) $\mathrm{Aut}F$ *is compact for every factor* F *of* G,

(b) 1. G *is compact or discrete,*

2. $\mathrm{Aut}G$ *is compact,*

(c) 1. G *is compact or discrete,*

2. G *is a minimax group,*

(d) G *is noetherian or discrete and of finite rank.*

Proof: Assuming (a) we obtain

$$G \cong \Delta_p{}^r \oplus \mathbb{H}_p{}^s \oplus \mathbb{Z}(p^\infty)^t \oplus E$$

with E finite by (3.1). Since $\mathrm{Aut}\,\mathbb{H}_p{}^s \cong GL(s, \mathbb{H}_p)$ is not compact for $s > o$ we know that $s = o$. Furthermore

$$\mathrm{Hom}(\Delta_p{}^r, \mathbb{Z}(p^\infty)^t) = \mathbb{Z}(p^\infty)^{rt}$$

is compact iff $rt = o$. Hence (b) holds.

If (b) holds we conclude (c) from (2.1) and (2.1)*. We already mentioned above that (d) is a consequence of (c).
If finally (d) holds by duality we may assume that G is noetherian. Then every factor of G is a finitely generated Δ_p-module hence of the form $\Delta_p{}^n \times E$ with E finite. This implies (a).

(3.5) **Corollary:** *For a locally compact abelian (p)-group* G *whose factors all have a locally compact group of automorphisms the following are equivalent:*

(a) $\mathrm{Aut}G$ *is compact*

(b) AutF *is compact for every factor* F *of* G .

Proof: Assume that (a) holds but not (b). Then G is neither compact nor discrete by (3.4). Hence, as a consequence of (3.3), there is a factor of G isomorphic to $\mathbb{H}_p$ or $\Delta_p \times \mathbb{Z}(p^\infty)$, q. e.a.

For an arbitrary group L in $\mathfrak{K}$, where $\mathfrak{K}$ denotes the class of all locally compact, totally disconnected groups in which every element lies in a compact subgroup, we denote by $L_{(p)}$ the set of all p-adic elements. From [B], Chap. III it follows that $L_{(p)}$ is a closed characteristic subgroup and that $G = \underset{p}{\Lambda} L_{(p)}$, where the compact-open subgroups $H_{(p)}$ of $L_{(p)}$ can be chosen arbitratily.

(3.6) **Theorem:** *For a group* $G \in \mathfrak{K}$ *the following statements are equivalent:*

(a) AutF *is locally compact for every factor* F *of* G

(b) $G_{(p)}$ *is a minimax group for every prime* p .

Proof: With (3.3) we deduce (b) from (a). If (b) holds find by (3.2) compact open subgroups $C_{(p)}$ of $L_{(p)}$ for which $\Omega(C_{(p)}, C_{(p)})$ is compact. By [B], p. 68, Théorème 1 we have

$$\mathrm{Aut}G = \mathrm{Aut}(\underset{p}{\Lambda}(G_{(p)}, C_{(p)})) = \underset{p}{\Lambda}(\mathrm{Aut}G_{(p)}, \Omega(C_{(p)}, C_{(p)}))$$

and by [B], p. 8, Prop. 1 this local direct product is locally compact. Since (b) is inherited by every factor of G we obtain (a) .

Literature.

[B] Braconnier, J.: Sur les groupes topologiques localement compacts. J. Math. Pures Appl., N.S. 27, 1 - 85 (1948).

[F] Fuchs, L.: Infinite abelian groups, New York 1970, 1973.

[GLM] Grosser, S. - O. Loos - M. Moskowitz: Über Automorphismengruppen lokal-kompakter Gruppen und Derivationen von Lie-Algebren.

Math. Z. 114, 321 - 339 (1970).

[GM] Grosser, S. - M. Moskowitz: Compactness conditions in topological groups. J. reine u. angew. Mathematik, 246, 1 - 40 (1973).

[HR] Hewitt, E. - K.A. Ross: Abstract harmonic analysis I, Berlin-Göttingen-Heidelberg 1963.

[P] Plaumann, P.: Abelian groups with locally compact automorphism groups, to appear.

ZUR KENNZEICHNUNG VON ELATIONEN

K. Faltings

Sei K ein Körper von Primzahlcharakteristik p und sei A ein Vektorraum über K. Sind U und V Unterräume von A mit $U \subseteq V$, so sei $E(U,V)$ die Gruppe aller Automorphismen δ von A mit $A(\delta - 1) \subseteq U$ und $V(\delta - 1) = 0$; bekanntlich ist $E(U,V)$ zu $\mathrm{Hom}_K(A/V,U)$ isomorph, also insbesondere abelsch, und wegen $\mathrm{Char}(K) = p > 0$ ist $E(U,V)$ sogar eine elementar abelsche p-Gruppe. Ist $P \subseteq A$ ein Punkt [= Unterraum vom Rang 1], so ist $E(P,P)$ die Gruppe aller <u>Elationen</u> mit dem <u>Zentrum</u> P, und ist $H \subseteq A$ eine Hyperebene, so ist $E(H,H)$ die Gruppe aller <u>Elationen</u> mit der <u>Achse</u> H. Sind U,V Unterräume mit $0 \neq U,V \neq A$, so ist genau dann $E(U,U) = E(V,V)$, wenn $U = V$ gilt; sind überdies U und V Punkte oder Hyperebenen, so ist genau dann $E(U,U) \cap E(V,V) \neq 1$, wenn $U = V$ gilt oder aber U und V ein Paar inzidenter Punkte und Hyperebenen bilden. In diesem Zusammenhang ist der folgende Satz zu sehen:

SATZ. *Sei K ein Schiefkörper mit $\mathrm{Char}(K) = p > 0$ und sei A ein K-Vektorraum mit $\mathrm{Rang}_K A = n > 1$. Sei Δ eine Untergruppe von $\Gamma = \mathrm{Aut}_K A$. Dann sind die folgenden Aussagen äquivalent:*

(i) (a) *Δ ist eine abelsche p-Untergruppe von Γ,*
und (b) $(C_\Gamma \Delta)[p] \subseteq \Delta$,
und (c) *ist $\delta \in \Delta$ mit $o(\delta) = p$, so gilt $C_\Gamma \delta \subseteq N_\Gamma \Delta$.*

(ii) (a) *Es gibt einen Punkt $P \subseteq A$ mit $\Delta = E(P,P)$.*
oder (b) *Es gibt eine Hyperebene $H \subseteq A$ mit $\Delta = E(H,H)$.*
oder (c) *Es ist $p = |K| = 3$, $n = 3$ und Δ ist eine der beiden weiteren maximalen Untergruppen einer 3-Sylowuntergruppe von Γ.*

oder (d) Es ist $p = |K| = 2$, $n = 3$ und Δ ist die dritte maximale Untergruppe einer 2-Sylowuntergruppe von Γ.

Die Bezeichnungen sind wie üblich; insbesondere ist $C_G U$ der Zentralisator, $N_G U$ der Normalisator von U in G, $G[p]$ ist die Menge aller $x \in G$ mit $x^p = 1$ und $o(x)$ ist die Ordnung von x. Sind U,V Unterräume von A mit $U \subseteq V$, so ist $S(U,V)$ die *Stabilitätsgruppe* der Kette $0 \subseteq U \subseteq V \subseteq A$; d. i. die Gruppe aller Automorphismen δ von A mit $U = U\delta$ und $V = V\delta$, die in jedem der Faktoren U, V/U und A/V den Einsautomorphismus induzieren.

Wir beginnen mit einigen mehr oder minder trivialen Bemerkungen, die der Leser selbst beweisen möge. - Seien γ, σ Endomorphismen von A mit $\gamma = 1 - \sigma$. Wegen $\mathrm{Char}(K) = p > 0$ gilt genau dann $\gamma^p = 1$, wenn $\sigma^p = 0$ ist; in diesem Fall ist $\gamma \in \Gamma$, und es gilt $\gamma^{-1} = 1 + \sigma + \cdots + \sigma^{p-1}$. Ist α ein weiterer Endomorphismus, so gilt genau dann $\alpha\gamma = \gamma\alpha$, wenn $\alpha\sigma = \sigma\alpha$ ist. Offensichtlich gilt

(1) Hilfssatz. *Seien $\sigma, \tau \in \mathrm{End}_K(A)$ und sei ξ der kanonische Isomorphismus von $A/\mathrm{Ker}(\sigma)$ auf $A\sigma$. Dann sind äquivalent:*

(i) $\sigma\tau = \tau\sigma$.

(ii) *Es gilt* $(A\sigma)\tau \subseteq A\sigma$, $(\mathrm{Ker}(\sigma))\tau \subseteq \mathrm{Ker}(\sigma)$ *und* $(\tau|A/\mathrm{Ker}(\sigma))\xi = \xi(\tau|A\sigma)$.

(2) Folgerung. *Sei $\sigma \in \mathrm{End}_K(A)$ mit $\sigma^2 = 0$ und sei $C = C_\Gamma(\sigma)$. Sei X einer der Faktoren $A\sigma$, $\mathrm{Ker}(\sigma)/A\sigma$ oder $A/\mathrm{Ker}(\sigma)$ von A. Dann wird jeder Automorphismus von X von einem Element aus C induziert.*

(3) Hilfssatz. *Seien $\sigma \in \mathrm{End}_K(A)$ und $\alpha \in \mathrm{Aut}_K(A)$. Dann gilt $A\alpha^{-1}\sigma\alpha = (A\sigma)\alpha$ und $\mathrm{Ker}(\alpha^{-1}\sigma\alpha) = (\mathrm{Ker}(\sigma))\alpha$.*

Sei $\gamma \in \Gamma = \mathrm{Aut}_K A$. γ heisst *multiplikativ* genau dann, wenn es zu jedem $x \in A$ ein $t \in K$ mit $x\gamma = xt$ gibt; d. h. also $P = P\gamma$ für jeden Punkt P. Ist $\mathrm{Rang}\, A > 1$, so ist dieses t ein durch γ eindeutig bestimmtes Element aus dem Zentrum von K; vgl. BAER[1; p. 43 und p. 201]. Da die Charakteristik von K eine Primzahl p ist, gilt:

(4) Hilfssatz. Ist γ multiplikativ und ist die Ordnung von γ eine Potenz von p, so gilt $\gamma = 1$.

Zum Beweis sei lediglich bemerkt, dass es in der multiplikativen Gruppe $K - \{0\}$ von K keine Elemente $\neq 1$ von p-Potenzordnung gibt. - Mit Hilfe von (3) zeigt man:

(5) Hilfssatz. Sei U ein Unterraum von A, sei $\Delta = E(U,U)$ und sei $\alpha \in \Gamma$. Dann ist $\alpha \in N_\Gamma \Delta$ genau dann, wenn $U\alpha = U$ gilt.

(6) Lemma. Sei U ein Punkt oder eine Hyperebene von A und sei $\Delta = E(U,U)$. Dann ist Δ maximal unter den abelschen p-Untergruppen von Γ, und aus $1 \neq \delta \in \Delta$ folgt stets $C_\Gamma(\delta) \subseteq N_\Gamma \Delta$.

Beweis. Ist P ein Punkt mit $P \subseteq U$ und ist H eine Hyperebene mit $U \subseteq H$, so gibt es eine Elation $1 + \sigma \in \Delta$ mit $P = A\sigma$ und $H = \mathrm{Ker}(\sigma)$. Ist also $\delta \in C_\Gamma \Delta$, so gilt $P = P\delta$ und $H = H\delta$ wegen (3). Nun ist jeder Unterraum von U Erzeugnis von Punkten, und jeder Unterraum von A/U ist Durchschnitt von Hyperebenen, so dass δ auf U und auf A/U multiplikativ wirkt. Ist nun δ von p-Potenzordnung, so folgt $\delta \in \Delta$ aus (4) und damit ist Δ als maximale (elementar-)abelsche p-Untergruppe von Γ erkannt. Sei nun $1 \neq \delta = 1 + \sigma \in \Delta$. Dann ist $U = A\sigma$ falls U ein Punkt, $U = \mathrm{Ker}(\sigma)$ falls U eine Hyperebene ist; ist also $\alpha \in C_\Gamma(\delta)$, so gilt $U = U\alpha$ wegen (3) und aus (5) folgt $\alpha \in N_\Gamma \Delta$, q.e.d.

(7) Hilfssatz. Sei Δ eine abelsche p-Gruppe und sei X ein Δ-K-Bimodul mit $\mathrm{Rang}_K X > 1$. Dann ist X reduzibel.

Beweis. Sei N der Kern der von Δ auf X induzierten Darstellung. Ist $N = \Delta$, so ist X reduzibel wegen $\mathrm{Rang}_K X > 1$; sei also $\Delta/N \neq 1$ und sei $\gamma \in \Delta/N$ mit $o(\gamma) = p$. Sei $\gamma = 1 - \sigma$; dann ist $\sigma \neq 0$ und es gilt $\sigma^p = 0$. Also gilt $0 \subset X\sigma \subset X$ und aus (3) folgt die Invarianz von $X\sigma$ unter Δ/N, also auch unter Δ, q.e.d.

(8) Lemma. <u>Sei</u> Δ <u>eine abelsche</u> p-<u>Untergruppe von</u> $\Gamma = \mathrm{Aut}_K A$, <u>sei</u> $\delta = 1 - \tau \in \Delta$ <u>mit</u> $\tau^2 = 0$ <u>und sei</u> $C_\Gamma(\delta) \subseteq N_\Gamma \Delta$. <u>Dann gilt</u> $\Delta \subseteq S(A\tau, \mathrm{Ker}(\tau))$. <u>Insbesondere gilt für alle</u> $\gamma \in \Delta$ <u>stets</u> $\tau(\gamma - 1) = 0 = (\gamma - 1)\tau$.

Beweis. Wegen $\delta \in \Delta$ und (3) sind $A\tau$ und $\mathrm{Ker}(\tau)$ invariant unter Δ, so dass die auftretenden Faktoren $A/\mathrm{Ker}(\tau)$, $\mathrm{Ker}(\tau)/A\tau$ und $A\tau$ sämtlich Δ-K-Bimoduln sind; sei X einer von diesen. Wir zeigen nun, dass jedes $\gamma \in \Delta$ auf X multiplikativ ist: wegen (4) ist γ dann die Eins auf X, und die Behauptung ist bewiesen. Ist nun $\mathrm{Rang}_K X \leq 1$, so ist dies trivial; sei also $\mathrm{Rang}_K X > 1$. Wegen (7) gibt es dann einen Unterraum Y von X mit $Y\Delta = Y$ und $0 \subset Y \subset X$. Sei $\alpha \in C = C_\Gamma(\delta) = C_\Gamma(\tau)$; wegen (3) wirkt α in natürlicher Weise auf X. Nach Voraussetzung über Δ gilt aber $\Delta = \alpha^{-1}\Delta\alpha$ und daraus folgt $(Y\alpha)\Delta = Y\alpha\alpha^{-1}\Delta\alpha = Y\Delta\alpha = Y\alpha$, so dass auch $Y\alpha$ invariant unter Δ ist. Wegen (2) wird aber jeder Automorphismus von X von einem Element aus C induziert; und folglich ist jeder Punkt P von X Durchschnitt von Unterräumen $Y\alpha$ mit $\alpha \in C$. Also gilt $P\Delta = P$ für jeden Punkt P von X und damit ist Δ als multiplikativ erkannt, q.e.d.

Beweis des Satzes. In (6) wurde bereits gezeigt, dass aus (ii.a) oder (ii.b) jeweils (i) folgt; und man überzeugt sich durch leichte Rechnungen (etwa mit Matrizen), dass dies auch für (ii.c) und (ii.d) gilt. Es bleibt zu zeigen, dass unter der Annahme (i) für Δ eine der Aussagen (a) - (d) von (ii) zutrifft. Dabei sind zwei Fälle zu unterscheiden:

Fall 1: Es gilt $(\delta - 1)^2 = 0$ für alle $\delta \in \Delta$. Wegen (i) ist dann (8) anwendbar und liefert $\sigma\tau = 0 = \tau\sigma$ für alle $\sigma, \tau \in \Delta - 1$ und dies besagt gerade $A\sigma \subseteq \mathrm{Ker}(\tau)$ für alle $\sigma, \tau \in \Delta - 1$. Ist also U das Erzeugnis aller $A\sigma$ mit $\sigma \in \Delta - 1$ und ist V der Durchschnitt aller $\mathrm{Ker}(\sigma)$ mit $\sigma \in \Delta - 1$, so gilt $U \subseteq V$ und offensichtlich ist $\Delta \subseteq E(U,V)$. Da aber $E(U,V)$ eine elementar abelsche p-Gruppe ist, folgt $\Delta = E(U,V)$ aus (i.b); und da aus $U \subset V$ folgte, dass $E(U,V) \subset E(U,U)$ gilt, ist sogar $U = V$, d. h. es gilt $\Delta = E(U,U)$. Wegen $\mathrm{Rang}_K A > 1$ und (i.b) folgt $\Delta \neq 1$, also $0 \neq U \neq A$; sei angenommen, es wäre $\mathrm{Rang}_K U > 1$ und $\mathrm{Rang}_K A/U > 1$. Dann gibt es einen Punkt P mit $P \subset U$ und es gibt eine Hyperebene H mit $U \subset H$; sei $\eta = 1 - \lambda$ eine Elation mit Zentrum $P = A\lambda$ und Achse $H = \mathrm{Ker}(\lambda)$. Dann ist $1 \neq \eta \in \Delta$ mit $o(\eta) = p$, so dass $C_\Gamma(\eta) \subseteq N_\Gamma \Delta$ aus (i.c) folgt. Wegen (2) gibt es aber ein $\alpha \in C_\Gamma(\eta)$ mit $U\alpha \neq U$, und dies ist ein Widerspruch zu (5). Also ist U ein Punkt oder eine Hyperebene und damit gilt (ii.a) oder (ii.b).

Fall 2: Es gibt ein $\delta \in \Delta$ mit $(\delta - 1)^2 \neq 0$. Sei $\tau = \delta - 1$; dann ist $\tau^2 \neq 0 = \tau^p$. Sei $i \in \omega$ mit $2 \leq i$ sowie $\tau^i \neq 0$ und $\tau^{i+1} = 0$ und sei $\sigma = \tau^i$. Offenbar gilt $\Delta \subseteq C_\Gamma(\tau) \subseteq C_\Gamma(\sigma)$ und daraus folgt $1 - \sigma \in (C_\Gamma \Delta)[p]$, so dass $1 - \sigma \in \Delta$ wegen (i.b) gilt. Nun ist $1 \neq 1 - \sigma$ so dass (i.c) anwendbar ist, und mit (8) folgt nun $\Delta \subseteq S(A\sigma, \mathrm{Ker}(\sigma))$. Sei $P \subseteq A\sigma$ ein Punkt, sei $H \supseteq \mathrm{Ker}(\sigma)$ eine Hyperebene und sei η eine Elation mit Zentrum P und Achse H. Dann ist $1 \neq \eta \in E(A\sigma, \mathrm{Ker}(\sigma))$ und $E(A\sigma, \mathrm{Ker}(\sigma))$ liegt im Zentrum von $S(A\sigma, \mathrm{Ker}(\sigma))$, so dass wir $1 \neq \eta \in (C_\Gamma \Delta)[p]$ haben. Wegen (i.b) ist dann $\eta \in \Delta$, und wegen (i.c) dürfen wir (8) auf η und Δ anwenden. Also gilt $\Delta \subseteq S(P,H)$. Insbesondere gilt $\tau^3 = 0$, und wegen $\tau^2 \neq 0$ ist $P \subset H$. Weiter ist $A\tau \subseteq H$ und $P \subseteq \mathrm{Ker}(\tau)$; wegen $A\tau^2 \subseteq P$ ist sogar $A\tau^2 = P$ da P ein Punkt ist und ebenso folgt $H = \mathrm{Ker}(\tau^2)$. Schliesslich ist $H\tau = P$ wegen $\tau^2 \neq 0$ so dass $\mathrm{Ker}(\tau)$ eine Hyperebene in H ist und daraus folgt nun $\mathrm{Rang}_K A\tau = 2$; sei Q ein Punkt mit $A\tau = P \oplus Q$.

Sei angenommen, es wäre $\mathrm{Rang}_K H/P > 1$. Dann ist $\mathrm{Ker}(\tau) = P \oplus L$ mit $L \neq 0$; sei R ein Punkt mit $L = R \oplus M$. Dann ist also $\mathrm{Ker}(\tau) = P \oplus R \oplus M$. Wegen $\tau^2 \neq 0$ ist $Q \nsubseteq \mathrm{Ker}(\tau)$ und daraus folgt $H = Q \oplus \mathrm{Ker}(\tau) = Q \oplus P \oplus R \oplus M$. Sei S ein dritter Punkt auf der Geraden $Q \oplus R$; dann gilt also $Q \oplus R = Q \oplus S = R \oplus S$. Wegen (2) gibt es ein $\alpha \in C_\Gamma(\eta)$ mit $Q\alpha = R$ und $R\alpha = S$, und wegen (i.c) ist $\alpha \in N_\Gamma \Delta$, so dass $\alpha^{-1}\delta\alpha\delta = \delta\alpha^{-1}\delta\alpha$ wegen (i.a) gilt. Folglich gilt auch $\alpha^{-1}\tau\alpha\tau = \tau\alpha^{-1}\tau\alpha$. Wegen (3) haben wir $\mathrm{Ker}(\alpha^{-1}\tau\alpha) = (\mathrm{Ker}(\tau))\alpha = (P \oplus R \oplus M)\alpha \supseteq P \oplus S$; wäre nun $Q \subseteq \mathrm{Ker}(\alpha^{-1}\tau\alpha)$, so folgte $Q\alpha = R \subseteq Q \oplus S \subseteq \mathrm{Ker}(\alpha^{-1}\tau\alpha) = \mathrm{Ker}(\tau)\alpha$, also $Q \subseteq \mathrm{Ker}(\tau)$, ein

Widerspruch. Also haben wir $Q \not\subseteq \mathrm{Ker}(\alpha^{-1}\tau\alpha)$, so dass $A\tau \not\subseteq \mathrm{Ker}(\alpha^{-1}\tau\alpha)$ und folglich $\tau\alpha^{-1}\tau\alpha \neq 0$ gilt. Andererseits ist $A\alpha^{-1}\tau\alpha\tau = A\tau\alpha\tau = (P \oplus Q)\alpha\tau = (P \oplus R)\tau = 0$ und daraus folgt der Widerspruch $\alpha^{-1}\tau\alpha\tau = 0$. Also ist $\mathrm{Rang}_K H/P = 1$ und $\mathrm{Rang}_K A = n = 3$.

Sei $a \in A - H$ und sei $b = a\tau$ und $c = b\tau$; dann ist $\{a,b,c\}$ eine Basis von A. Sei K^* die multiplikative Gruppe von K. Ist $t \in K^*$, so wird durch $a\alpha = a$, $b\alpha = bt$ und $c\alpha = c$ ein Automorphismus $\alpha \in \Gamma$ definiert. Wegen (1) gilt $\alpha \in C_\Gamma(\eta)$, so dass $\alpha \in N_\Gamma \Delta$ aus (i.c) folgt. Wegen (i.a) ist dann δ mit $\alpha^{-1}\delta\alpha$ vertauschbar, woraus wie zuvor die Vertauschbarkeit von τ und $\alpha^{-1}\tau\alpha$ folgt. Offenbar ist $b\alpha^{-1} = bt^{-1}$; nach kurzer Rechnung ergibt sich $a\alpha^{-1}\tau\alpha\tau = ct$ und $a\tau\alpha^{-1}\tau\alpha = ct^{-1}$, woraus $ct = ct^{-1}$, also $t = t^{-1}$, folgt. Also ist K^* eine elementar abelsche 2-Gruppe, und K ist ein kommutativer Körper. Bekanntlich ist dann jede endliche Untergruppe von K^* zyklisch, und daraus folgt $|K^*| = 2$ oder $|K^*| = 1$, also $p = |K| = 3$ oder $p = |K| = 2$. Offensichtlich gilt nun (ii.c) im ersten, und (ii.d) gilt im zweiten Falle, und damit ist der Beweis fertig.

LITERATUR:

1. BAER, R., Linear Algebra and Projective Geometry. New York 1952

EXTENSIONS OF ISOMORPHISMS BETWEEN SUBGROUPS

L. Fuchs

All groups in this note are abelian p-groups; the results can, however, be easily extended, mutatis mutandis, to arbitrary abelian groups.

The fundamental idea of proofs in proving the isomorphisms of two groups, A and C, (see Ulm's theorem or the generalization by Hill) is to extend a height-preserving isomorphism

$$\phi : G \to H$$

between a subgroup G of A and a subgroup H of C to an isomorphism of A with C. As is well-known, such an extension exists only under rather restrictive conditions.

If we wish to investigate the possibility of extension in general, under more general situations than it has been studied so far, then the first case to consider is obviously the case of "simple extensions", i.e. the step of adjoining to G an element $a \in A$ such that $pa \in G$ and of finding a suitable element $c \in C$ such that ϕ extends to a height-preserving isomorphism

$$(1) \qquad \phi^* : G^* = \langle G,a \rangle \to H^* = \langle H,c \rangle$$

satisfying $\phi^* a = c$. Our present purpose is to raise the question of simple extensions as a prelude to the foundation for a general extension theory.

For notation and unexplained terminology, we refer to [1].

1. Simple extensions.

The problem of simple extensions can be regarded to consist of two steps: first, one has to pick a $c \in C$ of the same height as $a \in A$ with $\phi(pa) = pc \in H$; the second difficulty lies in assuring that ϕ^* is likewise height-preserving. This can be overcome if we can select a and c to be of maximal heights in their cosets mod G and H, respectively. In fact, if $a \in A$ satisfies $h(a) \geqq h(a+g)$ for all $g \in G$, then $h(a+g) = \min(h(a), h(g))$ for all $g \in G$ guarantees adequate control on the heights which enables us to show that

(2) $$h(a+g) = h(c+\phi g) \qquad \text{for all } g \in G,$$

whenever $c \in C$ is of the same sort.

In the proof of Ulm's theorem and generalizations, the crucial step is the simple extension described in Kaplansky-Mackey's lemma [4]:

Lemma 1. If $a \in A$ is proper with respect to G and $pa \in G$, $h(a) = \sigma$, then for every $c \in C$ with $\phi(pa) = pc$, $h(c) = \sigma$, ϕ extends to a height-preserving isomorphism (1) carrying a into c.

The case when $p(a+g)$ has height $\leqq \sigma+1$, for all $g \in G$, is routine, because then any choice of c is suitable. If, however, $h(pa) > \sigma+1$, i.e. the indicator of a has a gap at σ, then a suitable $c \in C$ can be found only if the relative Ulm invariants match. The σth relative invariant of G in A is the following vector space over $\mathbb{Z}/(p)$:

$$U_\sigma(A,G) = p^\sigma A[p]/G(\sigma)$$

where $G(\sigma) = (G + p^{\sigma+1}A) \cap p^\sigma A[p]$ (we follow R. Warfield in viewing the vector space itself rather than its dimension as an invariant). Suppose that we are given a monomorphism

$$\alpha_\sigma : U_\sigma(A,G) \rightarrow U_\sigma(C,H).$$

If a is as indicated, then for some $b \in p^\sigma A$, $a-b$ represents an element in $U_\sigma(A,G)$, and using α_σ, a corresponding representative in $U_\sigma(C,H)$ can be found, leading to a suitable $c \in C$.

Manifestly, this method breaks down if the coset $a+G$ contains no element of maximal height. In this case, one is compelled to look for another method.

It is clear that $a+G$ contains no element of maximal height means that

(3) $$\sup_{g \in G} h(a+g) = \lambda = \text{limit ordinal}$$

and $h(a+g) < \lambda$ for every $g \in G$. Such a situation was considered in a special case by Fuchs-Toubassi [3]; the "simple extension" corresponding to that case can be phrased as follows:

Lemma 2. Suppose $a+G$, $c+H$ contain no elements of maximal height and (3) holds. If

$$a \in \bigcap_{\rho<\lambda} (g_\rho + p^\rho A) \quad \text{and} \quad c \in \bigcap_{\rho<\lambda} (\phi g_\rho + p^\rho C)$$

for $g_\rho \in G$, and if $\phi(pa) = pc$, then ϕ extends to a height-preserving isomorphism (1) such that $\phi^* a = c$.

In order to prove this, all what we have to do is to verify (2). We have $h(a+g) = \sigma < \lambda$. If $\sigma < \rho < \lambda$, then $h(a - g_\rho) \geqq \rho$ implies $h(g_\rho + g) = \sigma$. Hence $h(\phi g_\rho + \phi g) = \sigma$ for $\sigma < \rho < \lambda$. As $h(c - \phi g_\rho) \geqq \rho$, it follows that

$$h(c + \phi g) = \min(h(c - \phi g_\rho), h(\phi g_\rho + \phi g)) = \sigma,$$

as claimed.

The hypothesis of a being contained in the intersection amounts to saying that $g_\rho \xrightarrow[\lambda]{} a$, a is the λ-limit of $\{g_\rho\}$, in the sense that $h(a - g_\rho) \geqq \rho$ for every $\rho < \lambda$, while that $a+G$ contains no element of maximal height is equivalent to the fact that $\{g_\rho\}$ has no λ-limit in G. Thus in Lemma 2, ϕ is assumed to be limit-preserving: if $g_\rho \xrightarrow[\lambda]{} a$ in A, but has no λ-limit in G, then $\phi g_\rho \xrightarrow[\lambda]{} c$ for some $c \in C$ and has no λ-limit in H. This limit-preserving character of ϕ is, however, not sufficient for the applicability of Lemma 2; in fact, the λ-limits have to satisfy $\phi(pa) = pc$ as well. It takes but a moment to recognize that this holds if the stronger condition

$$(4) \qquad \phi(G \cap p[\bigcap_{\rho<\lambda} (g_\rho + p^\rho A)]) \subseteq p[\bigcap_{\rho<\lambda} (\phi g_\rho + p^\rho C)]$$

is satisfied.

2. The coset valuation.

In order to study the heights in the cosets $a+G$, we introduce a kind of valuation for them. It is defined in terms of the height function $h_A = h$ in A (we suppress the reference to the group if there is no danger of confusion), and has properties analogous to the usual valuation in groups.

Recall that $h(a) = \sigma$ if $a \in p^{\sigma}A \setminus p^{\sigma+1}A$, while $h(a) = \infty$ if a belongs to a divisible subgroup of A.

As before, let G be a subgroup of A and $a \in A$. Define the function

$$k(a+G) = \sup_{g \in G}\{h(a+g)+1\}.$$

(This definition is reminiscent of the length of a group A which is defined as $\ell(A) = \sup_{h(a)\neq\infty}\{h(a)+1\}$.) Manifestly, $k(a+G)$ is not a limit ordinal exactly if the coset $a+G$ contains an element of maximal height in A.

It is straightforward to verify that for all $a, b \in A$:

(i) $k(a+b+G) \geqq \min(k(a+G), k(b+G))$; equality holds if $k(a+G) \neq k(b+G)$;

(ii) $k(na+G) \geqq k(a+G)$ for all $n \in \mathbb{Z}$; equality holds if $(n,p) = 1$, and strict inequality if $k(a+G)$ is not a limit ordinal and p divides n.

From the definition it follows at once that

$$k(a+G) \leqq h_{A/G}(a+G) + 1$$

for all $a \in A$. Easy examples show that the right member can be arbitrarily large, even if all elements $\neq 0$ of A have finite heights, i.e. if $k(a+G) \leqq \omega$ for all $a \notin G$. However, in some cases, equality can be established.

<u>Lemma 3</u>. If σ is an ordinal such that $k(a+G)$ is never a limit ordinal $< \sigma$, for any $a \in A$, then $k(a+G) < \sigma$ implies

(5) $$k(a+G) = h_{A/G}(a+G) + 1.$$

To prove this, we use transfinite induction on $k(a+G)$, up to σ. First, let $k(a+G) = 1$. Then $h(a+g) = 0$ for each $g \in G$, and the claim is evident. Suppose the assertion is true for ordinals $< \rho$, and let $k(a+G) = \rho + 1$ $(< \sigma)$. If $h(a+G) > \rho$, then, by definition, some $b \in A$ satisfies $h(b+G) \geqq \rho$ and $pb + G = a + G$. As $h(a+g_0) = \rho$ for some $g_0 \in G, h(b+g) < \rho$ for every $g \in G$. Therefore, $k(b+G) \leqq \rho$, and by induction hypothesis, $h(b+G) < \rho$, a contradiction.

The preceding lemma can be improved slightly:

Lemma 4. For every $a \in A$, (5) holds true if $k(a+G)$ is never a limit ordinal for any $a \in p^{-1}G$.

We induct on the order of a mod G to show that $k(a+G)$ is not a limit ordinal for any $a \in A$. Hypothesis takes care of the case when this order is p. Suppose a mod G has order p^r $(r \geqq 2)$ and $k(a+G) = \lambda$ is a limit ordinal. Set $k(pa+G) = \sigma+1$, by induction hypothesis on pa; thus some $g_0 \in G$ satisfies $h(pa+g_0) = \sigma$. Clearly, $\lambda < \sigma$, so there is a $b \in p^{\lambda}A$ such that $pa+g_0 = pb$. Furthermore, $p(a-b) = g_0 \in G$ shows that $k(a-b+G) \neq \lambda$. But $k(a+G) = \lambda$ and $h(b) \geqq \lambda$ imply $k(a-b+G) = \lambda$, a contradiction.

It is immediate that k defines a valuation on the $\mathbb{Z}/(p)$-vectorspace $p^{-1}G/G$ (with values in the ordinals with ∞ adjoined), in the sense of [2]. The subspaces $(p^{\sigma}A[p] + G)/G$ are of special interest. Notice that

$$U_{\sigma}(A,G) \cong (G+p^{\sigma+1}A + p^{\sigma}A[p])/(G+p^{\sigma+1}A) \cong (p^{\sigma}A[p] + G)/[(G+p^{\sigma+1}A) \cap (p^{\sigma}A[p] + G)]$$

where the denominator corresponds to the subspace of $(p^{\sigma}A[p] + G)/G$ consisting of cosets with k-values $> \sigma+1$. This leads to the interpretation of the σth relative invariant of G in A as the quotient of $(p^{\sigma}A[p] + G)/G$ modulo a subspace defined in terms of the k-valuation.

This suggests the introduction of new relative invariants for G in A, corresponding to limit ordinals λ in $p^{-1}G/G$:

$$V_{\lambda}(A,G) = [\bigcap_{\rho<\lambda} (G+p^{\rho}A) \cap p^{-1}G]/[(G+p^{\lambda}A) \cap p^{-1}G].$$

It is readily checked that $a \in p^{-1}G$ represents an element $\neq 0$ of $V_{\lambda}(A,G)$ if and only if it is the λ-limit of a sequence $g_{\rho} \in G$ which has no limit in G. These V_{λ} are functorial in the sense that if $\phi : A \to C$ is a homomorphism carrying G onto H, then there is an induced homomorphism $\eta_{\lambda} : V_{\lambda}(A,G) \to V_{\lambda}(C,H)$ which is an isomorphism whenever η is one. Hence a height-preserving isomorphism $\phi : G \to H$ can extend to an isomorphism of A with C only if the maps

$$\phi_{\lambda} : V_{\lambda}(A,G) \to V_{\lambda}(C,H) \qquad \text{(for limit } \lambda)$$

induced by ϕ are isomorphisms.

From what has been said above about the elements of $V_\lambda(A,G)$, the following statement should be clear:

Lemma 5. ϕ is limit-preserving if and only if, for all limit ordinals λ, the induced maps ϕ_λ are isomorphisms.

Condition (4) has no natural equivalent in terms of V_λ.

Notice that if G is a nice subgroup of A, then all these $V_\lambda(A,G)$ vanish. Moreover, from Lemmas 3 and 4 we infer:

Corollary 6. A subgroup G of A is nice exactly if $V_\lambda(A,G) = 0$ for every limit ordinal λ.

Thus the invariants $V_\lambda(A,G)$ can be regarded as measures to what extent G is not nice in A. It looks reasonable to expect that the subgroups G with $V_\lambda(A,G) = 0$ for $\lambda > \text{cof}\ \omega$ are of special interest.

3. Abstract coset valuations.

In the last section, we have introduced a new sort of valuation k on the factor group A/G which comes from the height function on A. This is obviously the dual to the valuation on a group, studied extensively by R. Hunter, F. Richman, E. Walker, and others, which arises from the height function of a suitable group containing it. It seems desirable to formulate the properties of this coset-valuation, abstractly. [During the Conference in Oberwolfach, I have learnt that recently R. Hunter has also come to the idea of coset-valuation.]

A coset-valuation of a p-group X is a function k from X to the ordinals with ∞ adjoined such that

(i) $k(x) \geqq 1$ for every $x \in X$, $k(0) = \infty$;

(ii) $k(x+y) \geqq \min(k(x), k(y))$ for all $x,y \in X$;

(iii) $k(x) \leqq k(px)$ with strict inequality if $k(x)$ is not a limit ordinal;

(iv) if $1 \leqq \sigma < k(x)$, then $x = py$ for some y such that $k(y) \geqq \sigma$.

The coset-valuation can also be defined via filtration. For an ordinal σ, set

$$p(p^\sigma)X = (p)(p^\sigma)X = (p^{\sigma+1})X$$

and for limit ordinals λ,

$$\bigcap_{\rho<\lambda} (p^{\rho})X \geqq (p^{\lambda})X.$$

Here equality need not hold; moreover, the quotient of these two subgroups of X can be any group. Now the filtration is

$$X > (p)X > (p^2)X > \ldots > \bigcap_{n<\omega} (p^n)X \geqq (p^{\omega})X > \ldots > (p^{\tau})X = (p^{\infty})X$$

where each subgroup $(p^{\sigma})X$, for a non-limit σ, is p times its predecessor. Conversely, every filtration of this kind defines a coset-valuation in the obvious way.

It is natural to inquire if every abstract coset-valuation arises by representing the group as a factor group and defining the coset function k as we did in 2. The answer is in the affirmative; in fact, we have:

Theorem 7. Let X be a p-group with a function k satisfying (i)-(iv). Then there is a p-group A and a subgroup G such that

(a) there is an isomorphism $\phi : A/G \to X$;

(b) for every $a \in A$, $k(\phi(a+G)) = \sup_{g\in G}\{h(a+g) + 1\}$.

We are going to use Nunke's generalized Prüfer groups H_{σ} of length σ in the construction of A. For every $x \in X$ with $k(x) \neq$ limit, consider a copy A_x of $H_{k(x)-1+r}$ where p^r denotes the order of x, and set $p^{k(x)-1}A_x = \langle a_x\rangle$. For every $x \in X$ with $k(x) = \infty$, let $A_x \cong \mathbb{Z}(p^{\infty})$ and $a_x \in A_x$ an element of order p^r. For every $x \in X$ with $k(x) =$ limit ordinal, let A_x be the direct sum of copies of $H_{\rho+r}(\rho < k(x))$, and set $p^{\rho}H_{\rho+r} = \langle a_{x,\rho}\rangle$. Define

$$A = \bigoplus_{x\in X} A_x.$$

Map the subgroup B which is the direct sum of all $\langle a_x\rangle$ and all $\langle a_{x,\rho}\rangle$ onto X by sending a_x or $a_{x,\rho}$ upon the x from which it was constructed. Since this map does not decrease heights, furthermore, B is a nice subgroup of A and A/B is totally projective, it extends to an epimorphism $\eta : A \to X$. Setting $G = \operatorname{Ker} \eta$, it is readily checked that the isomorphism $\phi : A/G \to X$ induced by η satisfies (a) and (b). This completes the proof of the theorem.

Notice that the group A we constructed is totally projective.

The question of studying subclasses of coset-valuated groups presents itself. For instance, those corresponding to simply presented groups could lead to a more general structure theorem for p-groups.

4. Concluding remarks

A more satisfactory analysis of simple extensions is closely related to the study of successive, possibly transfinite extensions, as each step has to prepare the next one. So far, I have been unable to pin down manageable conditions for a simple extension in the general case which would guarantee repeated extensions.

It follows directly that $\phi^*: G^* \to H^*$ in (1) is limit-preserving if a and c are of maximal heights in their cosets mod G and H, respectively. Moreover, it is straightforward to check that the relative invariants do not change if we adjoin λ-limits a, c to G, H. In other words, the simple extensions covered by Lemmas 1 and 2 do not interfere, one can follow the other. However, the limit-preserving character of the map can change if λ-limits are adjoined, and the real difficulty lies in finding out in which cases one can control the limit-preserving nature of maps involved.

References

1. L. Fuchs, Infinite Abelian Groups, vol.1-2, Academic Press (1970 and 1973).
2. L. Fuchs, Vector spaces with valuations, Journ. Alg. 35 (1975), 23-38.
3. L. Fuchs and E. Toubassi, On rank one mixed abelian groups with $p^{\omega+n}$-projective p-components, Comm. Math. Univ. St. Pauli 29 (1980), 125-133.
4. I. Kaplansky and G. W. Mackey, A generalization of Ulm's theorem, Summa Brasil Math. 2 (1951), 195-202.

THE DUALS OF TOTALLY PROJECTIVE GROUPS

Franz Kiefer

1. Introduction . Defined by Nunke, the totally projective p-groups turned out to be in some sense the best possible class of p-groups to which Ulm's classical theorem can be extended . This result is mainly due to Hill, who used an alternative definition to the one given by Nunke . Crawley and Hales showed the validity of Ulm's theorem for reduced simply presented p-groups, and combining this with Hill's result, concluded that the above are just the totally projective p-groups . For details about the story of totally projective p-groups we may refer to the book of Fuchs [2], which will serve for us as background with regard to definitions and results concerning totally projective p-groups .

Although Pontrjaginduality may be viewed as a reduction of the study of compact abelian groups to that one of discrete groups, it might be interesting to have a look which class of compact groups one has treated as well, after having obtained good results for a particular class of discrete groups . The suggestion to study the duals of totally projective p-groups is given in Problem 65 of Fuchs [2, p. 106]. For doing this, the most natural way consists in trying to dualize the various concepts which are used to describe totally projective p-groups . The fundamental results concerning the Pontrijaginduali-zation-machinery are taken from Hewitt-Ross [3] .

For convenience, we use $\hat{G}$ to denote the charactergroup of a locally compact abelian group G. Subgroups are always assumed to be closed and, of course, group is used synomym to abelian group .

2. Some preliminary remarks . The totally projective p-groups are a special kind of reduced p-groups . Thus, being in particular discrete torsiongroups, their duals must be compact o-dimensional groups .

Braconnier [1] calls the compact abelian group G a (p)-group , if for every $x \in G$ there is a continuous homomorphism from J_p onto $cl(\langle x\rangle)$. Clearly a discrete group is a p-group exactly if its dual is a (p)-group . Similar to the discrete case one has the result that every compact o-dimensional group is topologically isomorphic to the direct product of its (p)-components .

If we call the compact (p)-groups G <u>coreduced</u>, if it contains no (proper) subgroup H such that G/H is torsionfree, then reduced and coreduced are in one-to-one correspondence via duality. This follows from the fact that the dualgroup of $Z(p^{\infty})$ is topologically isomorphic to J_p.

To sum up, the duals of totally projective p-groups must be found within the class of compact o-dimensional coreduced abelian (p)-groups.

From now on, let G denote a (discrete) p-group and C a compact (p)-group.

3. <u>Some dual concepts</u>. With G_{α} we simply mean $p^{\alpha}G$, which is defined in the usual way. Dually, for C and every ordinal α, $C_{-\alpha}$ is defined by

$$C_{-1} := C[p]$$

$$C_{-(\alpha+1)} := \{x \in C \mid px \in C_{-\alpha}\}$$

$$C_{-\alpha} := \mathrm{cl}\Big(\bigcup_{\beta<\alpha} C_{-\beta}\Big) \quad \text{if } \alpha \text{ is a limit ordinal.}$$

If H is a subgroup of G, then $A(\hat{G},pH) = A(\hat{G},H)[p]$ and for a family $(H_i)_{i\in I}$ of subgroups of G we have

$$(1) \qquad A\Big(\hat{G}, \bigcap_{i\in I} H_i\Big) = \mathrm{cl}\Big(\sum_{i\in I} A(\hat{G},H_i)\Big).$$

This two things together yield

$$(2) \qquad A(\hat{G},G_{\alpha}) = \hat{G}_{-\alpha} \quad \text{for every ordinal } \alpha.$$

For example, the anihilator of the first Ulm-subgroup of G in $\hat{G}$ coincides with the closure of the torsionsubgroup of $\hat{G}$, that is $A(\hat{G},U_1(G)) = A(\hat{G},G_{\omega}) = \hat{G}_{-\omega} = \mathrm{cl}(t\hat{G})$.

<u>Definition</u>: A subgroup F of C is called <u>smart</u> in C, if $F_{-\alpha} = F \cap C_{-\alpha}$ for every ordinal α.

While a subgroup N of G is nice in G, iff every coset of N contains an element of the same (generalized) height as the coset itself, F is smart in C iff the elements of F have the same <u>(generalized) order</u> in F as in C. For $x \in C$ the (generalized) order $o(x)$ of x in C is α if $x \in C_{-\alpha} \setminus \big(\bigcup_{\beta<\alpha} C_{-\beta}\big)$ and $o(x) = \infty$ if there is no such α.

(3) A subgroup N of G is nice in G exactly if $F := A(\hat{G},N)$ is smart in $\hat{G}$.

<u>Proof</u>: Let N be nice in G, α an ordinal, $\omega : G \longrightarrow G/N$ the natural map and f the topological isomorphism from $\widehat{G/N}$ onto $A(\hat{G},N)$ induced by ω . Since $F = f(\widehat{G/N})$, we have $F_{-\alpha} = f(\widehat{G/N})_{-\alpha} = f((\widehat{G/N})_{-\alpha})$. Using (2) we get $F_{-\alpha} = f(A(\widehat{G/N},(G/N)_{\alpha}))$ $= f(A(\widehat{G/N},(G_{\alpha}+N)/N))$. The last equality holds because N is nice in G . Regarding that f also establishes the topological isomorphism between $A(\widehat{G/N},(G_{\alpha}+N)/N)$ and $A(\hat{G},G_{\alpha}+N)$, we conclude $F_{-\alpha} = f(f^{-1}(A(\hat{G},G_{\alpha}+N))) = A(\hat{G},G_{\alpha}+n) = A(\hat{G},G_{\alpha}) \cap A(\hat{G},N)$, that is $F_{-\alpha} = \hat{G}_{-\alpha} \cap F$. Thus, F is smart in $\hat{G}$. To prove the converse, similar arguments can be applied .

The following definition takes into consideration that totally projective groups are distinguished by having a nice system .

<u>Definition</u>: A system $\underline{F}$ of smart subgroups of C is a <u>smart system</u> for C , if it contains C , is closed under taking arbitrary intersections, and if it satisfies :

(S) For $F \in$ and a subgroup K of C contained in F such that F/K is metrisable, there is $D \in \underline{F}, D \subset K$ such that F/D is metrisable .

Condition (S) is a topological one and can be viewed in a slightly different way: Let G be an arbitrary abelian topological group and $\underline{P}$ the set of all continuous invariant pseudometrics on G . Then $\underline{U} := \{U^d_{\varepsilon}(o) | d \in P, \varepsilon > o\}$ is a neighbourhood basis for the identy in G , where $U^d_{\varepsilon}(o) := \{x \in G | d(x,o) < \varepsilon\}$ (compare [3 ,(8.13)]). Furthermore let $\underline{M}$ denote the set of all subgroups M of G such that G/M is metrisable . The metric on G/M can even be taken invariant ([3 ,(8.3)]).Now, given $M \in \underline{M}$ and an invariant metric d_M on G/M , d'_M defined by $d'_M(x,y) := d_M(x+M,y+M)$ for $x,y \in G$ is an element of $\underline{P}$. If we call $\underline{B} \subset \underline{M}$ a "basis" for the topology of G if the pseudometrics induced by the elements of $\underline{B}$ generate the topology of G , we get

(4) A system $\underline{F}$ of subgroups of C satisfies (S) exactly if it contains a "basis" for the topology of F for every $F \in \underline{F}$.

<u>Proof</u>: We assume that $\underline{F}$ satisfies (S). Take $F \in \underline{F}$ and let d be an invariant continuous pseudometric on F. Then $K_d := \{x \in F \mid d(x,o) = o\}$ is a closed subgroup of F, since K_d is the preimage of 0 under the restriction of d to $F \times O^{\cdot}$. Let d_k denote the metric on F/K_d induced by d. Observe that d_k generates the "right" topology on F; since $id : F \longrightarrow (F,d)$ is continuous, the induced map $id' : F/K_d \longrightarrow (F/K_d, d_k)$ is also continuous. Moreover id' is a homeomorphism, because F/K_d is compact and $(F/K_d, d_k)$ is a Hausdorffspace. By our assumption, we can find $D \in \underline{F}$, $D \subset K_d$ such that F/D is metrisable. Chose an invariant metric d_D' on F/D which induces an invariant continuous pseudometric d' on F. It remains to prove, that $id : (F,d') \longrightarrow (F,d)$ is continuous. Since F/K_d is topologically isomorphic to $(F/D)/(F/K_d)$ we have a continuous homomorphism j from F/D onto F/K_d. The continuity of id follows therefore from the diagram

$$\begin{array}{ccc} (F,d') & \xrightarrow{id} & (F,d) \\ \downarrow & & \downarrow \\ F/D & \xrightarrow{i} & F/K_d \end{array}$$

The converse is proved with a similar reasoning.

Our next definition dualizes the concept of having a nice composition series.

<u>Definition</u>: A well-ordered strictly descending chain of smart subgroups of $C\,(C = F_o > F_1 > \ldots > F_\lambda > \ldots > F_\mu = 0)$, starting with C and ending with 0, is called <u>smart composition series</u> for C, if it satisfies

$$\text{(i)} \quad |F_\lambda : F_{\lambda+1}| = p \quad \text{for every} \quad \lambda < \mu$$

$$\text{(ii)} \quad F_\lambda = \bigcap_{\tau<\lambda} F_\tau \quad \text{if } \lambda \text{ is a limit ordinal}$$

As smart subgroups of $\hat{G}$ correspond to nice subgroups of G, there is a kind of excellent subgroups of $\hat{G}$ which correspond to isotype subgroups of G.

(5) A subgroups B of G is isotype in G iff for $D := A(\hat{G},B)$ the following holds:

(+) $(\hat{G}/D)_{-\alpha} = (\hat{G}_{-\alpha}+D)/D$ for every ordinal α.

This can be shown by applying similar arguments as those used in the proof of (3) .

For later purpose, we may call a subgroup H of C <u>cobalanced</u> in C , iff H is smart in C and satisfies the condition (+) as well .

4. <u>Simply given groups</u> . We regard the following construction of a compact group as a subgroup of a toral group .

Let I be a nonempty set and (I,K,f,p) a tupel with the properties:

a) $I, K \subset I$; $I \cap K = \emptyset$; $I \cup K = I$

b) $f : K \longrightarrow I$ is an arbitrary map

c) p is a prime

Let $T = \mathbb{R}/Z$ denote the one-dimensional torus . Then

$$C_{(I,K,f,p)} := \{ (x_i)_{i\in I} \in \prod_{i\in I} T \mid px_i = o \text{ if } i \in I \text{ and } px_i = x_{f(i)} \text{ if } i \in K \}$$

with the induced topology is a subgroup of $\prod_{i\in I} T$.

If, for a compact group C , there exists a tupel (I,K,f,p) with the above properties, such that C is topologically isomorphic to $C_{(I,K,f,p)}$, we may call C <u>simply given</u> .

5. <u>Some excellent compact groups</u> . We recall that the so-called generalized Prüfergroups H^α (for every ordinal α) play an important role in the theory of totally projective groups . For reasons out of duality, there is for every ordinal α a compact (p)-group K^α being defined unique up to (topological) isomorphism by the following conditions:

a) $K^o = O$ and $(K^\alpha)_{-\alpha} = K^\alpha$ for every α

b) $K^{\alpha+1}/(K^{\alpha+1})_{-\alpha} \cong Z(p)$ and $(K^{\alpha+1})_{-\alpha} \cong K^\alpha$

c) $K^\alpha \cong \prod_{\beta<\alpha} K^\beta$ if α is a limit ordinal

Of course, the process for constructing the generalized Prüfergroups can be dualized . We use this to have a look on $K^{\omega+1}$, the dual of the Prüfergroup $H^{\omega+1}$. First note that for any integer $n > o$, $K^n \cong Z(p^n)$. Furthermore $K^\omega \cong \prod_{n>o} Z(p^n)$. Now, $K^{\omega+1}$ can be regarded as pushout of the following diagram:

$$\begin{array}{ccc} \bigoplus_{n>o} Z(p^{n+1})_n & \longrightarrow & \bigoplus_{n>o} Z(p^{n+1}) \\ \Big\downarrow \nabla & & \Big\downarrow \\ Z(p) & \dashrightarrow & H^{\omega+1} \end{array}$$

(Recall $Z(p^{n+1})_n := p^n Z(p^{n+1})$) ∇ is the codiagonal map, i the inclusion

since pushout and pullback are in one-to-one correspondance via duality, we obtain $K^{\omega+1}$ as pullback of the following diagram:

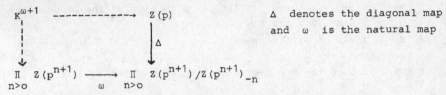

$$\begin{array}{ccc} K^{\omega+1} & \dashrightarrow & Z(p) \\ \Big\downarrow & & \Big\downarrow \Delta \\ \prod_{n>o} Z(p^{n+1}) & \xrightarrow{\ \omega\ } & \prod_{n>o} Z(p^{n+1})/Z(p^{n+1})_{-n} \end{array}$$

Δ denotes the diagonal map and ω is the natural map

Thus, $K^{\omega+1}$ is topologically isomorphic to the subgroup B of $\prod_{n>o} Z(p^n)$ consisting of those $(x_n)_{n>o} \in \prod_{n>o} Z(p^n)$ for which $x_i = x_j \pmod{p}$ for every $i,j > o$. The element $b_o := (1,1,\ldots,1,\ldots)$ of B has the property that it is not in $cl(tB)$ but pb_o is. This corresponds to the fact that $H^{\omega+1}$ contains an element of order p which is divisible by p^n for every $n > o$.

6. <u>The main result</u>: For a compact o-dimensional coreduced abelian (p)-group C the following conditions are equivalent

(i) C is the dual of a totally projective p-group

(ii) C has a smart system

(iii) C has a smart composition series

(iv) C is injective with respect to all cobalanced exact sequences of compact (p)-groups

(v) C is simply given

(vi) $(\mathrm{Ext}(A,C_{-\alpha}))_\alpha = O$ for every ordinal α and every compact group A

(vii) C is topologically isomorphic to a direct factor of a direct product of the K^{α}'s

(viii) C belongs to the smallest class $\underline{C}$ of compact groups which contains $Z(p)$, is closed under taking direct factors and direct products, and which contains a compact group A exactly if it contains both, $A_{-\alpha}$ and $A/A_{-\alpha}$ for every ordinal α.

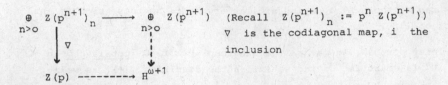

Proof: The equivalence of the above conditions will follow from the fact that each condition is equivalent to (i), using the corresponding results in the discrete case .

ad (ii): $\underline{F}$ is a smart system for C , iff $\underline{N} := \{A(\hat{C},F) \mid \in \underline{F}\}$ is a nice system for $\hat{C}$. If $\underline{F}$ is a smart system for C , then, because of (2), the elements of $\underline{N}$ are nice in $\hat{C}$. Moreover $0 = A(\hat{C},C)$ belongs to $\underline{N}$, and if $(A(\hat{C},F_i))_{i\in I}$ is a family of elements of $\underline{N}$, then $\sum_{i\in I} A(\hat{C},F_i) = A(\hat{C}, \bigcap_{i\in I} F_i)$ belongs to $\hat{C}$. Given $N := A(\hat{C},F) \in \underline{N}$ and a subgroup H of $\hat{C}$ containing N , such that H/N is countable, we can find $D \in \underline{F}$, $D \subset A(C,H)$ such that F/D is metrisable . Now, $A(\hat{C},D)$ contains H and $A(\hat{C},D)/N$ is countable, noticing that the dual of F/D is topologically isomorphic to $A(\hat{C},D)/A(\hat{C},F)$ and that a discrete group is countable iff its dual is metrisable . Thus, $\underline{N}$ is a nice system for $\hat{C}$. The converse is proved similarly .

ad (iii) We have to show, that $(C = F_o > F_1 > \ldots > F_\lambda > \ldots > F_\mu = o)$ is a smart composition series for C exactly if $(O = N_o < N_1 < \ldots < N_\lambda < \ldots < N_\mu = \hat{C})$ is a nice one for $\hat{C}$, taking $N_\lambda := A(\hat{C},F_\lambda)$ for every $\lambda \leq \mu$. This employs no other arguments as the one used above, except for showing that $|F_\lambda : F_{\lambda+1}| = p$ for $\lambda < y$ iff the some holds for the index of N_λ in $N_{\lambda+1}$; but this is obvious, because finite groups are selfdual.

ad (iv) Condition (iv) means that the following extension problem is always solvable:

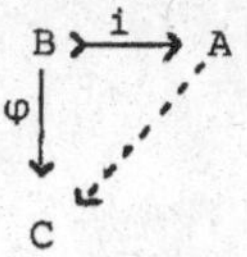

A,B compact (p)-groups
i(B) is cobalanced in A
$\underline{C}$ arbitrary continuous homomorphism

Applying the duality functor, we see that this is true iff the following lifting problem is always solvable .

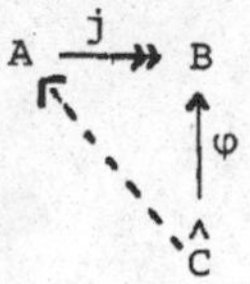

A,B p-groups
ker(j) balanced in A
φ arbitrary homomorphism

This is true iff $\hat{C}$ is totally projective

<u>ad (v)</u>: Every tupel (I,K,f,p) having the properties listed in 4. determines in an obvious way a simply presented p-group which turns out to be isomorphic to the dual of $C_{(I,K,f,p)}$. On the other hand, every simply presented p-group determines a tupel (I,K,f,p) by using a faithful presentation .

<u>ad (vi)</u>: We note that for discrete groups A and C , $\mathrm{Ext}(C,A)$ is as group isomorphic to $\mathrm{Ext}(\hat{A},\hat{C})$, if we take extensions in the category of locally compact abelian groups .

(vii) and (viii) are almost obvious .

7. <u>A last remark</u> . If G is a totally projective p-group then the α-th Ulm invariant if G is defined as the dimension of $P_{(\alpha)} := (G_\alpha[p])(G_{\alpha+1}[p])$, regarded as vectorspaces over $Z(p)$. The same invariants can be found for $\hat{G}$ by using $\hat{P}_{(-\alpha)} := (p\hat{G}+\hat{G}_{-(\alpha+1)})/(p\hat{G}+\hat{G}_{-\alpha})$ instead of $P_{(\alpha)}$.

R E F E R E N C E S

1. Braconnier,J. "Sur les groupes topologiques localement compacts", Jour. Math. Pure Appl. N.S. 27, (1948) 1-85

2. Fuchs, L. "Infinite abelian groups, Volume II" Academic Press, Juc. (1970)

3. Hewitt, E. and Ross K.A. "Abstract harmonic analysis" Springer-Verlag (1963)

ISOTYPE SUBGROUPS OF TOTALLY PROJECTIVE GROUPS

Paul Hill

I. Introduction.

About twelve years ago I published a paper [6] whose main result was that if G is a direct sum of countable, reduced, abelian p-groups and H is an isotype subgroup of countable length then H must also be a direct sum of countable groups. The length of an abelian p-group is the smallest ordinal λ such that $p^{\lambda}G = 0$; when G is reduced such a λ exists. Brief mention of some of the major known results for countable primary groups, direct sums of such groups, and finally the more general class of totally projective groups should give the casual reader some appreciation for the theorem cited above. At the same time, we will be providing the necessary preliminaries for the present paper with the hope that its purpose and meaning will then be clear. More details and information on the subject can be found, among other references, in [1], [2], [4], and [10].

The basic point of beginning of the theory is Ulm's uniqueness theorem published in 1933: a countable abelian p-group is completely determined (up to isomorphism) by certain cardinal numbers associated with the group. Actually, this is not the original version of Ulm's theorem but the interpretation by Kaplansky [10], and it would be more accurate to designate the Ulm invariants as the Kaplansky-Mackey numbers [11]. It was not until a quarter of a century later that G. Kolettis [12] succeeded in extending Ulm's theorem to an arbitrary direct sum of countable abelian p-groups. We shall assume henceforth that <u>all groups are reduced, p-primary abelian groups</u>. Observe that being a direct sum of countable groups in no way restricts the cardinality of the group itself. Moreover, unless it just happens to be presented that way, the question of whether or not a given group is in fact a direct sum of countable groups can be a difficult one. Yet, by Kolettis' theorem the Ulm invariants (Kaplansky-Mackey numbers) are a source of complete knowledge of the group if it is a direct sum of countable groups. Thus the question, difficult or not, of whether or not a group is a direct sum of countable groups is of high interest in abelian group theory. Following R. Nunke [14], we shall use the abbreviation "d.s.c." Even after Kolettis proved his theorem, virtually nothing was known apart from the obvious about subgroups of d.s.c.'s. It was Nunke [13] who first demonstrated in 1963 the unpleasant but not surprising feature that a subgroup of a d.s.c. need not again be a d.s.c. Subsequently, it developed that a typical d.s.c. has a preponderance of subgroups that are not d.s.c.'s. After giving the first example of a subgroup of a d.s.c. that is not a d.s.c. in [13], Nunke made a much deeper and more comprehensive study of d.s.c.'s and their subgroups in [14]. In this work Nunke explicitly considered the question: Under what conditions is a subgroup of a d.s.c. again a d.s.c.? While a complete answer was not seriously approached in [14], Nunke was able to provide several partial answers to the question. My theorem that I quoted in the beginning was, at least, a major initial step toward an answer for the special case of isotype subgroups. It has remained up to now as one of the most definitive positive statements about subgroups of d.s.c.'s.

One of the main purposes of the present paper is to give a necessary and sufficient condition for an isotype subgroup of a d.s.c to be itself a d.s.c. Thus we give here a complete answer to Nunke's question about subgroups of d.s.c.'s for isotype subgroups. For the sake of simplicity we have described our result for direct sums of countable groups, which is an important special case, but everything is actually done in the more general setting of totally projective groups. Not only are totally projective groups determined by their Ulm invariants but they are one of the most interesting classes of abelian groups. For example, totally projective groups admit three equivalent but distinctly different natural descriptions [2]. The one that we find most relevant for this paper is our own: the third axiom of countability.

Definition. A group G satisfies the third axiom of countability (or Axiom 3) if it has a collection $\mathcal{C}$ of nice subgroups that satisfy the following conditions.

(0) $0 \in \mathcal{C}$.

(1) $\mathcal{C}$ is closed with respect to arbitrary group unions, $\langle N_i \rangle_{i \in I} \in \mathcal{C}$ if $N_i \in \mathcal{C}$ for each i.

(2) If A is any countable subgroup of G, there exists a countable N in $\mathcal{C}$ that contains A

The subgroup N of G is nice if $p^\alpha(G/N) = \langle p^\alpha G, N \rangle / N$ for each α.

II. Separable Subgroups.

The concept of a nice subgroup has been transferred successfully to valuated groups [15] and to valuated vector spaces [3], [7]. Furthermore, we have shown that the more general notion of separability has an important role in valuated vector spaces [7]. The author believes that the same should be true for groups, and in fact the remainder of this paper will substantially support this view. Let us hasten to add that we use the term "separable" here a different way from its use in [2]. Throughout, $h_G(x)$ denotes the height of x in G; hence $h_G(x)$ is the ordinal α such that $x \in p^\alpha G$ but $x \notin p^{\alpha+1} G$. If $x \in p^\alpha G$ for each α, we set $h_G(x) = \infty$.

Definition 1. A subgroup H of G is separable in G if, for each $g \in G$,

$$\sup \{h_G(g+x) : x \in H\} = \sup \{h_G(g+x_n)\}_{n<\omega_0}$$

for some sequence $\{x_n\}$ in H.

Recall that the cofinality of an ordinal λ, abbreviated $\mathrm{cof}(\lambda)$, is the smallest ordinal β such that

$$\lambda = \sup \{\lambda_\alpha\}_{\alpha<\beta}$$

where $\lambda_\alpha < \lambda$ for each $\alpha < \beta$. It is understood that $\mathrm{cof}(\lambda) = 0$ if λ is isolated.

Proposition 1. A subgroup H of G is separable in G if and only if, for each $g \in G$, the set $\{h_G(g+x): x \in H\}$ has a largest element whenever its supremum is not cofinal with ω_o.

Proof. Suppose that H is separable in G and let

$$\lambda = \sup \{h_G(g+x): x \in H\} .$$

If $\mathrm{cof}(\lambda) \neq \omega_o$, then

$$\lambda = \sup \{h_G(g+x_n)\}_{n<\omega_o}$$

yields $h_G(g+x_o) = \lambda$ for some $x_o \in H$.

Conversely, if there always exists $x_o \in H$ such that

$$h_G(g+x_o) = \lambda = \sup \{h_G(g+x): x \in H\}$$

whenever $\mathrm{cof}(\lambda) \neq \omega_o$, then it follows at once that H is separable in G.

The following theorem establishes a necessary condition for an isotype subgroup to be totally projective.

Theorem 1. If H is an isotype subgroup of G, then H must be separable in G in order for H to be totally projective.

Proof. Suppose that the isotype subgroup H of G is totally projective. If H is not separable in G, there exist $g \in G$ and a limit ordinal λ whose cofinality is not ω_o such that $h_G(g+x) < \lambda$ for each $x \in H$ but

$$\lambda = \sup \{h_G(g+x): x \in H\} .$$

Since $H/p^\lambda H$ is totally projective and since $H/p^\lambda H \cong \langle H, p^\lambda G\rangle / p^\lambda G$ is isotype in $G/p^\lambda G$, there is no loss of generality in assuming that $p^\lambda G = 0$ because $h_G(g+x) = h_{G/p^\lambda G}(g+x+p^\lambda G)$ for each $x \in H$. Hence we assume that G has length λ. This means that H is the direct sum of subgroups having length less than λ since H is totally projective and λ is a limit ordinal; such a decomposition of H is a consequence of the existence and uniqueness theorems for totally projective groups [2]. Permitting $H_\alpha = 0$, we can write

$$H = \textstyle\sum_{\alpha<\lambda} H_\alpha , \text{ where } p^\alpha H_\alpha = 0 .$$

Now, for each $\mu < \lambda$, there exist $\beta < \lambda$ and $x \in \sum_{\alpha<\beta} H_\alpha$ such that $\mu < h_G(g+x) < \lambda$. Choose $\mu(0)$ arbitrarily but less than λ. Suppose that a finite sequence $\{x_n\}$, $1 \leq n \leq k$, of elements x_n in H have been chosen along with a strictly increasing sequence of ordinals

$$\mu(0) < \mu(1) < \cdots < \mu(k) < \lambda$$

so that the following conditions are satisfied for each $n \leq k$.

(i) $\mu(n-1) < h_G(g+x_n) \leq \mu(n)$,

(ii) $x_n \in \sum_{\alpha<\mu(n)} H_\alpha$,

(iii) $x_n = \sum_{\alpha<\mu(n)} x_{n,\alpha}$ where $x_{n,\alpha} \in H_\alpha$ and

$h_H(x_{n,\alpha}) > \mu(n-1)$ only if $x_{n,\alpha} = 0$.

We can certainly choose $x_{k+1} \in H$ and an ordinal $\mu(k+1) > \mu(k)$ so that conditions (i) and (ii) are satisfied for $n = k+1$. Letting

$$x_{k+1} = \sum_{\alpha<\mu(k+1)} x_{k+1,\alpha}$$

with $x_{k+1,\alpha} \in H_\alpha$, we can delete from the above representation of x_{k+1} all terms $x_{k+1,\alpha}$, if any, that belong to $p^{\mu(k)+1}H_\alpha$. Thereby we obtain condition (iii). Therefore, there exist an infinite sequence $\{\mu(n)\}$ of ordinals less than λ such that conditions (i) - (iii) are satisfied for all n .

Setting $\mu = \sup \{\mu(n)\}$, we observe that $\mathrm{cof}(\mu) = \omega_0$ and therefore $\mu < \lambda$ since $\mathrm{cof}(\lambda) \neq \omega_0$. Hence there exists $x \in H$ such that $h_G(g+x) \geq \mu$. We can write

$$x = \sum_{\alpha<\mu} x(\alpha) + \sum_{\alpha\geq\mu} x(\alpha),$$

where $x(\alpha) \in H_\alpha$ for each α (and where $x(\alpha) = 0$ for all but a finite number of α). Condition (i), together with $h_G(g+x) \geq \mu$, implies that

$$\mu(n-1) < h_G(x-x_n) \leq \mu(n) .$$

Since H is isotype in G, the preceding inequalities yield $\mu(n-1) < h_H(x-x_n) \leq \mu(n)$ for all $n \geq 1$. Since $x_n \in \sum_{\alpha<\mu} H_\alpha$ for each n, it follows at once that $h_H(x(\alpha)) \geq \mu$ if $\alpha \geq \mu$ because $h_H(x(\alpha)) > \mu(n-1)$ and $\mu = \sup \{\mu(n)\}$. Without loss of generality regarding the choice of x, we can therefore take $\sum_{\alpha\geq\mu} x_\alpha = 0$. Thus we can assume that $x = \sum_{\alpha<\mu} x(\alpha)$, and hence focus on the group $\sum_{\alpha<\mu} I_\alpha$ since it contains x as well as x_n for each n . Observe that $x \in \sum_{\alpha<\mu(k)} H_\alpha$ for some $k \geq 1$. If $n > k$, write

$$x_n = \sum_{\alpha<\mu(k)} x_{n,\alpha} + \sum_{\alpha\geq\mu(k)} x_{n,\alpha}$$

and notice that condition (iii) implies that $x_{n,\alpha} = 0$ for all $\alpha \geq \mu(k)$ because

$$h_H(x_{n,\alpha}) \geq h_H(x-x_n) > \mu(n-1) .$$

However, $x_n \in \sum_{\alpha<\mu(k)} H_\alpha$ for all n with k fixed is absurd since, if it is true,

$$\mu(n-1) < h_H(x_{n+1}-x_n) \leq \mu(n)$$

and $p^\alpha H_\alpha = 0$ imply, on one hand, that $x_n = x_{n+1}$ when $n > k$; whereas, on the other hand, $x_n = x_{n+1}$ is specifically precluded (by the condition

$h_H(x_{n+1}-x_n) \leq \mu(n))$. Therefore, H must be separable in G, and the theorem is proved.

Theorem 2 below shows that Theorem 1 has a partial converse, but unfortunately the converse holds only for relatively small totally projective groups. In order to prove Theorem 2 we need to introduce the concept of two subgroups of a given group being compatible.

Definition 2. If A and B are subgroups of G, then A is compatible with B in G if for each pair $(a,b) \in A \times B$ there exists c in $A \cap B$ such that $h_G(a+c) \geq h_G(a+b)$. If A is compatible with B, we write $A \| B$; it will be clear from the context what G is.

It should be observed that the compatible relation is symmetric. The following lemma established an important relationship between compatibility and separability.

Lemma 1. Suppose that B is a separable subgroup of G. If H is any subgroup of G, there exists a subgroup A of G such that:

(1) $A \supseteq H$,

(2) $|A| \subseteq |H|\aleph_0$

(3) $A \| B$.

Proof. For each element $x \in H$ choose a sequence $\{b_n^x\}$ in B so that

$$\sup \{h_G(x+b):b \in B\} = \sup \{h_G(x+b_n^x)\}_{n<\omega_o}.$$

Set $A_o = H$, and define $A_1 = \langle A_o, b_n^x \rangle_{n<\omega_o,\ x \in H}$. Suppose that $h_G(a_o+b) = \alpha$ where $a_o \in A_o$ and $b \in B$. Obviously there exists a positive integer n such that $h_G(a_o+b_n^x) \geq \alpha$ for $x = a_o \in A_o = H$. Notice that $b_n^x \in A_1 \cap B$. Moreover, $|A_1| \leq |H|\aleph_o$. Now, if we let A_1 replace $A_o = H$ and repeat the process, we obtain $A_2 \supseteq A_1 \supseteq A_0 = H$ where $|A_2| \leq |H|\aleph_o$ and has the property that for each pair $(a_1,b) \in A_1 \times B$ there exists c in $A_2 \cap B$ such that $h_G(a_1 + c) \geq h_G(a_1 + b)$. Continuing the process, we let $A = \bigcup_{n<\omega_o} A_n$. Since A satisfies conditions (1) - (3), the lemma is proved.

Theorem 2. Let G be a totally projective group of cardinality not exceeding $\aleph_1$ and let H be an isotype subgroup of G. A necessary and sufficient condition for H to be totally projective is that H is separable in G.

Proof. The necessity of the condition is a consequence of Theorem 1. Thus we need only show that H is totally projective if it is separable.

Since G is totally projective, it has a collection $\mathcal{C}$ of nice subgroups satisfying the third axiom of countability:

(0) $0 \in \mathcal{C}$.

(1) $\mathcal{C}$ is closed with respect to arbitrary group unions, $\langle N_i \rangle \in \mathcal{C}$ if $N_i \in \mathcal{C}$ for each i.

(2) If A is any countable subgroup of G, there exists a countable $N \in \mathcal{C}$ that contains A.

In order to show that H is totally projective, it suffices to prove that H has an ascending chain

$$0 = M_0 \subseteq M_1 \subseteq \cdots \subseteq M_\alpha \subseteq \cdots$$

of nice subgroups M_α of H meeting the following conditions:

(i) $M_{\alpha+1}/M_\alpha$ is countable.

(ii) $M_\beta = \bigcup_{\alpha<\beta} M_\alpha$ when β is a limit.

(iii) $H = \bigcup M_\alpha$, the union of all the M_α's.

For a proof of this assertion see [8] or [2].

By Lemma **1**, if A is any countable subgroup of G, there exists a countable subgroup B of G containing A with $B \| H$ since H is separable in G. Likewise, there of course exists a countable subgroup $C \supseteq B$ that belongs to the collection $\mathcal{C}$. By giving alternate consideration to these two properties, we construct a sequence

$$A \subseteq B_0 \subseteq C_0 \subseteq B_1 \subseteq C_1 \subseteq \cdots$$

of countable subgroups of G, where $B_i \| H$ and $C_i \in \mathcal{C}$. If N represents the union of the above sequence, then not only does N belong to $\mathcal{C}$ but N is also compatible with H. Since $|G| \leq \aleph_1$ it quickly follows that there exists an ascending chain

$$0 = N_0 \subseteq N_1 \subseteq \cdots \subseteq N_\alpha \subseteq \cdots$$

of countable subgroups leading up to G that satisfy the following conditions:

$$N_\beta = \bigcup_{\alpha<\beta} N_\alpha \text{ when } \beta \text{ is a limit,}$$

$$N_\alpha \in \mathcal{C} \text{ for each } \alpha, \text{ and}$$

$$N_\alpha \| H \text{ for each } \alpha .$$

If we let $M_\alpha = N_\alpha \cap H$, all we need to do is show that M_α is nice in H, for this will give us the desired chain of subgroups M_α of H mentioned earlier.

In order to show that M_α is nice in H, suppose for any $x \in H$ that

$$\sup \{h_H(x+m_\alpha) : m_\alpha \in M_\alpha\} = \lambda .$$

Immediately from this we have that

$$\sup\{h_G(x+n_\alpha): n_\alpha \in N_\alpha\} \geq \lambda .$$

Since N_α is nice in G, there is an element $a_\alpha \in N_\alpha$ such that $h_G(x+a_\alpha) \geq \lambda$. Since $x \in H$ and since $N_\alpha \| H$, there exists c_α in $M_\alpha = N_\alpha \cap H$ such that $h_G(x+c_\alpha) \geq \lambda$. But H being isotype in G implies that $h_H(x+c_\alpha) \geq \lambda$. This demonstrates that M_α is nice in H and completes the proof of the theorem.

The following corollary is of special interest. It provides a simple test for the total projectivity of an isotype subgroup of a d.s.c. of "normal" size.

Corollary 1. Let G be a d.s.c. with all (except a countable number) of its Ulm invariants not exceeding $\aleph_1$. An isotype subgroup H of G is totally projective if and only if H is separable in G.

We remark here that the condition on the Ulm invariants is necessary, but we shall later find a necessary and sufficient condition for H to be totally projective with this restriction removed.

III. Examples.

Example 1. We first construct an example of an isotype subgroup H of a d.s.c. group G so that H is not a d.s.c. itself, but in a certain sense is the closest possible in structure to a d.s.c. without being such. In fact G/H will be totally projective of length $\omega_1 + 1$.

For each countable ordinal α, let A_α denote a countable group of length $\alpha + 1$ such that

$$p^\alpha A_\alpha = \langle a_\alpha \rangle$$

is a cyclic group of order p. We use the groups A_α to construct an ascending chain, indexed by the countable ordinals, of countable groups B_λ defined as follows. Set $B_0 = A_0$, and suppose that B_μ has been defined for all $\mu < \lambda$ so that $p^\mu B_\mu = \langle b_\mu \rangle$ is cyclic of order p. As usual we distinguish two cases in the definition of B_λ.

Case 1: λ is a limit. Set $B_\lambda = \bigcup_{\mu<\lambda} B_\mu$.

Case 2: $\lambda - 1$ exists. Let $\mu = \lambda - 1$ and let B_λ be the pushout associated with the diagram

$$\begin{array}{ccc} Z(p) & \rightarrowtail & A_\lambda \\ \downarrow & & \downarrow \\ B_\mu & \rightarrowtail & B_\lambda \end{array}$$

where $Z(p)$ is mapped onto $p^{\mu}B_{\mu} = \langle b_{\mu}\rangle$ and likewise onto $p^{\lambda}A_{\lambda} = \langle a_{\lambda}\rangle$. Observe that

$$B_{\lambda} = (B_{\mu} \oplus A_{\lambda})/\langle(b_{\mu} - a_{\lambda})\rangle$$

contains B_{μ} and has the property that $P^{\lambda}B_{\lambda} = \langle\bar{b}_{\mu}\rangle = \langle\bar{a}_{\lambda}\rangle$ is cyclic of order p.

We have constructed the desired ascending chain of countable groups B_{λ} so that if c denotes b_{λ} then c has height exactly λ in B_{λ}. Furthermore,

$$B_{\lambda+1}/\langle c\rangle = B_{\lambda}/\langle c\rangle \oplus A_{\lambda+1}/\langle c\rangle.$$

Therefore, if $B = \bigcup_{\lambda<\omega_1} B_{\lambda}$ then $p^{\omega_1}B = \langle c\rangle$ and $B/\langle c\rangle$ is a d.s.c. Consider the exact sequence

$$H \rightarrowtail \sum \oplus B_{\lambda} \twoheadrightarrow B$$

where $\sum \oplus B_{\lambda} \longrightarrow B$ is the natural map associated with the inclusion maps $B_{\lambda} \rightarrowtail B$. In order to show that H is isotype in $\sum \oplus B_{\lambda}$, we let $G = \sum \oplus B_{\lambda}$ where the summation is over the countable ordinals λ. We want to show that $p^{\alpha}G \cap H = p^{\alpha}H$ for all α. This is accomplished by induction on α. Observe that the induction automatically survives limits, so assume that $p^{\alpha}G \cap H = p^{\alpha}H$ for $\alpha \leq \beta$ and let $h \in p^{\beta+1}G \cap H$. Choose $g \in p^{\beta}G$ so that $pg = h$ and write $g = \sum \oplus b_{\lambda}$ where $b_{\lambda} \in B_{\lambda}$. Since $g \in p^{\beta}G$, $b_{\lambda} \in p^{\beta}B_{\lambda}$ for each λ and therefore $b = \sum b_{\lambda} \in p^{\beta}B$. Observe that $pb = 0$ since $pg \in H$. Moreover, $b \in p^{\beta}B$ implies that $b \in p^{\beta}B_{\lambda}$ for some λ; there are two cases. If b is a multiple of c then $b \in p^{\beta}B_{\beta}$. If b is not a multiple of c, then the height of b in B_{λ} is the same as it is in $B_{\lambda+1}$ because

$$B_{\lambda+1}/\langle c\rangle = B_{\lambda}/\langle c\rangle \oplus A_{\lambda+1}/\langle c\rangle$$

where $p^{\lambda}B_{\lambda} = \langle c\rangle$. Thus $b \in p^{\beta}B_{\lambda}$ for any B_{λ} that contains b. Choose ν so that $b \in p^{\beta}B_{\nu}$ and beyond any of the components of g. Let $\tilde{b} \in G$ have only one nonzero component, namely the ν-component, and b for that component. Observe that $p(g-\tilde{b}) = h$ and $g - \tilde{b} \in p^{\beta}G \cap H$. Hence the induction hypothesis, $p^{\beta}G \cap H = p^{\beta}H$, implies that $h \in p^{\beta+1}H$, and H is isotype in G. According to Theorem 1, H cannot be a d.s.c. because H is not separable in G due to the fact that

$$\sup \{h_G(c+x) : x \in H\} = \omega_1 .$$

The reader should not confuse the example above with the well-known example where H is dense in G relative to the $p^{\omega}1$-topology. Our example is not an S-group; R. Warfield [17] has introduced and determined the structure of S-groups.

Example 2. Let A_α be a group of length not less than $\alpha + 1$ and not exceeding ω_1. We need to place a restriction on the cardinality of A_α depending on certain set-theoretic considerations. For convenience and simplicity, we shall assume the continuum hypothesis for the purpose of Example 2 only. However, this could be avoided by the approach used in the proof of Theorem 4 in [7]. With the assumption that $c = \aleph_1$, the only cardinality restriction on A_α is $|A_\alpha| \leq c$.

Letting ΠX_i denote the torsion product of a family of torsion groups X_i (since we are working exclusively in the category of torsion abelian groups), we define

$$B = \Pi A_\alpha \; , \; \alpha < \omega_1 \; .$$

Observe that $|B| = 2^c$, and that the subgroup C of B consisting of elements with countable support has cardinality only c; we mean that

$$C = \{(a_\alpha): a_\alpha = 0 \text{ for all but a countable number of } \alpha\} \; .$$

Let $\{B_i\}_{i\epsilon I}$ be the collection of all isotype subgroups of B that are countable, and consider the natural map $\sum \oplus B_i \twoheadrightarrow B$. It is well known [2, Lemma 81.8] that the kernel of this map is both nice and isotype. Thus if

$$H \rightarrowtail \sum \oplus B_i \twoheadrightarrow B$$

is the exact sequence associated with the epimorphism on the right, its kernel H is isotype in $G = \sum \oplus B_i$. We boldly claim that H is not a d.s.c. To avoid redundancies, the proof of this is deferred until Theorem 4 is proved. After Theorem 4 is proved, the fact that H is not a d.s.c. will follow from the fact that B does not satisfy the third axiom of countability with respect to separable subgroups; this axiom is discussed in the next section, where we prove a crucial theorem concerning it.

IV. The Third Axiom of Countability.

If H is a subgroup of G we shall say that G satisfies the third axiom of countability over H with respect to separable subgroups if there exists a collection $\mathcal{C}$ of separable subgroups $K \supseteq H$ of G satisfying the following conditions:

(0) $H \epsilon \mathcal{C}$

(1) $\langle K_i \rangle_{i\epsilon I} \epsilon \mathcal{C}$ if $K_i \epsilon \mathcal{C}$ for each $i \epsilon I$.

(2) If $H \subseteq L \subseteq G$ and L/H is countable, there exists $K \epsilon \mathcal{C}$ such that $K \supseteq L$ and K/H is countable.

If the subgroup H is clear from the context and it is clear that we are dealing with separable subgroups, as opposed for example to nice subgroups, we shall simply say that G satisfies Axiom 3. A closely related axiom but one that is apparently weaker is the following.

Axiom 3'. There exists a chain of separable subgroups

$$H = K_o \subseteq K_1 \subseteq \dots \subseteq K_\alpha \subseteq \dots , \alpha < \tau ,$$

satisfying the following conditions:

(i) $K_\beta = \bigcup_{\alpha<\beta} K_\alpha$ if β is a limit.

(ii) $K_{\alpha+1}/K_\alpha$ is countable.

(iii) $G = \bigcup_{\alpha<\tau} K_\alpha$.

The following theorem is crucial for the proof of our main result.

Theorem 3. Axiom 3 and Axiom 3' are equivalent.

Proof. It is immediate that Axiom 3 implies Axiom 3'. We shall show there exists a collection C of separable subgroups K of G containing H satisfying all the conditions of Axiom 3 when conditions (i) - (iii) are satisfied.

For convenience of notation, let T denote the initial segment of ordinals less than τ; note that T is the index set for the chain of separable subgroups K_α associated with Axiom 3'.

For each $\alpha \in T$, let $\{g_{\alpha,n}\}$ be a set of representatives for the nonzero cosets of K_α in $K_{\alpha+1}$. There is no loss of generality in assuming that $K_\alpha \neq K_{\alpha+1}$. Since $K_{\alpha+1}/K_\alpha$ is countable, we let n range over the positive integers or some nonempty finite subset of the positive integers depending on α . For convenience of notation, we denote

$$\sup \{h_G(g+h): h \in H\}$$

simply by $h_G(g+H)$; this should not be confused with $h_{G/H}(g+H)$, which is the height of $g+H$ in the group G/H . Since K_α is separable in G, when $h_G(g_{\alpha,n} + K_\alpha) = \mu$ is not cofinal with ω_o we can choose according to Proposition 1 the representative $g_{\alpha,n}$ of the coset $g_{\alpha,n} + K_\alpha$ so that $h_G(g_{\alpha,n}) = \mu$. We shall therefore assume such a choice has been made whenever $h_G(g_{\alpha,n} + K_\alpha)$ is not cofinal with ω_o .

Observe that each element $g \in G$ not in H has a unique representation of the form

$$(*)\ g = g_{\lambda(1),n(1)} + g_{\lambda(2),\, n(2)} + \dots + g_{\lambda(k),\, n(k)} + h ,$$

where $h \in H$ and $\lambda(1) > \lambda(2) > \dots > \lambda(k)$. This unique representation of g will be called its standard representation.

A subset S of T is <u>closed</u> if it satisfies the following conditions:

(a) If $g = \sum g_{\alpha(i),\, m(i)}$ has standard representation (*), then $\lambda(j)$ belongs to S for each $j \leq k$ provided $\alpha(i) \in S$ for each i.

(b) If $\alpha \in S$ and $h_G(g_{\alpha,n} + K_\alpha) = \mu$, then either $h_G(g_{\alpha,n}) = \mu$ or else for each $\mu' < \mu$ there exist $\lambda(i) \in S$, $1 \leq i \leq k$, and positive integers $n(i)$ such that $\alpha = \lambda(1)$, $n = n(1)$,

$$\lambda(1) > \lambda(2) > \lambda(3) > \ldots > \lambda(k),$$

and

$$h_G(\sum g_{\lambda(i),n(i)} + h) > \mu'$$

for some $h \in H$. In other words, there is an element $g \in G$ such that $h_G(g) > \mu'$ whose initial term in its standard representation is the given element $g_{\alpha,n}$ and all the indices belong to S.

We now observe that the union of any number of closed subsets of T is again closed. The proof that condition (a) is valid for the union S of a collection $\{S_\nu\}$ of closed sets follows by induction on the largest of the (finite number of) $\alpha(i)$'s involved in the given summation of $g = \sum g_{\alpha(i),\, m(i)}$. Suppose that $\alpha(i) = \beta$ for $i \leq j$ and that $\alpha(i) < \beta$ if $i > j$. We let

$$g_1 = \sum_{i \leq j} g_{\alpha(i),\, m(i)} .$$

Since $\alpha(i) = \beta$ for $i \leq j$, we know that $\alpha(i) \in S_\nu$ for some ν when $i \leq j$. Therefore, by condition (a) for S_ν, if

$$g_1 = \sum g_{\gamma(i),\, q(i)} + h_1$$

is the standard representation for g_1, we know that $\gamma(i) \in S_\nu \subseteq S$ for each i. In particular, $\gamma(1) \in S$. Notice that if $\gamma(1) < \beta$ then the induction hypothesis yields (a) for the set S because

$$g - h_1 = \sum_{i>j} g_{\alpha(i),m(i)} + \sum g_{\gamma(i),\, q(i)} ,$$

and the standard representation of $g - h_1$ differs from that of g only by $h_1 \in H$. Thus we may assume that $\gamma(1) = \beta$. If we let (*) $\sum g_{\lambda(i),\, n(i)} + h$ be the standard representation of

$$g - g_{\gamma(1),q(1)} - h_1 = \sum_{i>j} g_{\alpha(i),m(i)} + \sum_{i \geq 2} g_{\gamma(i),q(i)} ,$$

then $\lambda(i) \in S$ for each i by the induction hypothesis. Since $\lambda(1) < \beta = \gamma(1)$, it follows that

$$g_{\gamma(1),q(1)} + \sum g_{\lambda(i),n(i)} + h + h_1$$

is the standard representation of g. Since $\gamma(1) \in S$ and $\lambda(i) \in S$, condition (a) holds for S. It is immediate that condition (b) holds for S since each $\alpha \in S$ is contained in some S_ν. We have demonstrated that the union of any collection of closed subsets is again closed.

Since closure is an inductive property any countable subset of T is contained in a countable closed subset, for we can add successively the required elements to satisfy both of the conditions (a) and (b). In this connection, observe that if $\alpha(i)$ is restricted to a countable set, there are only a countable number of elements of the form $\sum g_{\alpha(i),m(i)}$ to be considered in condition (a). In regard to condition (b), if $h_G(g_{\alpha,n}) < \mu$, then $\mathrm{cof}(\mu) = \omega_o$ due to the choice of $g_{\alpha,n}$. Therefore, we can choose a sequence μ_j so that $\mu_j < \mu$ but $\mu = \sup\{\mu_j\}$. For each j, there exists $x_j \in K_\alpha$ such that $h_G(g_{\alpha,n} + x_j) > \mu_j$. Hence if S_o is a countable set and $\alpha \in S_o$, we adjoin to S_o all of the indices $\lambda(i)$ in T that are used in the standard representations of the x_j's associated with the various $g_{\alpha,n}$'s. As we have already indicated, for the benefit of condition (a) we also adjoin to S_o all of the indices $\lambda(i)$ in T that are used in the standard representations of elements of the form $\sum g_{\alpha(i),m(i)}$ with $\alpha(i) \in S_o$. Repeating the process, we obtain a sequence of countable sets

$$S_o \subseteq S_1 \subseteq \cdots \subseteq S_n \subseteq \cdots$$

such that $S = \bigcup S_n$ is closed and contains an arbitrary countable set S_o.

In order to show that G satisfies Axiom 3, we set

$$\mathcal{C} = \{K_S \subseteq G : K_S = \langle H, g_{\alpha,n} \rangle_{\alpha \in S}\}$$

where S ranges over the closed subsets of T. Naturally, we follow the conventio that the empty set ϕ is closed. Thus $\mathcal{C}$ satisfies all the condition of Axiom 3, but we need to show that K_S is separable in G whenever S is a closed subset of T. Let $g \in G$ and suppose that $h_G(g+K_S) = \mu$. To show that K_S is separable there is essentially nothing to do if $\mathrm{cof}(\mu) = \omega_o$, so assume that $\mathrm{cof}(\mu) \neq \omega_o$. Since H is separable by hypothesis, we may assume that $h_G(g+H) < \mu$. Letting g have the standard representation

$$(*) \quad g = g_{\lambda(1),n(1)} + g_{\lambda(2),n(2)} + \cdots + g_{\lambda(k),n(k)} + h,$$

we prove by induction on $\lambda(1)$ that $h_G(g+c) = \mu$ for some $c \in K_S$. We can without loss of generality assume that $\lambda(i) \notin S$ for each i because we do not change the coset $g + K_S$ if we delete from g the terms $g_{\lambda(i),n(i)}$ for which $\lambda(i) \in S$. With this choice of g and for any $\mu' < \mu$, choose $x \in K_S$ so that $h_G(g+x) > \mu'$. Let

$$x = g_{\alpha(1),m(1)} + g_{\alpha(2),m(2)} + \cdots + g_{\alpha(j),m(j)} + h_x$$

be the standard representation of x. Since $x \in K_S$, $\alpha(i) \in S$ for each i. Since $\lambda(1) \notin S$, $\alpha(1) \neq \lambda(1)$. Assume that $\alpha(1) > \lambda(1)$. Under this assumption, observe that

$$h_G(g_{\alpha(1),m(1)} + K_{\alpha(1)}) \geq h_G(g+x) > \mu' .$$

By condition (b) of a closed set, there exists $y \in K_S \cap K_{\alpha(1)}$ such that $h_G(g_{\alpha(1),m(1)} + y) > \mu'$. Therefore, if we set

$$z = x - g_{\alpha(1),m(1)} - y$$

then $z \in K_S \cap K_{\alpha(1)}$ and $h_G(g+z) > \mu'$, which means that we can replace x by z. Hence if x is chosen so that $\alpha(1)$ is minimal, it must be the case that $\lambda(1) > \alpha(1)$. Therefore $h_G(g+K_{\lambda(1)}) \geq \mu$ and consequently $h_G(g - g') \geq \mu$ for some $g' \in K_{\lambda(1)}$ because $K_{\lambda(1)}$ is separable. Since $g' \in K_{\lambda(1)}$, the initial term of its standard representation has a smaller index than $\lambda(1)$. Hence by the induction hypothesis there exists c in K_S such that $h_G(g'+c) \geq \mu$. But this implies that $h_G(g+c) \geq \mu$, and the theorem is proved.

V. A Necessary and Sufficient Condition.

In this section we establish a necessary and sufficient condition for an isotype subgroup of a totally projective group to be itself totally projective. This is our main result, and both Theorem 1 and Theorem 3 are required for the proof.

Theorem 4. Suppose that H is an isotype subgroup of a totally projective group G. Then H is totally projective if and only if G satisfies the third axiom of countability over H with respect to separable subgroups.

Proof. First assume that H is totally projective. Let $\mathcal{C}_G$ and $\mathcal{C}_H$, respectively, be collections of <u>nice</u> subgroups of G and H satisfying Axiom 3. Since H is totally projective, H is separable in G according to Theorem 1. We plan to show that G satisfies Axiom 3' over H with respect to separable subgroups. Toward this end, suppose that

(B) $0 = B_0 \subseteq B_1 \subseteq \dots \subseteq B_\alpha \subseteq \dots \quad , \alpha < \gamma ,$

is an ascending chain of subgroups of G that satisfy the following conditions for $\alpha < \gamma$.

(1) $B_\alpha \in \mathcal{C}_G$.

(2) $B_\alpha \cap H \in \mathcal{C}_H$.

(3) $B_\alpha \parallel H$.

(4) $|B_{\alpha+1}/B_\alpha| \leq \aleph_0$ whenever $\alpha + 1 < \gamma$.

(5) $B_\beta = \bigcup_{\alpha<\beta} B_\alpha$ whenever $\beta < \gamma$ is a limit.

If the subgroups B_α do not already lead to G, we want to define B_γ so that conditions (1) - (5) continue to hold for the extended chain. If γ is a limit, we simply set $B_\gamma = \bigcup_{\alpha<\gamma} B_\alpha$ as we must, and all the conditions are satisfied due to the fact that (1) - (3) are inductive properties. If $\gamma - 1$ exists, it is more difficult to find a proper extension B_γ of $B_{\gamma-1}$ having the desired properties, but we shall do that presently.

In the construction of B_γ, we temporarily set $B_{\gamma-1} = B$. Since B satisfies conditions (1) and (2), $\langle H,B\rangle/B \cong H/H \cap B$ is totally projective, and the same is true for G/B. Moreover, $\langle H,B\rangle/B$ is isotype in G/B. For if $h + B \in p^\lambda(G/B) = \langle p^\lambda G,B\rangle/B$ then $h + B = g + B$ where $g \in p^\lambda G$. Whence $h + b = g \in p^\lambda G$ where $b \in B$. In view of condition (3), this implies that $h + c \in p^\lambda G$ for some $c \in H \cap B$. But H being isotype in G yields $h + c \in p^\lambda H$, so $h + B \in p^\lambda(\langle H,B\rangle/B)$ and $\langle H,B\rangle/B$ is isotype in G/B. This means that we can replace H by $\langle H,B\rangle/B$ and G by G/B and yet retain the hypothesis: H is an isotype subgroup of G with H and G totally projective.

Obviously, the set

$$\mathcal{C}_{G/B} = \{ C/B: C \in \mathcal{C}_G \text{ and } C \supseteq B\}$$

is a collection of nice subgroups of G/B that satisfies the third axiom of countability. It is also easily verified that

$$\mathcal{C}_{\langle H,B\rangle/B} = \{\langle A,B\rangle/B: A \in \mathcal{C}_H \text{ and } A \supseteq H \cap B\}$$

is a collection of nice subgroups of $\langle H,B\rangle/B$ that satisfies Axiom 3; use the property $B \parallel H$, together with the property that H is isotype in G, to conclude that B is nice in $\langle H,B\rangle$. Then employ the isomorphism $\langle H,B\rangle/B \cong H/H \cap B$ to obtain the designated collection of nice subgroups of $\langle H,B\rangle/B$ from those in $\mathcal{C}_H$ containing $H \cap B$.

If C_o is any countable subgroup of G, there exists according to Lemma 1 a countable subgroup C of G containing C_o such that $\langle B,C\rangle/B \parallel \langle H,B\rangle/B$. In fact, we can choose C so that the following hold:

(a) $C \in \mathcal{C}_G$

(b) $\langle B,C\rangle/B \cap \langle H,B\rangle/B \in \mathcal{C}_{\langle H,B\rangle/B}$

(c) $\langle B,C\rangle/B \parallel \langle H,B\rangle/B$.

It is important to observe that (b) can be accomplished by choosing C so that $\langle B,C\rangle \cap H \in \mathcal{C}_H$, and we shall do this. Since B and C belong to $\mathcal{C}_G$ then so does $\langle B,C\rangle$. Thus if we define $B_\gamma = \langle B,C\rangle$ then B_γ satisfies conditions (1) and (2). Moreover, since $B = B_{\gamma-1}$ and since C is countable, $B_\gamma/B_{\gamma-1}$ is countable. Therefore, condition (4) continues to hold for the extended chain of subgroups B_α. Because condition (5) is not relevant, it remains only to show that $B_\gamma \parallel H$, which is condition (3). Suppose that $h_G(b_\gamma + h) = \lambda$ where $b_\gamma \in B_\gamma = \langle B,C\rangle$ and $h \in H$. Using (c), we know that

$$h_{G/B}(b_\gamma + a + B) \geq \lambda$$

for some $a \in \langle B,C\rangle \cap \langle H,B\rangle$. But $\langle B,C\rangle \cap \langle H,B\rangle = \langle\langle B,C\rangle \cap H, B\rangle$ means that we can choose $a \in \langle B,C\rangle \cap H$. Moreover, B being nice implies that $h_G(b_\gamma + a + b) \geq \lambda$ for some $b \in B$. This leads to $h_G(a - h + b) \geq \lambda$. Since $a \in H$ and since $B \parallel H$, there exists $d \in B \cap H$ such that $h_G(d + b) \geq \lambda$. We have now $h_G(b_\gamma + a - d) \geq \lambda$ and $a - d \in \langle B,C\rangle \cap H = B_\gamma \cap H$. Thus $B_\gamma \parallel H$.

It has been shown that there exists a chain (B) satisfying conditions (1) - (5) and such that $G = \bigcup_{\alpha<\gamma} G_\alpha$, for we can continue to extend such a chain until we reach G. With the chain (B) at our disposal, we set $K_\alpha = \langle H,B_\alpha\rangle$ and consider the chain

$$(K) \quad H \doteq K_0 \subseteq K_1 \subseteq \dots \subseteq K_\alpha \subseteq \dots, \alpha < \gamma .$$

Since condition (4) for (B) implies that $K_{\alpha+1}/K_\alpha$ is countable and condition (5) yields $K_\beta = \bigcup_{\alpha<\beta} K_\alpha$ whenever β is a limit, all we need to do is demonstrate that K_α is separable in G in order to prove that G satisfies Axiom 3' over H with respect to separable subgroups. To show that $K_\alpha = \langle H,B_\alpha\rangle$ is separable in G, suppose $h_G(g+K_\alpha) = \mu$ where $\mathrm{cof}(\mu) \neq \omega_0$. Letting $B = B_\alpha$ for convenience, we observe that

$$h_{G/B}(g + \langle H,B\rangle/B) \geq h_G(g + \langle H,B\rangle) = \mu .$$

As previously shown, $\langle H,B\rangle/B$ is isotype in G/B, and $\langle H,B\rangle/B$ is totally projective. By Theorem 1, there exists $h \in H$ such that $h_{G/B}(g + h + B) \geq \mu$. Since B is nice in G, $h_G(g + h + b)$ for some $b \in B$, which shows that $K_\alpha = \langle H,B\rangle$ is separable in G. According to Theorem 3, G must satisfy the third axiom of countability over H with respect to separable subgroups because G satisfies Axiom 3' over H.

Conversely, suppose that H is an isotype subgroup of the totally projective group G and suppose that G satisfies the third axiom of countability over H with respect to separable subgroups. Let $\mathcal{C}$ be a collection of nice subgroups of G satisfying Axiom 3 and let $\mathcal{D}$ be a collection of separable subgroups of G satisfying Axiom 3 over H. In particular, $H \in \mathcal{D}$ is a separable subgroup of G. Let us examine the consequences of a subgroup B of G satisfying the conditions:

(i) $B \in \mathcal{C}$.

(ii) $\langle H,B\rangle \in \mathcal{D}$.

(iii) $B \parallel H$.

Since $\langle H,B\rangle$ is separable, there exists a countable subgroup C of G not contained in $\langle H,B\rangle$ unless $G = \langle H,B\rangle$ such that

$$C \in \mathcal{C},$$

$$\langle H,C\rangle \in \mathcal{D},$$

$$C \parallel \langle H,B\rangle.$$

We claim that

$$\langle B,C\rangle \in \mathcal{C},$$
$$\langle H,B,C\rangle \in \mathcal{D},$$
$$\langle B,C\rangle \parallel H .$$

The first two conditions are immediate, and the third is a direct consequence of $C \parallel \langle H,B\rangle$ and $B \parallel H$. It follows that there exists an ascending chain

$$0 = B_0 \subseteq B_1 \subseteq B_2 \subseteq \dots \subseteq B_\alpha \subseteq \dots$$

of subgroups of G satisfying conditions (i) - (iii) such that $B_{\alpha+1}/B_\alpha$ is countable, $B_\beta = \bigcup_{\alpha<\beta} B_\alpha$ when β is a limit, and G is the union of all the B_α's. Set $A_\alpha = H \cap B_\alpha$, and consider the corresponding chain

$$0 = A_0 \subseteq A_1 \subseteq A_2 \subseteq \dots \subseteq A_\alpha \subseteq \dots$$

of subgroups of H. Since $A_\beta = \bigcup_{\alpha<\beta} A_\alpha$ when β is a limit, since $H = \bigcup A_\alpha$ and $A_{\alpha+1}/A_\alpha$ is countable, it is enough to show that A_α is nice in H in order to show that H is totally projective [2]. Assume, for some $x \in H$, that $h_H(x + A_\alpha) = \lambda$. Clearly, $h_G(x+B_\alpha) \geq \lambda$ which implies that $h_G(x+b) \geq \lambda$ for some $b \in B_\alpha$ since B_α in $\mathcal{C}$ is nice in G. Finally, $B_\alpha \parallel H$ means that $h_G(x+a) \geq \lambda$ for some $a \in A_\alpha$. But H isotype in G yields $h_H(x+a) \geq \lambda$, and therefore A_α is nice in H. This completes the proof of the theorem.

Using Theorem 4, we can easily prove that in Example 2 the subgroup H of $G = \sum \oplus B_i$ is not a d.s.c. Recall that H is both nice and isotype in G; subgroups satisfying these two conditions are called balanced [2]. Since H is nice in G, if $H \subseteq K \subseteq G$ then K is separable in G if and only if K/H is separable in $B = G/H$. Hence it suffices to prove that $B = \Pi A_\alpha$ does not satisfy the third axiom of countability (over zero) with respect to separable subgroups. Assume that B does satisfy Axiom 3 for separable subgroups. Since the continuum hypothesis has been invoked for Example 2, there must exist a separable subgroup K of B having cardinality c and containing the subgroup

$$C = \{(a_\alpha) \in \Pi A_\alpha : a_\alpha = 0 \text{ except for a countable number of } \alpha\}$$

because C has only c elements. However, if we choose $a_\alpha \in p^\alpha A_\alpha[p]$ different from zero, then $\Pi\langle a_\alpha\rangle \subseteq B$ has cardinality 2^c. Hence there exists $a \in \Pi\langle a_\alpha\rangle$ not in K. But

$$h_B(a+K) \geq h_B(a+C) = \omega_1$$

shows that K is not separable after all.

In connection with Example 2, we mention that Richman and Walker [15] were the first to show the existence of a balanced subgroup of a d.s.c. that is not itself a d.s.c. There are common ideas in the two examples, but their approach is different and not as direct as ours. Moreover, we have actually given a broad class of such examples.

The concluding corollary should serve as a reminder that the true significance of Theorem 4 is due to the uniqueness theorem for totally projective groups.

Corollary. Let H and H' be isotype subgroups of totally projective groups G and G', respectively. If G and G' satisfy the third axiom of countability over H and H' with respect to separable subgroups, then H and H' are isomorphic provided they have the same Ulm invariants.

REFERENCES

1. L. Fuchs, Infinite Abelian Groups, Vol. I, Academic Press, New York, 1970.

2. ______, Infinite Abelian Groups, Vol. II, Academic Press, New York, 1973.

3. ______, Vector spaces with valuations, Jour. of Algebra 35 (1975), 23-38.

4. P. Griffith, Infinite Abelian Groups, Chicago Lectures in Mathematics, Chicago and London, 1970.

5. P. Hill, On the classification of Abelian groups, photocopied manuscript, Houston, 1967.

6. ______, Isotype subgroups of direct sums of countable groups, Illinois Jour. Math. 13 (1969), 281-290.

7. ______, Criteria for freeness in Abelian Groups and valuated vector spaces, Lecture Notes in Mathematics # 616, 140-157. Springer-Verlag, Berlin, Heidelberg, New York, 1977.

8. ______, The third axiom of countability for Abelian groups, Proc. Amer. Math. Soc. (to appear).

9. P. Hill and C. Megibben, On direct sums of countable groups and generalizations, Studies on Abelian Groups, 183-206, Paris, 1968.

10. I. Kaplansky, Infinite Abelian Groups, University of Michigan Press, Ann Arbor, Michigan, 1969.

11. I. Kaplansky and G. Mackey, A generalization of Ulm's theorem, Summa Brasil. Math. 2 (1951), 195-202.

12. G. Kolettis, Direct sums of countable groups, Duke Math. Jour. 27 (1960) 111-125.

13. R. Nunke, On the structure of Tor, Proc. Colloq. Abelian Groups 115-124, Budapest 1964.

14. ______, Homology and direct sums of countable abelian groups, Math. Zeitschr. 101 (1967), 182-212.

15. F. Richman and E. Walker, Valuated groups, Journal of Algebra 56 (1979),145-167.

16. H. Ulm, Zur Theorie der Abzählbarunendlichen abelschen Gruppen, Math. Ann. 107 (1933), 774-803.

17. R. Warfield, A classification theorem for Abelian p-groups, Trans. Amer. Math. Soc. 210 (1975), 149-168.

CLASSIFICATION THEORY OF ABELIAN GROUPS, II:

LOCAL THEORY

Robert B. Warfield, Jr.[1]

This paper is a somewhat rewritten and shortened version of a paper first written many years ago but never published. The paper introduces a class of modules over a discrete valuation ring, gives some characterizations of this class, defines new invariants, and gives a complete classification theorem. The classification theorem is a generalization of Ulm's classification theorem for countable Abelian p-groups, and includes Kaplansky's and Mackey's generalization of Ulm's theorem to modules of torsion-free rank one, as well as Hill's generalization to totally projective p-groups. The new invariants are obtained by applying a suitable Krull-Schmidt theorem to an additive category associated with the category of modules.

This research was done primarily between 1969 and 1971 and the results originally reported in [30] and [31]. The results were written up as a paper (with the same title) in 1976, but publication was further delayed because of an error discovered by Fred Richman. (Fortunately, Hunter, Richman, and Walker had already given an alternative proof in [9] of the result which the author had proved incorrectly.) Recently, Hunter and Richman proved in [8] an alternative version of the original incorrect lemma, which makes the approach taken here again a valid one. While the results in this paper have in the interim become "well known," the original proofs have never appeared, and it seemed appropriate that they should, in that they represent a different point of view from the alternative versions provided by others. However, some of the results in the 1971 and 1976 accounts have been removed from this paper because new proofs are now available which are substantial improvements, and it no longer seemed worth while to publish the original ones.

---

1. Research supported in part by grants from the National Science Foundation.

In the following, R will be a discrete valuation ring, p a generator of its maximal ideal, Q the quotient field of R, and all modules will be R-modules. A module is simply presented if it can be defined in terms of generators and relations in such a way that the relations all have the form $px = 0$ or $px = y$. The modules classified in this paper are exactly those modules which are summands of simply presented modules.

Simply presented Abelian p-groups were introduced in [4] by Crawley and Hales, who called them T-groups. They were able to extend Ulm's theorem to these groups and to show that the reduced T-groups coincided with the totally-projective p-groups introduced by Nunke [16] and classified by Hill [6]. The original description of these groups given by Nunke is not very useful in the classification theory, and Hill was forced to give a new description of the totally projective groups, characterizing them as exactly those reduced p-groups satisfying what we will call *Hill's condition.* We postpone a description of Hill's condition until section 1, but we remark that Hill called it "the third axiom of countability," and in the terminology of [5], a group satisfied this condition if and only if it "has a nice system." Hill's condition can be used to give another description of the modules in our class: a module is a summand of a simply presented module if and only if it is a summand of a direct sum of modules of (torsion-free) rank one which satisfy Hill's condition. (The rank of a module M is the Q-dimension of $Q \otimes M$.)

The above description of summands of simply presented modules is the most useful technically, though it initially appears somewhat ad hoc, because of the rank one condition. In fact, the most important elementary observation about simply presented modules is that they are direct sums of modules whose torsion-free rank is one. (Conversely, any countable generated module of torsion-free rank one is a summand of a simply presented module, but it is not necessarily simply presented.) The fact that rank one modules appear in the discussion immediately suggests how one should attempt to classify these modules, using the ideas developed by Kaplansky to classify countably generated modules of rank one over a complete discrete valuation ring.

In order to define the invariants we need to classify simply presented modules and their summands, we introduce some terminology. For any ordinal α, and any module M, we define $p^\alpha M$ inductively, setting $p^{\alpha+1}M = p(p^\alpha M)$ and, if α is a limit ordinal, $p^\alpha M = \cap_{\beta<\alpha} p^\beta M$. The *height* of x, $h(x)$, is α if $x \in p^\alpha M$ and $x \notin p^{\alpha+1}M$. We define $p^\infty M = \cap p^\alpha M$, where the intersection is over all ordinals α. (Equivalently, $p^\infty M = p^\alpha M$, where α is the smallest ordinal such that $p^{\alpha+1}M = p^\alpha M$.) If $x \in p^\infty M$, then we write $h(x) = \infty$. $p^\infty M$ is the maximal divisible submodule of M, and is a summand of M. M is said to be *reduced* if $p^\infty M = 0$. The *Ulm functors* are the functors associating to M the vector spaces $U_\alpha(M) = (p^\alpha M)[p]/(p^{\alpha+1}M)[p]$ and $U_\infty(M) = (p^\infty M)[p]$, and the *Ulm invariants* are the dimensions of these spaces, as vector spaces over R/pR,

$$f(\alpha,M) = \dim U_\alpha(M), \quad f(\infty,M) = \dim U_\infty(M).$$

It should be clear that the Ulm invariants only give information about the torsion part of the module M.

An *Ulm sequence* is a sequence $\{\alpha_i\}$, $0 \le i < \omega$, such that all of the α_i are ordinals or ∞ and such that (i) if $\alpha_i = \infty$ then $\alpha_{i+1} = \infty$, and (ii) if $\alpha_{i+1} \ne \infty$ then $\alpha_{i+1} > \alpha_i$. If M is a module and $x \in M$, the *Ulm sequence* of x, $U(x)$, is the sequence $\{\alpha_i\}$ where $\alpha_i = h(p^i x)$. Two Ulm sequences $\{\alpha_i\}$ and $\{\beta_i\}$ are *equivalent* if for some positive integers n and m, $\alpha_{i+n} = \beta_{i+m}$ for all $i \ge 0$. If x and y are elements of infinite order in a module M of torsion-free rank one, then $U(x)$ and $U(y)$ are equivalent, so the equivalence class is an invariant of M, written $U(M)$. Ulm sequences and the invariant $U(M)$ were invented by Kaplansky. Using the techniques from Mackey's proof of Ulm's theorem, Kaplansky showed in [13] that two countably generated modules M and N of rank one over a complete discrete valuation ring are isomorphic if and only if they have the same Ulm invariants and $U(M) = U(N)$. (The completeness requirement was removed by Megibben [15].)

If M is simply presented and $M = \oplus_{i\in I} M_i$, where the M_i all have rank one, we would like to define $g(e,M)$ for any equivalence class e of Ulm sequences, to be the number of summands M_i such that $U(M_i) = e$. It is not obvious that this is an invariant of M, or that it makes sense for summands of simply presented modules, which are not necessarily direct sums of modules with torsion-free rank one. Both of these problems are overcome by using a Krull-Schmidt theorem in a category H whose objects are modules, but in which the modules of torsion-free rank one are indecomposable and small. (This program is carried out in section 4, and the classification theorem proved in section 5.)

The Ulm sequences also give us a characterization of summands of simply presented modules as relatively projective modules. If $u = \{\alpha_i\}$ is an Ulm sequence and M a module, we define

$$uM = \{x \in M : h(p^i x) \geq \alpha_i \text{ for all } i \geq 0\}.$$

(Kaplansky showed in [12] that if M is a countably generated torsion module then any fully invariant submodule of M is of the form uM for some Ulm sequence u.) We say a short exact sequence

(1) $$0 \to A \to B \to C \to 0$$

of modules is *sequentially-pure* if for any Ulm sequence u, the sequence

$$0 \to uA \to uB \to uC \to 0$$

is exact. A module P is *sequentially-pure-projective* if for any sequentially-pure short exact sequence (1), the map $\mathrm{Hom}(P,B) \to \mathrm{Hom}(P,C)$ is surjective. We show in section 3 that *a module is sequentially-pure-projective if and only if it is a summand of a simply presented module.*

The first three sections of this paper are the same as those in the paper with the same title written in 1976, except for a slight shortening of section 1. In section 4, the previous incorrect lemma is replaced by the correct lemma proved by Hunter and Richman, and the proof of the classification theorem in section 5 is slightly modified to use the new lemma. The 1976 paper contained in section 6 a long and essentially independent treatment of direct sums of countably generated modules. Subsequent methods have made it possible to vastly simplify the proofs of these results, so the original treatment has been omitted and there is just a brief discussion. Section 7 of the 1976 paper contained a discussion of open problems, most of which have now been solved, so it has been replaced by a brief discussion of significant recent work on simply presented groups and related classes of groups.

1. Modules satisfying Hill's condition.

1.1. Definition. If M is a module and N a submodule, then N is nice in M if for all ordinals α, $p^{\alpha}(M/N) = (p^{\alpha}M + N)/N$.

1.2. Lemma. (i) Any finitely generated torsion submodule is nice. (ii) If the discrete valuation ring R is complete, then any finitely generated submodule of an R-module is nice. (iii) If $M = \oplus_{i\in I} M_i$ and $S = \Sigma_{i\in I} S_i$, where $S_i \leq M_i$, then S is nice if and only if S_i is nice in M_i for each i. (iv) If M has submodules N and S with $N \geq S$, and S is a nice submodule of M, then N is a nice submodule of M if and only if N/S is a nice submodule of M/S.

Proof. [34, lemmas 1.3-1.6]. All of these are essentially trivial to verify except the statement about modules over a complete discrete valuation ring, which follows from the linear compactness of the ring.

1.3. Definition. If M is a module over a discrete valuation ring R, then M is said to satisfy Hill's condition if M has a family $\mathcal{C}$ of submodules, such that

(i) If $C \in \mathcal{C}$, then C is a nice submodule.

(ii) If $C_i (i \in I)$ are in $\mathcal{C}$, then $\Sigma_{i\in I} C_i \in \mathcal{C}$.

(iii) If $C \in \mathcal{C}$ and X is a countable subset of M, then there is a $C' \in \mathcal{C}$ such that $C' \geq C$, $C' \geq X$, and C'/C is countably generated.

(iv) $\{0\} \in \mathcal{C}$.

1.4. Lemma. If M is a module of torsion-free rank one and N a cyclic submodule, then N is nice.

Remark: This is contained in [29], but we give a separate proof for completeness. Our proof uses the truth of the statement when the ring R is complete.

Proof. Let R^* be the completion of R, and $M^* = R^* \otimes M$. Let x be an element of M and N the submodule generated by x. By Lemma 1.2 (i), we may assume x has infinite order. To show that N is a nice submodule of M, we need to show that the natural map $M \to M/N$ takes $p^{\alpha}M$ onto $p^{\alpha}(M/N)$ for all ordinals α.

We can identify M/N with M^*/N^* where N^* is the cyclic R^*-submodule generated by x. Since N^* is a nice submodule of M^*, every element of $p^\alpha(M/N)$ is the image of an element of $p^\alpha M^*$. When we regard M as an R-module of M^*, the torsion submodule of M^* is contained in M and for all ordinals α, $p^\alpha M = M \cap p^\alpha M^*$. (This follows from the fact that R^*/R is a torsion-free divisible R-module.) Therefore, if $p^\alpha M^*$ is torsion, then $p^\alpha M^* = p^\alpha M$, so every element of $p^\alpha(M/N)$ is in the image of $p^\alpha M$. Suppose then that $p^\alpha M^*$ is not torsion. Then for some integer n, $p^n x \in p^\alpha M^*$, and $M^* = M + [p^n x]^*$. If $y \in p^\alpha M^*$, then in this expression we can write $y = w + rp^n x$, for some $w \in M$ and $r \in R^*$. Since $rp^n x \in N^*$, it follows that an element of M^*/N^* which is in the image of $p^\alpha M^*$ is also in the image of $p^\alpha M$. This completes the proof that N is a nice submodule of M.

We now recall the basic theorem on extending isomorphisms defined on nice submodules. (The basic idea of this theorem is contained in the work on the classification theorem for torsion groups, done simultaneously in 1967 by Hill and Crawley and Hales, [4,6]. Hill's version is more generally applicable. Walker's version [27] is actually the first not to require that all modules be torsion. The version we need is stated precisely in [34, 1.11].) If M and N are modules, and S and T nice submodules, and $f : S \to T$ an isomorphism which is height preserving (i.e. $f(S \cap p^\alpha M) = T \cap p^\alpha N$ for all ordinals α), if M/S and N/T are torsion modules satisfying Hill's condition and M and N have the same Ulm invariants, then f extends to an isomorphism of M onto N if and only if the "relative Ulm invariants" of S in M are equal to those of T in N. We now define the relative Ulm invariants, and give some pertinent examples. If α is any ordinal, we let $S_\alpha = S \cap p^\alpha M$ and let $S_\alpha^* = \{x \in S_\alpha : px \in S_{\alpha+2}\}$. There is an injective map $S_\alpha^*/S_{\alpha+1} \to U_\alpha(M)$ defined as follows: if $x \in S_\alpha^*$, let $px = py$ for some $y \in S_{\alpha+1}$. If $\eta_\alpha(x) = s - y + (p^{\alpha+1}M)[p]$, then η_α is the desired injective homomorphism $S_\alpha^*/S_{\alpha+1} \to (p_\alpha M)[p]/(p^{\alpha+1}M)[p] = U_\alpha(M)$. The image of this map is denoted $I_\alpha(S)$, and the corresponding relative Ulm invariant is the codimension of $I_\alpha(S)$ in $U_\alpha(M)$. Similarly, we define $I_\infty(S) = S \cap (p^\infty M)[p]$. The existence of a height preserving isomorphism from S to T implies the isomorphism between $I_\alpha(S)$ and $I_\alpha(T)$, but for the map to extend to an isomorphism of M onto N it is clearly also necessary that the codimensions be equal. We note that if S and T are finitely generated, this will be true automatically, since $I_\alpha(S)$ and $I_\alpha(T)$ will be finite dimensional subspaces. We note also that if [x] is a cyclic submodule of M, and u(x) is its Ulm sequence, then $I_\alpha([x]) \neq 0$ if and only if for some non-negative integer n, $h_p(p^n x) = \alpha$, and $h_p(p^{n+1}x) > \alpha + 1$. In this situation, the Ulm sequence is said to have a *gap at* α.

1.5 Theorem. Let M and N be modules of torsion-free rank one satisfying Hill's condition. Then $M \simeq N$ if and only if $U(M) = U(N)$ and M and N have the same Ulm invariants.

Proof. We can easily find elements x and y of infinite order in M and N such that $U(x) = U(y)$. The function taking x to y therefore induces a height preserving isomorphism from [x] to [y]. Since [x] and [y] are nice submodules (1.4), we infer from the extension theorem [34, 1.11] that the isomorphism from [x] to [y] extends to an isomorphism of M onto N. (In view of the discussion preceeding this theorem, the condition in [34, 1.11] concerning relative Ulm invariants is not relevant in this case, since [x] and [y] are finitely generated submodules.)

Remarks. This contains the Kaplansky-Mackey [13] and Megibben [15] classification theorems for countably generated modules of rank one. In [29], Wallace proves (in effect) that if M and N are modules whose torsion submodules satisfy Hill's condition, and if M and N have torsion-free rank one, then $M \simeq N$ if and only if $U(M) = U(N)$ and M and N have the same Ulm invariants. (Wallace also obtains a global form of this theorem, which does not concern us in this paper.) We note that if M is such a module, and x is an element of infinite order in M, then M satisfies Hill's condition, since it is a countable extension of a module satisfying Hill's condition. It follows that Wallace's class of modules is included in ours. The converse is false however, since there are modules in our class whose torsion submodules do not satisfy Hill's condition, (see [33]).

2. Simply presented modules. In this section we establish the basic properties of modules with simple presentations. We show that they are direct sums of modules of torsion-free rank one, and that they satisfy Hill's condition. An existence theorem is proved which establishes that lots of large simply presented modules exist. In passing, some other structural properties of simply presented modules are obtained.

Our approach is, of course, closely related to the treatment of simply presented p-groups given by Crawley and Hales [4], [5, Ch. 12]. Lemma 2.1 was suggested by a presentation of the results of [4] given in a seminar by R. S. Pierce.

We remind the reader that a module is *simply presented* if it can be defined in terms of generators and relations in such a way that the only relations are of the form $px = 0$ or $px = y$. In different terms, we start with a set X, take the free module on this set, $F(X)$, and take a subset L of $F(X)$ consisting only of elements of the form px or $px-y$ (with x and y in X) and consider the module $M = F(X)/[L]$. A presentation of a module M is given by a set X, a subset L of $F(X)$ of the required form, and a homomorphism $\phi : F(X) \to M$ which is surjective and has kernel $[L]$. We say such a presentation, given by X, L, and ϕ, is a *standard presentation* if (i) if $x \in X$ then $\phi(x) \neq 0$, (ii) if x and y are distinct elements of X then $\phi(x) \neq \phi(y)$, (iii) if $x \in X$ and $p\phi(x) \neq 0$ then $p\phi(x) = \phi(y)$ for some $y \in X$.

2.1 Lemma. If M is a simply presented module then M has a standard presentation: $\{X, L, \phi\}$. If $Y = \phi(X)$ then (i) Y generates M, (ii) $0 \notin Y$, (iii) if $y \in Y$, and $py \neq 0$ then $py \in Y$, and (iv) for any $y \in Y$, if Z is the set of elements $z \in Y$ such that $p^n z \neq y$ for all $n \geq 0$, then $y \notin [Z]$. Conversely, if M is an R-module and Y is a subset of M satisfying these conditions, then M is simply presented and there is a standard presentation $\{X, L, \phi\}$ of M such that $Y = \phi(X)$.

Proof: We prove the second part of the lemma first. Suppose that M is a module with such a subset Y. Let N be the module defined by taking the set Y as a set of generators with the relations $px = 0$ and $px = y$ whenever the corresponding equations are valid in M. There is clearly a natural surjective homomorphism from N onto M. Let F be the free module on the set Y, K the kernel of the homomorphism of F onto N and K' the kernel of the natural homomorphism of F onto M. Choose, if possible, an element of K' not in K whose expression in terms of the generators has minimal length. It is clear that more than one generator must be involved. We may also assume that in the expression $r_1x_1 + \ldots + r_nx_n$ (in F) that all of the coefficients are units, since otherwise we can alter the element by elements of K to bring this about. Multiplying by a unit, we may in fact assume that $r_1 = 1$. Clearly, for any $i > 1$ and any integer $n > 0$, $p^nx_i \neq x_1$, (in M) since otherwise we could obtain a shorter expression (an element in K' not in K and of shorter length), by replacing the r_ix_i by $(r_i + p^n)x_i$ and omitting x_1. But in this case, the equation $x_1 = -(r_2x_2 + \ldots + r_nx_n)$ in M contradicts condition (iv).

We now show that a simply presented group has a subset satisfying the conditions (i) - (iv). Starting with an arbitrary simple presentation $\{X, L, \phi\}$ we let Y be all nonzero elements of M of the form $p^n\phi(x)$ for some $x \in X$. We show that Y satisfies conditions (i) - (iv). This will show (by the previous argument) that standard presentations exist. Also, it is clear that if the presentation $\{X, L, \phi\}$ were a standard presentation, we would have $Y = \phi(X)$, so the first part of the theorem will be proved.

It is clear that the set Y satisfies (i), (ii), and (iii). We prove (iv) by starting with an element $y \in Y$, and constructing a homomorphism σ' such that $\sigma'(Z) = 0$ and $\sigma'(y) \neq 0$. Let K be the module given by generators and relations as follows: the generators form a sequence $\{w_i : 0 < i < \omega\}$, and the relations are $pw_1 = 0$ and $pw_{i+1} = w_i$, $0 < i < \omega$. (If Q is the quotient field of R, then clearly $K \cong Q/R$.) Define a homomorphism σ from $F(X)$ to K as follows: $\sigma(x) = w_{n+1}$ if $x \in X$ and $p^n\sigma(x) = y$, and $\sigma(x) = 0$ otherwise. It is a routine computation to show $\sigma(L) = 0$, so this defines a homomorphism $\sigma' : M \to K$ such that $\sigma'(y) = w_1 \neq 0$ and $\sigma'(Z) = 0$.

Hereafter we will not be concerned with the presentation of a simply presented-module M but only with a subset satisfying the conditions of Lemma 2.1. Such a subset will be called a T-*basis of* M.

2.2 Lemma. A simply presented-module is a direct sum of submodules of torsion-free rank at most one.

Proof. Let X be a T-basis of M and say that two elements x and y of X are *equivalent* if for some positive integers n and m, $p^n x = p^m y$. Let Λ be the set of equivalence classes, and for any $\lambda \in \Lambda$, let M_λ be the submodule of M generated by the elements of λ. From Lemma 2.1 or from the nature of the relations in a presentation of M, it is clear that $M = \oplus_{\lambda\in\Lambda} M_\lambda$. If λ_o is the equivalence class consisting of the torsion elements of X, then M_{λ_o} is torsion, and if $\lambda \neq \lambda_o$, then M_λ has torsion-free rank one.

2.3 Lemma. If M is a simply presented module and X is a T-basis of M, and Y is a subset of X, then $[Y]$ and $M/[Y]$ are simply presented modules.

Proof. $M/[Y]$ is simply presented since it can be defined by the same generators as M with some additional relations of the form $y = 0$. $[Y]$ is simply presented with the set $X \cap [Y]$ as a T-basis since $X \cap [Y]$ satisfies the conditions of Lemma 2.1.

2.4 Lemma. If M is a simply presented module and X is a T-basis for M, and Y a subset of X, then $[Y]$ is a nice submodule.

Proof. We first note that we may use the decomposition of lemma 2.2 to restrict ourselves to the case where M has torsion-free rank at most one, and where either M is torsion or every element of X is of infinite order. If M is torsion, the proof of 2.4 is exactly the same as it is for p-groups, as in [4] or [5, Ch. 12]. We note that in those proofs, it is shown that any element can be expanded in the form $r_1x_1 + \ldots + r_nx_n$ where the x_i are distinct elements of X and $0 < r_i < p$. The expansion so obtained is uinque. In our general situation, one simply chooses some set of coset representatives of R/pR in R, to replace the integers between 0 and p, and the proof goes through as before. We note that the fact that the module is torsion is essential for such an expansion to exist. If one wished to carry out this argument directly for mixed modules, one would have to content oneself with weaker results, as was done in the notes [31].

In the general case, we let $y \in Y$, and note that by 2.3, $M/[y]$ is simply presented. By the previous case, since $M/[y]$ is torsion, the submodule $[Y]/[y]$ is nice in $M/[y]$, and since $[y]$ is nice in M by 1.4, it follows by 1.2, (iv), that $[Y]$ is nice in M.

2.5. Theorem. If u is any Ulm sequence, there is a simply presented module T_u of torsion-free rank one such that T_u contains an element x_u of infinite order such that $U(x_u) = u$. If α is any ordinal, there is a simply presented torsion module P_α such that $p^\alpha P$ is cyclic of order p.

Remark. The proof here is based on Walker's construction in [23], and the modules P_α are those constructed by Walker.

Proof. We first consider the torsion case. We let the generators of the module be finite sequences of ordinals $(\alpha_1, \ldots, \alpha_n)$ with $\alpha_n = \alpha$ and $\alpha_{i+1} > \alpha_i$. The idea is that the height of $p^k(\alpha_1, \ldots, \alpha_n)$ should be $\alpha_k (k \le n)$, and to make this true, we impose the relations

$$p(\alpha_1, \ldots, \alpha_n) = (\alpha_2, \ldots, \alpha_n)(n > 1)$$

and $p(\alpha) = 0$. An easy induction shows that this construction is correct and that if P_α is the resulting group, then $p^\alpha P$ is the cyclic subgroup generated by (α).

For the group T_u we let the Ulm sequence be $\{u_i : i \ge 0\}$. If $u_1 = \infty$, then the module Q is our example. If $u_{n+1} = \infty$ but $u_n < \infty$, then we let $T_u = P_{u_n} \oplus Q$, and choose the element x_u to be $(u_1, \ldots, u_n) + 1$, where $(u_1, \ldots, u_n)$ is one of the generators of P_{u_n} and $1 \in Q$. Finally, if all of the u_i are ordinals, we let T_u be the module with generators consisting of finite sequences of ordinals $(\alpha_1, \ldots, \alpha_n)$, (for various n, $n \ge 1$) such that $\alpha_{i+1} > \alpha_i$ and α_n is one of the ordinals u_i, $(i \ge 0)$. We impose the relations

$$p(\alpha_1, \ldots, \alpha_n) = (\alpha_2, \ldots, \alpha_n)(n > 1),$$

and

$$p(u_i) = (u_{i+1}).$$

This clearly gives a module of the desired sort, with $x_u = (u_0)$.

2.6. Theorem. The following conditions are equivalent for a module M of rank one:

(a) M satisfies Hill's condition,

(b) M is a summand of a simply presented module of rank one.

Proof. Lemma 2.4 shows that a simply presented module satisfies Hill's condition, so (b) implies (a). Conversely, if M satisfies Hill's condition, then there is a simply presented module N of rank one such that $U(N) = U(M)$ by 2.5. We can find a torsion simply presented module T whose Ulm invariants are so large that $M \oplus T$ and $N \oplus T$ have the same Ulm invariants. It follows from 1.6 that $M \oplus T \cong N \oplus T$, since $U(M \oplus T) = U(M) = U(N) = U(N \oplus T)$, and $M \oplus T$ and $N \oplus T$ both satisfy Hill's condition. M is therefore a summand of a simply presented module of rank one.

2.7. Example. A module of torsion-free rank one satisfying Hill's condition is not necessarily simply presented. We give an example which is actually countably generated. Let $u = \{\alpha_i\}$ be the Ulm sequence defined by $\alpha_i = 2i$. By the Rotman-Yen existence theorem [19], we can construct a countably generated module M of torsion-free rank one, such that $U(M)$ is the equivalence class of the sequence u, $f(2i, M) = 1 (i \geq 0)$, $f(2i + 1, M) = 0$, $(i \geq 0)$, $f(\omega, M) = 1$, and $f(\alpha, M) = 0$ for all $\alpha > \omega$. In any decomposition of M, one of the summands has bounded order, by [12, Lemma 22]. If M were simply presented, then by the construction of Lemma 2.2, we would have a decomposition $M = S \oplus N$, such that N had torsion-free rank one, and such that N was simply presented with a T-basis all of whose elements had infinite order. Since the module S has bounded order, $p^\omega M = p^\omega N$. If we let y be an element of the T-basis for N, then $N/[y]$ is a torsion simply presented group. By [5, Ch. 12, p. 97, (f)], if x is an element of the T-basis for N, the height of $x + [y]$ in $N/[y]$ is the height of x in N. Hence $N/[y]$ has a T-basis consisting of elements with finite height. It follows that $p^{\omega}(N/[y]) = 0$, by [5, Ch. 12, p. 96, (c)]. Hence, $p^{\omega}N \leq [y]$, which is impossible, since $[y]$ is torsion-free and $p^{\omega}N$ is torsion. This example shows that Theorem 1.5 is (slightly) more general than a classification for simply presented modules of rank one. (A classification theorem for simply presented modules of rank one was, in fact, proved by Hales in unpublished work done prior to the work reported here.)

3. A projective characterization. In this section, we find a family of short exact sequences such that the simply presented modules and their summands are exactly the projectives with respect to these sequences. We recall that if u is an Ulm sequence then $uM = \{x \in M: h_p(p^n x) \geq u_n, \ n \geq 0\}$. A short exact sequence $0 \to A \to B \to C \to 0$ is *sequentially pure* if for every Ulm sequence u, the sequence $0 \to uA \to uB \to uC \to 0$ is exact. A module M is *sequentially-pure-projective* if for every such sequence, the natural mapping $\mathrm{Hom}(M,B) \to \mathrm{Hom}(M,C)$ is surjective.

3.1. Thoerem. A module is sequentially pure projective if and only if it is a summand of a simply presented module. If C is any module, there is a sequentially pure short exact sequence $0 \to K \to P \to C \to 0$ such that P is sequentially pure projective.

Proof. We first note that a sequence $0 \to A \to B \to C \to 0$ is sequentially pure if and only if (i) the map $uB \to uC$ is surjective for all Ulm sequences u, and (ii) A is isotype in B--that is, for all ordinals α, $p^{\alpha}A = A \cap p^{\alpha}B$. We now recall from [34, Lemma 2.1] that if the sequence satisfies condition (i), then A is isotype if and only if for every ordinal α, $(p^{\alpha}C)[p] = \nu((p^{\alpha}B)[p])$, where $\nu : B \to C$ is the natural map. We use these facts to construct for any module C a sequentially pure short exact sequence $0 \to K \to P \to C \to 0$ such that P is simply presented. We let the generators of P be a set X such that there is a bijective map $\phi : X \to C$. We impose the relation $px = y$ if and only if $p\phi(x) = \phi(y)$. ϕ clearly extends to a homomorphism $\phi : P \to C$. An easy induction shows that if

$x \in X$ and $\phi(x) = c$, then $h(x) = h(c)$, where x is regarded as an element of the module P. Also, if $x \in X$ and $\phi(x)$ has order p, then so does x. If $\phi(x) = c$, then it is clear that if we regard x as an element of P, that $u(x) = u(c)$. Hence, by the above criterion, the sequence $0 \to K \to P \to C \to 0$, where the map $P \to C$ is induced by ϕ, is sequentially pure. This proves the required statement.

It follows immediately from this construction that any sequentially pure projective module is a summand of a simply presented module. To prove the converse, we must show directly that a simply presented module is sequentially pure projective. We first note that if $0 \to A \to B \to C \to 0$ is sequentially pure, then the induced sequence

$$0 \to p^{\alpha}A \to p^{\alpha}B \to p^{\alpha}C \to 0$$

is exact for all ordinals α (using the Ulm sequence u where $u_i = \alpha + i$.) Hence a sequentially pure sequence is balanced in the sense of [34], so a torsion module satisfying Hill's condition is sequentially pure projective, since it is a balanced projective. It will therefore suffice to show that a simply presented module of rank one is a sequentially pure projective module.

We let M be a simply presented module of torsion-free rank one, and $x \in M$ an element of infinite order. Let $0 \to A \to B \to C \to 0$ be a sequentially pure sequence, and $f : M \to C$ a homomorphism. If $\nu : B \to C$ is the map appearing in the sequence, then $f(x) \le u(x)C = \nu(u(x)B)$, by the sequential purity. There is therefore an element $y \in u(x)B$ such that $\nu(y) = f(x)$. The map $g : [x] \to [y]$ taking x to y is therefore a homomorphism which does not decrease heights, (i.e. for all α, $g([x] \cap p^{\alpha}M) \le p^{\alpha}B$.) Since, by 1.2, $[x]$ is a nice submodule, g extends to a homomorphism $g : M \to B$, by [29, 1.13]. Since $f - \nu g$ is a map from M to C which has $[x]$ in its kernel, it may be regarded as a map $M/[x] \to B$. $M/[x]$ is a torsion module satisfying Hill's condition (2.4), and hence, by the earlier part of the proof, is sequentially pure projective. There is therefore a homomorphism $g' : M \to B$ such that $g'(x) = 0$ and $\nu g' = f - \nu g$. It follows that $f = \nu(g + g')$, which proves the projectivity property of M. This completes the proof of 3.1.

3.2. Example. A sequentially pure projective module need not be a direct sum of modules of torsion-free rank one. In [19, p. 251] Rotman and Yen give an example of a countably generated module M of torsion-free rank two over a complete discrete valuation ring which is not a direct sum of modules of torsion-free rank one, (in fact, if $M = A \oplus B$ then either A or B is of bounded order) but such that for a suitable countably generated torsion module S, $M \oplus S$ is a direct sum of modules of torsion-free rank one. From 2.6 it is clear that M is sequentially pure projective.

4. The category $\mathcal{H}$. If M is a simply presented module, then M is a direct sum of a torsion module and modules of torsion-free rank one, (2.2), and each of these summands belongs to a class of modules with a classification theorem. It is certainly plausible, therefore, to suppose that by suitably collecting the data which classified these summands, one could classify all simply presented modules. It would clearly be too much to write down what each of the summands is in terms of invariants, because this data would clearly not be invariant--some other decomposition would yield different data. We therefore try the following: for each equivalence class e of Ulm sequences, let $g(e,M)$ be the number of summands in the above decomposition $M = \oplus M_i$, such that $U(M_i) = e$. One might expect that the cardinal numbers $g(e,M)$ plus the Ulm invariants would classify the simply presented modules. This is indeed the case, but one first has the non-obvious task of showing that the numbers $g(e,M)$ are, in fact, invariant. This will be shown below. A more difficult problem comes up when we try to extend the classification theorem to summands of simply presented modules, since these modules are not necessarily direct sums of modules of rank at most one. The solution of this problem, as well as the proof of the invariance of the $g(e,M)$, is obtained by introducing a new category $\mathcal{H}$. The objects of $\mathcal{H}$ are the R-modules, but the homomorphism groups are changed, so that two nonisomorphic modules may be isomorphic in $\mathcal{H}$. We show, for example, that two modules, M and N, of torsion-free rank one, are isomorphic in $\mathcal{H}$ if and only if $U(M) = U(N)$. We also show that a summand of a simply presented module is isomorphic in $\mathcal{H}$ to a direct sum of modules of rank one.

This allows us to extend our invariants, and our classification theorem, to summands of simply presented modules--i.e. to all sequentially-pure-projective modules.

If M and N are R-modules and A is a submodule of M, we define $H_A(M,N)$ to be the submodule of $\mathrm{Hom}(A,N)$ consisting of homomorphisms f such that for all $x \in A$, $h(x) \leq h(f(x))$, where $h(x)$ is computed in M. If B is another submodule of M and $B \subsetneq A$ there is a natural restriction map $H_A(M,N) \to H_B(M,N)$. We define

$$\mathcal{H}(M,N) = \varinjlim H_A(M,N)$$

where the limit is taken over all submodules A of M such that M/A is torsion. Finally, we define $\mathcal{H}$ to be the category whose objects are R-modules and whose morphisms are the modules $\mathcal{H}(M,N)$.

It is clear that if M and N are modules then M and N are isomorphic in $\mathcal{H}$ if and only if there are free subgroups F and G of M and N such that M/F and N/G are torsion and an isomorphism $\phi: F \to G$ such that if $x \in F$ then $h(x) = h(\phi(x))$, where the heights are computed in M and N. In particular, if the rank of M is finite or countable, then M and N are isomorphic in $\mathcal{H}$ if and only if they have the same "invariant" as described by Rotman and Yen [19] and Bang [2]. The category $\mathcal{H}$ therefore provides a natural setting for theorems such as those of Rotman-Yen and Bang, who showed, in effect, that if M and N are countably generated modules over a *complete* discrete valuation ring, then $M \cong N$ if and only if M and N are isomorphic in $\mathcal{H}$ and have the same Ulm invariants.

4.1. Definition. If M is a module and K a submodule, then K^o is the submodule of M consisting of all elements $x \in M$ such that for some $r \in R$, $r \neq 0$, $rx \in K$. We note that K^o always includes the torsion submodule of M.

4.2. Definition. An additive category with kernels and infinite direct sums satisfies a weak Grothendieck condition if for every object with a direct sum decomposition $M = \oplus_{i\in I} M_i$, and every nonzero subobject S of M, there is a finite subset $J \subseteq I$ such that $S \cap (\oplus_{i\in J} M_i) \neq 0$.

4.3. Lemma. $\mathcal{H}$ is an additive category with infinite direct sums and kernels, and satisfies a weak Grothendieck condition. A module is small as an object of $\mathcal{H}$ if and only if its torsion-free rank is finite.

A proof of this is in [34, 3.2]. We remark that if M and N are modules, A a submodule of M such that M/A is torsion, $f \in H_A(M,N)$, and K is the kernel of f (regarding f as an element of Hom(A,N)), then K^o, with its natural imbedding into M, is a kernel for the element of $\mathcal{H}(M,N)$ represented by f.

We regard it as obvious from the definition that two modules of torsion-free rank one are isomorphic in $\mathcal{H}$ if and only if they determine the same equivalence class of Ulm sequences. We have therefore classified the "rank one" objects in our category, and now proceed to their direct sums.

4.4. Theorem. Let M be a module which is isomorphic in $\mathcal{H}$ to a direct sum of modules of torsion-free rank one. If we choose such a decomposition and, for any equivalence class e of Ulm sequences, we let g(e,M) be the number of summands corresponding to e, then the numbers g(e,M) are independent of the choice of decomposition and are therefore invariants of M. Furthermore, any summand of M is again isomorphic in $\mathcal{H}$ to a direct sum of modules of torsion-free rank 1.

Proof. By [26, Theorem 3] if M is an object in an additive category with kernels and infinite direct sums, and $M = \oplus_{i\in I} M_i = \oplus_{j\in J} N_j$, where the endomorphism rings of the objects M_i and N_j are all local rings, then there is a bijective map $\phi: I \to J$ such that $M_i \cong N_{\phi(i)}$, $i \in I$, (the Azumaya theorem). If, in addition, the objects M_i are small and the category satisfies a weak Grothendieck condition, then [26, Theorem 4] implies that if N is a summand of M, there is a subset $J \subseteq I$ such that $N \cong \oplus_{i\in J} M_i$, (the Crawley-Jónsson theorem). Theorem 4.4 will follow from these results if we show that the $\mathcal{H}$-endomorphism ring of a module of torsion-free rank one is local. If M is $\mathcal{H}$-isomorphic to Q then the $\mathcal{H}$-endomorphism ring of M is isomorphic to Q. In all other cases, the $\mathcal{H}$-endomorphism ring is isomorphic to the discrete valuation ring R. To see this, let f be an $\mathcal{H}$-endomorphism of M, and let A be a submodule such that M/A is torsion and such that there is an $f' : A \to M$ representing f. A may be assumed to be cyclic of infinite order, $A = [x]$. If $f'(x) = y$, then since M has torsion-free rank one, there are units u and v and integers n and m such that $p^n ux = p^m vy$. We clain that $n \geq m$, since otherwise $h(p^n x) < h(p^m x) \leq h(p^m y)$ which contradicts the above equation. (Here we use the fact that f' does not decrease heights.) This shows that there is a unit w and a nonnegative integer $k = n - m$ such that $f'(p^m x) = (wp^k)p^m x$, so that if B is the submodule generated by $p^m x$, f' restricted to B is just multiplication by wp^k. This proves the result.

Theorem 4.4 contains the main point of this section, in that it shows that the numbers $g(e,M)$ are invariants and that they can be defined for summands of simply presented modules. For technical purposes it will be desirable to reformulate the result in terms of certain kinds of subsets of M, using an idea of Rotman's, [18, 19]. If X is a subset of a module M, we let $[X]$ denote the submodule generated by X. We let $T(M)$ denote the maximal torsion submodule of M.

4.5. Definition. If M is an R-module, and X is a subset of M, then S is a *decomposition set* if (i) the elements of X are independent, (ii) $[X] \cap T(M) = 0$, and (iii) if $x_1, \ldots, x_n$ are in X and $r_1, \ldots, r_n$ in R, then $h(\Sigma r_i x_i) = \min\{h(r_i x_i)\}$. X is a *decomposition basis* if it is a decomposition set and $M/[X]$ is torsion.

If M is a direct sum of modules of torsion-free rank one, $M = \oplus_{i\epsilon I} M_i$, and if we choose in each M_i an element x_i of infinite order, then the set $X = \{x_i : i \in I\}$ clearly forms a decomposition basis for M. Conversely, if X is a decomposition basis for M then M is H-isomorphic to $\oplus_{x\epsilon X}[x]^o$, so we have proved the following lemma.

4.6 Lemma. If M is a module with decomposition basis, then any summand of M has a decomposition basis. Furthermore, if we choose a decomposition basis X for M and for any equivalence class e of Ulm sequences, we let $g(e,M)$ be the number of elements $x \in X$ such that $U(x) \in e$, then the numbers $g(e,M)$ are independent of the choice of X and are therefore invariants of M.

This is just a reformulation of Theorem 4.4. For modules of finite torsion-free rank, the second part of this result (concerning the invariants) was previously proved by Rotman and Yen in [19].

To prove a classification theorem, we not only need to know that summands of simply presented modules have decomposition bases (as they do by 4.6) but also we need decomposition bases which generate submodules with additional properties. If X and Y are decomposition bases of a module M, we say that Y is *subordinate* to X if for every $y \in Y$, there is an $x \in X$ and a nonnegative integer n such that $p^n x = y$. The main result is that for the groups we are interested in, any decomposition basis has a subordinate basis with additional good properties. In the author's original notes (1971) and the original version of this paper (1976), a lemma was used which has since been discovered to be false. (For a discussion and a counterexample see [8].) Here we will use instead the following lemma which follows from the recent work of Hunter and Richman, and resurrects this approach to the theory.

4.7 Lemma. Let M be a module and X and Y two decomposition bases of M, and suppose that X is a nice submodule of M. Then there is a decomposition basis Z subordinate to Y such that $[Z] \leq [X]$ and $[Z]$ is nice. If, in addition, $M/[X]$ is simply presented, then there is a decomposition basis Z subordinate to Y such that $[Z] \leq [X]$, $[Z]$ is nice, and $M/[Z]$ is simply presented.

Proof. See [8, Theorem 9.3 and Corollary 9.4].

5. <u>The classification theorem</u>. Suppose that we have two modules M and N such that (i) M and N are both summands of simply presented modules, (ii) M and N have the same Ulm invariants, and (iii) for every equivalence class e of Ulm sequences, $g(e,M) = g(e,N)$. We want to find an isomorphism of M onto N. We may clearly start by finding decomposition bases X and Y for M and N (using 4.6) and a bijection $\phi: X \to Y$ such that $U(x) = U(\phi(x))$ for all $x \in X$. To apply the extension theorems, we need the submodules $[X]$ and $[Y]$ to be nice, and (as we shall see) this can be arranged by an application of 4.7, which will also

guarantee that $M/[X]$ and $N/[Y]$ can be chosen to be simply presented. We would be all set to apply the extension theorem discussed in section 1 (before Theorem 1.6), if the relative Ulm invariants of $[X]$ and $[Y]$ were equal. This is not necessarily the case, and the next task is to see how to replace X and Y by new decomposition bases which will have this property.

Definition. If M is an R-module, a decomposition basis X of M is a *lower decomposition basis* if for every ordinal α, either $I_\alpha([X])$ is finite dimensional, or

$$\dim(U_\alpha(M)) = \dim(U_\alpha(M)/I_\alpha([X])).$$

We recall that an Ulm sequence $\{\alpha_i\}$ has a *gap at* α if for some $i \geq 0$, $\alpha_i = \alpha$ and $\alpha_{i+1} > \alpha + 1$. It is easy to see that if X is a decomposition basis for M, the dimension of $I_\alpha([X])$ is just the number of elements of X whose Ulm sequences have gaps at α.

5.1 Lemma. If M is a module with decomposition basis X, there is a decomposition basis X' subordinate to X which is a lower decomposition basis.

Proof. By a standard transfinite induction we can find disjoint countable subsets X_λ, $\lambda \in \Lambda$, of X such that X is the union of the X_λ, and such that for each $\lambda \in \Lambda$, and each ordinal α, if $I_\alpha([X])$ is infinite dimensional, then $I_\alpha([X_\lambda])$ is either 0 or infinite dimensional. We define a new decomposition basis X' by defining a new decomposition set X'_λ associated to each of the X_λ, $\lambda \in \Lambda$. For each $\lambda \in \Lambda$, let $h(\lambda)$ be the set of ordinals α, such that $I_\alpha([X_\lambda])$ is infinite dimensional. If $h(\lambda)$ is empty we let $X'_\lambda = X_\lambda$. Otherwise, we claim that we can choose disjoint, infinite subsets X^α_λ of X_λ, one for each $\alpha \in h(\lambda)$, such that if $x \in X^\alpha_\lambda$, then $U(x)$ has a gap at α. We postpone the proof of this till the next paragraph and assume that it has been done. If $x \in X^\alpha_\lambda$, there is an integer $k(x)$ such that $p^{k(x)}x$ has height greater than α. Let X'_λ be the set of elements $p^{k(x)}x$ for all $x \in X^\alpha_\lambda$, $\alpha \in h(\lambda)$, and those elements $x \in X_\lambda$ which are not in any of the sets X^α_λ, $\alpha \in h(\lambda)$. Let X' be the union of the sets X'_λ, $\lambda \in \Lambda$. By construction, if α is any ordinal such that $I_\alpha([X])$ is infinite dimsional, the codimension of $I_\alpha([X'])$ in $I_\alpha([X])$ is at least as great as the dimension of $I([X'])$, from which it follows that X' is a lower decomposition basis.

For the proof that X_λ can be partitioned in the indicated way, let $X_\lambda = Y$ and let the elements of $h(\lambda)$ be α_i, $0 < i < \omega$. (The case where the set $h(\lambda)$ is finite is easier.) Let Y_{11} be an infinite subset of Y, such that if $x \in Y_{11}$ then $U(x)$ has a gap at α_1. Let x_{11} be an element of Y_{11}. If Y_{11} has only a finite number of elements whose Ulm sequence has a gap at α_2, we let $Y_{12} = Y_{11}$. Otherwise, let Y'_{12} be an infinite subset of Y_{11}, such that if $x \in Y'_{12}$ then $U(x)$ has a gap at α_2, and such that $Y_{11} - Y'_{12}$ is infinite and contains x_{11}. Let $Y_{12} = Y_{11} - Y'_{12}$, and choose an element x_{12} in Y_{12} such that $x_{12} \neq x_{11}$. We proceed in this way, finally letting $Y_1 = \cap_{i=1}^{\infty} Y_{1i}$, where Y_1 is still infinite since it contains all of the elements x_{1i}. We now apply the same process to $Y - Y_1$, using the ordinal α_2, and forming the set Y_2. We have, in the end, disjoint, infinite sets Y_i, such that if $x \in Y_i$ then $U(x)$ has a gap at α_i.

5.2 Lemma. If M is a summand of a simply presented module, then M has a decomposition basis X such that for every decomposition basis X' subordinate to X, $[X']$ is a nice submodule of M and $M/[X']$ is simply presented.

Proof. There is a module K and a simply presented module L such that $M \oplus K = L$. According to 4.6, M and K have decomposition bases W and Z and we let $Y = W \cup Z$. Let B be a decomposition basis for L which is a subset of a T-basis. Using 5.1 and the fact that every decomposition basis subordinate to B is also a subset of the same T-basis, we may assume that B is a lower decomposition basis. We choose a decomposition basis Y' subordinate to Y with the properties that (i) $Y' \leq [B]$, (ii) Y' is a lower decomposition basis, and (iii) there is a bijective map $\phi : Y' \to B$ such that $U(y)$ and $U(\phi(y))$ are equivalent and $U(y) \geq U(\phi(y))$ for all $y \in Y'$. Using 4.7 and passing to a further subordinate (which does not change any of the previous conditions) we may assume that $[Y']$ is nice and that $L/[Y']$ is simply presented.

We now replace B by a subordinate basis B' so that if $p^n\phi(y) \in B'$, then $U(y) = U(p^n\phi(y))$. We therefore have a height preserving isomorphism

$f : [B'] \to [Y']$. This implies that $I_\alpha([B']) \cong I_\alpha([Y'])$ for all ordinals α, and hence that if $I_\alpha([B'])$ and $I_\alpha([Y'])$ are finite dimensional, that $U_\alpha(L)/I_\alpha([B']) \cong U_\alpha(L)/I_\alpha([Y'])$. If $I_\alpha([B'])$ and $I_\alpha([Y'])$ are infinite dimensional, then since B' and Y' are lower decomposition bases, $U_\alpha(L) \cong U_\alpha(L)/I_\alpha([B']) \cong U_\alpha(L)/I_\alpha([Y'])$ so that the hypotheses of the extension theorem [34, 1.11] are satisfied, and we can conclude that f extends to an isomorphism of L onto itself. Since the image of a T-basis under an automorphism is again a T-basis, Y' is a subset of a T-basis for L.

We let $X = Y' \cap M$. Since Y' is a decomposition basis for L, subordinate to Y, X is a decomposition basis for M, subordinate to W. Since X is a subset of a T-basis for L, every decomposition basis X' for M subordinate to X is also a subset of the same T-basis for L, so $[X']$ is a nice submodule of M (according to 2.4 and 1.2). If $Z' = Y' \cap K$, then $M/[X']$ is a summand of the torsion simply presented module $L/[X' \cup Z']$, and since a summand of a torsion simply presented module is simply presented, $M/[X']$ is simply presented. This shows that the decomposition basis X of M has all of the properties required.

5.3. Theorem. Let M and N be summands of simply presented modules. Then $M \cong N$ if and only if $f(\infty,M) = f(\infty,N)$ and for all ordinals α and all equivalence classes e of Ulm sequences, $f(\alpha,M) = f(\alpha,N)$ and $g(e,M) = g(e,N)$.

Proof. Using 5.2, we may choose decomposition bases X and Y for M and N such that for all decomposition bases X' and Y' subordinate to X and Y, $[X']$ and $[Y']$ are nice submodules of M and N, and $M/[X']$ and $N/[Y']$ are simply presented. Passing to subordinate bases if necessary, we may assume that X and Y are lower decomposition bases of M and N, and that there is a bijection $\phi : X \to Y$ such that for all $x \in X$, $U(x) = U(\phi(x))$. It is clear that ϕ extends to a height preserving isomorphism $\phi' : [X] \to [Y]$. Using the extension theorem as in the proof of the previous lemma, we see that ϕ' extends to an isomorphism of M onto N.

6. Countably generated modules. If we specialize our previous results to modules which are direct sums of countably generated modules, then we obtain immediately a classification theorem for modules which are both (i) direct sums of countably generated modules and (ii) summands of direct sums of modules of torsion-free rank at most one. In this section in the original papers (1971 and 1976) we proved that for modules which are direct sums of countably generated modules, this class coincides with the apparently larger class of modules with decomposition bases--i.e. modules which are H-isomorphic to direct sums of modules of rank one. The theorems are the following:

6.5. Theorem. Let M and N be modules over the discrete valuation ring R, such that both are direct sums of countably generated modules and both have decomposition bases. Then $M \cong N$ if and only if $f(\infty,M) = f(\infty,N)$ and for every ordinal α and every equivalence class e of Ulm sequences, $f(\alpha,M) = f(\alpha,N)$ and $g(e,M) = g(e,N)$.

6.6. Corollary. If M is a direct sum of countably generated R-modules and M has a decomposition basis, then there is a torsion module T such that $M \oplus T$ is a direct sum of countably generated modules with torsion-free rank one. In particular, M is a summand of a simply presented module.

The slight of hand of Richman and Walker [17] reduces the first of these to the countable case. The author's proof of this was independent of the earlier part of the paper and proceeded by first proving the result for modules over a complete discrete valuation ring and then doing the general case. All of this has become redundant because of subsequently developed methods. In [9], Theorem 33], Hunter, Richman, and Walker show that any countable decomposition basis of a module has a subordinate which is nice. If the module is countably generated we can immediately call upon our previous methods to prove the result without further ado, so these theorems are now easy. In fact, the Hunter-Richman-Walker result allows an immediate generalization. It shows that a module of countable rank is a summand of a simply presented module if and only if it has a decomposition basis and satisfies

Hill's condition. Whether this generalizes any further I still do not know (see problem 15 of [35]).

There were some earlier special cases of these results and related results in the literature which are suggestive. In [19], Rotman and Yen show that if M is a countably generated module over a *complete* discrete valuation ring, then M is a summand of a finite direct sum of modules of rank one if and only if M is *H*-isomorphic to such a sum. In [2, 3], Bang shows that for direct sums of countably generated modules over a complete discrete valuation ring, $M \cong N$ if and only if M and N have the same Ulm invariants and are isomorphic in *H*. (In all of these papers, *H*-isomorphism is described in terms of a very complicated "invariant." The category *H* gives a much more natural setting for these results.) Stratton [25] shows that the completeness is essential for the results of Rotman-Yen and Bang. It is possible that an analogue of the Rotman-Yen-Bang results could be obtained without completeness using instead of the category *H*, the category *Walk*, discussed in [35].

7. Recent progress and the global theory. A great deal of progress has been made in the questions considered in this paper since it was written in 1976. We review in this section (A) the progress in the local theory through 1980, (B) the present status (1980) of the global theory, and (C) some prospects for the further development of the theory of mixed groups.

A. The local theory. The invariants used in this paper are obtained from a counting argument--a Krull-Schmidt theorem in a suitable category. In [20], Stanton gave a functorial description of these invariants in a way that made them defined for all modules. (These general invariants are now frequently called "Warfield invariants"--an injustice with historical precedent, since the "Ulm invariants" were first defined in their modern form by Kaplansky and Mackey.) An alternative way of proving the results here was developed by Hunter, Richman, and Walker in [9], using ideas from the theory of groups with valuations. They also used this point of view to develop an existence theory for the local case in [10], in which it is also shown that a summand of a simply presented module is the direct sum of a module of countable rank and a simply presented module. A remaining problem in the local theory was that the torsion subgroup of a summand of a simply presented group is not necessarily simply presented. The author developed in [33] a theory of a class of p-groups called "S-groups" for which a classification theorem was proved. Hunter showed in [7, 5.2] that if M is a module which is a summand of a simply presented module, then the torsion submodule of M is a summand of an S-module. (Some of this work was done for p-groups, but the results hold without exception for modules over any discrete valuation ring.) Finally, Stanton showed in [23] that a summand of an S-group is an S-group, solving a long standing problem and completing at the same time the classification of the torsion submodules of the modules considered in this paper. An alternative treatment of this result is given in [11].

B. The global theory. The global analogue of the theory developed here concerns simply presented groups (groups defined by generators and relations in which each relation involves at most two generators) and their summands. The study of these groups was initiated in [32]. It is easy to carry out a classification program for groups of rank one which are summands of simply presented groups, ([32] and [35, Thm. 1]), to give a projective characterization of these groups similar to that in section 3 above, and to show that if M and N are two simply presented groups with the same Ulm invariants which are direct sums of groups of rank one in such a way that the rank one summands are pairwise $\mathcal{H}$-isomorphic, then M and N are isomorphic [32]. However, it was also shown in [32] that the Krull-Schmidt theorem fails in $\mathcal{H}$, that is, a simply presented group can be isomorphic in $\mathcal{H}$ to a direct sum of rank one groups in two different ways in which rank one summands are not pairwise isomorphic. The problem posed by this anomaly was resolved by Stanton [21] who showed that you can extract from a decomposition into rank one summands in $\mathcal{H}$ enough invariant data to provide an isomorphism classification for direct sums of rank one objects in $\mathcal{H}$. The second step in the global theory was to show that every summand of a simply presented group has a decomposition basis. This was done by Stanton [22] and by Arnold, Hunter, and Richman [1], who give an elegant categorical treatment of this entire aspect of the theory in a form that may well prove useful in other investigations. Finally, in [8], Hunter and Richman did the group theory necessary to obtain from this structure theory in $\mathcal{H}$ the actual classification theory for summands of simply presented groups. (An earlier proof in [22] is insufficient because it relies on [21, Thm. 13], which is incorrectly proved.) Hunter and Richman also provide the existence theory for the global case. Stanton [24] has subsequently obtained a more general classification theory including both simply presented groups and S-groups, extending previous work of Wick's [36, 37].

C. Further developments. In [35] the author gave a list of 27 open problems in the theory of mixed groups. The work reported above has taken care of most of the problems dealing with classification theory (problems 11, 13, 16, 17, 18, 19, 23), which still leaves 20 more. An obvious kind of question is to find alternative characterizations of summands of simply presented groups. In [9, Thm. 33], Hunter, Richman, and Walker show (in effect) that if M is a module of countable rank over a discrete valuation ring, then M is a summand of a simply presented module if and only if M has a decomposition basis and satisfies Hill's condition. This is not known in general, [35, problem 15] except for certain modules over a complete discrete valuation ring, [34, 4.4].

In general, mixed groups can be expected to behave at least as badly as torsion-free groups, but there are some indications that methods which have been successful in the study of torsion-free groups of finite rank should have a great deal to say about mixed groups of finite rank also, [35]. In this connection, it may be worth while to point out a categroical way of describing the invariants used in the global classification theory of summands of simply presented groups. If $\mathcal{H}$ is the category defined previously (for groups, as in [34], not for modules), then we define $\mathcal{H}_{(p)}$ to be the category with the same objects as $\mathcal{H}$, but with $\mathcal{H}_{(p)}(G,H)$ defined to by $(\mathcal{H}(G,H))_{(p)}$, where this is ordinary localization of Abelian groups at the prime p. It is easy to see that if G and H are isomorphic in $\mathcal{H}$ to direct sums of groups of rank one, then G and H have the same invariants (as defined by Stanton [21], or Arnold, Hunter, and Richman [1]) if and only if G and H are isomorphic in $H_{(p)}$ for all primes p. If one likes, one can tie together the different primes by defining another category N, with $N(G,H) = \Pi_p H_{(p)}(G,H)$. For torsion-free groups of finite rank, isomorphism in the category N is the same as "near-isomorphism" in the sense of [14]. The fact that near-isomorphism turns out to to of significance for mixed groups in this way suggests that a great many of the ideas of the theory of torsion-free groups might have more general settings in the theory of mixed groups.

REFERENCES

1. D. Arnold, R. Hunter, and F. Richman, Global Azumaya Theorems in Additive Categories, *J. Pure Appl. Alg.* 16(1980), 223-242.

2. C. Mo Bang, Countably generated modules over complete discrete valuation rings, *J. Alg.* 14(1970), 552-560.

3. C. Mo Bang, Direct sums of countably generated modules over complete discrete valuation rings, *Proc. Amer. Math. Soc.* 28(1971), 381-388.

4. P. Crawley and A. W. Hales, The structure of Abelian p-groups given by certain presentations, *J. Alg.* 12(1969), 10-23.

5. L. Fuchs, "Infinite Abelian Groups," 2 vols., Academic Press, New York, 1970 and 1973.

6. P. Hill, On the classification of Abelian groups, lecture notes.

7. R. Hunter, Balanced subgroups of Abelian groups, *Trans. Amer. Math. Soc.* 215 (1976), 81-98.

8. R. Hunter and F. Richman, Global Warfield Groups, to appear, *Trans. Amer. Math. Soc.*

9. R. Hunter, F. Richman, and E. Walker, Warfield modules, in "Abelian Group Theory" (Proceedings of the 2nd New Mixico State University Conference, 1976), *Lecture Notes in Mathematics* 616, Springer-Verlag, Berlin, 1977, pp. 87-123.

10. R. Hunter, F. Richman, and E. Walker, Existence theorems for Warfield groups, *Trans. Amer. Math. Soc.* 235(1978), 345-362.

11. R. Hunter and E. Walker, S-groups revisited, (preprint).

12. I. Kaplansky, "Infinite Abelian Groups," revised edition, Ann Arbor, 1969.

13. I. Kaplansky and G. W. Mackey, A generalization of Ulm's theorem, *Summa Brasil. Math.* 2(1951), 195-202.

14. E. Lady, Nearly isomorphic torsion-free Abelian groups, *J. Alg.* 35(1975), 235-238.

15. C. Megibben, Modules over an incomplete discrete valuation ring, *Proc. Amer. Math. Soc.* 19(1968), 450-452.

16. R. Nunke, Homology and direct sums of countable Abelian groups, *Math. Zeit.* 101 (1967), 182-212.

17. F. Richman and E. A. Walker, Extending Ulm's theorem without group theory, *Proc. Amer. Math. Soc.* 21(1969), 194-196.

18. J. Rotman, Mixed modules over valuation rings, *Pac. J. Math.* 10(1960), 607-623.

19. J. Rotman and Ti Yen, Modules over a complete discrete valuation ring, *Trans. Amer. Math. Soc.* 98(1961), 242-254.

20. R. O. Stanton, An invariant for modules over a discrete valuation ring, *Proc. Amer. Math. Soc.* 49(1975), 51-54.

21. R. O. Stanton, Decomposition bases and Ulm's theorem, in "Abelian Group Theory" (Proceedings of the 2nd New Mexico State University Conference, 1976) *Lecture Notes in Mathematics* 616, Springer-Verlag, Berlin, 1977, pp. 39-56.

22. R. O. Stanton, Almost affable Abelian groups, *J. Pure Appl. Alg.* 15(1979), 41-52.

23. R. O. Stanton, S-groups, preprint.

24. R. O. Stanton, Warfield groups and S-groups, preprint.

25. A. E. Stratton, Mixed modules over an incomplete discrete valuation ring, *Proc. London Math. Soc.* 21(1970), 201-218.

26. C. Walker and R. B. Warfield, Jr., Unique decomposition and isomorphic refinement theorems in additive categories, *J. Pure Appl. Alg.* 7(1976), 347-359.

27. E. A. Walker, Ulm's theorem for totally projective groups, *Proc. Amer. Math. Soc.* 37(1973), 387-392.

28. E. A. Walker, The Groups P_β, Symposia Mathematica XIII, (Gruppi Abeliani), Academic Press, London, 1974, 245-255.

29. K. Wallace, On mixed groups of torsion-free rank one with totally projective primary components, *J. Alg.* 17(1971), 482-488.

30. R. B. Warfield, Jr., Classification theorems for p-groups and modules over a discrete valuation ring, *Bull. Amer. Math. Soc.* 78(1972), 88-92.

31. R. B. Warfield, Jr., Invariants and a classification thoerem for modules over a discrete valuation ring, University of Washington notes, 1971.

32. R. B. Warfield, Jr., Simply Presented Groups, in Proceedings of the Special Semester on Abelian Groups, Spring 1972, University of Arizona, Tucson.

33. R. B. Warfield, Jr., A classification theorem for Abelian p-groups, *Trans. Amer. Math. Soc.* 210(1975), 149-168.

34. R. B. Warfield, Jr., Classification theory of Abelian groups I: Balanced projectives, *Trans. Amer. Math. Soc.*, 222(1976), 33-63.

35. R. B. Warfield, Jr., The structure of mixed Abelian groups, in "Abelian Group Theory" (Proceedings of the 2nd New Mexico State University Conference, 1976) *Lecture Notes in Mathematics* 616, Springer-Verlag, Berlin, 1977, pp. 1-38.

36. B. D. Wick, A projective characterization for SKT-modules, *Proc. Amer. Math. Soc.* 80(1980), 39-43.

37. B. D. Wick, A classification theorem for SKT-modules, *Proc. Amer. Math. Soc.* 80(1980), 44-46.

VALUATED p-GROUPS

Roger Hunter and Elbert Walker

1. INTRODUCTION

Our goal is a comprehensive theory of simply presented valuated p-groups. This goal looks attainable, but there are many obstacles. First we must better understand the theory of simply presented, or totally projective, groups. Particularly, that theory must be examined in the light of possible extensions to valuated p-groups. For example, there are several characterizations of totally projective groups. These characterizations are not equivalent for valuated p-groups, and relevant counterexamples must be provided and understood in order to put these characterizations in proper perspective in the new theory. Second, in the course of developing a theory for valuated p-groups, new mathematical entities are bound to arise which demand to be explored in their own right as well as to advance the theory under development. Valuated trees are a case in point since they, almost by definition, will be the key structural component. Finally, we must sort out the proper characterizing invariants and prove the necessary structure and existence theorems. This is, of course, our ultimate goal anyhow. We believe such a theory of simply presented valuated p-groups is within reach and will be provided by the abelian group theory community in the next few years. It is hoped that this paper will contribute to that end.

AMS (MOS) subject classification (1970), Primary 20K10; secondary 20K99.
The authors were supported by NSF grant MCS 80-03060

2. PRELIMINARIES

Let p be a prime and G a p-local group. A p-valuation on G is a function v from G to ordinal numbers and the symbol ∞ such that

(a) $v(x + y) \geq \min\{v(x), v(y)\}$;

(b) $v(px) > v(x)$;

(c) $v(nx) = v(x)$ if p does not divide n.

We adopt the convention that $\infty > \alpha$ for any ordinal α and that $\infty > \infty$. A group G with such a function is called a valuated group. The notion of a valuated Abelian group was introduced in [RicWa], and studied there from a categorical point of view. A map, or homomorphism, from one valuated group G to another is a group homomorphism f such that $vf(x) \geq v(x)$ for all x in G. A group becomes a valuated group via the ordinary p-height function on it, and a subgroup H of a group G becomes a valuated group by restricting the p-height function of G to the subgroup H. In fact, all valuated groups arise in this way [RicWa, Theorem 1]. In general, restricting the valuation on a valuated group G to a subgroup H of G makes H into a valuated group, and we say that H is a valuated subgroup of G. Of particular importance are nice valuated subgroups. They are those valuated subgroups H of a valuated group G such that every coset $g + H$ has an element of maximum value in G. An element g is proper with respect to H if it is of maximum value in $g + H$.

If G is a valuated group and α is an ordinal, then

$$G(\alpha) = \{g \in G : v(g) \geq \alpha\}.$$

The length of a reduced valuated group G is $\min\{\alpha : G(\alpha) = 0\}$. Multiplication by p induces a natural map

$$G(\alpha)/G(\alpha + 1) \rightarrow G(\alpha + 1)/G(\alpha + 2).$$

The kernel and cokernel of this map are vector spaces $U_G(\alpha)$ and $D_G(\alpha)$.

Their dimensions are denoted $f_G(\alpha)$ and $g_G(\alpha + 1)$ and are the *α-th Ulm invariants* and the *α-th derived Ulm invariants*, respectively, of G. For limit ordinals α, $g_G(\alpha)$ is defined as the dimension of

$$G(\alpha)/ \bigcap_{\beta<\alpha} (G(\alpha + 1) + G(\alpha) \cap pG(\beta)).$$

If H is a valuated subgroup of the valuated group G, there is a natural map $U_H(\alpha) \to U_G(\alpha)$ the dimension of whose cokernel is denoted $f_{G,H}(\alpha)$ and called the *α-th Ulm invariant of G relative to H*. Relationships between f_G, $f_{G,H}$, and g_G are spelled out in [HRWal, Section 3].

The subgroup generated by a subset X of a group is denoted $\langle X \rangle$. The symbol □ marks the end of a proof.

3. NICE SYSTEMS AND p-BASES

Fundamental in proving Ulm's theorem for totally projective groups was a new characterization of these groups in terms of systems of nice subgroups [Hill1]. Similar to this new characterization by Hill are subsequent ones given by Griffith [Griff] and Fuchs [Fuchs]. A proof, which avoids the use of Ulm's theorem, that these three characterizations are indeed equivalent was given by Hill [Hill2]. Embedded in that proof is another characterization reminiscent of p-bases. A fifth characterization is the existence of a p-basis, and a sixth is being projective in the category of p-groups relative to nice isotype short exact sequences. These characterizations of totally projective p-groups all make sense when stated in the category of valuated p-groups, and we will determine their relationships there.

DEFINITION 3.1. A valuated p-group G *satisfies condition* (H) if it has a set Ψ of nice valuated subgroups such that

(a) the subgroup generated by any subset of Ψ is in Ψ;

(b) if S is a countable subgroup of G then there is a countable $N \in \Psi$ with $S \subset N$.

A valuated p-group satisfying condition (H) will be said to have a nice system, and the set Ψ will be called a nice system. Condition (H) was Hill's characterization of totally projective groups [Hill1], and Griffith formulated the following characterization [Griff].

DEFINITION 3.2. A valuated p-group G satisfies condition (G) if it has a set Ψ of nice valuated subgroups such that

(a) the subgroup generated by any chain in Ψ is in Ψ;

(b) if S is a countable subgroup of G and $N \in \Psi$, then there exists $M \in \Psi$ with $S + N \subset M$ and M/N countable.

It was Fuchs who noted that condition (F) for groups in the following definition characterizes totally projectives.

DEFINITION 3.3. A valuated p-group G satisfies condition (F) if it is the union of a well ordered chain $\{G_\beta\}$ of nice valuated subgroups such that

(a) $G_0 = 0$;

(b) $G_{\beta+1}/G_\beta$ is of order p;

(c) if β is a limit ordinal, then $G_\beta = \cup_{\alpha<\beta} G_\alpha$.

Condition (F) is expressed by saying that the valuated p-group G has a nice composition series, or is Fuchsian. The fourth condition, which is extracted from [Hill2] is in the next definition.

DEFINITION 3.4. A valuated p-group G satisfies condition (W) if it has a subset X such that

(a) every element in G is uniquely expressible in the form

$\sum_{x \in X} n_x x$ with $0 \le n_x < p$;

(b) for such an expression, $v(\Sigma n_x x) = \min\{v(n_x x)\}$.

The set X is called a pseudo p-basis. If $px \in X \cup \{0\}$ for all $x \in X$, then X is a p-basis. A valuated group with a p-basis is said to be simply presented. This concept is due to Crawley and Hales who defined a p-group to be simply presented if it is given by a set X with relations of the form $px = y$ and $p^n x = 0$ [CrHal]. They then showed that such a group has a p-basis. Having a p-basis is the more immediately useful concept, and we take it as the definition of simply presented.

For p-bases in groups, condition (b) is redundant. However, condition (b) is not redundant for pseudo p-bases of groups because every p-group G has a set X satisfying (a). Simply let $\{C_\alpha\}$ be any composition series of G and take $x_\alpha \in C_{\alpha+1} \setminus C_\alpha$. The x_α satisfy (a). Various properties of pseudo p-bases for groups noted by Hill in [Hill2] carry over for valuated groups. They include the following lemmas.

LEMMA 3.5. If $\{G_\alpha\}$ is a composition series of nice valuated subgroups of the valuated group G, and if $x_\alpha \in G_{\alpha+1} \setminus G_\alpha$ with x_α proper with respect to G_α, then $\{x_\alpha\}$ is a pseudo p-basis of G.

PROOF. Clearly, every element in G can be written in the form $\Sigma_\alpha n_\alpha x_\alpha$ with $0 \le n_\alpha < p$. If $\Sigma n_\alpha x_\alpha = \Sigma m_\alpha x_\alpha$ with $0 \le m_\alpha < p$ and $0 \le n_\alpha < p$, then $\Sigma(m_\alpha - n_\alpha)x_\alpha = 0$. If β is the largest ordinal such that not both m_β and n_β are 0, then $(m_\beta - n_\beta)x_\beta \in G_\beta$, whence $m_\beta = n_\beta$. Hence every element in G can be written uniquely in the form $\Sigma n_\alpha x_\alpha$ with $0 \le n_\alpha < p$. To complete the proof we need to show that $v(\Sigma n_\alpha x_\alpha) = \min\{v(n_\alpha x_\alpha)\}$. Let β be the largest ordinal such that $n_\beta \ne 0$. Then $v(\Sigma n_\alpha x_\alpha) = v(n_\beta x_\beta + \Sigma_{\alpha<\beta} n_\alpha x_\alpha) = \min\{v(n_\beta x_\beta), v(\Sigma_{\alpha<\beta} n_\alpha x_\alpha)\}$ since $n_\beta x_\beta$ is proper with respect to G_β. The result follows by induction on the number of non-zero n_α's. Thus $\{x_\alpha\}$ is a pseudo p-basis of G. □

DEFINITION 3.6. **Let X be a pseudo p-basis of the valuated group G. A subset Y of X is a <u>sub-psuedo p-basis</u> if whenever $y \in Y$, then the support of py is contained in Y.**

It is fairly obvious that any infinite subset of a pseudo p-basis X is contained in a sub-pseudo p-basis of the same cardinality, and that unions of sub-pseudo p-bases are sub-pseudo p-bases.

LEMMA 3.7. **Sub-pseudo p-bases generate nice subgroups.**

PROOF. Let C be generated by a sub-pseudo p-basis. For $g \in G$, we need an element of maximum value in $g + C$. Write $g = \sum_{x\in X} n_x x$ with $0 \leq n_x < p$. The element of maximum value is $\sum n_x x$, where the sum is taken over those terms with $x \notin C$. □

Now it is easy to prove the equivalence of conditions (F), (G), (H), and (W). The proof is almost identical to that given by Hill for groups [Hill2].

THEOREM 3.8. **(H) ⇔ (G) ⇔ (F) ⇔ (W) for valuated p-groups.**

PROOF. To see that (H) implies (G) it is enough to check condition (b) of 3.2 for the nice system Ψ of 3.1. If S is countable and $N \in \Psi$, then (b) of 3.1 gives a countable $N' \in \Psi$ with $S \subset N'$. By condition (a) of 3.1, $M = N + N' \in \Psi$ and it is easy to see that M/N is countable and $S + N \subset M$. Condition (G) implies that the group is the union of an ascending chain $\{G_\alpha\}$ of nice valuated subgroups such that $G_{\alpha+1}/G_\alpha$ is countable and $G_\beta = \cup_{\alpha<\beta} G_\alpha$ for limit ordinals β. Since finite extensions of nice valuated subgroups are nice, the chain $\{G_\alpha\}$ can obviously be filled out to the required composition series. Thus (G) implies (F).

Assume (F), and let $\{G_\alpha\}$ be a composition series of nice valuated subgroups. Pick $x_{\alpha+1} \in G_{\alpha+1} \setminus G_\alpha$ with $x_{\alpha+1}$ proper with respect to G_α. Then $\{x_\alpha\}$ is a pseudo p-basis by LEMMA 3.5. Thus (F) implies (W).

To get (W) implies (H), let Ψ be the system of nice subgroups generated by all sub-pseudo p-bases of a pseudo p-basis. □

As we just saw in the proof that (W) implies (H), there is a canonical way to get a nice system from a pseudo p-basis. Simply take those subgroups generated by sub-pseudo p-bases of X. There is no canonical way to go from a nice system to a pseudo p-basis. However, there is a tighter relationship between an arbitrary nice system and an arbitrary pseudo p-basis than one might expect. First a lemma.

LEMMA 3.9. Let G be a valuated p-group satisfying condition (H). Then the intersection of countably many nice systems is a nice system.

PROOF. Let C be a countable subset of G, and let $\Psi_1, \Psi_2, \Psi_3, \cdot \cdot \cdot$ be the nice systems under consideration. There is a chain $C \subset N_{11} \subset N_{21} \subset N_{12} \subset N_{22} \subset N_{32} \subset N_{13} \subset N_{23} \subset N_{33} \subset N_{43} \subset \cdot \cdot \cdot$ with $N_{ij} \in \Psi_i$ and countable. The union of this chain is in $\cap_{i<\omega}\Psi_i$ and is countable. The rest is clear. □

COROLLARY 3.10. If Ψ is a nice system of a valuated p-group G, and X is a pseudo p-basis of G, then Ψ has a nice subsystem each member of which is generated by a sub-pseudo p-basis of X.

PROOF. The desired subsystem is $\Psi \cap \Psi(X)$, where $\Psi(X)$ is composed of the subgroups generated by sub-pseudo p-bases of X. □

COROLLARY 3.11. If Ψ is a nice system of a simply presented

p-group G, and if X is a p-basis of G, then Ψ has a nice subsystem every member of which is simply presented and has as a p-basis a subset of X. □

Since a p-basis is a pseudo p-basis, valuated groups with p-bases have nice systems. Countable valuated p-groups obviously have nice systems, so all countable valuated p-groups have pseudo p-bases. However, not even finite valuated p-groups need have p-bases [HRWa2, page 132]. Let $G = \langle x \rangle \oplus \langle y \rangle \oplus \langle z \rangle$, where the orders of x, y, and z are p^2, p^4, and p^5, respectively. Then the valuated subgroup H of G generated by $x + py$ and $p^2y - p^3z$ does not have a p-basis, as is easily checked. Thus for valuated p-groups, the existence of a p-basis is a stronger condition than the existence of nice systems. That these two conditions are equivalent for p-groups is quite non-trivial. In its proof, one needs to show that groups of limit length with nice systems are direct sums of groups of shorter length. This seems to involve both Ulm's and Zippin's theorems for groups with nice systems. Specifically, one needs that two such groups are isomorphic if they have the same Ulm invariants, that admissible functions of limit length are sums of admissible functions of shorter length, and that given an admissible function, there exists a given group with a nice system whose Ulm function is that admissible function. (Fuchs' proof [Fuchs, Theorem 83.5] that a nice system implies a p-basis is faulty because at that point he has not shown that totally projective groups of limit length are direct sums of shorter ones. This can be remedied by doing his Theorem 83.6 first.)

Having a nice system seems to be a condition of an entirely different character than that of being simply presented. As further evidence of this, it is relatively easy to show that summands of groups with nice systems have nice systems. In fact, there is a nice proof of this in [Hill1] that goes through for valuated groups. It is hard to show that summands of simply presented groups are simply presented. The equivalence of groups with nice systems and groups with p-bases is needed. It is trivial that a simply presented group of limit length is a direct sum of simply presented shorter ones. It is hard to prove this fact for groups with nice systems. Thus, although the classes of simply presented valuated p-groups and of valuated groups with nice systems are closely enough related to coincide for p-groups, these two classes may be of entirely different characters.

THEOREM 3.12. Direct sums and direct summands of valuated p-groups with nice systems have nice systems.

PROOF. If X_i is a pseudo p-basis of the valuated p-group X_i, then clearly $\cup X_i$ is a pseudo p-basis of $\oplus G_i$. Thus direct sums of valuated p-groups with nice systems have nice systems.

Now suppose that Ψ is a nice system of the valuated p-group $A \oplus B$. Let $\Psi(A)$ be those valuated subgroups M of A such that $N = M \oplus (N \cap B)$ for some $N \in \Psi$. It is clear that $\Psi(A)$ is a family of nice valuated subgroups of A satisfying (a) of 3.1. Let S be a countable subgroup of A. Choose a countable N_1 in Ψ with $S \subset N_1$. Then there are valuated subgroups X_1 and Y_1 of A and B, respectively, such that $N_1 \subset X_1 \oplus Y_1$ with $X_1 \oplus Y_1$ countable. Now choose a countable $N_2 \in \Psi$ with $X_1 \oplus Y_1 \subset N_2$. Next choose X_2 and Y_2 valuated subgroups of A and B, respectively, such that $X_1 \subset X_2$, $Y_1 \subset Y_2$, $N_2 \subset X_2 \oplus Y_2$ and $X_2 \oplus Y_2$ is countable. Thus we get chains of countable valuated subgroups N_i, X_i and Y_i with $N_i \in \Psi$ and $N_i \subset X_i \oplus Y_i \subset N_{i+1}$. Then $M = \cup_{i<\omega} N_i$ is countable, is in Ψ, and satisfies $M = (M \cap A) \oplus (M \cap B)$. Clearly $S \subset M \cap A$. □

It is obvious that simply presented valuated p-groups of limit length are direct sums of simply presented valuated p-groups of shorter length. However, the question of summands remains open.

OPEN QUESTION. Are summands of simply presented valuated p-groups simply presented?

The answer is yes for p-groups, but as indicated, the known proofs involve structure theory which is not available for valuated p-groups. The answer is also yes for finite simply presented valuated p-groups, and also for simply presented valuated p-groups with no elements of infinite value.

In view of our earlier discussion, it is also natural to ask whether a valuated group satisfying (H) and of limit length is a direct sum of valuated groups of shorter length. However, this is not the case, as there are examples of indecomposable countable valuated groups of length ω.

Now we turn to homological aspects. A p-group is simply presented, and hence has a nice system, if and only if it is projective with respect to the class of nice isotype, or balanced, short exact sequences of p-groups. This is the appropriate homological characterization although this class of p-groups was originally discovered by Nunke as a subclass of the projectives of other relative homological algebras [Nunke].

In the category of valuated p-groups, nice isotype short exact sequences are those $0 \to A \to B \to C \to 0$ with A a nice valuated subgroup of B. There are no non-trivial valuated p-groups that are projective with respect to such sequences. However, the valuated p-groups of homological dimension 1 in that category are precisely the valuated p-groups with nice systems [RicWa, Theorem 13]. Thus valuated p-groups with nice systems are of homological heritage.

In [RicWa, Theorem 3], a functor T defined on valuated groups is constructed with the properties that $T(G)$ is a group, G is a nice valuated subgroup of $T(G)$, and $T(G)/G$ is a simply presented p-group.

THEOREM 3.13. A valuated p-group G has a nice system if and only if T(G) is simply presented.

PROOF. Assume that G has a nice system. Since T(G)/G is simply presented, it has a nice system, whence a nice composition series for G can be run up to one of T(G). The group T(G) then is simply presented.

If T(G) is simply presented, then since T(G)/G is also simply presented, they both have dimension 1, whence so does G. Therefore G has a nice system. □

Note that THEOREM 3.13 shows that if G has a nice system, then it is a nice valuated subgroup of a simply presented group, namely of T(G). Nice valuated subgroups of simply presented groups do not have to have nice systems, however, since there exist nice isotype subgroups of simply presented groups which are not simply presented [RicWa, page 158].

A short direct proof of THEOREM 3.13 which avoids the homological algebra of the category of valuated groups would be instructive.

4. SIMPLY PRESENTED p-GROUPS AND THEIR p-BASES

In this section we look more closely at the relationship between a simply presented p-group and its p-bases. In particular, we want information about extending valuated p-bases of valuated subgroups of a simply presented p-group to p-bases of that group. In the process we will give a direct and relatively simple proof that countable p-groups have p-bases.

It is convenient at this point to discuss valuated trees, and in particular to show exactly how they give rise to simply presented valuated p-groups. Considering a p-basis of a simply presented valuated group as an entity in its own right gives rise to a structure called a valuated tree. It will be a fundamental object in the study of simply presented valuated groups.

DEFINITION 4.1. A <u>tree</u> is a set X with a distinguished element 0 (the <u>root</u> of X) that admits a multiplication by p satisfying

(a) $p0 = 0$;

(b) for each $x \in X$, $p^n x = 0$ for some positive integer n.

The smallest integer n in (b) is called the <u>order</u> of x. If G is a p-group, then G with the usual multiplication of its elements by p is a tree. A height function is defined on a tree X exactly as the p-height function is defined on a group. If X is a p-basis of a simply presented valuated p-group, then $X \cup \{0\}$ is a tree. But the elements of X come with values on them, and this gives rise to the more general concept of valuated trees.

DEFINITION 4.2. A <u>valuated tree</u> is a tree X with a function v on X such that

(a) $v(x)$ is an ordinal or ∞;

(b) $v(px) > v(x)$.

A tree with its height function is a valuated tree. A p-basis together with the 0 element of a simply presented valuated p-group, with the value function inherited from the valuated group, is a valuated tree. Now we show exactly how to associate with a valuated tree X a valuated group $S(X)$ in which X, with its root removed, is a p-basis. This will give a straightforward technique for obtaining simply presented valuated p-groups.

Let X be a valuated tree and let $F_X = \oplus Z[x]$ be the free Abelian group on the elements of X. Let R_X be the subgroup of F_X

generated by elements of the form

$$p[x] - [px], \qquad x \in X,$$

and

$$[0]$$

and set $S(X) = F_X/R_X$. The next lemma allows us to place an appropriate valuation on this quotient group.

LEMMA 4.3. Each element s of $S(X)$ has a unique representative whose coordinates are in $\{0,...,p-1\}$. The unique representative of $\sum n_i[x_i] + R_X$ involves only the non-zero $[x_i]$'s and their non-zero p-power multiples.

PROOF. Let $\sum n_i[x_i]$ represent an element $s \in S(X)$. We first show that each n_i may be assumed non-negative. Since every element of X has finite order, there is an m so that $p^m[x_i] = [0] \in R_X$ and $|n_i| < p^m$. Then $(n_i + p^m)[x_i] + R_X = n_i[x_i] + R_X$ so we may replace n_i by the positive coefficient $n_i + p^m$.

Now if $n_k \geq p$, we write

$$\begin{aligned} s &= n_k[x_k] + \Sigma_{i\neq k} n_i[x_i] + R_X \\ &= p[x_k] + (n_k - p)[x_k] + \Sigma_{i\neq k} n_i[x_i] + R_X \\ &= [px_k] + (n_k - p)[x_k] + \Sigma_{i\neq k} n_i[x_i] + R_X \\ &= \sum n_i'[x_i] + R_X. \end{aligned}$$

The sum of the coefficients of the new representative $\sum n_i'[x_i]$ is less than the sum of the coefficients in the original representative. Continuing, we obtain a representative $\sum u_i[x_i]$ for s with each u_i in $\{0,...,p-1\}$. It remains to show that the u_i's are unique. Suppose $\sum t_i[x_i]$ is another representative for s with $0 \leq t_i < p$. Then the difference $t = \sum t_i[x_i] - \sum u_i[x_i]$ is in R_X so we may write

$$t = \sum(t_i - u_i)[x_i] = \sum s_j(p[x_j] - [px_j]).$$

For each k, the coefficients t_k and u_k are both non-negative and less than p, so $|t_k - u_k| < p$. On the right hand side, however, the coefficient of $[x_k]$ is divisible by p, which implies $t_k = u_k$.

Hence this representation is unique. □

It should be noted that the uniqueness of representation is lost when trees with elements of infinite order are allowed. Now we can define the value of a non-zero element of $S(X)$: if $s = \sum u_i[x_i] + R_X$ with $0 < u_i < p$, then set $vs = \min\{vx_i\}$. We set $v(0) = \infty$.

THEOREM 4.4. $S(X)$ is a valuated group with p-basis

$$\{[x] + R_X : 0 \neq x \in X\}.$$

PROOF. If we can show that $S(X)$ is a valuated group under the given valuation, it will follow from LEMMA 4.3 that $\{[x] + R_X : x \in X\}$ is a p-basis. Let s, t be elements of $S(X)$. Since the canonical representative for $s + t$ involves only p power multiples of the $[x_i]$'s appearing in the canonical representatives for s and t, it follows that $v(s + t) \geq \min\{vs, vt\}$. Similarly, the canonical representative of ps contains only p^n multiples, $n \geq 1$, of these elements, so $vps > vs$, and so $S(X)$ is a valuated group. □

We now provide techniques for enlarging finite trees within groups. Let G be a valuated group. We say that a valuated tree X which is a subset of G is a <u>tree in</u> G if $X \setminus \{0\}$ is a p-basis of the valuated subgroup $\langle X\rangle$ generated by X. If $\langle X\rangle$ is nice in G, we say that X is a <u>nice tree</u> in G. The fundamental fact we will use is this: if X is a tree in G, if $g \in G\setminus\langle X\rangle$, if $pg \in X$ and if g is proper with respect to $\langle X\rangle$, then $X \cup \{g\}$ is a tree in G.

Let X be a tree in G. Let $x = \sum n_i x_i$ be an element of $\langle X\rangle$ with $0 < n_i < p$. The <u>value support</u> of x is the set

$$\text{vspt}(x) = \{x_i : vx_i = vx\}.$$

The next two Lemmas follow readily from definitions.

LEMMA 4.5. Let x and y be elements of $\langle X\rangle$ such that $v(x + y) >$

$v(x)$. Then $\mathrm{vspt}(x) = \mathrm{vspt}(y)$. □

LEMMA 4.6. Let $a \in \langle X \rangle$ and suppose there is a $b \in \langle X \rangle$ such that $v(a + pb) > va$. Then if $x \in \mathrm{vspt}(a)$, there is a $x' \in X$ with $px' = x$ and $vx' \geq vb$. □

LEMMA 4.7. Let X be a finite tree in a group G and let x be an element of X with value α. If $\beta < \alpha$ then there is an element x' of G of value at least β such that $px' = x$ and $X \cup \{x'\}$ is a tree in G.

PROOF. If α is not a limit ordinal, there is a g in G of value $\alpha - 1$ such that $pg = x$. If g is not proper with respect to $\langle X \rangle$, let $s \in \langle X \rangle$ be chosen so that $v(g + s) > vg$. But then $v(pg + ps) = v(x + ps) > vx$ and LEMMA 4.5 tells us the required x' is already in X. If g is proper with respect to $\langle X \rangle$, we set $x' = g$. Now suppose α is a limit ordinal. Since X is finite, we may find $g \in G$ so that $pg = x$ with $vg > \beta$ and so that vg is not an element of $\{\gamma : vx = \gamma$ for some $x \in X\}$. Then g is clearly proper with respect to $\langle X \rangle$ and we again set $x' = g$. □

THEOREM 4.8. Let X be a finite tree in a p-group G and let g be an element of G. Then there is a finite extension X' of X which is a tree in G such that $g \in \langle X' \rangle$.

PROOF. We may assume g is not in $\langle X \rangle$, that g is proper with respect to $\langle X \rangle$, and that $pg \in \langle X \rangle$. Let $pg = \sum a_i x_i$ with $0 < a_i < p$. LEMMA 4.7 allows us to enlarge X to X_1 so that, for each x_i, the equation $px_i' = x_i$ can be solved in X with $vx_i' \geq vg$, and if $vpg > vg + 1$, then $vx_i' > vg$. Set $g_1' = g - \sum a_i x_i'$. There are two cases.

CASE 1: g_1' is proper with respect to $\langle X_1 \rangle$. Since $pg_1' = 0$, the tree $X_1 \cup \{g_1'\}$ is the required extension.

CASE 2: g_1' is not proper with respect to $\langle X_1 \rangle$. Then we must have

$vpg = vg + 1$. Choose g_1 proper with respect to $\langle X_1 \rangle$ and so that $g_1 + \langle X_1 \rangle = g_1' + \langle X_1 \rangle = g + \langle X_1 \rangle$. Then $pg_1 \in \langle X \rangle$ and $vpg_1 > vpg$. Now repeat the above process with g_1 and X_1 in place of g and X. If we do not fall out in case 1, we obtain a g_2 and X_2 so that $pg_2 \in \langle X \rangle$, $vpg_2 > vpg_1$, and $g_2 + \langle X_2 \rangle = g + \langle X_2 \rangle$. Continuing, we obtain a sequence $\{(g_k, X_k)\}$ with the properties

1) each X_k is a finite tree in G
2) $X \subset X_k$
3) $g + \langle X_k \rangle = g_k + \langle X_k \rangle$
4) g_k is proper with respect to $\langle X_k \rangle$
5) $pg_k \in \langle X \rangle$
6) $vpg_k > vpg_{k-1}$.

Since the value support of $\langle X \rangle$ is finite, conditions 5) and 6) ensure that this sequence will eventually terminate and we will drop out through case 1. □

COROLLARY 4.9. Every finite valuated subgroup of a group is contained in a finite simply presented valuated subgroup of that group.

PROOF. Starting with $X = \{0\}$, apply THEOREM 4.8 over and over. □

COROLLARY 4.10. A countable valuated subgroup H of a p-group G is contained in a countable simply presented valuated subgroup of G.

PROOF. Let $x_1, x_2, \ldots$ be a list of the elements of H, and let X_0 be the zero tree. Then THEOREM 4.8 shows there is a sequence $X_0 \subset X_1 \subset \ldots$ of finite trees such that $x_i \in \langle X_i \rangle$ for each i, so the tree $X = \cup X_i$ is a p-basis for a countable valuated subgroup of G containing H. □

COROLLARY 4.11. A countable p-group is simply presented. □

Of course COROLLARY 4.11 is well known, but there does not seem to be a direct proof in the literature. Indeed, Crawley and Hales [CrHal] use both Ulm's and Zippin's theorems for simply presented p-groups for this result.

COROLLARY 4.12. Every finite tree in a countable p-group G extends to a p-basis of G. □

COROLLARY 4.13. Let Ψ be a nice system of a p-group G. Then every countable subset of G is contained in a countable simply presented member of Ψ.

PROOF. Let C be a countable subset of G. Then C is contained in a countable member N_1 of Ψ. Order N_1 and subsequent N_i with order type ω. There is a finite tree X_1 such that the first element of N_1 is in $\langle X_1 \rangle$. There is a countable member N_2 of Ψ containing $\langle N_1, X_1 \rangle$. Extend X_1 to a finite tree X_2 with $\langle X_2 \rangle$ containing the first two elements of both N_1 and N_2. There is a countable member N_3 of Ψ containing $\langle N_2, X_2 \rangle$. Extend X_2 to a finite tree X_3 with $\langle X_3 \rangle$ containing the first three elements of N_1, of N_2, and of N_3. Then $\cup_{i<\omega} X_i = X$ is a countable tree, and X is the countable member $\cup_{i<\omega} N_i$ of Ψ. □

Notice that none of the proofs of these corollaries uses the structure theory of simply presented groups. In particular, the proof of 4.13 does not use the fact that G is simply presented.

Although every finite tree in a countable group extends to a generating tree of the whole group, the following example shows that this is the most we can hope for without imposing additional conditions. That is, nice trees of a countable group do not necessarily extend to trees of the whole group.

EXAMPLE. Consider the tree

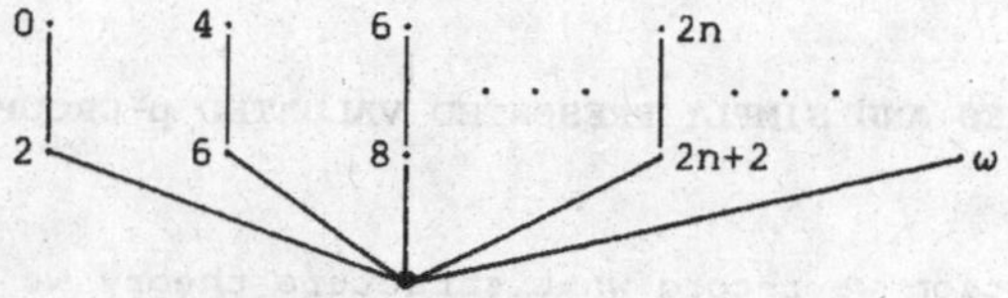

The countable fine embedding theorem [HRWal,Theorem 12] says that we may embed this tree nicely in a group with no relative Ulm invariants. Thus, the tree exists as a nice subgroup of a group G with the same Ulm invariants and no relatives. There is no way to extend this tree to a generating tree of G, since the attachment of anything to the element of order p and value ω introduces a relative Ulm invariant. □

There are necessary and sufficient conditions in terms of relative and derived invariants for extending a nice tree in a simply presented group to a p-basis of that group. Their establishment involves the structure theorem for simply presented p-groups, however.

THEOREM 4.14. Let G be a simply presented p-group and X a nice tree in G. Then X extends to a p-basis of G if and only if $f_{G,\langle X\rangle}$ dominates $f_{G,\langle X\rangle} + g_{\langle X\rangle}$.

PROOF. Suppose $f_{G,\langle X\rangle}$ dominates $f_{G,\langle X\rangle} + g_{\langle X\rangle}$. By [HRWal, Theorem 3], X can be extended to a tree Y such that $f_{\langle Y\rangle,\langle X\rangle} = f_{G,\langle X\rangle}$. Then $\langle Y\rangle \cong G$ with the isomorphism extending the identity on $\langle X\rangle$.

The converse follows immediately from [HRWal, Theorem 1]. □

Noticing that if X is finite, then $\langle X\rangle$ is nice and $f_{G,\langle X\rangle}$ dominates $f_{G,\langle X\rangle} + g_{\langle X\rangle}$ yields

COROLLARY 4.15. Every finite tree in a simply presented p-group G extends to a p-basis of G. □

5. TREES AND SIMPLY PRESENTED VALUATED p-GROUPS

In this section we record what structure theory we know for simply presented valuated p-groups. A beginning has been made in [HRWa2], where a satisfactory structure theory is given for _finite_ simply presented p-groups. That theory will be outlined here, but first some preliminaries are needed. We refer to 4.2 for the definition of _valuated tree_.

DEFINITION 5.1 Let X and Y be valuated trees. A _map_ $f : X \to Y$ is a function from X to Y such that

(a) $f(px) = pf(x)$;

(b) $vf(x) \geq v(x)$.

With this definition, we arrive at the category $\underline{T}$ of valuated trees. The direct sum (= coproduct) of a set of valuated trees is their disjoint union with the roots identified. Thus a valuated tree is _indecomposable_ if it has at most one element of order p. A valuated tree is partially ordered by $y \geq x$ if $p^n y = x$ for some n. The _valuated tree at x_ is the set

$$T_x = \{y \in X : y \geq x \text{ or } y \leq x\}.$$

A valuated tree X is thus the coproduct of the valuated trees T_x with x in X and of order p.

With a valuated tree X is associated a simply presented valuated group S(X), as described in section 4. The functor S is actually the adjoint of the forgetful functor from valuated p-groups to valuated trees. It is easy to see that if G is a reduced simply presented valuated p-group, then $G = \oplus S(X_\alpha)$ with X_α indecomposable.

However, $S(X)$ and $S(Y)$ can be isomorphic valuated groups without X and Y being isomorphic trees, even if X and Y are indecomposable.

DEFINITION 5.2. **A** **retraction** **of a valuated tree X is a map $r : X \to X$ such that $r^2 = r$.**

A decomposable tree obviously has non-trivial retractions, while an indecomposable one may or may not. However, a non-zero retraction of an indecomposable valuated tree preserves order. Also, it is easy to see that if X is indecomposable and $r(X)$ is finite irretractable, then any other retraction of X onto a finite irretractable is isomorphic to $r(X)$. Clearly, a tree must be bounded in order to have such a retraction. In [HRWa2], irretractable trees are used to construct examples of indecomposable simply presented valuated p-groups that have arbitrary finite rank. On the other hand, it is shown there that any indecomposable simply presented valuated p-group must be finite unless all values are ∞.

A basic fact [HRWa2, Theorem 7] is that if X is a finite valuated tree that has no non-trivial retractions, and if $S(X)$ is isomorphic to $S(Y)$, then X is isomorphic to Y. This is instrumental in the fundamental structure theorem for finite simply presented p-groups, which follows.

THEOREM 5.3. [HRWa2] Let X be a finite valuated tree. Then there exists a set $X_1, X_2, \ldots, X_n$ of irretractable trees, unique up to isomorphism, such that $S(X) = S(X_1) \oplus S(X_2) \oplus \ldots \oplus S(X_n)$. Each $S(X_i)$ is indecomposable, and any direct decomposition of $S(X)$ may be refined to this one. □

Therefore, a simply presented p-group G, which is $S(X)$ for any p-basis X of G, is determined by the unique set of X_i's in THEOREM

5.3. If G were a direct sum of cyclic valuated p-groups, then the $S(X_i)$ would simply be appropriate cyclic summands of G. In particular, this is the case if G were a finite p-group. So the usual invariants for a finite p-group are generalized from the number of summands of a given order, or what is the same thing, from a finite set of trees with one branch each, to a finite set of irretractable valuated trees. THEOREM 5.3 says that there is a canonical one-to-one correspondence between finite simply presented valuated p-groups and finite sets of irretractable valuated trees.

A countable group with no elements of infinite height is a direct sum of cyclics. More generally, any simply presented group with no elements of infinite height is a direct sum of cyclics. We now prove a corresponding result for valuated groups with p-bases.

LEMMA 5.4. Suppose $n < \omega$. Then there are finitely many irretractable valuated trees in which the maximum value of a nonzero element is at most n.

PROOF. We use induction on n. If $n = 0$ there is clearly only one such tree (it has two elements). Assume the result for $n - 1$ and let X be an irretractable tree whose nonzero elements have maximum value at most n. Since X is irretractable there is only one element in X of value n. Let X_α be the valuated tree gotten from X by setting $px_\alpha = 0$ if $px_\alpha = x$ in X. To avoid a retraction of X corresponding to an embedding $X_\alpha \to X_\beta$, we must, at the very least, have distinct X_α's. Thus the number of possibilities for X is bounded by the number of subsets of the set of irretractable trees whose non-zero elements have value at most $n - 1$. Our induction assumption ensures this number is finite. □

THEOREM 5.5. Let X be an indecomposable valuated tree of finite length. Then X has a retraction r such that:

(a) rX is finite irretractable;

(b) rX is unique up to isomorphism;

(c) X/rX is a direct sum of finite irretractables.

PROOF. We induct on the length n. If $n = 1$, then X has exactly two elements and is irretractable. We assume the result for all k less than n. Thus the unique order p element x of X may be assumed to have value n. Let $\{x_\alpha\}$ be the set of elements of X such that $px_\alpha = x$ and, for each x_α, let X_α be the tree whose elements are $\{x_\alpha\} \cup \{y \in X \mid p^m y = x_\alpha$ for some $m\}$ and such that $px_\alpha = 0$. All other relations remain as in X. Then, by the induction hypothesis, X_α has a retraction r_α with properties (a) - (c). The collection of r_α's yields a retraction r' of X in the natural way:

$$r'(y) = \begin{cases} r_\alpha y & \text{if } y \in X_\alpha \\ x & \text{if } y = x. \end{cases}$$

Clearly $X/r'X$ is a direct sum of finite irretractables, while $r'X$ is the tree made up of the set of finite irretractables $r_\alpha X_\alpha$ sitting over x. By LEMMA 5.4, there is a finite set of distinct representatives of $\{r_\alpha X_\alpha\}$, from which it follows that $r'X$ has a retraction r'' onto a finite irretractable with $r'X/r''r'X$ a direct sum of finite irretractables. Let $r = r''r'$. Then X/rX is the direct sum of $X/r'X$ and $r'X/rX$. □

THEOREM 5.6. Let G be a simply presented valuated p-group. If G has no elements of infinite value then $G = \oplus G_\alpha$ where each G_α is a finite indecomposable simply presented valuated p-group. Any decomposition of G refines to this one.

PROOF. We may assume $G \cong S(X)$ where X has a unique element of maximum value $n < \omega$. THEOREM 5.5 shows that X may be written as a direct sum $\oplus X_\alpha$ of finite irretractrables X_α, and hence G may be written as the direct sum $S(\oplus X_\alpha) \cong \oplus S(X_\alpha)$ of finite indecomposables. Azumaya's theorem provides the isomorphic refinement property. □

THEOREM 5.7. If G is a simply presented valuated p-group of length $\omega + 1$, then G is a direct sum of countable simply presented valuated p-groups.

PROOF. We can assume $G \cong S(X)$, where X has only one element x of value ω. By THEOREM 5.5, we can assume that each T_y, where $py = x$, is finite irretractable. By LEMMA 5.4, there are only countably many distinct T_y. Thus there is a retraction r of X onto a countable subset which includes x. Thus $G \cong S(X) \cong S(rX) \oplus S(X/rX)$. But X/rX has no elenents of infinite value, so we are done by THEOREM 5.6. □

A finite tree X is irretractable if and only if $S(X)$ is cyclic. Thus, COROLLARY 4.11 shows that every finite p-group is a direct sum of cyclics. Although this, together with THEOREM 5.3 yields the structure theorem for finite groups, the proof of THEOREM 5.3 involves Azumaya's theorem. If G is a simply presented group with no elements of infinite height, then THEOREM 5.6 shows that G is a direct sum of cyclics. In particular, COROLLARY 4.11 and THEOREM 5.6 show that a countable p-group with no elements of infinite height is a direct sum of cyclics. Thus our results give an alternative development of some of the well known and fundamental facts for Abelian p-groups.

REFERENCES

[CrHa] P. Crawley and A. Hales. The structure of Abelian p-groups given by certain presentations. *J. Algebra* 12(1969), 10-23.

[Fuchs] L. Fuchs. *"Infinite Abelian Groups"*. Volume 2, Academic Press. New York, 1970.

[Griff] P. Griffith. *"Infinite Abelian group theory"*. The University of Chicago Press. Chicago, 1970.

[Hill1] P. Hill. On the classification of Abelian groups. *preprint*.

[Hill2] P. Hill. The third axiom of countability for Abelian groups. to appear.

[HRWa1] R. Hunter. F. Richman, and E. Walker. Existence theorems for Warfield groups. *Trans. Amer. Math. Soc.* 235(1978), 345-362.

[HRWa2] R. Hunter. F. Richman, and E. Walker. Simply presented valuated p-groups. *J. Algebra* 49(1977), 125-133.

[Nunke] R. J. Nunke. Homology and direct sums of countable abelian groups. *Math Z.* 101(1967), 182-212.

[RicWa] F. Richman and E. Walker. Valuated groups. *J. Algebra* 56(1979), 145-167.

MIXED LOCAL GROUPS

Fred Richman

1. Introduction. Any abelian group may be viewed as an extension of a torsion group by a torsion-free group. This observation reduces the study of abelian groups to that of torsion groups, torsion-free groups, and how these two kinds of groups can fit together. Unfortunately the biggest problem seems to be the fitting together. More fruitful has been the idea of viewing an abelian group G as an extension of a torsion-free subgroup H by a torsion group, where the heights in G of the elements of H are recorded, that is, H is considered as a valuated group [RW]. The theory of summands of simply presented groups (Warfield groups), viewed as nice extensions of direct sums of cyclic valuated groups by simply presented torsion groups, is the prototype example of this point of view [HRW1].

If the torsion group G/H is simply presented, as it is if G is countable, and the torsion-free subgroup H is required to be nice, then the Hill-Walker theorem [WALK] says that there is only one way to fit the two together with specified relative Ulm invariants. The choice of relative Ulm invariants $f_{G,H}(\alpha)$ is limited by the equation

$$f_{G,H}(\alpha) + g_H(\alpha + 1) = f_{G/H}(\alpha),$$

where $f_{G/H}(\alpha)$ is the αth Ulm invariant of G/H, and $g_H(\alpha+1)$ is the $\alpha+1^{st}$ derived Ulm invariant of H (see [HRW2; p. 348]). Exactly what relative Ulm invariants may be specified in the countable case is the subject of Theorem 9.4, the "fine existence theorem".

The success of this approach in the case of Warfield groups suggests considering valuated groups other than direct sums of cyclics to serve as the torsion-free subgroups. Such a class is introduced in this paper and the theory of the corresponding mixed groups is developed. Cyclics are replaced by pseudo-cyclics (Section 7) which are valuated groups satisfying certain finiteness conditions and having local endomorphism rings in a suitable category. Theorem 8.1 is an isomorphism theorem for these generalized Warfield groups while Theorem 9.4 is an existence theorem for the countable case. Closure under summands, in the finite rank case, is established by Theorem 10.8. The general approach provides a complete classification of countable local groups, with no elements of infinite height and reduced torsion-free part, that "almost" split (Theorem 5.2).

All groups considered will actually be modules over $Z_{(p)}$, the integers localized at the prime p. Accordingly, the phrase "cyclic group" will denote a cyclic $Z_{(p)}$-module, "subgroup" will mean a $Z_{(p)}$-submodule, etc. A valuated group is value-reduced, or simply reduced, if zero is the only element of value ∞. By and large we are interested only in reduced valuated groups, but a blanket restriction to such groups would change the meaning of G/H. So I have tried to invoke this hypothesis only when it is needed or convenient.

2. Warfield categories. In dealing with summands of simply presented groups, Warfield [WARF] constructed a category in which the objects were abelian groups and the maps were height increasing homomorphisms defined on full-rank subgroups (subgroups H of G such that G/H is torsion). The category of valuated groups provides a more natural starting point than the category of groups as explicit reference to height may be omitted. In fact the construction works in a quite general setting, and we will use it in three slightly different situations.

Let CAT be an additive category with kernels, and KER a class of "proper" kernels containing the identity maps and closed under composition and pullbacks. Using set-theoretic notation this means that

$A \subset A$ is a proper kernel; that if $K \subset B$ is a proper kernel, and f is a map from A to B, then $f^{-1}K \subset A$ (exists and) is a proper kernel; and if $K \subset H$ and $H \subset A$ are proper kernels, then $K \subset A$ is a proper kernel. We construct a category W(CAT,KER) by defining a map from A to B to be a map from A' to B, where $A' \subset A$ is a proper kernel. Two maps in W(CAT,KER) are equal if they agree on some proper kernel. Composition of maps $f : A' \to B$ and $g : B' \to C$ is achieved by composing g with f restricted to $f^{-1}B'$. It is easily verified that W(CAT,KER) is an additive category with kernels.

The usefulness of W(CAT,KER) resides in the characterization of ismorphism in that category.

THEOREM 2.1. **Let A and B be objects of CAT. Then A is isomorphic to B in W(CAT,KER) if and only if there are kernels $H \subset A$ and $K \subset B$ in KER such that H is isomorphic to K in CAT.**

PROOF. The "if" part is clear. Conversely suppose that $H' \subset A$ and $K' \subset B$ are in KER, and the maps $f : K' \to A$ and $g : H' \to B$ are mutual inverses in W(CAT,KER). Then there is $H'' \subset H'$ in KER so that fg is the identity on H'', and $K'' \subset K'$ in KER so that gf is the identity on K''. Let

$$H = H'' \cap g^{-1}K'' \quad \text{and} \quad K = K'' \cap f^{-1}H.$$

As KER is closed under pullbacks, both $H \subset H''$ and $K \subset K''$ are in KER. Now $gH \subset gg^{-1}K'' \subset K''$ and $gH \subset gH'' \subset f^{-1}H$, so $gH \subset K$. Clearly $fK \subset H$ and f and g are mutual inverses when restricted to K and H. □

We will be interested in three different kinds of categories of the form W(CAT,KER), namely: W, W_α and C_α.

W: CAT is the category VAL of valuated groups, and KER consists of the full-rank (valuated) subgroups. This is the generalized Warfield category. Two valuated groups are isomorphic in W if and only if they have isomorphic full-rank (valuated) subgroups. Warfield's original category, called H by him and <u>Warf</u> by others, is the full subcategory of W whose objects are the groups.

W_α: CAT is the category VAL_α of valuated groups G such that $G(\alpha) = 0$, and KER consists of the full-rank subgroups H such that $(G/H)(\alpha) = 0$. Two valuated groups in VAL_α are isomorphic in W_α if and only if they have isomorphic full-rank subgroups whose cokernels in VAL are in VAL_α.

C_α: CAT is the category VAL_α and KER consists of those kernels that have nonzero intersection with every cyclic subgroup of length α. Two valuated groups in VAL_α are isomorphic in C_α if and only if they have isomorphic subgroups whose cokernels in VAL are in VAL_α, and which have nonzero intersection with every cyclic subgroup of length α.

3. <u>Generalized KM-modules</u>. Kaplansky and Mackey [KM] gave a complete set of invariants for countably generated modules of torsion-free-rank 1 over complete discrete valuation rings. Completeness of the ring turns out to be a superfluous condition: it guarantees that cyclic submodules are nice, but Wallace [WALL] showed this to be automatic in the rank 1 case.

Rotman and Yen generalized these results to finite rank modules which they called <u>KM-modules</u>. Here completeness is essential to insure niceness of finitely generated submodules, that is, the "coset property" [RY; Lemma 2]. Their isomorphism theorem for KM-modules [RY; Theorem 1] is generalized by Theorem 3.4 below. Completeness of the ring and finiteness of the torsion-free rank are replaced by postulating a nice full-rank torsion-free subgroup H with finite Ulm invariants and easily described nice subgroups. The countability assumption is replaced by requiring that G/H be simply presented, and that torsion quotients of H have nice composition series. The condition on nice subgroups can be phrased in terms of the topology given by the neighborhood base $\{H(\theta+1) : \theta < \text{length } H\}$.

DEFINITION 3.1. If H is a valuated group of length α, and A is a subgroup of H, then we say that A is closed if $(H/A)(\alpha) = 0$.

Clearly every nice subgroup of a reduced valuated group H is closed. For some H the converse holds. One obvious example is when all subgroups of H are nice. Another is when H has length ω. This latter case can be extended by the notion of packed length.

If H is a valuated group, then $\{vx : x \in H \text{ and } vx \neq \infty\}$ is a well-ordered set which is order isomorphic to a unique ordinal α. This ordinal α is called the packed length of H. We shall be particularly concerned with valuated groups of packed length ω. Clearly reduced infinite cyclics have packed length ω.

THEOREM 3.2. If H is a valuated group of packed length ω, then every closed subgroup of H is nice.

PROOF. Let α be the length of H and let A be a subgroup of H. If $x + A$ has no element of maximum value, then $\sup \{v(x+a) : a \in A\} = \alpha$ so x represents a nonzero element of $(H/A)(\alpha)$. □

COROLLARY 3.3. Let H_1 and H_2 be reduced valuated groups of length α and packed length ω. Then the following are equivalent:

a) H_1 and H_2 are isomorphic in C_α

b) H_1 and H_2 are isomorphic in W_α

c) H_1 and H_2 have isomorphic nice full-rank subgroups.

PROOF. The kernels in VAL_α are those subgroups $A \subset H$ such that $(H/A)(\alpha) = 0$. The proper kernels in C_α are those that intersect every cyclic subgroup of length α. But since H_1 and H_2 have packed length ω, these are exactly the full-rank kernels. Thus a) and b) are equivalent. That b) and c) are equivalent follows from Theorems 2.1 and 3.2. □

Recall that a valuated group is called Fuchsian if it has a nice composition series [HR], and that a group is Fuchsian exactly when it is simply presented torsion [FUC2; page 82].

THEOREM 3.4. **Let G be a reduced group and H a torsion-free subgroup of G such that**

i) Every closed subgroup of H is nice,

ii) All Ulm invariants of H are finite,

iii) Every nice torsion image of H is Fuchsian,

iv) H is nice in G,

v) G/H is simply presented torsion.

If G' and H' are another such pair, then G is isomorphic to G' if and only if

a) G and G' have the same Ulm invariants, and

b) H and H' have isomorphic nice full-rank subgroups.

PROOF. Suppose G is isomorphic to G'. Then certainly G and G' have the same Ulm invariants. As G/H and G'/H' are torsion, it is clear that length H = length H' = α. As $G/G(\alpha)$ and $G'/G'(\alpha)$ are isomorphic, and are isomorphic in W_α to H and H', we have H and H' isomorphic in W_α. Thus H and H' have isomorphic closed, hence nice, full-rank subgroups.

Conversely suppose there are isomorphic nice full-rank subgroups K of H and K' of H'. Since all Ulm invariants of K and K' are finite, the Ulm invariants of G relative to K are the same as the Ulm invariants of G' relative to K'. Since H/K and H'/K' are Fuchsian, G/K and G'/K' are simply presented torsion. Hence by the Hill-Walker theorem, G and G' are isomorphic. □

Any finitely generated module over a complete discrete valuation ring is nice. Hence conditions i) and iv) hold for a KM-module G where H is any full-rank torsion-free submodule. If H is finitely generated, then H satisfies ii) and iii), and if G is countably generated, then v) holds. Thus if G is a KM-module, and H is any finitely generated, full rank, torsion-free submodule of G, then G and H satisfy the hypotheses of Theorem 3.4.

4. Torsion-free groupification. The valuated group H in Theorem 3.4 contains much more information than the torsion-free group G/G_t, where G_t is the maximal torsion subgroup of G. We shall now show that it contains at least as much. In fact, every valuated group has a unique torsion-free group envelope - and the envelope of H is G/G_t.

THEOREM 4.1. **Let H be a valuated group and let K be the subgroup of $Q \otimes H$ defined by**

$$K = \cup_{n<\omega} p^{-n}(1 \otimes H(n))$$

and $f : H \to K$ the map taking x to $1 \otimes x$. Then K is a torsion-free group, f is a map of valuated groups, and any map from H to a torsion-free group factors uniquely through f. Moreover $K/f(H)$ is torsion and

$$\text{ht } f(x) = \sup \{m : vp^n x \geq n + m \text{ for some } n\}$$

where the supremum is ∞ if the set is unbounded.

PROOF. Note that the subgroups $p^{-n}(1 \otimes H(n))$ of $Q \otimes H$ form an ascending chain, so K is indeed a group. As $Q \otimes H$ is torsion-free, so is K. If $n < \omega$ and $x \in H(n)$, then $f(x) \in p^n K$ so f is a map of valuated groups as K is torsion-free. Suppose g maps H to a torsion-free group F. Then g induces a unique map g^* from $Q \otimes H$ to $Q \otimes F$ taking $1 \otimes H(n)$ into $p^n F$. Hence $g^*(K) \subset F$. Clearly $g = g^* f$.

It is immediate that $K/f(H)$ is torsion whence the map g^* is unique. To see that the computation of $\mathrm{ht}\, f(x)$ is correct, suppose $vp^n x \geq n + m$. Then $\mathrm{ht}\, p^n f(x) \geq n + m$, so $\mathrm{ht}\, f(x) \geq m$ as K is torsion-free. Conversely suppose $\mathrm{ht}\, f(x) \geq m$. Then

$$1 \otimes x = f(x) \in p^m K = \cup_{n<\omega} p^{m-n}(1 \otimes H(n)).$$

Thus $1 \otimes p^n x = 1 \otimes p^m h$ for some n and $h \in H(n)$. So, for some positive integer r, we have $p^r p^n x = p^r p^m h \in H(r+n+m)$. □

The group K in Theorem 4.1, together with the map $f : H \to K$, is clearly unique up to isomorphism over H. We shall refer to it as the <u>torsion-free group envelope</u> of H. It arises whenever H is a full-rank subgroup of some group.

THEOREM 4.2. Let H be a full-rank subgroup of a group G. Then G/G_t is the torsion-free group envelope of H.

PROOF. Let K and f be as in Theorem 4.1 and let g be the natural projection from G to G/G_t restricted to H. Then there is a unique map $g^* : K \to G/G_t$ such that $g^* f = g$. If $g^* f(x) = 0$, then $g(x) = 0$ so $x \in G_t$ whence $f(x) = 0$. But $K/f(H)$ is torsion, so g^* is one-to-one. To see that g^* is onto, let $y \in G$. Then

$p^n y \in H(n)$ for some $n < \omega$. Hence $x = p^{-n}(1 \otimes p^n y) \in K$ and $g^*(p^n x) = g(p^n y)$ so $g^*(x) = g(y)$. □

We may characterize those groups G such that G/G_t is reduced in terms of their cyclic subgroups. Call a valuated group gapless if $vpx = vx + 1$ for every nonzero element x. Note that a valuated group is gapless if and only if all of its Ulm invariants are zero.

COROLLARY 4.3. If G is a group, then G/G_t is reduced if and only if every infinite cyclic subgroup of G contains a nozero gapless subgroup of length ω.

PROOF. Let f be the natural map from G to G/G_t and let $x \in G$. As G/G_t is the torsion-free group envelope of G by Theorem 4.2, we have $\text{ht}\, f(x) = \sup \{m : vp^n x \geq n + m$ for some $n\}$ by Theorem 4.1. Hence $\text{ht}\, f(x)$ is finite exactly when the subgroup generated by x contains a nonzero gapless subgroup of length ω. □

COROLLARY 4.4. If G is a group of torsion-free rank 2 such that $p^\omega G = 0$ and G/G_t is reduced, then G has a nice full-rank torsion-free subgroup.

PROOF. By Corollary 4.3 the group G contains a nonzero gapless subgroup H. The p-adic closure of H in G is also gapless for if $\text{ht}(g - h) > \text{ht}\, h$ for some h in H, then $\text{ht}(pg - ph) > \text{ht}\, h + 1 = \text{ht}\, ph$ so $\text{ht}\, pg = \text{ht}\, ph = \text{ht}\, g + 1$. Thus we may assume that H is closed in G. If H is rank 2 we are done. If H is rank 1, then G/H has torsion-free rank 1, so by [WALL] any cyclic subgroup is nice. The preimage of any infinite cyclic subgroup of G/H in G is the desired full-rank nice torsion-free subgroup of G. □

What groups have full-rank nice torsion-free subgroups? Corollary 4.4 gives a sufficient but by no means necessary condition. Indeed we can embed any torsion-free valuated group as a full-rank nice subgroup of a group via the crude embedding theorem of [RW]. The problem has not been intensively studied, but no example of a group that does not contain a full-rank nice torsion-free subgroup has yet been established.

5. W-split groups. If the maximal torsion subgroup G_t of a reduced nontorsion group G is a summand, i.e. G splits, then G contains a nice torsion-free gapless subgroup of length ω, and G/G_t is reduced. However this is not a sufficient condition for G to split; rather it characterizes groups that split in W.

THEOREM 5.1. Let G be a reduced group with maximal torsion subgroup G_t, and f the natural map from G to G/G_t. Then the following are equivalent:

1) There is a map $g : G/G_t \to G$ in W such that $fg = id$ in W.

2) The map f is a W-isomorphism.

3) G is W-isomorphic to a torsion-free group.

4) G contains a gapless full-rank subgroup H with $H(\omega) = 0$.

If, moreover, $p^{\omega}G = 0$, then the subgroup in (4) can be taken to be nice.

PROOF. If (1) holds, then $fgf = f$ so $f(gf - id) = 0$. But $\ker f = 0$ in W so f is monic, whence $gf = id$. Clearly (2) implies (3), and (3) implies (4) by Theorem 2.1 and the fact that any subgroup K of a torsion-free group is gapless with $K(\omega) = K(\infty)$. Now suppose G contains a gapless full-rank subgroup H with $H(\omega) = 0$. As H is gapless, we cannot raise heights of elements of H by adding torsion elements. But $H(\omega) = 0$ so f maps H isomorphically onto a full-rank subgroup of G/G_t. Finally, if $p^{\omega}G = 0$, then the p-adic closure of H in G is also gapless (see proof of Corollary 4.4). □

Note that condition (1) may be thought of as stating that G splits in W. If we look ahead to Theorem 9.4 we can get a complete classification theorem for countable mixed (local) groups, with no elements of infinite height, that split in W.

THEOREM 5.2. There is a correspondence between

1) Countable groups G with no elements of infinite height such that G_t is unbounded and G splits in W,

2) Pairs (S,T) where S is a countable gapless valuated group, $S(\omega) = 0$, and T is an unbounded countable direct sum of torsion cyclic groups,

given by letting G correspond to (S,T) if G_t is isomorphic to T and G contains a full-rank nice subgroup isomorphic to S. We have

i) To every G there corresponds a pair (S,T),

ii) To every pair (S,T) there corresponds a G,

iii) If G corresponds to (S,T), and G' corresponds to (S',T'), then G is isomorphic to G' if and only if T is isomorphic to T', and S is isomorphic to S' in C_ω.

PROOF. Part i) is Theorem 5.1. Part iii) follows from Theorem 3.4. Part ii) follows from Theorem 9.4. □

COROLLARY 5.3. Let F be a countable reduced torsion-free group, and T_1 and T_2 unbounded countable direct sums of torsion cyclics. Then there is a natural correspondence between isomorphism classes of W-split extensions of T_1 by F and isomorphism classes of W-split extensions of T_2 by F.

PROOF. "Natural correspondence" is meant in the same sense that there is a natural correspondence between countable 5-groups and countable 17-groups. Let G_1 be a W-split extension of T_1 by F, and let S be a full-rank gapless nice subgroup of G_1. Define G_2 to be a group corresponding to (S,T_2). We may assume that T_2 is the maximal torsion subgroup of G_2 and that S is a full-rank gapless nice subgroup of G_2. The only issue is whether G_2/T_2 is isomorphic to F, but each is a torsion-free group envelope of S. □

If G is a reduced group that splits in W, then G/G_t is reduced. The converse does not hold. In fact, by Theorems 4.1 and 4.2 the group G/G_t is reduced precisely when $p^\omega G \subset G_t$ and every cyclic subgroup of G has only finitely many Ulm invariants. An example of such a group that is not W-split is constructed as follows.

EXAMPLE 5.4. Let P be the p-adic integers, Z the rational p-adics and B an unbounded direct sum of cyclic torsion groups. Let x be an irrational in P and y a nontorsion element of the completion $\hat{B}$ of B. Let G be a subgroup of $P \oplus \hat{B}$ containing $x + y$ and $Z \oplus B$ such that $G/(Z \oplus B)$ is rank-one torsion-free divisible. It is readily verified that every cyclic subgroup of G has only finitely many Ulm invariants, but every rank 2 subgroup of G has nonzero Ulm invariants. □

6. p-rank. Some authorities, e.g. [ARN], define the p-rank of a torsion-free abelian group H to be the dimension of H/pH. Equivalently, the p-rank of H is the least cardinal n such that $H/p^m H$ can be generated by n elements for each $m < \omega$. We define a generalization of this notion for valuated groups.

DEFINITION. Let H be a valuated group and n a cardinal number. If $H/H(\theta+1)$ can be generated by n elements for each $\theta <$ length H, then we say that p-rank $H \leq n$. The p-*rank* of H is the least n such that p-rank $H \leq n$.

A few comments on this definition are in order:

1) The only situation we are seriously interested in is when n is finite and H is torsion-free reduced, and hence of limit length. For n an infinite cardinal some other definition might be more useful.

2) The point of using $\theta+1$ instead of θ is to assign p-rank one to a cyclic group of order p.

3) If H is a torsion-free group, then p-rank $H = \dim H/pH$ since length $H = \omega$ and $H(m) = p^m H$.

4) A finitely generated valuated group K can be generated by n elements, for finite n, if and only if $\dim K/pK \leq n$.

Examples of valuated groups of finite p-rank are provided by the following two theorems.

THEOREM 6.1. If H is a valuated group of packed length ω, then p-rank $H \leq \dim H/pH$.

PROOF. As H has packed length ω, the quotient $H/H(\theta+1)$ is bounded for each $\theta <$ length H. Thus $H/H(\theta+1)$ can be generated by $\dim H/pH$ elements. □

To see the necessity for the packed length condition in Theorem 6.2, let H be a direct sum of the p-adic integers and a cylic valuated group of length greater than ω. Then $\dim H/pH = 2$ while the p-rank of H is uncountable.

THEOREM 6.2. Let H be a valuated group. If K is a subgroup of H, or if K is a nice quotient of H, then p-rank $K \leq$ p-rank H.

PROOF. If K is a subgroup of H, then $K/K(\theta+1) \subset H/H(\theta+1)$ for each $\theta <$ length $K \leq$ length H. But for any cardinal number n, every subgroup of a group generated by n elements is generated by n elements. If K is a nice quotient of H, then $H/H(\theta+1)$ maps onto $K/K(\theta+1)$ for each $\theta <$ length $K \leq$ length H. □

We shall show that a valuated group H of finite p-rank is small in an appropriate category. First we show that for such H the generators of $H/H(\theta+1)$ can be chosen independently of θ.

THEOREM 6.3. Let H be a valuated group of length λ. Let n be a nonnegative integer. Then p-rank $H \leq n$ if and only if there is a subgroup S of H, generated by n elements, such that $H = S + H(\theta+1)$ for each $\theta < \lambda$.

PROOF. Suppose such an S exists. If $\theta < \lambda$, then $H/H(\theta+1)$ is generated by n elements, so p-rank $H \leq n$. Conversely if p-rank $H \leq n$, choose $\alpha < \lambda$ maximizing the number of elements required to generated $H/H(\alpha+1)$, and let S be a subgroup of H with n generators such that $S + H(\alpha+1) = H$. If $\theta \leq \alpha$, then $S + H(\theta+1) = H$. If $\alpha < \theta < \lambda$, then the finitely generated group $H/H(\theta+1)$ maps onto $H/H(\alpha+1)$ so, by the choice of α we have $pH + H(\alpha+1) = pH + H(\theta+1)$ whereupon $S + pH + H(\theta+1) = H$. By the Nakayama Lemma it follows that $S + H(\theta+1) = H$. □

THEOREM 6.4. Let H be a valuated group of length λ and finite p-rank. Then any map from H into a direct sum of reduced valuated groups of lengths at most λ goes into a finite subsum.

PROOF. Suppose $\{K_i\}$ is a family of valuated groups with $K_i(\lambda) = 0$ and f maps H into the direct sum of the K_i. Choose S as in Theorem 6.3. Then S maps into a finite subsum $\Sigma_{i\in J} K_i$. If $i \notin J$, then the induced map from $H/H(\theta+1)$ to $K_i/K_i(\theta+1)$ is zero for each θ

$< \lambda$. Since $K_i(\lambda) = 0$, the induced map from H to K_i is zero. □

7. Pseudo-cyclic valuated groups. Cyclic valuated groups are finitely generated and have local endomorphism rings in various categories. These properties are abstracted in the notion of a pseudo-cyclic valuated group.

DEFINITION. A reduced valuated group H of length α is pseudo-cyclic if:

1) H has finite p-rank
2) H has packed length ω
3) H has a local endomorphism ring in C_α.

As we shall see later, we might as well require a pseudo-cyclic to be torsion-free. The prototype pseudo-cyclic groups are finitely generated subgroups of the p-adic integers that contain 1 but no irrational algebraic number. The endomorphism ring of such a group is $Z_{(p)}$, the integers localized at p. Other examples are achieved by revaluating these groups with an increasing sequence of ordinals. More generally we have the following theorem.

THEOREM 7.1. Let H be a finitely generated torsion-free reduced valuated group of packed length ω and length α. If the endomorphism ring of H is $Z_{(p)}$, the integers localized at p, then H is pseudo-cyclic. Moreover the endomorphism ring of H is $Z_{(p)}$ in W and in C_α.

PROOF. As H is finitely generated it clearly satisfies (1). We have assumed that H satisfies (2). To establish (3), let f be a map from a full-rank subgroup of H to H. As H is finitely generated, we may take this subgroup to be p^nH. Then fp^n is a map from H to

H, so $fp^n = up^m$ where u is a unit in $Z_{(p)}$. As f increases values we have $n \leqslant m$ so f extends uniquely to the map up^{m-n} on all of H. Hence the endomorphism ring of H in W and in C_α is equal to $Z_{(p)}$, so (3) holds. □

Examples of pseudo-cyclic groups that are not subgroups of the p-adic integers are provided by a construction from [MOOR]. By the value sequence Vx of an element x we mean the sequence of ordinals vp^nx. We write $Vx \leqslant Vy$ if $vp^nx \leqslant vp^ny$ for all n. Note that if f is a map, then $Vx \leqslant Vf(x)$.

EXAMPLE 7.2. For each positive integer m, there is a direct sum of cyclic valuated groups with basis $x_1, \cdots, x_m$ such that

$$V(\Sigma a_i x_i) \leqslant V(\Sigma b_i x_i) \text{ if and only if } a_i \text{ divides } b_i \text{ for all } i$$

[MOOR; Example 5.4]. Let H be the subgroup of this group consisting of $\{\Sigma a_i x_i : \Sigma a_i = 0\}$. We shall show that the endomorphism ring of H is $Z_{(p)}$. Suppose f is an endomorphism of H. Then $V(x_i - x_j) \leqslant Vf(x_i - x_j)$ so $f(x_i - x_j) = ax_i + bx_j$. But $a + b = 0$, so $f(x_i - x_j) = r_{ij}(x_i - x_j)$. Moreover because

$$f(x_i - x_k) = r_{ik}(x_i - x_k) = r_{ij}(x_i - x_j) + r_{jk}(x_j - x_k),$$

so $r_{ij} = r_{jk}$, all the r_{ij} are equal. Finally note that H is generated by the elements $x_i - x_j$. □

THEOREM 7.3. Let H be a reduced valuated group of finite p-rank. Then all the Ulm invariants of H are finite, and each nice torsion image of H is finite.

PROOF. The quotient $H/H(\theta+1)$ is finitely generated for each $\theta <$

length H. Thus the Ulm invariants $f_H(\theta) \subset H/H(\theta+1)$ are finite. If K is a nice torsion image of H, then K has finite p-rank by Theorem 6.2, and length K $\leq$ length H. If S is a subgroup of the socle of $K/K(\theta+1)$, then dim S $\leq$ p-rank K. Thus the socle of K is finite, so K is finite. □

We shall consider two pseudo-cyclic valuated groups to be equivalent if they have isomorphic full-rank nice subgroups. From Theorem 3.2 we see that two pseudo-cyclics of length α are equivalent if and only if they are isomorphic in C_α (or in W_α). The torsion subgroup of a pseudo-cyclic H is finite, so $H(\theta)$ is torsion-free for some $\theta <$ length H, whence H is equivalent to the torsion-free pseudo-cyclic valuated group $H(\theta)$.

While a cyclic valuated group is equivalent to any of its nonzero subgroups, the situation is quite different for subgroups of the p-adic integers. To see this we first characterize the full-rank nice subgroups of such valuated groups.

LEMMA 7.4. Let H be a reduced valuated group of finite p-rank, length λ, and packed length ω. Let K be a subgroup of H. Then K is full-rank nice if and only if K contains $H(\alpha)$ for some $\alpha < \lambda$. Moreover, if H has p-rank one, then K must equal some $H(\alpha)$.

PROOF. Suppose K contains $H(\alpha)$ for some $\alpha < \lambda$. As H has packed length ω, we have $H(\alpha) \supset p^n H$ for some n, so $H/H(\alpha)$ is bounded whence K is full-rank. As H has finite p-rank, $H/H(\alpha)$ is finite so $K/H(\alpha)$ is nice in $H/H(\alpha)$. Since $H(\alpha)$ is nice in H, so is K.

Conversely suppose K is nice and full-rank. Then length H/K $\leq \lambda$, and H/K is finite by Theorem 7.3. Since λ is a limit,

$$\alpha = \max\{vx : 0 \neq x \in H/K\} + 1 < \lambda,$$

and K contains $H(\alpha)$. If, moreover, H has p-rank one, then $H/H(\alpha)$ is cyclic, so $K = H(\theta)$ for some $\theta \leq \alpha$. □

THEOREM 7.5. **Let H be a subgroup of the p-adic integers, with the induced valuation, generated by n elements and containing both nonzero rational and irrational numbers. If H is equivalent to p^iH for some $i > 0$, then H contains an irrational algebraic number of degree n or less.**

PROOF. The p-adic integers are injective in the category of valuated groups so any isomorphism between $H(m)$ and $(p^iH)(m)$ is induced by multiplication by some p-adic integer r. This r is a unit and is algebraic of degree n or less. If r is irrational, then $(p^iH)(m)$ contains an irrational algebraic number of degree n or less since $H(m)$ contains a nonzero rational. If r is rational, we can take r to be 1. But $H(m) \neq (p^iH)(m)$ because H contains an irrational and a nonzero rational, and therefore an irrational of height 0 and value greater than m. □

8. Generalized Warfield groups. In this section we introduce a generalization of the notion of a local Warfield group, and prove that two such groups are isomorphic if they have the same invariants.

We say that $\sum H_i$ is a Warfield decomposition of G if $\sum H_i \subset G$ is a valuated direct sum with torsion cokernel. A generalized Warfield group is one that admits a nice Warfield decomposition into pseudo-cyclics with simply presented cokernel. The invariants of such a group are its Ulm invariants and the equivalence classes (with multiplicities) of the pseudo-cyclics in the Warfield decomposition.

THEOREM 8.1. **Let G and G' be reduced groups that admit nice**

Warfield decompositions into pseudocyclics $\Sigma_{i\in I} H_i$ and $\Sigma_{j\in J} H_j'$ with simply presented cokernels. Then G and G' are isomorphic if and only if they have the same Ulm invariants, and there is a one-to-one correspondence $u : I \to J$ such that H_i is equivalent to $H_{u(i)}'$.

PROOF. If G and G' are isomorphic, then they certainly have the same Ulm invariants. Let λ be a limit ordinal. As $G/G(\lambda)$ and $G'/G'(\lambda)$ are isomorphic in C_λ, the Azumaya theorem [WW] in C_λ provides a one-to-one correspondence

$$u : \{i\in I : \text{length } H_i = \lambda\} \to \{j\in J : \text{length } H_j' = \lambda\}$$

such that H_i is equivalent to $H_{u(i)}'$. To apply the Azumaya theorem we must invoke Theorems 6.4 and 6.2.

Conversely, suppose G and G' have the same Ulm invariants and $u : I \to J$ is as indicated. We may assume that H_i is isomorphic to $H_{u(i)}'$. By the Hill-Walker theorem it suffices to arrange that G and G' have the same Ulm invariants relative to $H = \Sigma H_i$ and $H' = \Sigma H_j'$ respectively. As $f_G(\alpha) = f_H(\alpha) + f_{G,H}(\alpha)$, this is automatic if $f_G(\alpha)$ is finite. Let

$$X = \{\alpha : f_G(\alpha) \neq 0\} \quad \text{and} \quad F_i = \{\alpha : f_{H_i}(\alpha) \neq 0\}.$$

Then F_i is countable, so by [HRW1; Lemma 9] there is a function $e : I \to X$ such that $e(i) \in F_i$ and

$$\text{card}\,\{i : e(i) = \alpha\} = \text{card}\,\{i : \alpha \in F_i\}$$

whenever the latter is infinite. Redefine H_i to be $H_i(e(i)+1)$ and $H_{u(i)}'$ to be $H_{u(i)}'(e(i)+1)$. Then $f_G(\alpha) = f_{G,H}(\alpha)$ whenver $f_G(\alpha)$ is infinite. □

9. A fine embedding theorem. The question of what generalized Warfield groups exist hinges on finding out when a given direct sum of pseudo-cyclics can be nicely embedded in a group with specified relative Ulm invariants and simply presented torsion cokernel. In this section we give a complete answer to the question of when a given valuated group can be nicely embedded in a countable reduced group with specified relative Ulm invariants and torsion cokernel. Recall that the derived Ulm invariants of a reduced valuated group H are defined by the vector spaces

$$g_H(\lambda) = H(\lambda)/\cap_{\theta<\lambda}(H(\lambda+1) + H(\lambda)\cap pH(\theta))$$

which are filtered by

$$J_H(\alpha,\lambda) = \text{image of } H(\lambda)\cap pH(\alpha) \text{ in } g_H(\lambda).$$

By $pH(\theta)$ we mean $p(H(\theta))$, not $(pH)(\theta)$.

DEFINITION. A nontrivial coset C of $pH(\theta)$ in $H(\theta+1)$ is bad if every element x in C, whose value is sufficiently close to the length of C, represents zero in $g_H(vx)$.

That is, there is $\beta \leq vc$ for some c in C such that if $x \in C$ and $vx \geq \beta$, then for each $\alpha < vx$ there is y_α in $H(\alpha)$ with $v(x - py_\alpha) > vx$.

DEFINITION. An ordinal λ is bad for H if λ is the length of a bad coset of $pH(\theta)$ in $H(\theta+1)$ for some θ.

Note that such a λ must be a limit ordinal. Bad ordinals cause relative Ulm invariants in the following sense.

THEOREM 9.1. **If H is nice in a reduced group G, and λ is bad for H, then for each $\beta < \lambda$ there is α such that $\beta \leq \alpha < \lambda$ and $f_{G,H}(\alpha) \neq 0$.**

PROOF. Let C be a bad coset of $pH(\theta)$ in $H(\theta+1)$ of length λ. We may assume that $\beta \geq \theta$ and that if $x \in C$ and $vx > \beta$, then x represents zero in $g_H(vx)$. Choose g in G so that $pg \in C$ and $vg \geq \beta$. Let g' be an element of maximum value in $g + H(\theta)$ and set $\alpha = vg' \geq \beta$. Since $pg' \in C$ we have $\alpha < \lambda$. If $vpg' > \alpha + 1$, then g' represents a nonzero element of $f_{G,H}(\alpha)$. If $vpg' = \alpha + 1$, then, since pg' represents zero in $g_H(\alpha + 1)$, there is h in $H(\alpha)$ such that $v(pg' + ph) > \alpha + 1$, so $g' + h$ represents a nonzero element of $f_{G,H}(\alpha)$. □

An example of a valuated group H such that ω is bad for H is provided by taking an irrational p-adic integer ξ and letting H be the subgroup of the p-adic integers generated by 1 and $p\xi$. The coset $C = p\xi + pH \subset H(1)$ is bad, for if $x = p\xi + ph$, then we can find a positive integer n such that $x - p(n+h) = p(\xi - n)$ has arbitrarily high value and (so) $n + h \in H(vx - 1)$. This example is cited in [HRW2; page 359] as a valuated group that cannot be embedded nicely in a group with relative Ulm invariants, yet satisfies the admissibility conditions given in that paper. Theorem 9.1 provides an explanation of this phenomenon.

The following technical lemma is used to show that bad ordinals are inherited by nice extensions of finite index.

LEMMA 9.2. **Let K be a reduced valuated group and H a nice subgroup of index p. Let $K = H + \langle z\rangle$ where z is of maximum value in $z + H$. If x in H represents a nonzero element of the kernel of**

the map from $g_H(vx)$ to $g_K(vx)$, then $vx = vz + 1$.

PROOF. For $vx = \alpha + 1$ this follows from the exact sequence

$$f_{K/H}(\alpha) \to g_H(\alpha+1) \to g_K(\alpha+1).$$

In general, since $H(vz + 1) = K(vz + 1)$, if $vx > vz + 1$, then $g_H(vx) = g_K(vx)$. Thus we may assume that $vx \leq vz$. If x represents zero in $g_K(vx)$, then for each $\alpha < vx$ there is k_α in $K(\alpha)$ such that $v(x + pk_\alpha) > vx$. We can write $k_\alpha = n_\alpha z + h_\alpha$ with h_α in H. Hence $h_\alpha = k_\alpha - n_\alpha z \in H(\alpha)$ and

$$v(x + ph_\alpha) = v(x + pk_\alpha - n_\alpha pz) > vx,$$

so x represents zero in $g_H(vx)$. □

LEMMA 9.3. Let K be a reduced valuated group and H a nice subgroup of finite index. If λ is bad for K, then λ is bad for H.

PROOF. We may assume that H is of index p in K. Let $x + pK(\theta)$ be a bad coset of $pK(\theta)$ of length λ. There is x' in H such that $x - x'$ has maximum value in $x + H$, so $v(x - x') \geq \lambda$. It is easy to see that $x' + pK(\theta)$ is a bad coset of $pK(\theta)$ of length λ. We may assume that $pK(\theta) \neq pH(\theta)$, so

$$x' + pK(\theta) = C_1 \cup C_2 \cup \cdots \cup C_p$$

where C_i is a coset of $pH(\theta)$. Some C_i must have length λ, and Lemma 9.2 implies that this C_i is a bad coset of $pH(\theta)$. □

DEFINITION. Let f be a function from ordinals to cardinals, and H a valuated group. We say that H <u>admits</u> f if, whenever $\lambda \geq \theta + \omega$, we have

1) If λ is bad for H, then there is α in (θ,λ) such that $f(\alpha) \neq 0$,

2) If $f(\lambda) \neq 0$ or $g_H(\lambda) \neq 0$, then there is α in $(\theta,\theta+\omega)$ such that $f(\alpha) \neq 0$ or $J(\alpha,\alpha+n) \neq 0$ for some $n < \omega$.

It is readily seen that this definition is equivalent to the one in [HRW2] provided no ordinals are bad for H, the only case of interest there.

THEOREM 9.4. **Let H be a countable valuated group such that vx is a countable ordinal for each nonzero x in H. Let f be a function from ordinals to countable ordinals, vanishing beyond some countable ordinal. Then H admits f if and only if H can be embedded as a nice subgroup of a countable reduced group G with G/H torsion and $f_{G,H} = f$.**

PROOF. The group G is constructed as the union of valuated groups $H = H_0 \subset H_1 \subset H_2 \subset \cdots$ where H_n is of index p in H_{n+1} and $f_{Hn,H} \leq f$. The proof proceeds as in [HRW2; Theorem 12] with the following modifications. Instead of showing $pH_n(\theta)$ is nice for every θ, we invoke Lemma 9.3 to show that if λ is bad for H_n, then it is bad for H. Given x in H_n and $\beta < vx = \alpha$ we may assume either that x represents a nonzero element of $g_{Hn}(\alpha)$, as in [HRW2; Theorem 12], or that $x + pH_n(\beta)$ is a bad coset. If $x + pH_n(\beta)$ is a bad coset, then we may assume that $f_{Hn,H}(\beta) = 0$ and that $f(\beta) \neq 0$. Then H_{n+1} is constructed by adjoining y such that $vy = \beta$ and $py = x$. □

10. Summands. Let H and K be subgroups of length α of a valuated group G. We say that H permeates K if for each k in

K there is h in H such that $v(k - h) > vk$.

If $H \subset K$ and H permeates K we say that H is <u>dense</u> in K. This corresponds to Fuchs' "s-dense" for valued vector spaces [FUC1]. A <u>closure</u> of a subgroup H of a valuated group G is a nice subgroup of G containing H as a dense subgroup. Unlike the situation for p-bounded valuated groups [FUC1], not every subgroup of a valuated group has a closure, nor need the closure be unique.

EXAMPLE 10.1. Let G be a p^2-bounded direct sum of cyclics with basis $x_1, x_2, \cdots$ where $vx_n = n$ and $vpx_n = \omega$. Let H be generated by the elements $x_i - x_{i+1}$. Any subgroup of G properly containing H contains px_1, so H is not dense in such a subgroup. On the other hand, the coset $x_1 + H$ contains no element of maximum value. □

EXAMPLE 10.2. Let K be the subgroup of the p-adic integers generated by 1 and an irrational p-adic integer ξ. Let $G = K \oplus \langle x \rangle$ where $vx = \omega$. If $H = \langle 1 \rangle$, then H is dense in K and in $K' = \langle 1, \xi + x \rangle$, and both K and K' are summands of G.

If H and K are subgroups of length α of a reduced valuated group G, then we say that H and K are <u>super close</u> if

$$H + G(\alpha) = K + G(\alpha).$$

Note that K and K' in Example 10.2 are super close.

THEOREM 10.3. **If H is super close to K, then H is isomorphic to K. If, moreover, H is nice, then K is nice.**

PROOF. For each h in H there is a unique k in K such that $v(h - k) \geq \alpha = \text{length } H$. This is the isomorphism.

If H is nice, let x be an element of G. We may assume that

$$v(x + K) \subset vK, \text{ so}$$

$$v(x + K) = v(x + K + G(\alpha)) = v(x + H + G(\alpha)),$$

which has a maximal element. □

THEOREM 10.4. Let H be a subgroup of a reduced valuated group G. If H has packed length ω, then any two closures of H in G are super close. If, moreover, length G = length H, then H has precisely one closure in G.

PROOF. Let α be the length of H. We first show that if $G(\alpha) = 0$, then there is precisely one closure of H. Let $K = \cap_{\theta<\alpha}(H + G(\theta))$. Clearly H is dense in K. To see that K is nice, let $g \in G$. If $g + H$ has an element of maximum value, this element has maximum value in $g + K$. If $g + H$ has no element of maximum value, then $g \in K$ because H has packed length ω. Clearly K contains any subgroup in which H is dense, while any nice subgroup containing H must contain K.

In the general case, let K and M be closures of H. If $k \in K$, then there is $m \in M$ maximizing the value of $k + m$. As $H \subset M$, and H is dense in K, we must have $v(k + m) \geqslant \alpha$. Hence $K \subset M + G(\alpha)$. Similarly $M \subset K + G(\alpha)$. □

LEMMA 10.5. Let $N_1, \cdots, N_n$ be nice subgroups of a reduced valuated group G. Let α_j be the length of N_j and suppose that if $x \in N_j$, then $vx \geqslant \alpha_{j-1}$, for $j = 2,\cdots,n$. Then $\sum N_j$ is nice and the sum is direct.

PROOF. The sum is direct because the value sets are disjoint. If $j < n$, then, as all values of elements of N_n are bigger than those of N_j, the image of N_j in G/N_n is nice and isomorphic to N_j. Thus we are done by induction on n. □

LEMMA 10.6. If a reduced group G admits a nice Warfield

decomposition into a finite number of pseudo-cyclics, then so does any summand of G.

PROOF. Let $K_1 \oplus \cdots \oplus K_m$ be a nice Warfield decomposition of G with each K_i a finite direct sum of pseudo-cyclics of the same length α_i. Set $\alpha_0 = 0$. We may assume that, for $i = 1, 2, \cdots, m$, the value of any element of K_i is at least α_{i-1}. Suppose $G = A \oplus B$, and let L_j be the projection of $A \cap (K_j \oplus \cdots \oplus K_m)$ on K_j. Then A permeates L_j, so A permeates the closure $\bar{L}_j$ of L_j in K_j. Thus any element of the projection of $\bar{L}_j$ on B has value at least α_j. Let N_j be the projection of $\bar{L}_j$ on A. Then N_j is super close to $\bar{L}_j$. Thus each N_j is nice and isomorphic to $\bar{L}_j$ by Theorem 10.3 so the values of elements in N_j are at least as big as the length of N_{j-1}. Therefore $\sum N_j$ is nice and the sum is direct by Lemma 10.5. To see that $A/\sum N_j$ is torsion we prove that if $va \geq \alpha_{j-1}$, then there is a positive integer r and x in N_j so that $v(p^r a - x) \geq \alpha_j$. Indeed choose r so that

$$p^r a \in K_1 \oplus \cdots \oplus K_m \quad \text{so} \quad p^r a \in K_j \oplus \cdots \oplus K_m;$$

choose y in L_j so that $v(p^r a - y) \geq \alpha_j$, and let x be the projection of y on A.

The Azumaya theorem in $C\alpha_j$ shows that N_j is isomorphic in $C\alpha_j$ to a finite direct sum $\sum H_i$ of pseudo-cyclics. Therefore there exist isomorphic nice full-rank subgroups $S \subset \sum H_i$ and $S' \subset N_j$. By Lemma 7.4 we can take S to be $\sum H_i(\theta)$ for some $\theta < \alpha_j$, so N_j contains a nice full-rank direct sum of pseudo-cyclics. □

It remains to show that the cokernel of the Warfield decomposition of the summand in Lemma 10.6 is simply presented if the original group was a generalized Warfield group. By the <u>rank</u> of a generalized Warfield group we mean the number of pseudo-cyclics in its Warfield decomposition.

LEMMA 10.7. **Let G be a reduced finite-rank generalized Warfield group, and M a nice full-rank subgroup of G of finite p-rank. Then G/M is simply presented.**

PROOF. Let K_i and α_i be as in the proof of Lemma 10.6 with the additional property that $G/(K_1 \oplus \cdots \oplus K_m)$ is simply presented (possibly $m = 0$). First we show that we can assume that $G(\alpha_m) = 0$.

Note that for $\alpha = \alpha_m$ and $K = K_1 \oplus \cdots \oplus K_m$ we have

$$G(\alpha) \cong \frac{G}{K}(\alpha) \text{ is Fuchsian and } \frac{G(\alpha)}{M(\alpha)} \cong \frac{G}{M}(\alpha)$$

As M has finite p-rank, and $M(\alpha)$ is torsion, $M(\alpha)$ is finite so $\frac{G}{M}(\alpha)$ is Fuchsian. Thus it suffices to show that $\frac{G}{M + G(\alpha)}$ is simply presented. Since $M/M(\alpha)$ has finite p-rank, and is a nice full-rank subgroup of the finite rank generalized Warfield group $G/G(\alpha)$, we may assume $G(\alpha) = 0$.

If $m = 0$, then $\alpha_m = 0$ so $G = 0$. If $m > 0$, then $H = M \cap K_m$ is a nice full-rank subgroup of K_m. As K_m has finite p-rank, H is of finite index in K_m by Theorem 7.3. So G/H is a finite rank generalized Warfield group, and M/H is a nice full-rank subgroup of G/H of finite p-rank. By induction on m, $\frac{G/H}{M/H} \cong G/M$ is simply presented. □

THEOREM 10.8. **A summand of a reduced finite-rank generalized Warfield groups is a finite-rank generalized Warfield group.**

PROOF. Lemma 10.6 provides a nice Warfield decomposition into a finite number of pseudo-cyclics. Lemma 10.7 says that the cokernel of this decomposition is a summand of a simply presented group. □

11. Leftovers. The definition of a pseudo-cyclic valuated group H has been continually changing during the past couple of years. Two conditions which have dropped out of the definition are

a) H/pH be finite, and

b) H have a local endomorphism ring in W.

Conditions a) and b), together with

c) H has a local endomorphism ring in C_α,

imply that H is pseudo-cyclic in the sense used in this paper. This implication was important when a), b) and c) constituted the definition of pseudo-cyclic, but became quite irrelevant later. However the tools developed in the proof of this implication are of some interest in their own right and have found other applications, so we include them here.

We say that a valuated group has the equal length property if any two nonzero subgroups have the same length. For torsion-free reduced valuated groups H such that dim H/pH is finite, this condition is equivalent to having packed length ω.

THEOREM 11.1. Let H be a reduced valuated group such that dim H/pH is finite. Then H has the equal length property if and only if H is torsion-free of packed length ω, or H is finite and all elements in H of order p have the same value.

PROOF. If H is torsion-free of packed length ω, or H is finite and all elements of order p have the same value, then H has the equal length property. Conversely, suppose H has the equal length property. If $\alpha <$ length H, then $H/H(\alpha)$ is torsion. Since H/pH is finite, $H(\alpha)$ is of finite index in H, so there are only finitely many ordinals less than α in vH. If H has limit length, then H is torsion-free and has packed length ω. Otherwise H is torsion, hence finite, with all elements of order p having value length $H - 1$. □

LEMMA 11.2. Let H be a valuated group such that $\dim H/pH$ is finite. Then there is x in H such that length $\langle x \rangle$ = length H.

PROOF. Passing to $H/H(\infty)$ we may assume H is reduced. If $H = 0$, choose $x = 0$. If H is torsion, then H is finite and we let x be a nonzero element of maximum value. Otherwise let y be an element of infinite order in H and set α = length $\langle y \rangle$. If α = length H, set $x = y$. If $\alpha <$ length H, Let $K/H(\alpha)$ be the torsion subgroup of $H/H(\alpha)$. Then $K/H(\alpha)$ is finite, so K is nice in H. Since $y \notin K$ and H/K is torsion-free reduced, $\dim K/pK < \dim H/pH$. Hence, by induction on $\dim H/pH$, there is x in K such that length $\langle x \rangle$ = length K = length H. □

THEOREM 11.3. Let H be a reduced valuated group such that $\dim H/pH$ is finite. Then H is isomorphic in W to a direct sum of valuated groups K with the equal length property such that $\dim K/pK$ is finite. Hence if H is torsion-free and has a local endomorphism ring in W, then H has packed length ω.

PROOF. As the torsion subgroup of H is finite, $p^n H$ is torsion-free for some n, so we may assume H is torsion-free. We may assume $H \neq 0$. If $K = \{x \in H : x = 0$ or length $\langle x \rangle$ = length $H\}$, then K is a nonzero (Lemma 11.2) pure subgroup of H. If $K = H$ we are done. Otherwise let L be any subgroup of H maximal with respect to $L \cap K = 0$. Then L is pure in H, so $\dim L/pL$ is finite, and $L \neq 0$. Thus by Lemma 11.2 there is y in L such that

$$\alpha = \text{length } L = \text{length } \langle y \rangle < \text{length } K = \text{length } H.$$

Then $K(\alpha) \oplus L$ is a decomposition of H in W. Clearly $K(\alpha)$ has the equal length condition and, since $K/K(\alpha)$ is torsion and $\dim K/pK$ is finite, $\dim K(\alpha)/pK(\alpha)$ is finite. But L is isomorphic to a direct sum of valuated groups with the equal length property by induction on length. □

Theorem 11.3, together with Theorem 6.1, show that if a reduced torsion-free valuated group H satisisfies conditions a), b) and c), then H is pseudo-cyclic.

REFERENCES

[ARN] Arnold, D., A duality for torsion-free modules of finite rank over a discrete valuation ring, Proc. London Math. Soc. 24(1972), 204-216.

[FUC1] Fuchs, L., Vector spaces with valuations, J. Algebra, 35(1975), 23-38.

[FUC2] ________, Infinite abelian groups, Academic Press, 1970

[HR] Hunter R., and F. Richman, Global Warfield groups, Trans. Amer. Math. Soc. (to appear)

[HRW1] Hunter, R., F. Richman, and E. Walker, Warfield modules, Abelian group theory, Springer Lecture Notes 616, 1977, 87-123.

[HRW2] ____________________________________, Existence theorems for Warfield groups, Trans. Amer. Math. Soc. 235(1978), 345-362.

[MOOR] Moore, J., Warfield groups and related topics, Ph.D thesis, New Mexico State University, 1980.

[RW] Richman, F., and E. Walker, Valuated groups, J. Algebra 56(1979), 145-167.

[RY] Rotman, J., and T. Yen, Modules over a complete discrete valuation ring. Trans. Amer. Math. Soc. 98(1961), 242-254.

[WALK] Walker, E., Ulm's theorem for totally projective groups, Proc. Amer. Math. Soc. 37(1973), 387-392.

[WALL] Wallace, K., On mixed groups of torsion-free rank one with totally projective primary components, J. Alg. 17(1971), 482-488.

[WARF] Warfield, R. B. Jr., Classification of abelian groups I, Balanced projectives, Trans. Amer. Math. Soc. 222(1976), 33-63.

[WW] Walker C., and R. B. Warfield, Jr., Unique decomposition and isomorphic refinement in additive categories, J. Pure Appl. Algebra 7(1976), 347-359.

The author was supported by NSF grant MCS 80-03060

NICE SUBGROUPS OF VALUATED GROUPS

Judy H. Moore

1. INTRODUCTION. All groups considered in this paper will be p-local abelian groups; that is, modules over Z_p, the integers localized at p. We will use valuated groups throughout, adopting the convention that a subgroup of a valuated group will always carry the induced valuation. The category V_p of p-local valuated abelian groups was studied by Richman and Walker in [4] and we will use their notation and definitions.

The desire to know when a subgroup of a direct sum of valuated cyclics is nice motivated the study reported here. Noting that every subgroup of a valuated cyclic is nice (Corollary 2), we prove that every subgroup of a finite direct sum of valuated cyclics is nice by proving that the class of valuated groups where every subgroup is nice is closed under finite direct sums (Theorem 1). Rotman and Yen [5] proved that for finite rank countably generated modules over a complete discrete valuation ring, every finitely generated submodule is nice. For rank one modules over a discrete valuation ring, Wallace [6] showed that a full rank submodule, and hence every finitely generated submodule, is nice. This result was generalized by Hunter and Richman [2], who proved that a finite subset of a decomposition basis generates a nice subgroup. We use their result to prove that finitely generated subgroups of valuated groups with decomposition bases are nice (Corollary 4). The remainder of the paper is devoted to a characterization of the class of valuated groups where every subgroup is nice.

The following terminology and notation will be used. For a valuated group A and a value α, let $A(\alpha) = \{a \in A \mid v(a) \geq \alpha\}$. We say that A is a <u>reduced</u> if $A(\infty) = 0$. The rank of a group A, denoted by $r(A)$, will refer to the torsion free rank of A. The torsion subgroup of A will be denoted by A_t. If $G = A \oplus B$, then we will use π_A and π_B to denote the projections of G onto A and B, respectively.

2. DIRECT SUMS OF VALUATED CYCLICS. For a subgroup A of a valuated group B, we say that A is nice in B if every coset of A has an element of maximal value. As a corollary to the following theorem, we see that every subgroup of a finite direct sum of valuated cyclics is nice.

THEOREM 1. Let A and B be valuated groups. If every subgroup of A is nice and every subgroup of B is nice, then every subgroup of $A \oplus B$ is nice.

PROOF. Let C be a subgroup of $A \oplus B$ and K a coset of C. Let $\lambda = \sup\{v(a,b) \mid (a,b) \in K\}$. Then we must show that $\lambda = v(a,b)$ for some (a,b) in K.

We first show that for each (a,b) in K, there is an (a^*,b^*) in K so that $v(a^*,b^*) \geq v(a,b)$ and if (c,d) is in K, then $v(c) \geq v(a^*)$ implies $v(d) \leq v(b^*)$ and $v(d) \geq v(b^*)$ implies $v(c) \leq v(a^*)$. Let

$$X = \{x \mid (x,y) \in C \text{ for some } y \text{ and } v(y) \geq v(b)\}.$$

Since X is nice in A, there exists $(x',y') \in C$ so that $v(a + x') \geq v(a + x)$ for each $x \in X$, and $v(y') \geq v(b)$. Set $a' = a + x'$ and $b' = b + y'$. Notice that $v(a',b') \geq v(a,b)$. Now let

$$Y = \{y \mid (x,y) \in C \text{ for some } x \text{ and } v(x) \geq v(a')\}.$$

Since Y is nice in B, there exists $(x'',y'') \in C$ so that $v(b' + y'') \geq v(b' + y)$ for each $y \in Y$ and $v(x'') \geq v(a')$. Set $a^* = a' + x''$ and $b^* = b' + y''$. Notice that $v(a^*,b^*) \geq v(a',b') \geq v(a,b)$. If (c,d) is in K with $v(c) \geq v(a^*)$, then $v(c) \geq v(a')$ so that $v(c - a') \geq v(a')$. Since $(c - a',d - b') \in C$, we know that $d - b'$ is an element of Y. Thus $v(d) = v(b' + d - b') \leq v(b^*)$. Similarly, if $v(d) \geq v(b^*)$, then $v(d) \geq v(b)$ so that $v(d - b) \geq v(b)$. Since $(c - a, d - b) \in C$, we know that $c - a \in X$. Thus $v(c) = v(a + c - a) \leq v(a') \leq v(a^*)$.

Let $K' = \{(a^*,b^*) \mid (a,b) \in K\}$. Then $\lambda = \sup\{v(a,b) \mid (a,b) \in K'\}$. Define

$$L = \{(a,b) \mid (a,b) \in K' \text{ and } v(a) \leq v(b)\} \text{ and}$$
$$U = \{(a,b) \mid (a,b) \in K' \text{ and } v(a) \geq v(b)\}.$$

Choose $(a',b') \in L$ so that $v(b') = \min\{v(b) \mid (a,b) \in L\}$. If (a,b) is in L, then $v(b') \leq v(b)$ which implies $v(a) \leq v(a')$. Thus $v(a,b) = v(a) \leq v(a') = v(a',b')$. Therefore $v(a',b') = \sup\{v(a,b) \mid (a,b) \in L\}$. Similarly, we can find (a'',b'') in U so that $v(a'',b'') = \sup\{v(a,b) \mid (a,b) \in U\}$. Therefore $\lambda = \max\{v(a',b'), v(a'',b'')\}$. □

COROLLARY 2. Every subgroup of a finite direct sum of valuated cyclics is nice.

PROOF. Since every subgroup of a valuated cyclic group is nice, the result follows by induction. □

COROLLARY 3. Let A and B be valuated groups. If every finitely generated subgroup of A is nice and every finitely generated subgroup of B is nice, then every finitely generated subgroup of

$A \oplus B$ is nice.

PROOF. Let C be a finitely generated subgroup of $A \oplus B$. Because $\pi_A(C)$ and $\pi_B(C)$ are finitely generated subgroups of A and B, respectively, $\pi_A(C) \oplus \pi_B(C)$ is nice in $A \oplus B$. Thus it suffices to prove that C is nice in $\pi_A(C) \oplus \pi_B(C)$. But every subgroup of $\pi_A(C)$ is finitely generated and thus nice, and every subgroup of $\pi_B(C)$ is finitely generated and thus nice. Hence C is nice. □

Before giving the next result, we recall a definition. A *decomposition basis* X of a valuated group G is a basis for a direct sum of valuated cyclics in G so that $G/\langle X\rangle$ is torsion.

COROLLARY 4. Every finitely generated subgroup of a valuated group with a decomposition basis is nice.

PROOF. Let G be a valuated group with a decomposition basis X. Let Y be a finitely generated subgroup of G. Since $G/\langle X\rangle$ is torsion and Y is finitely generated, there exists a finite subset X' of X and an n so that $nY \subset \langle X'\rangle$. By Corollary 2, nY is nice in $\langle X'\rangle$. But any finite subset of a decomposition basis is nice [2], so that nY must be nice in G. Since Y/nY is finite, we know that Y/nY is nice in G/nY. Hence Y is nice in G. □

3. VALUATED GROUPS WHERE EVERY SUBGROUP IS NICE. Theorem 1 says that the class of valuated groups where every subgroup is nice is closed under finite direct sums. A natural problem to consider at this point is the characterization of the valuated groups in this class. If G is a valuated group, then every subgroup of G is nice if and only if every subgroup of $G/G(\infty)$ is nice. Hence we

need only consider reduced valuated groups. Let B be a basic subgroup of a reduced valuated group G. Since G/B is divisible, either $B = G$ or B is not a nice subgroup of G. Thus if every subgroup of G is nice, G has only one basic subgroup which means that G is the group direct sum of a bounded group and a finite rank free group [1, Theorem 35.5]. Now we need information about the valuations allowed on such groups. The first lemma observes that we must have infinitely many values to produce a subgroup which is not nice. In fact, we will show that in the bounded case, every subgroup is nice precisely when there are only finitely many values.

LEMMA 5. Let G and H be valuated groups. If $v(H)$ is finite and $H \subset G$, then H is nice in G.

PROOF. Let β be the largest value of a nonzero element of H. By induction, we may assume $H/H(\beta)$ is nice in $G/H(\beta)$. Thus it suffices to show that $H(\beta)$ is nice in G. Let $g \in G$. We may assume $v(g) = \beta$ and that there exists an $h \in H$ with $v(h) = \beta$ so that $v(g + h) > \beta$. If $h' \in H(\beta)$ and $h' \neq h$, then

$$\begin{aligned} v(g + h') &= v(g + h + h' - h) \\ &= v(h' - h) \\ &= \beta < v(g + h). \end{aligned}$$

Hence $H(\beta)$ is nice in G. □

LEMMA 6. Let G be a reduced valuated p-group. If there exists an n so that $G[p^n]$ has infinitely many distinct values, then G has a subgroup which is not nice.

PROOF. By induction on n we may assume that $G[p^n]$ has infinitely many distinct values and $G[p^{n-1}]$ has only finitely many distinct values. Choose $\{x_i\} \subset G[p^n]$ with distinct values and no values in common with $G[p^{n-1}]$. We may assume that $i < j$ implies $v(x_i) < v(x_j)$. Define $X = \langle\{x_1 - x_i\}\rangle$. Now $v(x_1 + X) \geq \sup\{v(x_i)\}$. However, we will show that $v(x_1 + y) \leq v(x_i)$ for each $y \in X$. For y in X write $y = ax_1 + \sum r_i x_i$ where $\sum r_i = -a$. If $1 + a$ is a unit, then $v(x_1) < v(r_i x_i)$ for each i so $v(x_1 + y) = v(x_1)$. If $1 + a$ is not a unit, then a is a unit so $I = \{i \mid r_i \text{ is a unit}\}$ is not empty. Let m be the smallest element of I. Since $(1 + a)x_1 + \sum_{i \notin I} r_i x_i$ is in $G[p^{n-1}]$, we have $v(x_1 + y) \leq v(\sum_{i \in I} r_i x_i) = v(r_m x_m) = v(x_m)$. Thus X is not nice in G. □

PROPOSITION 7. Let G be a reduced valuated p-group. Then every subgroup of G is nice if and only if $v(G)$ is finite.

PROOF. There exists an n so that $G = G[p^n]$. If every subgroup of G is nice, then G has finitely many distinct values by Lemma 6. The converse is Lemma 5. □

Next we consider finitely generated torsion free valuated groups where every subgroup is nice. The following proposition gives a characterization in a special setting. Following Richman [3], we say that a valuated group has <u>packed length</u> ω if the order type of the set of values of nonzero elements is ω.

PROPOSITION 8. Let G be a finitely generated torsion free valuated group with packed length ω. Then every subgroup of G is nice if and only if the value topology on G is equivalent to the height topology on G.

PROOF. Since G has packed length ω, we may assume that $G(\omega) = 0$.

To prove necessity, let H be a subgroup of G. For $g \in G\backslash H$, there exists a k so that $ht(g + h) \leq k$ for all $h \in H$. Hence $g + H$ is in $G\backslash p^{k+1}G$. Since the value topology is equivalent to the height topology, there exists an n so that $g + H \in G\backslash G(n)$. Thus H is nice in G since $G(\omega) = 0$.

To prove the converse, let $\{x_1, x_2, \ldots, x_m\}$ be an independent set of generators of the group G. For each i, define $G_i = \langle x_1, \ldots, x_{i-1}, x_{i+1}, \ldots, x_m\rangle$. We must show that for every k, there exists an n so that $G(n) \subset p^kG$ or equivalently that $G\backslash p^kG \subset G\backslash G(n)$. Let $n_i = v(p^{k-1}x_i + G_i)$ and set $n = 1 + \max\{n_i\}$. If x is in G with $ht(x) = \alpha < k$, then writing $x = \Sigma r_i x_i$, we must have that $ht(r_j x_j) \geq \alpha$ for all j, and there exists an i so that $ht(r_i x_i) = \alpha$. Thus

$$v(x) = v(r_i x_i + \Sigma_{j\neq i} r_j x_j) \leq v(p^\alpha x_i + G_i) \leq n_i < n$$

so that $x \in G\backslash G(n)$ as desired. □

Observe from the proof of the proposition that for a finitely generated torsion free valuated group G with packed length ω, if the finite collection of subgroups $\{G_i\}$ defined above are nice in G, then every subgroup of G is nice.

The class of finite rank free groups with valuations so that all subgroups are nice contains the subgroups of valuated groups with finite decomposition bases. However, it is important to note that the class must be larger, as Example 10 will show. The following lemma gives a necessary condition for a valuated group to be a subgroup of a direct sum of valuated cyclics.

LEMMA 9. Let G be a finite direct sum of valuated cyclics and let y and z be elements of G. Then there exists an M so that

$v(ay + bz) = \min\{v(ay), v(bz)\}$ whenever $|ht(a) - ht(b)| > M$.

PROOF. Let $G = \oplus_{i=1}^{n}\langle x_i\rangle$ and let $y = \sum r_i x_i$ and $z = \sum s_i x_i$. Let $M = \max\{|ht(r_i) - ht(s_i)|\}$. If $v(a\pi_i y) = v(b\pi_i z)$ for some i, then

$$ht(a) + ht(r_i) = ht(b) + ht(s_i)$$

so that

$$|ht(a) - ht(b)| = |ht(r_i) - ht(s_i)| \leq M.$$

Therefore if $|ht(a) - ht(b)| > M$, then $v(a\pi_i y + b\pi_i z) = \min\{v(a\pi_i y), v(b\pi_i z)\}$ for every i. Thus

$$\begin{aligned} v(ay + bz) &= \min\{v(a\pi_i y + b\pi_i z)\} \\ &= \min\{\min\{v(a\pi_i y), v(b\pi_i z)\}\} \\ &= \min\{v(ay), v(bz)\}. \ \square \end{aligned}$$

EXAMPLE 10. A rank two torsion free valuated group which is not a subgroup of a valuated group with a finite decomposition basis.

Let $G = \langle y,z\rangle$ with valuation defined by

$$\begin{aligned} v(p^n y) &= 2n \\ v(p^n z) &= 4n, \text{ and} \\ v(p^n y + p^m z) &= 2n + 1 \text{ if } n = 2m. \end{aligned}$$

Every subgroup of G is nice by Proposition 8 and the fact that $G(4m) \subset p^m G \subset G(2m)$ for each m. However, if an integer M is given, $v(p^{2M+2}y + p^{M+1}z) = 4M + 5 > \min\{v(p^{2M+2}y), v(p^{M+1}z)\} = 4M + 4$ even though $|2M + 2 - (M + 1)| = M + 1 > M$. Hence Lemma 9 implies that ry and sz are not elements of a finite direct sum of valuated cyclics for any r and s in Z_p. Thus G cannot be a subgroup of a group with a finite decomposition basis. □

As we have mentioned before, Hunter and Richman proved that a finite subset of a decomposition basis generates a nice subgroup. Their argument centers around the proof that if x is an element of a decomposition basis then $\langle x\rangle$ is nice. Lemma 11 is a generalization of this.

LEMMA 11. Let $K \subset H \subset G$ where K is torsion free, packed length ω, and every subgroup of K is nice. If K is nice in H and G/H is torsion, then K is nice in G.

PROOF. Let $g \in G$. Since G/H is torsion, there exists an n so that $p^n g \in H$, and because K is nice in H, there exists a $k^* \in K$ so that $v(p^n g + k^*) = v(p^n g + K) = \alpha$. Let $S = \{v(g + k) \mid k \in K\}$. If there is an element k' of K so that $v(g + k') \notin v(K)$, then $v(g + k) \leqslant v(g + k')$ for all $k \in K$. Hence we may assume that $S \subset v(K)$. Since K has packed length ω, it suffices to find $\lambda \in v(K)$ so that λ is an upper bound for S.

Since α is an upper bound for S, we may assume that $K(\alpha) = 0$. Also, if $v(p^n k) \neq v(k^*)$ for any $k \in K$, then $v(k^*)$ is an upper bound for S. Thus we will assume that $v(k^*) = v(p^n k')$ for some $k' \in K$. Let $\beta = v(g + k')$. If $K(\beta) = 0$, then for each $k \in K$,

$$v(g + k) = v(g + k' + k - k') = v(k - k') < v(g + k').$$

So we will assume that $K(\beta) \neq 0$.

By proposition 8, the value topology and that height topology on K are equivalent, so there exists a τ so that $K\backslash K(\beta + 1) \subset K\backslash p^\tau K$

and a λ so that $K\backslash p^{\tau+n}K \subset K\backslash K(\lambda)$. If $v(g + k) > \beta$ for some k, then $v(k - k') = \beta$. Hence $k - k' \in K\backslash p^{\tau}K$ and $p^n(k - k') \in K\backslash p^{\tau+n}K$. Therefore $v(g + k) < v(p^n(k - k')) < \lambda$. □

LEMMA 12. Let H be a subgroup of a valuated group G so that every subgroup of H is nice and so that G/H is torsion with $v(G/H)$ finite. Then every subgroup of G is nice.

PROOF. By the previous lemma, H is nice in G. Let β be the largest value of a nonzero element of G/H. Since $G(\beta)$ is nice in G, we have that $(H + G(\beta))/G(\beta)$ is nice in $G/G(\beta)$. Also, if $S/G(\beta) \subset (H + G(\beta))/G(\beta)$, then $G(\beta) \subset S \subset H + G(\beta)$ so that $S/G(\beta)$ is nice in $(H + G(\beta))/G(\beta)$. Thus every subgroup of $(H + G(\beta))/G(\beta)$ is nice. If $v(g + H) = \beta$, then $v(g + H + G(\beta)) = \infty$ and if $v(g + H) < \beta$ then $v(g + H + G(\beta)) = v(g + H)$. Hence $v[(H + G(\beta)/G(\beta)] = v(G/H)\backslash\{\beta\}$. By induction on the cardinality of $v(G/H)$, it suffices to prove the lemma in the case where every nonzero element of G/H has the same value, say β.

Let $S \subset G$. Since $S \cap H$ is nice in H and H is nice in G, we have that S is nice in G precisely when $S/(S \cap H)$ is nice in $G/(S \cap H)$. Therefore we may assume that $S \cap H = 0$ which implies $S \subset G_t$. Hence we need only prove that G_t has finitely many values.

For each $g \in G\backslash H$, there exists an $h_g \in H$ so that $\beta = v(g + h_g) \geq v(g)$. If $g \in G_t$, then there exists an n so that $ng = 0$ and thus $v(n(g + h_g)) = v(nh_g) > \beta$. Therefore $h_g + H(\beta + 1)$ is in $[H/H(\beta + 1)]_t$. Since every subgroup of $H/H(\beta + 1)$ is nice, there are only finitely many values in $[H/H(\beta + 1)]_t$. But $v(g)$ is in $v[(H/H(\beta + 1))_t] \cup \{\beta\}$ so G_t has finitely many values as required. □

We are now ready to characterize those valuated groups where every subgroup is nice.

THEOREM 13. Every subgroup of a reduced valuated group G is nice if and only if G is an extension of a finite direct sum of valuated groups $\{G_i\}$ by a torsion group with a finite value set where

i) each G_i is a finitely generated torsion free group,

ii) each G_i has packed length ω, and

iii) the value topology on G_i is equivalent to the height topology on G_i.

PROOF. As was pointed out earlier, there is a bounded subgroup C and a finite rank free subgroup D so that $G = C + D$. But D is isomorphic, in the category W, to a direct sum of valuated groups of packed length ω [3]. Thus, there exists a finite number of finitely generated torsion free valuated groups G_i with packed length ω, so that $\oplus G_i \subset D$ and $D/(\oplus G_i)$ is finite. Since $G/\oplus G_i$ is torsion and every subgroup is nice, $G/\oplus G_i$ has only finitely many values by Proposition 7. It is clear that G_i satisfies conditions i) and ii). Proposition 8 implies that G_i satisfies iii).

To prove sufficiency, first note that every subgroup of G_i is nice by Proposition 8. By Theorem 1, every subgroup of $\oplus G_i$ is nice. Finally by Lemma 12, every subgroup of G is nice. □

BIBLIOGRAPHY

1. L. Fuchs, Infinite Abelian Groups, Vol. I, Academic Press, New York, 1970.

2. R. Hunter and F. Richman, "Global Warfield Groups", Trans. Amer. Math. Soc. (to appear).

3. F. Richman, "Mixed Local Groups", this volume.

4. F. Richman and E.A. Walker, "Valuated groups", J. Algebra, 56 (1979), 145 - 167.

5. J. Rotman and T. Yen, "Modules over a complete discrete valuation ring", Trans. Amer. Math. Soc. 98(1961), 242 - 254.

6. K. Wallace, "On mixed groups of torsion-free rank one with totally projective primary components", J. Algebra, 17(1971), 482 - 488.

DIAGRAMS OVER ORDERED SETS: A SIMPLE MODEL OF ABELIAN GROUP THEORY

Michael Höppner and Helmut Lenzing

Introduction. Diagrams on an ordered set I with values in the category R-Mod of modules over some ring R are usually investigated in the framework of category theory [16]. As was shown by Mitchell, a category of diagrams $\mathcal{D}$ = [I, R-Mod] satisfies nearly all the properties of a category of modules: it is therefore natural to consider $\mathcal{D}$ as the category of modules over a *ring with several objects* [17]. For investigations of diagram categories in this spirit of holomogical algebra and general module theory we refer to [17, 18, 3, 4, 5, 7, 12].

The aim of the present notes is to show that in case I = $\mathbb{Z}$ is the ordered set of integers and R = F denotes an arbitrary field, the category $\mathcal{D}$ = [$\mathbb{Z}$, F-Mod] inherits nearly all the properties of abelian group theory [9, 10, 15], more specifically of the category of *modules over a complete, discrete valuation ring* R. Therefore, in section 1 we give a short discription of the translation procedure from R-modules to diagrams. As an application, we consider Whitehead's problem for diagrams in section 3. As may be expected from the case of modules over a complete, discrete valuation ring the difficulties of Whitehead's problem, well-known in the group theory case [8, 20], will not appear in the case of diagrams.

Sections 2 and 4 are of a homological nature. In section 2 the case of global dimension 1, in section 4 flatness and injectivity as well as the determination of the weak global dimension are considered for diagram categories.

1. Vector space valued diagrams on the ordered set of integers

We want to consider diagrams

$$M : \cdots \longrightarrow M_{n-1} \xrightarrow{d} M_n \xrightarrow{d} M_{n+1} \longrightarrow \cdots$$

on the ordered set $\mathbb{Z}$ of integers with values in a category F-Mod of vector spaces over some fixed field F. Therefore, M consists of an infinite sequence of F-vector spaces together with a sequence of F-linear maps $d_n : M_n \to M_{n+1}$ $(n \in \mathbb{Z})$. It is convenient to denote all these maps by the same letter d and also by dM the subdiagram consisting of the subspaces $dM_{n-1} \subset M_n$. Similarly, Ker(d) denotes the subdiagram consisting of the subspaces $\mathrm{Ker}[M_n \xrightarrow{d} M_{n+1}]$ of M_n.

There is an obvious notion of morphism $u : M \to M'$ between diagrams. There results an abelian category $\mathcal{D} = [\mathbb{Z}, \text{F-Mod}]$ sharing all the usual properties with module categories (cf. [16]). Moreover, we want to indicate how the usual notions and properties of abelian group theory (cf. [9], [10], [15]), more precisely of the *theory of modules over a complete discrete valuation ring* R may be transferred to the category $\mathcal{D}$. As was shown by Ringel [19], in the case of tame, hereditary, finite dimensional F-algebras, the category of modules behaves in a similar fashion.

The following list gives a short description of the transfer from R-modules to diagrams.

R-modules		diagrams
R-module	R	$S_k(F): \cdots 0 \to F = F = F = \cdots$ = diagram constant with value F for indices $i \geq k$ and 0 elsewhere
Cyclic R-module	R/p^nR	$S_k(F)/d^nS_k(F) = S_{k,k+n}(F)$ = diagram constant with value F on the interval $[k,k+n-1]$ and 0 elsewhere
quotient field	Q	$\Delta(F) : \cdots = F = F = F \cdots$
Prüfer module	Q/R	$T_k(F) = \Delta(F)/S_{k+1}(F)$
divisible module		dM = M: diagram of epimorphisms

R-modules	diagrams
torsion-free module	Ker d = 0: diagram of monomorphisms
torsion module	$tM = \cup \operatorname{Ker}(d^n)$
module of bounded order	$d^nM = 0$ for some n
pure submodule	$U \subset M$ pure $\iff$ $d^kM \cap U = d^kU$, all k.
hight of an element	$ht(x) = \sup \{n \mid x \in d^nM\}$
p-adic topology	filtration $M_n \supset (dM)_n \supset (d^2M)_n \supset \dots$ on every M_n
complete, reduced R-module	$M \to \varprojlim_n M/d^nM$ is an isomorphism

With the aid of this translation between R-modules and diagrams it is easy to translate theorems 2 to 23 (with the exception of thms. 15, 18, 19) of Kaplansky's book [14] as well as their proofs almost verbatim to the case of diagrams.

In particular, the $S_n(F)$, $n \in \mathbb{Z}$, are a generating set of finitely generated projectives in $\mathcal{D}$. Since any subdiagram of $S_n(F)$ is of the form $S_k(F)$ for some $k \geq n$, $\mathcal{D}$ is noetherian and hereditary, i.e. subdiagrams of projectives [quotient diagrams of injectives] are projective [injective, resp.]. We refer to section 2 for the determination of all "hereditary" ordered sets I. Every projective diagram P is free, i.e. $P = \bigoplus_{n\in\mathbb{Z}} S_n(P_n)$ for some projective modules P_n. (According to [12] this is also true in arbitrary categories of diagrams.) With the aid of Baer's test [2] divisibility coincides with injectivity. Further, any injective diagram decomposes as a direct sum of $T_k(F)_s'$ and $\Delta(F)$'s. Torsion-free diagrams are exactly the flat diagrams, where flatness [and also purity] are defined via tensor-products (see section 4 for further information).

Moreover, any subdiagram of a direct sum of cyclics is again of this form.

We refer to [6] for an alternative proof of Kulikov's theorem in a categorical setting.

In order to illustrate the translation from R-modules to diagrams we give explicit proofs for two theorems in Kaplansky's book [15]: diagrams of bounded order and indecomposable diagrams. We feel free, however, to use some shortcuts due to the diagram situation.

1.1. Proposition. *Let* $M : \mathbb{Z} \to F\text{-Mod}$ *be a diagram of bounded order. Then* M *is a direct sum of cyclic diagrams. Consequently,* M *is* $\sum$*-algebraically compact.*

Proof. We denote by $\mathcal{D}^{(n)}$ the full subcategory of $\mathcal{D}$ consisting of all diagrams M satisfying $d^n M = 0$. Since $S_{k,n+k}(F)$, $k \in \mathbb{Z}$, is a generating set of finitely generated projectives for $\mathcal{D}^{(n)}$, we work in a noetherian situation. From

$$\mathrm{Hom}(M, S_{k,n+k}(F)) = \mathrm{Hom}(M_{k+n-1}, F), \quad \text{for } M \in \mathcal{D}^{(n)}$$

we further deduce the $\mathcal{D}^{(n)}$-injectivity of $S_{k,n+k}(F)$ for each k.

Consequently, M is the direct sum of suitably chosen $S_{k,n+k}(F)$'s and some diagram N in $\mathcal{D}^{(n-1)}$. Via induction, this proves the first claim. As a result, any diagram in $\mathcal{D}^{(n)}$ is algebraically compact [1], hence algebraically compact in $\mathcal{D}$ by the splitting criterion below. □

If $M : I \to F\text{-Mod}$ is a diagram, there is an obvious notion of F-dual

$$M^* : I^{op} \to F^{op}\text{-Mod}, \quad i \mapsto \mathrm{Hom}_F(M_i, F) .$$

It belongs to the folklore of the subject that M is algebraically compact if and only if the canonical embedding $M \to M^{**}$ splits. This immediately implies the following proposition.

1.2. Proposition. *Let* I *be an ordered set,* F *a field and* $M : I \to F\text{-Mod}$. *Suppose either*

(i) $M = \Delta_J(X)$ *for some subset* J *of* I *and* $X \in F\text{-Mod}$, *or*

(ii) every M_i *is finite dimensional.*

Then M *is algebraically compact.*

Here, $\Delta_J(X)$ denotes the diagram which is constant on J with value X (and identity maps) and 0 elsewhere. A special case for $J = \{k \mid k \geq i\}$ is the diagram $S_i(X)$. Since $(\Delta_J(X))^{(L)} = \Delta_J(X^{(L)})$, $\Delta_J(X)$ is even $\sum$-algebraically compact.

<u>1.3. Proposition.</u> *Any diagram* $0 \neq M : \mathbb{Z} \to$ F-Mod, F *a field, contains an indecomposable direct summand of the form* $S_k(F)$, $S_{k,k+n}(F)$, $T_k(F)$, *or* $\Delta(F)$.

<u>Proof.</u> We may assume that M is reduced. If $T = tM$ denotes the torsion diagram of M, T is pure in M since M/T is flat.

If $T \neq 0$, we have $T \neq dT$ since T is reduced. So there is an element $x \in \operatorname{Ker} d$ with $x \in d^nT \setminus d^{n+1}T$ for some n. If $y \in T$ with $d^ny = x$, the subdiagram U generated by x is pure in T, therefore also pure in M. By algebraic compactness, U is a direct summand of M.

If $T = 0$, M is torsion-free and reduced and possesses a cyclic direct summand by 3.2. □

We conclude this section with the remark that a diagram is complete if and only if it is algebraically compact and reduced. Further, the following example fits into the framework of algebraic compactness.

<u>1.4. Example.</u> *The diagram* $M = \prod_{n\in\mathbb{Z}} S_n(X_n) / \bigoplus_{n\in\mathbb{Z}} S_n(X_n)$ *is injective for every family* (X_n) *in* F-Mod.

2. The hereditary case

The following proposition which extends a result of Brune [5] includes a description of the hereditary categories [I, R-Mod].

<u>2.1. Proposition.</u> *The following are equivalent for any nondiscrete ordered set* I

(i) gl.dim[I, R-Mod] = 1 + gl.dim R *for every ring* R.

(ii) gl.dim[I, R-Mod] = 1 + gl.dim R *for some commutative, noetherian ring of finite global dimension.*

(iii) I *does neither contain* $\overline{2} \times \overline{2}$ *nor* $(\omega+1)^{op}$.

Here $\overline{2}$ denotes the ordered set $\{1 < 2\}$ and, as usual, ω denotes the ordered set of natural numbers.

Proof. (i) ⇒ (ii) is obvious. (ii) ⇒ (iii): If I contains $\overline{2} \times \overline{2}$, then $\overline{2} \times \overline{2}$ is a retract of I, and

$$\text{gl.dim } [I, R\text{-Mod}] \geq \text{gl.dim } [\overline{2} \times \overline{2}, R\text{-Mod}]$$

follows [17]. But $[\overline{2} \times \overline{2}, R\text{-Mod}] = [\overline{2},[\overline{2}, R\text{-Mod}]]$, and every step increases the global dimension about 1. Consequently, gl.dim [I, R-Mod] ≥ 2 + gl.dim R, a contradiction.

If $(\omega+1)^{op}$ is contained in I, it is a retract of I. Therefore, we may assume $I = (\omega+1)^{op}$ without loss of generality. If X is a finitely generated R-module of projective dimension n = gl. dim R, we deduce from Lemma 2.2 that the diagram

$$\Delta_{\mathbb{N}^{op}}(X) : (\omega+1)^{op} \to R\text{-Mod}$$

has projective dimension n+1. Since $S_\omega(X)$ has projective dimension n, we conclude that $S_\omega(X)/\Delta_{\mathbb{N}^{op}}(X)$ has projective dimension n+2 , a contradiction.

(iii) ⇒ (i): We only consider the case R = F is a field and refer to [5] for the general result. Since any subdiagram of $S_i(F)$, for i in I, has the form $U = \Delta_J(F)$ for some right open subset J, $i \leq J$, of I we conclude from (iii) that every connected component J_p of J contains a smallest element k_p. Hence, $U = \bigoplus_p S_{k_p}(F)$ is projective, and [I, F-Mod] is hereditary. □

2.2. Lemma. *Suppose* R *is a commutative, noetherian ring of finite global dimension* n *and* X *is a finitely generated* R*-module with* proj dim X = n. *Then* $\text{proj dim } \Delta_{\mathbb{N}^{op}}(X) = n+1$ *in* $[\mathbb{N}^{op}, R\text{-Mod}]$.

Proof. From $\Delta_{\mathbb{N}^{op}}(X) = \varinjlim_{n\in\mathbb{N}} S_n(X)$ we deduce proj dim $\Delta_{\mathbb{N}^{op}}(X) \leq 1 +$ proj dim X. In order to prove the converse inequality, we first reduce to the case that R is a regular local ring. If $F = R/m$ denotes the residue class field of R, F is an R-submodule of X [14]. If $Y = X/F$, it suffices to consider proj dim $\Delta(F)$ in $[\mathbb{N}^{op}, R\text{-Mod}]$ due to the exactness of

$$0 \to \Delta(F) \to \Delta(X) \to \Delta(Y) \to 0 .$$

From a variant of the well-known change of rings theorem of Kaplansky [14] we deduce

$$\text{proj dim } \Delta(X) \geq \text{proj dim } \Delta(F) = n+1.$$ □

We want to point out that in 2.1.(ii) some restriction on the ring R is necessary. For example, if R is a countable, von Neumann regular, non semi-simple ring, then gl.dim R = 1 and gl.dim $[(\omega+1)^{op}, R\text{-Mod}] = 2$, as follows from (4.3) together with the countability of R (compare [13]). But gl. dim $[(\omega+1)^{op}, \mathbb{Z}\text{-Mod}] = 3$ essentially as a consequence of (2.1). Similar effects on the "global dimension" of finite ordered sets have been observed by Spears [21] and Mitchell [17].

3. Whitehead's problem for diagrams

As was pointed out in section 1, diagrams $M : \mathbb{Z} \to F\text{-Mod}$, F a field, behave rather similar to modules over a complete discrete valuation ring. Here, we prove further results in this direction concerning the Whitehead problem.

M is called a *W-diagram* if it satisfies condition

$$\text{(W)} \quad \text{Ext}^1(M, S_n(F)) = 0 \quad \text{for each } n \in \mathbb{Z}.$$

Since we are working in a hereditary (cf. 2.1), noetherian situation, we immediately get the following properties:

(1) Any finitely generated W-diagram is projective.

(2) Any subdiagram of a W-diagram is a W-diagram, too.

(3) Any W-diagram is flat (= torsion-free).

Consequently, with $X = \bigcap_{n\in\mathbb{Z}} M_n$ we get for any W-diagram M

(4) $M = \Delta(X) \oplus N$, where N is a W-diagram satisfying $\bigcap_{n\in\mathbb{Z}} N_n = 0$.

3.1. Propositon. $M : \mathbb{Z} \to F\text{-Mod}$, F *a field, is a* W*-diagram if and only if* M *is torsion-free.*

Proof. Suppose, M is torsion-free. Since $S_n(F)$ is algebraically compact and M is flat, every exact sequence

$$0 \to S_n(F) \to D \to M \to 0$$

splits. □

3.2. Lemma. *Suppose* M *is torsion-free and reduced. Then any* $x \in M_n$ *is contained in a finitely generated direct summand* D *of* M.

Proof. By (4) $\bigcap M_n = 0$. We may therefore assume that $0 \neq x \in M_{k+1} \setminus M_k$ for some $k \in \mathbb{Z}$. Now the cyclic diagram D generated by x is pure in M and therefore a direct summand by (1.2). □

As an obvious consequence we get

3.3. Proposition. *If* M *is torsion-free, reduced and countably generated, then* M *is free.*

Alternatively, one may prove (3.3) by establishing Pontrjagin's criterion for freeness [8,9] by induction on rank(M).

3.4. Example. $M = \prod_{n\in\mathbb{Z}} S_n(F)$ *is a reduced* W*-diagram which is not free.*

Proof. M_0 is not countably generated as a vector space over F. Therefore, M is not countably generated as a diagram and therefore not free, since $M/dM = \bigoplus_{n\in\mathbb{Z}} S_n(F)/d\,S_n(F)$ is countably generated. □

For further information concerning Whitehead's problem for diagrams $M : I \to F\text{-Mod}$ on arbitrary hereditary ordered sets I (cf. 2.1) we refer to [11].

4. A general criterion for flatness and injectivity

It has been shown in [12] that a diagram $M : I \to R\text{-Mod}$ (for I an arbitrary ordered set and R an arbitrary ring) is projective if and only if M is free, i.e. $M = \bigoplus_{i \in I} S_i(P_i)$ for some projective R-modules P_i. We now present a general characterization of flatness. Recall that M is *flat*, if the tensor product functor (see [17]) $- \otimes M$: $[I^{op}, R^{op}\text{-Mod}] \to Ab$, $X \mapsto X \otimes M$ is exact. Equivalently, every map $F \to M$, with F finitely presented, admits a factorization through a (finitely generated) projective or even free diagram [22]. The following proposition extends a result of Cheng and Mitchell [7].

4.1. Proposition. *The following are equivalent for* $M : I \to R\text{-Mod}$.

(i) M *is flat.*

(ii)(a) M_i *is flat for each* i *of* I.

(b) The canonical map $\varinjlim_J M \to M_i$ *is a pure monomorphism for each* i *of* I *and every left open set* J *of predecessors of* i.

(iii) M *is a diagram of monomorphisms such that*

(a) M_i *is flat for each* i *of* I.

(b) If $i_1,\dots,i_n \le i$, *and* K *consists of all* $k \le i_1$ *s.t.* $k \le i_s$ *for some* $s = 2,\dots,n$ *then*

$M_{i_1} + \dots + M_{i_n}$ *is pure in* M_i *and*

$$M_{i_1} \cap (M_{i_2} + \dots + M_{i_n}) = \sum_{k \in K} M_k .$$

Proof. (i) $\Rightarrow$ (ii): M is a direct limit of free diagrams [22].
(ii) $\Rightarrow$ (iii): We only need to show condition (b). Denoting by J' (J'' resp.) the set of all predecessors of i_1 (of i_2 or ... or i_n, resp.) and setting $J = J' \cup J''$ we obtain an obvious exact sequence

$$(1)\quad 0 \to \Delta_K(N) \to \Delta_{J'}(N) \oplus \Delta_{J''}(N) \to \Delta_J(N) \to 0$$

in $[I^{op}, R^{op}\text{-Mod}]$. Tensoring with M we get the exactness of

$$0 \to \varinjlim_K M \to \varinjlim_{J'} M \oplus \varinjlim_{J''} M \to \varinjlim_J M \to 0$$

In view of (ii)(b) this reduces to the exactness of

$$0 \to \sum_{k\in K} M_k \to M_{i_1} \oplus (M_{i_2} + \ldots + M_{i_n}) \to M_{i_1} + \ldots + M_{i_n} \to 0,$$

from which we get (b). The proof (iii) ⇒ (i) is divided into three steps:

Step 1. If $J \leq i$ is the left open set generated by $i_1,\ldots,i_n$, an obvious induction proves $\varinjlim_J M = M_{i_1} + \ldots + M_{i_n}$. Therefore,

(2) $\qquad \varinjlim_J M \to M_i$

is a pure monomorphism. If $\Delta_J(X) : I^{op} \to R^{op}\text{-Mod}$ denotes the diagram, constant on J with value X and 0 elsewhere, we prove by induction on n that

(3) $\qquad \mathrm{Tor}_1(\Delta_J(X),M) = 0.$

$n = 1$ amounts to prove $\mathrm{Tor}_1(S_i(X), M) = \mathrm{Tor}_1(X,M_i) = 0$, which is obvious.

We now assume, that (3) is proved for (n-1)-generated left open subsets of I. In the notation of (ii) ⇒ (iii) we get from (1) with the aid of the induction hypothesis the exact sequence

$$0 \to \mathrm{Tor}_1(\Delta_J(X),M) \to X \otimes \varinjlim_K M \xrightarrow{\varphi} (X \otimes \varinjlim_{J'} M) \oplus (X \otimes \varinjlim_{J''} M)$$

where φ is a monomorphism because of (2), thus proving (3).

Step 2. We first observe that (3) holds for any left open subset J of I. Now, for $K \subset J \leq i$ left open subsets of I, define $S_{J,K}(X)$ by the exactness of

(4) $\qquad 0 \to \Delta_K(X) \to \Delta_J(X) \to S_{J,K}(X) \to 0.$

Tensoring with M we get the exactness of

$$0 \to \mathrm{Tor}_1(S_{J,K}(X),M) \to X \otimes \varinjlim_K M \to X \otimes \varinjlim_J M,$$

from which we deduce with the aid of (2) that

(5) $\qquad \mathrm{Tor}_1(S_{J,K}(X),M) = 0.$

Step 3. Any cyclic and finitely presented diagram $D : I^{op} \to R^{op}\text{-Mod}$ has the form $D = S_i(X)/U$, where $U = \Delta_{J_1}(X_1) + \ldots + \Delta_{J_r}(X_r)$ for suitably chosen submodules $X_1,\dots,X_r$ of X and left open subsets $J_1,\dots,J_r \leq i$. Since $n(U) = \# \{U_k \mid k \in I\}$ is finite, we may chose a maximal member, say V, among the U_k's. If K (L, resp.) denotes the set of elements k of I with $U_k = V$ ($U_k = X$, resp.), then $S_{K,L}(X/V)$ is contained in D and the quotient $\overline{D} = D/ S_{K,L}(X/V)$ has the form $\overline{D} = S_i(X)/\overline{U}$, where $\overline{U} = U + \Delta_K(X)$, and consequently $n(\overline{U}) < n(U)$.

Continuing in this fashion, we therefore obtain a filtration

$$0 = D_o \subset D_1 \subset \dots \subset D_p = D$$

of D, with all quotients D_s/D_{s-1} of the form $S_{J,K}(Y)$ for suitably chosen J, K and Y. Consequently, $\mathrm{Tor}_1(-,M)$ vanishes on all diagrams which are cyclic and finitely presented. Hence, M is flat. □

We now consider an ordered set I, and define

$$d_R(I) = 1 + \sup_J \text{flat dim } \Delta_J(R) ,$$

where the sup is taken over all left-open and right-bounded subsets J of I and flat dimension is considered in $[I^{op}, R^{op}\text{-Mod}]$.

<u>Proposition 4.2.</u> *Let* I *be a non-discrete ordered set and* R *be any ring, then*

$$\max(\text{w.gl.dim } R, d_R(I)) \leq \text{w.gl.dim } [I, R\text{-Mod}] \leq \text{w.gl.dim } R + d_R(I).$$

<u>Proof.</u> We only have to show the right hand inequality. Let $D : I \to R\text{-Mod}$ be a diagram, $d = d(I)$, $n = \text{w.gl.dim } R$, and

(1) $0 \to M \to P_{d-1} \to \dots \to P_o \to D \to 0$

(2) $0 \to K \to Q_{n-1} \to \dots \to Q_o \to M \to 0$

exact sequences with P_p und Q_q projective ($1 \leq p \leq d-1$, $1 \leq q \leq n-1$). If $S_{i,J}(R)$ is defined by the exactness of

(3) $0 \to \Delta_J(R) \to S_i(R) \to S_{i,J}(R) \to 0$

in $[I^{op}, R^{op}\text{-Mod}]$, then by definition of $d(I)$ we get

(4) $\mathrm{Tor}_p(S_{i,J}(R),M) = \mathrm{Tor}_{d+p}(S_{i,J}(R),D) = 0$ for each $p \geq 0$ and also

(5) $\mathrm{Tor}_1(S_{i,J}(R),K) = \mathrm{Tor}_{d+n+1}(S_{i,J}(R),D) = 0$

by the usual shifting argument.

Therefore, $\varinjlim_J K \to K_i$ is a monomorphism (see the proof of 4.1), and $K_i/\varinjlim_J K = S_{i,J}(R) \otimes K$ is flat, because (2) remains exact after tensoring with $S_{i,J}(R)$ as a consequence of (4). Notice further that all the $S_{i,J}(R) \otimes Q_q$ are flat and $n = \text{w.gl.dim } R$. By definition of K, K_i is flat for each i in I. Therefore K itself is flat as a consequence of 4.1. □

Corollary 4.3. *Let* I *be a non-discrete ordered set and* R *a von Neumann regular ring, then*

$$\text{w.gl.dim } [I, R\text{-Mod}] = d_R(I).$$

We refer to Brune [3] for the determination of all diagram categories of weak global dimension one.

Injective diagrams admit a description which is somewhat dual to 4.1.

4.4. Proposition. *Suppose either* I *satisfies the ascending chain condition or* R *is left noetherian. Then the following are equivalent for* $M : I \to R\text{-Mod}$

(i) M *is injective.*

(ii) (a) M_i *is injective for every* i *of* I.

(b) The canonical map $M_i \to \varprojlim_J M$ *is a direct epimorphism for any right open subset* J *in* I *with* $i \leq J$.

Proof. We first observe that (ii) is equivalent to the condition

(*) $\quad \text{Ext}^1(S_{i,J}(X), M) = 0$

for every left R-module X and $i \leq J$, J right open in I. Thus (ii) is satisfied by every injective diagram.

We now assume that (*) holds. By an obvious transfinite induction it is sufficient to prove that every diagram $D \neq 0$ has a subdiagram $S_{i,J}(X)$ for suitably chosen i,J and $X \neq 0$.

Let i be a maximal member of $\{k \in I \mid D_k \neq 0\}$, if I satisfies acc. Denoting by $E_i(D_i)$ the diagram with value D_i at i and 0 elsewhere, $E_i(D_i)$ is a subdiagram of D, which proves the claim in this case.

Now, let R be left noetherian. We may assume that D is cyclic, i.e. $D = S_p(R)/U \neq 0$, where $U = (U_i)_{i \in I}$ is a family of left ideals of R satisfying $U_i \subset U_k$ for $i \leq k$ in I. By hypothesis, U_i may be chosen to be maximal with respect to $U_i \neq R$. Denoting by J the set of all $k \geq i$ with $U_k = R$, we obtain that $S_{i,J}(R/U_i)$ is contained in D.

REFERENCES

[1] Baer, D.: Zerlegungen von Moduln und Injektive über Ringoiden. Archiv Math., to appear.

[2] Baer, R.: Abelian groups that are direct summands of every containing abelian group. Bull. Amer. Math. Soc. 46 (1940), 800-806.

[3] Brune, H.: Flache Darstellungen von geordneten Mengen. Manuscripta Math. 26 (1978), 141-154.

[4] ________ : Some left pure semisimple ringoids which are not right pure semisimple. Comm. Alg. 7 (1979), 1795-1803.

[5] ________: On projective representations of ordered sets. To appear.

[6] ________: On a theorem of Kulikov for artinian rings. To appear.

[7] Cheng, C.C. and B. Mitchell: Posets of homological dimension one J. pure appl. Alg. 13 (1978), 125-137.

[8] Eklof, P.C.: Whitehead's problem is undecidable. Amer. Math. Monthly 83 (1976), 775-788.

[9] Fuchs, L.: Infinite Abelian Groups I, II. Academic Press, New York 1970, 1973.

[10] Griffith, P.A.: Infinite Abelian Group Theory . The University of Chicago Press, Chicago, 1970.

[11] Höppner, M.: Homological properties of ordered sets.
In preparation.

[12] Höppner, M. and H. Lenzing: Projective diagrams over partially ordered sets are free.
J. pure appl. Alg. 20 (1981), 7-12.

[13] Jensen, C.U.: On homological dimension of rings with countably generated ideals.
Math. Scand. 18 (1966), 97-105.

[14] Kaplansky, I.: Fields and Rings. The University of Chicago Press, Chicago, 1965.

[15] ___________ : Infinite Abelian Groups. The University of Michigan Press, Ann. Arbor, 1971.

[16] Mitchell, B.: Theory of Categories.
Academic Press, New York, 1969.

[17] ___________: Rings with several objects.
Advances Math. 8 (1972), 1-161.

[18] ___________: Some applications of module theory to functor categories.
Bull. Amer Math. Soc. 84 (1978), 867-885.

[19] Ringel, C.M.: Infinite dimensional representations of finite dimensional herditary algebras.
Symposia Math. 23 (1979), 321-412.

[20] Shelah, S.: Infinite abelian groups, Whitehead's problem , and some constructions.
Israel J. Math. 18 (1974), 243-256.

[21] Spears, W.T.: Global dimension in categories of diagrams.
J. Alg. 22 (1972), 219-222.

[22] Stenström, B.: Purity in functor categories. J. Alg. 8 (1968), 352-361.

Kawada's theorem

Claus Michael Ringel

Kawadas's theorem solved the Köthe problem for basic finite-dimensional algebras: It characterizes completely those finite-dimensional algebras for which any indecomposable module has squarefree socle and squarefree top, and describes the possible indecomposable modules. This seems to be the most elaborate result of the classical representation theory (prior to the introduction of the new combinatorical and homological tools: quivers, partially ordered sets, vectorspace categories, Auslander-Reiten sequences). However, apparently his work was not appreciated at that time.

These are the revised notes of parts of a series of lectures given at the meeting on abelian groups and modules in Trento (Italy), 1980. They are centered around the second part of Kawada's theorem: the shapes of the indecomposable modules over a Kawada algebra.

1. Köthe algebras and algebras of finite representation type

Recall the following important property of abelian groups, thus of $\mathbb{Z}$-modules: every finitely generated module is a direct sum of cyclic modules. Köthe showed that the only commutative finite-dimensional algebras which have this property are the uniserial ones, and he posed the question to classify also the non-commutative finite-dimensional algebras with this property [11]. An algebra for which any finitely generated left or right module is a direct sum of cyclic modules, is now called a Köthe-algebra, and a classification of these algebras seems to be rather difficult. In fact, for a solution one would need a classification of all algebras of finite representation type, as well as some further insight into the structure of the modules over a given algebra of finite representation type.

(1.1) Notation. Let k be a (commutative) field, and A a finite-dimensional k-algebra (associative, with 1). We want to investigate the representations of A, thus we consider A-modules (usually, we will deal with finite-dimensional left A-modules and call them just modules). Always, homomorphisms will be written on the opposite side as scalars, thus the composition of $f : {}_AX \to {}_AY$ and $g : {}_AY \to {}_AZ$ will be denoted by fg. Given any module M, we denote by $\mathrm{rad}M$ the radical of M, it is the intersection of all maximal submodules, and call $M/\mathrm{rad}M =: \mathrm{top}M$ the top of M. If $\mathrm{rad}M = 0$, then M is called semisimple. Also, let $\mathrm{soc}M$ be the socle of M, it is the sum of all simple submodules of M. Any finite-dimensional A-module M has a composition series

$$0 = M_o \subset M_1 \subset \ldots \subset M_\ell = M ,$$

with M_i/M_{i-1} simple, for all $1 \le i \le \ell$. The M_i/M_{i-1} are called composition factors, and the number ℓ is called the length of M, denoted by $|M|$. (The module $grM := \bigoplus_{i=1}^{\ell} M_i/M_{i-1}$ will be called the graded module corresponding to M; we will need this construction later.) We choose a fixed ordering $S(1),\ldots,S(n)$ of the simple A-modules, and denote by $(\underline{\dim}M)_i$ the number of composition factors of M isomorphic to $S(i)$, this number is independent of the given composition series (theorem of Jordan-Hölder). In this way, we obtain an n-tupel $\underline{\dim}M$, called the dimension type of M. If M is semisimple and $(\underline{\dim}M)_i \leq 1$ for all i, then M is called squarefree. A module is semisimple and squarefree if and only if it is the direct sum of pairwise non-isomorphic simple modules. Again, assuming M to be finite-dimensional, then we can write M as a direct sum $M = \bigoplus_{i=1}^{m} M_i$ of indecomposable modules, and such a decomposition is unique up to isomorphism (theorem of Krull-Schmidt). In order to know all finite-dimensional modules, we therefore may restrict to the indecomposable ones. Note that a finite-dimensional module M is indecomposable if and only if its endomorphism ring $End(M)$ is local. In particular, we always have the indecomposable direct summands of the left module ${}_AA$, we denote representatives of their isomorphism classes by $P(1),\ldots,P(n)$, where $topP(i) = S(i)$, for $1 \leq i \leq n$. Thus, ${}_AA = \bigoplus_{i=1}^{n} P(i)^{p(i)}$ for some $p(i) \in \mathbb{N}$ (here, M^m denotes the direct sum of m copies of M), Note that we can calculate $\underline{\dim}M$ for any module M as follows:

$$(\underline{\dim}M)_i = |_{EndP(i)}Hom_A(P(i),M)| .$$

The projective modules are the direct sums of various $P(i)$, they are the modules with the usual lifting property. For any module M, there exists (uniquely up to isomorphism) an epimorphism $\varphi : P \to M$ with P projective and with kernel contained in $radP$, it is called the projective cover. If $\varphi: P \to M$ is a projective cover, then $topP \approx topM$. The (left) A-module ${}_AM$ is called cyclic provided it is an epimorphic image of ${}_AA$. Note that for a cyclic module M, we have $|M| \leq |_AA|$.

(1.2) The module M is cyclic if and only if $(\underline{\dim}\ topM)_i \leq p(i)$ for all i. Namely, let $P \to M$ be a projective cover of M. Then $P = \bigoplus_{i=1}^{n} P(i)^{m(i)}$, where $m(i) = (\underline{\dim}\ topM)_i$, since $topM = topP$. Now if $m(i) \leq p(i)$ for all i, then P is a direct summand of ${}_AA$, thus M is an epimorphic image of ${}_AA$. Conversely, if there exists an epimorphism ${}_AA \to M$, then P is isomorphic to a direct summand of ${}_AA$, thus $m(i) \leq p(i)$ for all i.

(1.3) The algebra A is said to be of finite representation type provided there are only finitely many indecomposable A-modules. (In this case, even the infinite-dimensional modules are direct sums of those finite-dimensional modules [17]). For example, the algebra $A = k[T]/(T^n)$, with $k[T]$ being the polynomial ring in one variable T and (T^n) the ideal generated by T^n, for some n, is of finite representation type: the only indecomposable modules being the modules $k[T]/(T^i)$, where $1 \leq i \leq n$. On the other hand, the three-dimensional algebra $k[T_1,T_2]/(T_1^2,T_1T_2,T_2^2)$ is not of finite representation type. There is a general theorem due to Rojter [18] which asserts that a finite-dimensional algebra with a bound on the length of the indecomposable modules, is necessarily of finite representation type. In particular, any Köthe algebra A has to be of finite representation type (here, $|_A A|$ is a bound for the length of the indecomposable A-modules).

(1.5) Conversely, one may ask when an algebra A of finite representation type actually is a Köthe algebra. Let $M_1,\ldots,M_m$ be the indecomposable left A-modules. As we have seen above, M_j is cyclic if and only if $(\underline{\dim}\,\mathrm{top}M_j)_i \le p(i)$, for all i. For any (left) module M, let $M^* = \mathrm{Hom}_k(M,k)$ be its dual module, it is a right A-module. Note that $M_1^*,\ldots,M_m^*$ are the indecomposable right modules, and it follows that M_j^* is cyclic if and only if $(\underline{\dim}\,\mathrm{soc}M_j)_i \le p(i)$, for all i. Let $q_A(i)$ be the maximum of all $(\underline{\dim}\,\mathrm{top}M_j)_i$ and all $(\underline{\dim}\,\mathrm{soc}M_j)_i$, where $1 \le j \le m$. Then, A is a Köthe algebra if and only if $q_A(i) \le p(i)$, for all i. If we replace A by a Morita equivalent algebra A', then there is a canonical bijection between the A-modules and the A'-modules. In particular, we may index the simple A-modules and the simple A'-modules in the same way. With A, also A' is of finite representation type, and $q_A(i) = q_{A'}(i)$. However, the numbers $p(i) = p_A(i)$ can be changed arbitrarily, by choosen an appropriate Morita equivalent algebra. For example, for the ring $M(d,A)$ of all $d \times d$-matrices over A, we have $p_{M(d,A)}(i) = dp_A(i)$, for all i. As a consequence, we see: Any algebra of finite representation type is Morita equivalent to a Köthe algebra.

(1.6) If $p_A(i) = 1$ for all i, then A is called a basic algebra. For any algebra A, there exists (uniquely up to isomorphism) a basic algebra A_o which is Morita equivalent to A. The following conditions now obviously are equivalent for an algebra A:

(i) A_o is a Köthe algebra.

(ii) Any algebra Morita equivalent to A is a Köthe algebra.

(iii) Any indecomposable A-module has squarefree top and squarefree socle.

An algebra A satisfying these conditions will be called a Kawada algebra.

2. The work of Kawada

These algebras which we now call Kawada algebras, were thoroughly investigated by Y. Kawada around 1960. He both gave a characterization of these algebras in terms of their indecomposable projective modules, as well as a full classification of the possible indecomposable modules.

(2.1) In 1960, Kawada reported his results at a meeting of the Mathematical Society of Japan, and a survey appeared in 1961 in two parts [1] : "The purpose of this paper is to announce that Köthe's problem mentioned above is completely solved for the case of self-basic algebras." This survey contains a set of 19 conditions which characterize Kawada algebras, as well as the list of the possible indecomposable modules.

One may formulate these two results separately, as Kawada did it in his survey. His proof however derives both results at the same time. This proof is published in a series of three papers [1] amounting altogether to 255 pages, and devoted just to this one theorem.

(2.2) The 19 conditions. These conditions are formulated in terms of the indecomposable projective A-modules and their submodules and factor modules. Let us give some examples: Condition VI has the shortest formulation (we use the notation introduced in 1): For any primitive idempotent e , the A-module Ae(radA)e is serial. Some of the conditions are, however, rather clumsy. We quote condition X:

> **X.** Assume that $Ae_{\lambda_1}g_1$ is a module such that $Ne_{\lambda_1}g_1 = Ae_\kappa te_{\lambda_1}g_1 + Ae_\rho we_{\lambda_1}g_1$ where $Ae_\kappa te_{\lambda_1}g_1$ is uni-serial, $Ae_\kappa te_{\lambda_1}g_1 \frown Ae_\rho we_{\lambda_1}g_1 = N^m e_\kappa te_{\lambda_1}g_1 = Ae_\sigma uwe_{\lambda_1}g_1 \neq 0 (m \geqq 1)$, $Ne_\rho we_{\lambda_1}g_1 = Ae_\sigma uwe_{\lambda_1}g_1 \oplus Ae_\tau vwe_{\lambda_1}g_1$ where $Ae_\tau vwe_{\lambda_1}g_1$ is uni-serial, and $S(Ae_{\lambda_1}g_1) = Ae_\sigma uwe_{\lambda_1}g_1 \oplus N^k e_\tau vwe_{\lambda_1}g_1 (k \geqq 0)$. Assume that $Ae_{\lambda_2}g_2$ is a non-simple module whose socle is isomorphic to $N^k e_\tau vwe_{\lambda_1}g_1$. Let φ be an isomorphism which maps $S(Ae_{\lambda_2}g_2)$ onto $N^k e_\tau vwe_{\lambda_1}g_1 + Ae_\kappa te_{\lambda_1}g_1 / Ae_\kappa te_{\lambda_1}g_1$ considered as a submodule of $Ae_{\lambda_1}g_1 / Ae_\kappa te_{\lambda_1}g_1$. Then φ is extendable; more precisely, either φ is extendable to a monomorphism $\Phi_1 : Ae_{\lambda_2}g_2 \to Ae_{\lambda_1}g_1 / Ae_\kappa te_{\lambda_1}g_1$, or φ^{-1} is extendable to a monomorphism $\Phi_2 : Ae_{\lambda_1}g_1 / Ae_\kappa te_{\lambda_1}g_1 \to Ae_{\lambda_2}g_2$.

(Here, the elements e_* are primitive idempotents, N = radA, and S(M) denotes the socle of M.)

Of course, one may reformulate these conditions in terms of the quiver with relations which defines A , at least in case the base field is algebraically closed. Then the conditions are more easy to visualize. For example, it is clear that any vertex a can have at most 4 neighbors, with at most two arrows having a as endpoint, and at most two arrows having a as starting point. Namely, otherwise, we obtain a subquiver of type D_4 with one of the orientations

or

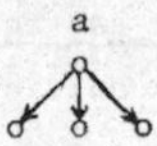

;

in the first case, we obtain an indecomposable module with socle $S(a)^2$, namely with dimension type $\begin{smallmatrix} 1 & 1 & 1 \\ & 2 & \end{smallmatrix}$, in the second case, we obtain an indecomposable module with top $S(a)^2$. Also, we see that we have to expect a rather long list of conditions. For example, we have to exclude subquivers of the form E_6 (with no relation) with all possible orientations. This is easy to formulate if one can use a diagrammatic language, however, it amounts to a large number of awkward conditions in terms of idempotents and serial modules.

(2.3) The possible indecomposable modules. The second part of Kawada's theorem describes completely the shape of the indecomposable modules over a Kawada algebra. Kawada first devides the indecomposable projective modules into 5 different types and then lists 38 possibilities of forming indecomposable modules as amalgamations of indecomposable projective modules. We want to present this list in a slightly different form. In order to do this, we first introduce the notion of the shape of a module.

3. The shape of a module

In order to define the shape of a module, we have to develop some of the machinery presently available in representation theory. Given a finite-dimensional algebra A, we will make use of its Auslander-Reiten-species $\Gamma(A)$, and the universal covering $\tilde{\Gamma}(A)$ of $\Gamma(A)$, as defined by Gabriel and Riedtmann. Since for an algebra A of finite representation type, $\tilde{\Gamma}(A)$ is the Auslander-Reiten-species of some "locally finite-dimensional algebra", we always have to take into account certain infinite-dimensional algebras (such an algebra will not contain a unit element).

(3.1) Locally finite-dimensional algebras. The k-algebra C is said to be locally finite-dimensional provided there exists a set $\{e_i \mid i \in I\}$ of orthogonal idempotents e_i of C such that $C = \bigoplus_{i,j\in I} e_iCe_j$, with Ce_i and e_iC finite-dimensional for every $i \in I$. For a C-module ${}_CM$, we require $CM = M$, or, equivalently, $M = \bigoplus_{i\in I} e_iM$. All modules considered will be assumed to be finite-dimensional over k. Note that we may and will assume that the idempotents e_i all are primitive, so that the left modules Ce_i are indecomposable. In case Ce_i and Ce_j are isomorphic as left C-modules only for $i = j$, we call C basic. As for finite-dimensional algebras, two locally finite-dimensional algebras will be called Morita equivalent in case their module categories are equivalent. And, given C, locally finite-dimensional, there exists a basic locally finite-dimensional algebra C_o which is Morita equivalent to C.

Assume now that C is locally finite-dimensional. For any module M, we define its support algebra $C(M)$ as the factor algebra of C modulo the ideal $\langle e \mid eM = 0,\ e^2 = e\rangle$ generated by all idempotents e with $eM = 0$. With M also $C(M)$ is finite-dimensional over k. Note that M is a sincere $C(M)$-module (i.e. no idempotent $\neq 0$ annihilates M). The $C(M)$-modules will be considered as C-modules, and the set of $C(M)$-modules is closed under submodules, factor modules and extensions. Clearly, the modules $P(i) = Ce_i$ are indecomposable and projective, and the modules $I(i) = (e_iC)^*$ indecomposable and injective. There is a categorial equivalence ν between the category of (finite-dimensional) projective modules and the category of (finite-dimensional) injective modules, with $\nu P(i) = I(i)$, the Nakayama functor (see [5]). Namely, maps $P(i) = Ce_i \to Ce_j = P(j)$ are given by right multiplication with elements from C of the form e_ice_j. But left multiplication by e_ice_j also gives a map $e_jC \to e_iC$, thus let $\nu(\cdot e_ice_j) = (e_ice_j\cdot)^*$. For any C-module M, there exists a projective cover $P(M) \twoheadrightarrow M$, and an injective envelop $M \hookrightarrow I(M)$.

We always have $\nu P = I(\text{top}P)$ for P projective.

Let us now describe the Auslander-Reiten translation: Let M be a C-module, and

$$P_1 \xrightarrow{p} P_o \longrightarrow M \longrightarrow 0$$

the first terms of a minimal projective resolution of M. Then, by definition, $\tau M = \text{Ker}\ \nu(p)$, thus we have the exact sequence

$$0 \longrightarrow \tau M \longrightarrow \nu P_1 \xrightarrow{\nu(p)} \nu P_o \quad .$$

Let $X = P_o \oplus P_1 \oplus \nu P_1 \oplus \nu P_o$. Then all modules involved in the construction of $\tau M = \tau_C M$ are in fact C(X)-modules, and we see that we have $\tau_{C(X)}M = \tau M$. Now assume that M is indecomposable and not projective. For the finite-dimensional algebra C(X), we have an Auslander-Reiten sequence

$$0 \longrightarrow \tau M \longrightarrow E \longrightarrow M \longrightarrow 0$$

and we claim that this sequence has the usual lifting properties with respect to all C-modules, not only the C(X)-modules. Namely, given a C-module Y, then we also may consider $C(X \oplus Y)$. Since $\tau_{C(X \oplus Y)}M = \tau M$, the Auslander-Reiten sequence ending with M, in the category of $C(X \oplus Y)$-modules, must be the given sequence (since it is characterized as being a socle element of ${}_{\text{End}(M)}\text{Ext}^1(M,\tau M)$. Thus, we may call the given sequence an <u>Auslander-Reiten sequence</u> in the category of C-modules.

Finally, we also will need the <u>dimension type</u> of a C-module. Let $\{P(i) \mid i \in I_o\}$ be a complete set of pairwise non-isomorphic indecomposable projective modules. The dimension type <u>dim</u>M of the C-module M will be an I_o-tuple, with

$$(\underline{\dim} M)_i = |{}_{\text{End}(P(i))}\text{Hom}_C(P(i),M)| \quad ,$$

or, equivalently, the number of composition factors of M of the form top $P(i)$.

(3.2) <u>Translation species.</u> We first need the notion of a translation quiver. A quiver (Γ_o,Γ_1) is given by a set Γ_o of "vertices" and a set Γ_1 of "arrows", to any arrow being assigned its starting point and its end point in Γ_o. A translation quiver (Γ_o,Γ_1,τ) is given by a locally finite quiver (Γ_o,Γ_1) without loops $\circ\!\circlearrowleft$ or multiple arrows $\circ\rightrightarrows\circ$ and with an injective function $\tau : \Gamma'_o \longrightarrow \Gamma_o$ where Γ'_o is a subset of Γ_o, such that for any $y \in \Gamma'_o$, we have $y^- = (\tau y)^+$. Here, y^- denotes the set of starting points of arrows with end point y, and y^+ the set of end points of arrows with starting point y. We denote by Γ'_1 the set of arrows $\alpha : x \to y$ with $y \in \Gamma'_o$. For every arrow $\alpha : x \to y$ in Γ'_1, there is a unique arrow $\tau y \to x$, and it will be denoted by $\sigma\alpha$. A translation species $(\Gamma_o,\Gamma_1,F,N,\tau,\chi)$ is given by the following data: (Γ_o,Γ_1,τ) is a translation quiver. For every $y \in \Gamma_o$, there is given a division ring $F(y)$, and for every arrow $x \xrightarrow{\alpha} y$ in Γ_1, there is given an $F(x)$-$F(y)$-bimodule $N(\alpha) = N(x,y)$, finite-dimensional on either side. Next, for every $y \in \Gamma'_o$, there also is given an isomorphism $\tau_y : F(y) \to F(\tau y)$. Finally, let $\alpha : x \to y$ be in Γ_1' , we may consider $N(\sigma\alpha)$ as an $F(y)$-$F(x)$-bimodule using the isomorphism τ_y . There is given a non-degenerate bilinear form

$$\chi_\alpha = \chi_{xy} : {}_{F(x)}N(\alpha)_{F(y)} \otimes {}_{F(y)}N(\sigma\alpha)_{F(x)} \to {}_{F(x)}F(x)_{F(x)} .$$

Note that $N(\sigma\alpha)$ is isomorphic as an $F(y)$-$F(x)$-bimodule to the left-dual $\mathrm{Hom}_{F(x)}(N(\sigma\alpha),F(x))$ of $N(\alpha)$, using the bilinear form χ_α . [However, in general it may not be possible to identify the division rings in an α-orbit in such a way that all maps τ_y are identity maps, since the τ-orbit may be closed. Similarly, it may not be possible to identify $N(\sigma\alpha)$ with $\mathrm{Hom}_{F(x)}(N(\alpha),F(x))$ for all α , so that the bilinear forms χ_α are the evaluation maps.] Also, the bilinear form χ_{xy} determines a unique element c_{xy} in $N(\tau y,x) \otimes_{F(x)} N(x,y)$ as follows: let $n_1,\dots,n_t$ be a basis of ${}_{F(x)}N(x,y)$ and $\varphi_1,\dots,\varphi_t$ the dual basis with respect to χ_{xy} , then $c_{xy} = \Sigma\, \varphi_i \otimes n_i$. This element c_{xy} is called the canonical element [2] .

Given a translation species $\Gamma = (\Gamma_o,\Gamma_1,F,N,\tau,\chi)$, we can construct the tensor category $\otimes\Gamma$ over Γ . It has Γ_o as set of objects. Given a pair $x,y \in \Gamma_o$, let $W(x,y)$ be the set of (oriented) paths from x to y in (Γ_o,Γ_1) , and if $w = (\underset{x}{o} \xrightarrow{\alpha_1} o \xrightarrow{\alpha_2} o \dots o \xrightarrow{\alpha_r} \underset{y}{o})$ is a path in $W(x,y)$, let $N(w) = N(\alpha_1) \otimes N(\alpha_2) \otimes \dots \otimes N(\alpha_r)$, the tensor products being taken with respect to the various $F(*)$; note that for the constant path w_x at the point x we have $N(w_x) = F(x)$. Now, for $x,y \in \Gamma_o$ define as set of homomorphisms from x to y the set $\bigoplus_{w\in W(x,y)} N(w)$, the composition being given by the tensor product. The <u>mesh category</u> $\diamond\Gamma$ is defined as the factor category of $\otimes\Gamma$ modulo the ideal generated by the elements $\sum_{x\in y^-} c_{xy}$ with $y \in \Gamma_o'$.

(3.3) <u>The universal covering of translation species</u>. Let (Γ_o,Γ_1,τ) be a translation quiver. A covering of (Γ_o,Γ_1,τ) is given by a translation quiver (Δ_o,Δ_1,τ) and a quiver map $\pi : (\Delta_o,\Delta_1) \to (\Gamma_o,\Gamma_1)$ which is compatible with τ , with $\pi^{-1}(\Gamma_o') = \Delta_o'$ and such that the induced maps $x^+ \to (\pi x)^+$ and $x^- \to (\pi x)^-$ are bijective, for any $x \in \Delta_o$, see [6] . Now assume a translation species $(\Gamma_o,\Gamma_1,F,N,\tau,\chi)$ is given, and (Δ_o,Δ_1,τ) is a covering of (Γ_o,Γ_1,τ) with covering map π . Then we can construct a translation species $(\Delta_o,\Delta_1,F,N,\tau,\chi)$ with underlying translation quiver (Γ_o,Γ_1,τ) as follows: For $y \in \Delta_o$, let $F(y) = F(\pi y)$; for $\alpha \in \Delta_1$, let $N(\alpha) = N(\pi\alpha)$; for $y \in \Delta_o'$, let $\tau_y = \tau_{\pi y} : F(y) = F(\pi y) \to F(\tau\pi y) = F(\pi\tau y) = F(\tau y)$, and for α in Δ_1' , let $\chi_\alpha = \chi_{\pi\alpha}$. Thus, all the data of a translation species are lifted back via π . We call $(\Delta_o,\Delta_1,F,N,\tau,\chi)$ a <u>covering</u> of $(\Gamma_o,\Gamma_1,F,N,\tau,\chi)$.

Any translation quiver (Γ_o,Γ_1,τ) has a universal covering, as Gabriel and Riedtmann (see [6]) have shown; it will be denoted by $(\widetilde{\Gamma}_o,\widetilde{\Gamma}_1,\tau)$. Given a translation species

$\Gamma = (\Gamma_o,\Gamma_1,F,N,\tau,\chi)$, we therefore can consider $\tilde{\Gamma} = (\tilde{\Gamma}_o,\tilde{\Gamma}_1,F,N,\tau,\chi)$ and call it the universal covering of Γ . Note that there exists a group G of automorphisms of $(\tilde{\Gamma}_o,\tilde{\Gamma}_1,\tau)$ such that $(\Gamma_o,\Gamma_1,\tau) = (\tilde{\Gamma}_o,\tilde{\Gamma}_1,\tau)/G$. By construction, G is also a group of automorphisms of the translation species $\tilde{\Gamma}$, and clearly $\Gamma = \tilde{\Gamma}/G$.

Note that for the universal covering $(\tilde{\Gamma}_o,\tilde{\Gamma}_1,\tau)$ of a translation quiver, neither any τ-orbit of $\tilde{\Gamma}_o$, nor any σ-orbit of $\tilde{\Gamma}_1$ is closed. Therefore, for the universal covering $(\tilde{\Gamma}_o,\tilde{\Gamma}_1,F,N,\tau,\chi)$, we can assume that all maps τ_y for $y \in \tilde{\Gamma}'_o$ are identity maps (we choose one representative y in any τ-orbit of $\tilde{\Gamma}_o$, fix $F(y)$, and identify the division rings corresponding to the remaining vertices in this τ-orbit with $F(y)$, using the maps τ_*). Similarly, in any σ-orbit of $\tilde{\Gamma}_1$, we select one arrow α , fix the corresponding bimodule $N(\alpha)$, and replace the bimodules corresponding to the remaining arrows in this σ-orbit by suitable dualized forms of $N(\alpha)$.

(3.4) Auslander-Reiten-species. Let C be a locally finite-dimensional algebra. We denote by $\Gamma(C) = (\Gamma_o,\Gamma_1,F,N,\tau,\chi)$ its Auslander-Reiten-species, which is defined as follows: Γ_o is a fixed set of representatives of the isomorphism classes of indecomposable C-modules. For any $X \in \Gamma_o$, let $F(X) = \mathrm{End}(X)/\mathrm{rad}\,\mathrm{End}(X)$, the residue division ring of $\mathrm{End}(X)$, and τX its Auslander-Reiten translate. For $X,Y \in \Gamma_o$, let $N(X,Y) = \mathrm{rad}(X,Y)/\mathrm{rad}^2(X,Y)$ the bimodule of irreducible maps [15], it is an $F(X)$-$F(Y)$-bimodule; given $X,Y \in \Gamma_o$, there is an arrow $X \to Y$ in Γ_1 if and only if $N(X,Y) \neq 0$; also, Γ'_o is the set of non-projective modules in Γ_o , and τ is the Auslander-Reiten translation.

In order to be able to define the isomorphisms τ_Y and the bilinear forms χ_{XY} , we have to consider Auslander-Reiten sequences. Let $Y \in \Gamma'_o$. Since Y is indecomposable and non-projective, there exists an Auslander-Reiten sequence

$$0 \longrightarrow \tau Y \xrightarrow{(f_{X,i})} \bigoplus_{X\in Y^-} X^{d(X)} \xrightarrow{(g_{X,i})} Y \longrightarrow 0 ,$$

here, both $(f_{X,i})$ and $(g_{X,i})$ are indexed by $X \in Y^-$ and $1 \leq i \leq d(X)$. We fix some $X \in Y^-$ and write $f_i = f_{X,i}$, $g_i = g_{X,i}$. Then $(f_i)_i$ gives modulo $\mathrm{rad}^2(\tau Y,X)$ a basis of $N(\tau Y,X)_{F(X)}$, and $(g_i)_i$ gives modulo $\mathrm{rad}^2(X,Y)$ a basis of ${}_{F(X)}N(X,Y)$, see [15]. Any automorphism of Y lifts to an automorphism of the middle term $\bigoplus_{X\in Y^-} X^{d(X)}$, and therefore induces an automorphism of τY . In this way, we define an isomorphism $\tau_Y : F(Y) \to F(\tau Y)$. More precisely, given $h \in F(Y)$ and g_i , there are elements $h_{ij} \in F(X)$ satisfying $\sum_j h_{ij}g_j = g_i h$, and then $\tau(h)\cdot f_j = \sum_i f_i h_{ij}$. Now define for $u_i,v_i \in F(X)$

$$\chi_{XY}\Big(\sum_i u_i g_i , \sum f_i v_i\Big) = \sum_i u_i v_i ,$$

and note that this factors over the tensor product $N(X,Y) \otimes_{F(Y)} N(\tau Y,X)$. Namely, if

$h \in F(Y)$, then

$$\chi_{XY}(g_i h, f_j) = \chi_{XY}(\sum_r h_{ir} g_r, f_j) = h_{ij}$$

and

$$\chi_{XY}(g_i, \tau(h) f_j) = \chi_{XY}(g_i, \sum_r f_r h_{rj}) = h_{ij} .$$

Thus, χ_{XY} is a bilinear form on ${}_{F(X)}N(X,Y) \underset{F(Y)}{\otimes} N(\tau Y,X)_{F(X)}$ with values in $F(X)$.

(3.5) Coverings of an Auslander-Reiten species. Given a finite-dimensional algebra A of finite representation type, we may consider coverings Δ of the Auslander-Reiten species $\Gamma(A)$ of A. Gabriel and Riedtmann [6] have shown (at least in the case when k is algebraically closed) that Δ is again the Auslander-Reiten species of a suitable locally finite-dimensional algebra. In fact, one can construct such an algebra $R(\Delta)$ as follows: First, consider the mesh category $C = \diamond\Delta$ over Δ. Given any (not necessarily finite) family J of objects in C, let $\mathrm{End}_o(J)$ be the ring of all row-and-column finite matrices indexed by J, with entries in the x-y-position (where $x,y \in J$) from $\mathrm{Hom}_C(x,y)$, and with matrix multiplication, using the composition in C. Now, let $R(\Delta) = \mathrm{End}_o(\Delta_o \setminus \Delta'_o)$. (Note that $\Gamma_o \setminus \Gamma'_o$ is a complete set of indecomposable projective A-modules, and we want that $\Delta_o \setminus \Delta'_o$ becomes a complete set of indecomposable projective $R(\Delta)$-modules.) The result of Gabriel and Riedtmann can be formulated as follows:

Proposition: Let A be a finite-dimensional algebra of finite representation type. Let Δ be a covering of the Auslander-Reiten species $\Gamma(A)$ of A. Then $R(\Delta)$ is locally finite-dimensional, and $\Gamma(R(\Delta)) = \Delta$.

Two special cases of this proposition are of particular interest. First of all, let $\Delta = \Gamma(A)$. Note that the algebra A and $R(\Gamma(A))$ are not necessarily isomorphic, not even in case k is algebraically closed. Algebras of the form $A = R(\Gamma(A))$ have been called standard, and for any algebra A of finite representation type, there is associated $R(\Gamma(A))$, its standard form. (Note that for algebraically closed k, the algebra $R(\Gamma(A))$ is nothing else than Kupisch's "Stamm-Algebra" [14] of the Auslander algebra of A.)

The other special case we are interested in is the case of $\Delta = \tilde{\Gamma}(A)$, the universal covering of $\Gamma(A)$. We will denote $R(\tilde{\Gamma}(A))$ by $\tilde{A}$. Note that $\tilde{A}$ usually will not be finite-dimensional. This was the reason for considering the more general class of locally finite-dimensional algebras from the beginning. Also note that $\tilde{A}$ can be shown to be the only basic locally finite-dimensional algebra satisfying $\Gamma(\tilde{A}) = \tilde{\Gamma}(A)$.

(3.6) The type of an $\tilde{A}$-module. Let A be a finite-dimensional algebra and $\tilde{A}$ the basic locally finite-dimensional algebra with $\Gamma(\tilde{A}) = \tilde{\Gamma}(A)$. Recall that for any $\tilde{A}$-module X, the support algebra of X is denoted by $\tilde{A}(X)$. Now $\tilde{A}(X)$ is a finite-dimensional algebra, of finite representation type, since with $\tilde{A}$ also $\tilde{A}(X)$

is of bounded representation type. Also the Auslander-Reiten quiver of $\tilde{A}(X)$ has no oriented cycles, since otherwise we would obtain an oriented cycle in $\tilde{\Gamma}(A)$. As a consequence, we can apply the results of [8]. In particular, we see that if X,Y are indecomposable $\tilde{A}$-modules with $\dim X = \dim Y$, then X,Y are isomorphic. Also, we have the following: By definition, the $\tilde{A}(X)$-module X is always sincere. Thus, if X is indecomposable, then X is a faithful $\tilde{A}(X)$-module. Both results rest on the fact that for X indecomposable, the algebra $\tilde{A}(X)$ is a tilted algebra in the sense of [8]. In fact, we have the following proposition [8]:

Proposition. Let X be an indecomposable $\tilde{A}$-module. Then there exists a basic, hereditary, finite-dimensional algebra H, a tilting module T_H, and a primitive idempotent e of H such that $\tilde{A}(X) = \mathrm{End}(T_H)$, ${}_{\tilde{A}(X)}X = {}_{\tilde{A}(X)}Te$ and moreover He is the only simple projective H-module. Also, H, T_H, and e are uniquely determined by X.

The algebra H is called the type of X.

Let us indicate in which way H, T_H, e are constructed: One constructs a "complete slice" $\mathcal{T}$ in the Auslander-Reiten quiver of $\tilde{A}(X)$ with X being the unique sink of $\mathcal{T}$, and defines $T = \bigoplus_{T_i \in \mathcal{T}} T_i$, $H = \mathrm{End}({}_{\tilde{A}(X)}T)$, with e being the projection onto the direct summand X of T.

(3.7) The shape of an A-module. Again, let A be a finite-dimensional algebra, and $\tilde{A}$ the basic locally finite-dimensional algebra with $\tilde{\Gamma}(A) = \Gamma(\tilde{A})$. Also, denote by π the covering map $\tilde{\Gamma}(A) \to \Gamma(A)$, note that π is surjective, and let G be the group of automorphisms of $\tilde{\Gamma}(A)$ such that $\tilde{\Gamma}(A)/G = \Gamma(A)$.

Now let M be an indecomposable A-module. Since π is surjective, there exists some $\tilde{M}$ in $\tilde{\Gamma}_o$ such that $\pi(\tilde{M}) = M$. Let $\tilde{A}(\tilde{M})$ be the support algebra of $\tilde{M}$, and consider $\tilde{M}$ as an $\tilde{A}(\tilde{M})$-module. The pair $(\tilde{A}(\tilde{M}),\tilde{M})$, or also the pair $(\tilde{A}(\tilde{M}),\dim\tilde{M})$ will be called the shape of M. Note that the shape of M is independent of the choice of $\tilde{M}$. For, any other inverse image of M under π is of the form $\tilde{M}^g$, with $g \in G$, and clearly g defines an isomorphism between the corresponding support algebras and also between the $\tilde{A}(\tilde{M})$-module $\tilde{M}$ and the $\tilde{A}(\tilde{M}^g)$-module $\tilde{M}^g$.

(3.8) Example. Let A be the matrix algebra

$$A = \left\{ \begin{pmatrix} a & b & 0 \\ 0 & c & 0 \\ 0 & d & c \end{pmatrix} \mid a \in \mathbb{R} ; b,c,d \in \mathbb{C} \right\} .$$

There are two simple A-modules S(1) and S(2), with $\mathrm{End}\, S(1) = \mathbb{R}$, $\mathrm{End}\, S(2) = \mathbb{C}$. The A-modules which do not split off a copy of S(1) are given by a $\mathbb{C}$-vectorspace V endowed with an endomorphism φ satisfying $\varphi^2 = 0$, and an $\mathbb{R}$-subspace U contained in the kernel of φ, thus we will use the notation (V,U,φ). There are the indecomposable projective modules

$$P(1) = (\mathbb{C}, \mathbb{R}, 0) \ , \ P(2) = (\mathbb{C}\mathbb{C}, 0, \begin{pmatrix} 0 & 1 \\ 0 & 0 \end{pmatrix}) \ ,$$

the indecomposable injective modules

$$I(1) = S(1) \ , \ I(2) = (\mathbb{C}\mathbb{C}, \ \mathbb{C}0, \begin{pmatrix} 0 & 1 \\ 0 & 0 \end{pmatrix}) \ ,$$

and three additional indecomposable modules of length > 1

$$M_1 = (\mathbb{C}\mathbb{C}\mathbb{C}, \mathbb{R}\,\mathbb{R}\,0, \begin{pmatrix} 0 & 0 & 1 \\ 0 & 0 & i \\ 0 & 0 & 0 \end{pmatrix}) \ , \ M_2 = (\mathbb{C}, \mathbb{C}, 0) \ , \ M_3 = (\mathbb{C}\mathbb{C}, \mathbb{R}0, \begin{pmatrix} 0 & 1 \\ 0 & 0 \end{pmatrix}) \ .$$

The Auslander-Reiten species of A is of the form

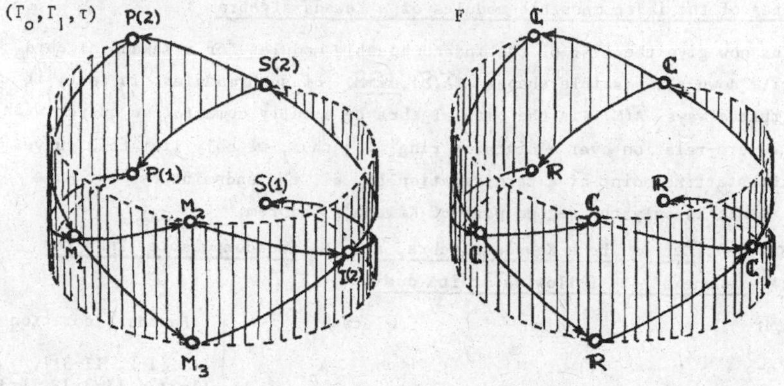

The arrows $\alpha : x \to y$ are always endowed with the bimodule $N(\alpha) = {}_{F(x)}\mathbb{C}_{F(y)}$, the bilinearforms ${}_{\mathbb{C}}\mathbb{C}_{F(y)} \otimes {}_{F(y)}\mathbb{C}_{\mathbb{C}} \to {}_{\mathbb{C}}\mathbb{C}_{\mathbb{C}}$ are the multiplication map, those of the form ${}_{\mathbb{R}}\mathbb{C}_{\mathbb{C}} \otimes {}_{\mathbb{C}}\mathbb{C}_{\mathbb{R}} \to {}_{\mathbb{R}}\mathbb{R}_{\mathbb{R}}$ are given by a projection ${}_{\mathbb{R}}\mathbb{C}_{\mathbb{R}} \to {}_{\mathbb{R}}\mathbb{R}_{\mathbb{R}}$.

The universal covering $\widetilde{\Gamma}(A)$ is of the form

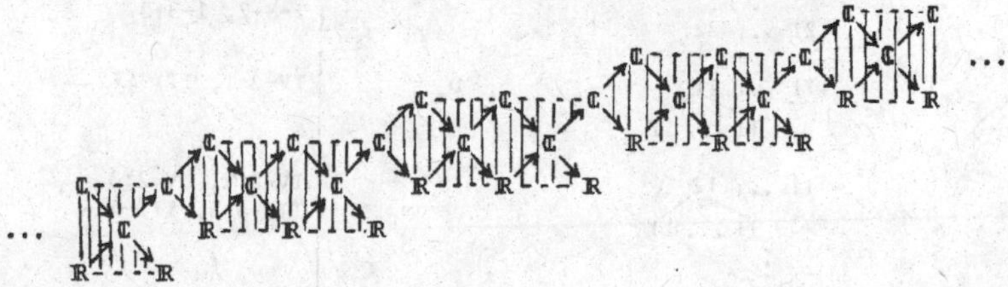

with the same description of the bimodules and bilinear forms. Finally, $\widetilde{A}$ is given by the species

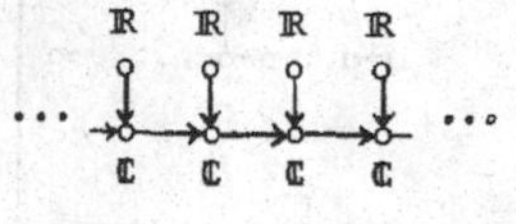

again with bimodules ${}_{\mathbb{R}}\mathbb{C}_{\mathbb{C}}$ and ${}_{\mathbb{C}}\mathbb{C}_{\mathbb{C}}$, and with all compositions being zero.

Let us determine the shape of the A-module M_1. An inverse image of M_1 under the covering map is given by $\tilde{M}_1$ with dimension type $\cdots\begin{smallmatrix}0&2&0&0\\0&2&1&0\end{smallmatrix}\cdots$, thus $\tilde{A}(\tilde{M})$ is the hereditary algebra

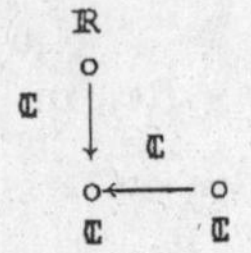

and $\tilde{A}(\tilde{M}_1)\tilde{M}_1$ is the indecomposable representation of dimension type $\begin{smallmatrix}2&\\2&1\end{smallmatrix}$.

4. The shapes of the indecomposable modules of a Kawada algebra

(4.1) Let us now give the list of the indecomposable modules for a Kawada algebra. We will write down all possible shapes $(\tilde{A}(\tilde{M}), \underline{\dim}\tilde{M})$ of such modules. In fact, it turns out that always $\tilde{A}(\tilde{M})$ is the path algebra of a fully commutative quiver with at most one zero-relation over a division ring F ; thus, we only list this quiver and mark the starting point of a zero-relation by ● , the endpoint by ■ . Note that this result is only the second part of Kawada's theorem.

Theorem (Kawada). Let A be a Kawada algebra, and M indecomposable. Then the shape of M is one of the following or its dual:

$\tilde{A}(\tilde{M})$	$\underline{\dim}\tilde{M}$	type	Kawada's notation
o-o-o ... o-o	111 ... 11	A_n	I-1, I-2, II-3·1, II-3·2, II-3·3, II-3·4
o←o←o ... o←o←o (with o attached below)	111 ... 111 / 1	D_n	I-4·1, II-2·1, II-1·1, I-3·1
	111 ... 121 / 1 ⋮ 121 ... 221 / 1		I-4·2 II-2·2, II-2·3 II-1·2 I-3·2, I-3·3
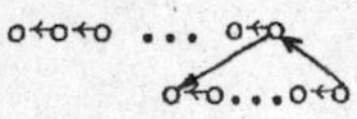	111 ... 11 / 11 ... 11	D_n	IV-2·1, III-4·2
	111 ... 12 / 11 ... 11 ⋮ 122 ... 22 / 11 ... 11		IV-2·2, IV-2·3 III-4·3, III-4·4
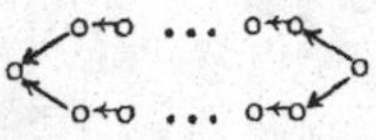	11 ... 11 / 1 1 / 11 ... 11	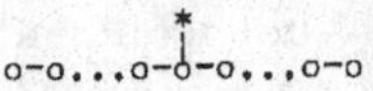	III-4·1
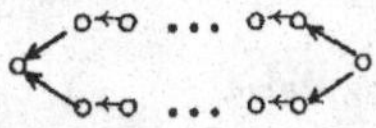	11 ... 11 / 1 1 / 11 ... 11	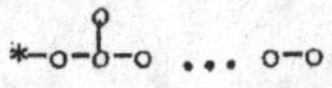	V-1

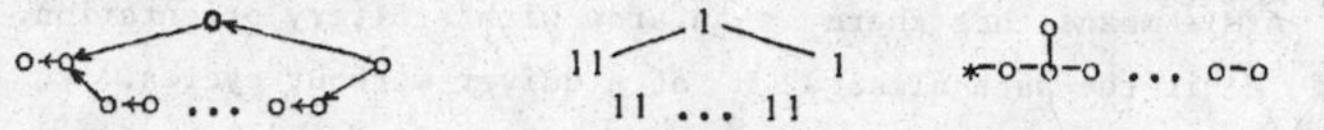	III-3·1, III-1
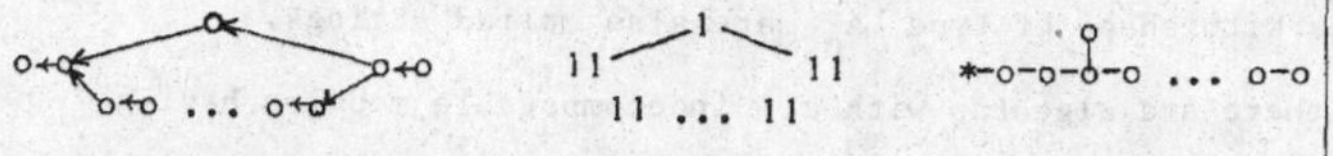	III-3·1
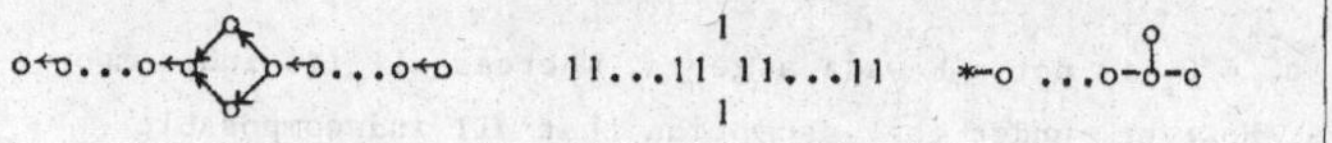	III-1, III-2
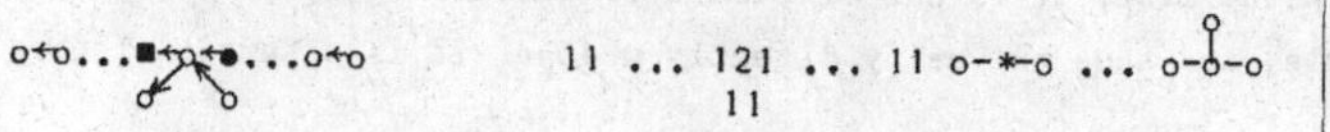	II-2·4
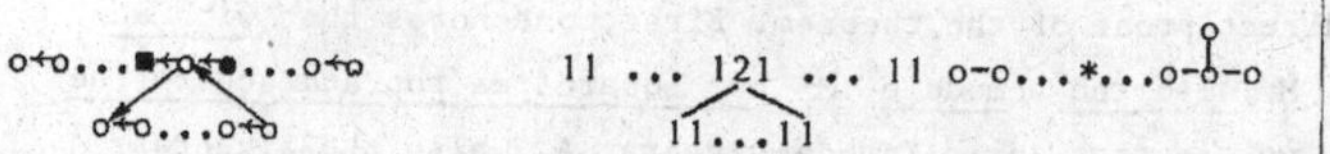	IV-2·4
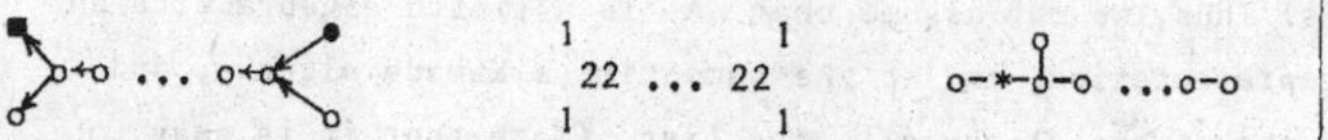	II-1·3
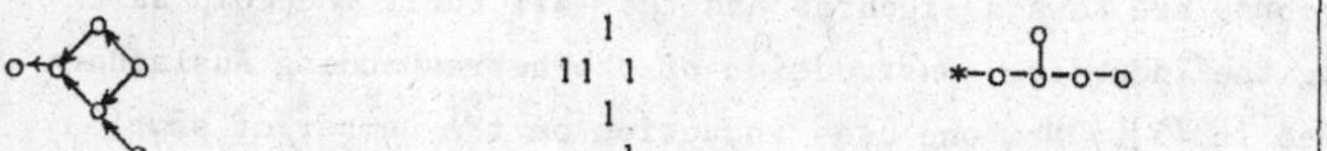	III-3·2, IV-1·1
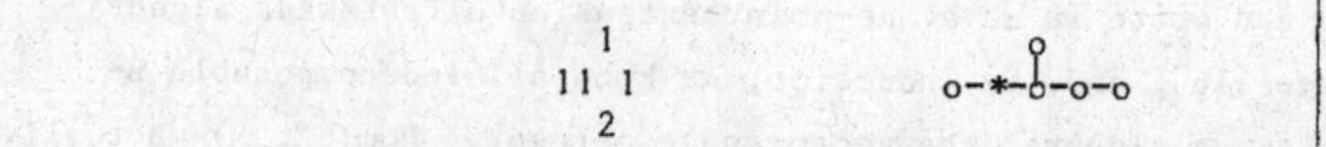	III-3·3, IV-1·2
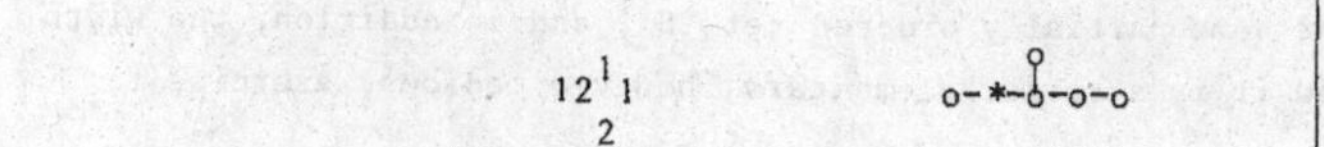	III-3·4, IV-1·3
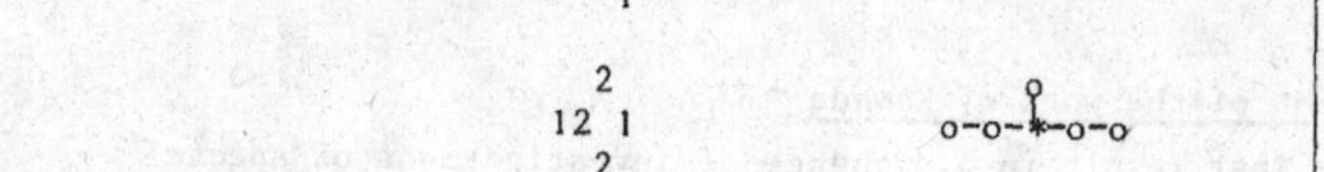	III-3·5, IV-1·4
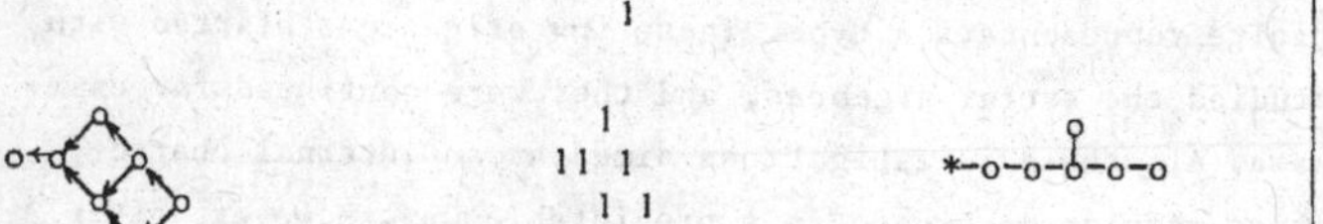	V
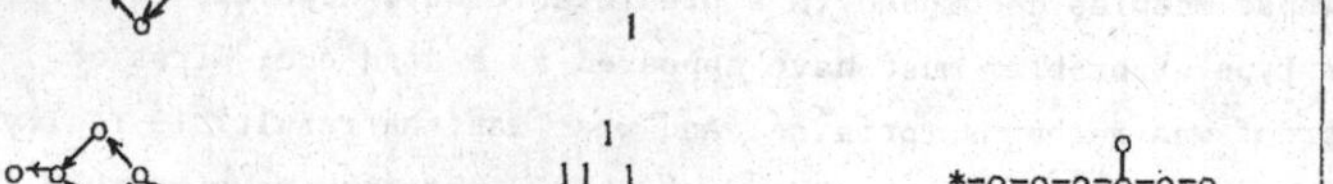	V

An edge in the quiver of $\tilde{A}(\tilde{M})$ means that there is an arow with arbitary orientation. In all cases, the type of $\tilde{M}$ is the path algebra H of a quiver without cycles. We have listed these quivers, the unique sink being marked by * (except in the first cases). Note that modules with shape of type A_n are also called strings.

(4.2) Note however that there are algebras with all indecomposable modules having shapes as in the list, without being a Kawada algebra. For example, the path algebra of $\circ \longrightarrow \circ \circlearrowleft \alpha$ with $\alpha^2 = 0$ is not a Kawada algebra, whereas all its indecomposable modules are strings. However, under the assumption that all indecomposable modules have shapes as in the list, it is not difficult to check for any of these modules both top and socle, and thus to verify directly whether it is a Kawada algebra or not.

(4.3) Let us outline a direct proof of the theorem. First, one notes that with an indecomposable A-module M also the $\tilde{A}$-module $\tilde{M}$ has squarefree top and squarefree socle. As a consequence, we see that for a Kawada-algebra A, also the algebras $\tilde{A}(\tilde{M})$ are Kawada algebras. Thus, we may assume that A is a tilted algebra with an indecomposable sincere representation, and at the same time a Kawada algebra, and have to show that A is one of the algebras in the list. (Note that it is easy to check that all these algebras are Kawada algebras and that all their indecomposable modules are listed, using the inductive construction of the corresponding Auslander-Reiten quiver, as outlined in [5].) Now one uses induction on the number of simple A-modules: Given A, we can write it as a one-point-extension of a Kawada algebra B by a B-module ${}_BX$, see [16], and, by induction, we know all indecomposable B-modules. Since A is a Kawada algebra, the vectorspace category $\mathrm{Hom}({}_BX, {}_BM)$ actually is of the form $\mathrm{add}S$ for some partially ordered set S, and in addition, the width of S must be ≤ 2. Now it is a rather elementary, however tedious, exercise to check all possibilities.

5. Appendix: The reception of the work of Kawada

Kawada's theorem was the last result in a sequence of investigations of special classes of algebras of finite representation type. These investigations started with Köthe and Nakayama who studied the serial algebras, and they were continued for example by Yoshii and Tachikawa. All these investigations aimed at an internal characterization of algebras whose modules decompose in a predictable way. However, after the work of Kawada, this type of problem must have appeared as a dead end: First of all, the length of his proof was rather surprising. And what was the result? 19 really horrible conditions which are difficult to check and which did not seem to give much insight into the problem. As a consequence, for a long time, there were no further attempts to deal with algebras of finite representation type, the work of Kawada was forgotten.

Some of Kawada's results were rediscovered later, and usually not in a simpler form. His methods involve a large number of different ways of amalgamation of modules in order to form large indecomposable modules, and also different ways of splitting off certain types of modules in order to decompose a given module. Several of these techniques were needed later by different authors and had to be introduced again. In particular, the decomposition of modules which are direct sums of strings has been investigated thoroughly (strings also have been called V-modules [13]), they play a rather dominant role in representation theory. We note however that not all algebras of finite representation type with only strings as indecomposable modules are Kawada algebras (see 4.2).

The most important Kawada algebras are perhaps the blocks of group algebras with cyclic defect group (in particular, the group algebras of groups with cyclic p-Sylow group over a field of characteristic p). These algebras were investigated by Dade, Janusz and Kupisch. Using deep character theoretical results of Dade, both Janusz [9] and Kupisch [12,13] determined the structure first of the indecomposable projective modules, they are of shape

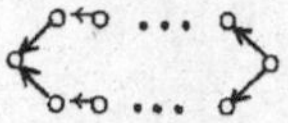 or o←o←o ... o←o ,

and then of the remaining modules: they are strings. After having derived the structure of the indecomposable projective modules, one could have applied Kawada's theorem.

A special class of Kawada algebras (which includes the blocks of group algebras with cyclic defect group) have been considered recently [4]: algebras of distributive module type. Recall that a module is said to be distributive in case its lattice of submodules is distributive. Note that a module M over a finite-dimensional algebra is distributive if and only if for every pair of submodules $0 \subseteq U \subseteq V \subseteq M$ with V/U semisimple, this module V/U is squarefree. The finite dimensional algebra A is said to be of distributive module type provided any indecomposable module is distributive. Clearly, algebras of distributive module type are Kawada algebras. Thus, we can apply Kawada's theorem. Note that the shape of a distributive module is again a distributive module, and the only quivers with relations occuring in Kawada's list for which all indecomposable representations are distributive, are

o-o-o ... o-o

and the commutative quiver

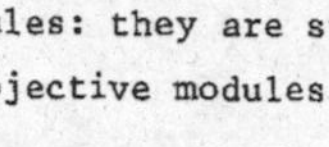 .

There also is a recent survey on the Köthe problem (which there is called the σ-cyclic problem, and correspondingly Köthe rings there are called σ-cyclic rings), with a "look to the future". It was presented at the 1978 annual AMS-meeting and then also published. This survey does have a reference to the papers [1] of Kawada, but it refers to them as follows: "Kawada gave a determination of a very special case of the σ-cyclic problem (e. g. radical square zero, and every indecomposable cyclic embeds in R), but even then some 19 conditions were deemed necessary and sufficient." The number of conditions is the right one, but everything else is pure fantasy (actually, under the mentioned assumptions, the problem would be very easy [10]). On the other hand, the author poses the problem to do what Kawada actually did: "Call a ring property P Morita stable if every ring Morita equivalent to a ring with P also has P . ... It would be a reasonable conjecture that any semi-perfect Morita stable σ-cyclic ring is uniserial." At least Nakayama gave a counter example to such a conjecture, and we have seen above the large variety of possible shapes of modules found by Kawada. A look to the past is sometimes valuable.

References.

[1] Kawada, Y.: On Köthe's problem concerning algebras for which every indecomposable module is cyclic. I-III. Sci. Rep. Tokyo Kyoiku Daigaku 7 (1962), 154-230; 8 (1963), 1-62; 9 (1964), 165-250. Summery I-II: Proc. Japan Acad. 37 (1961), 282-287; 288-293.

[2] Dlab, V., Ringel, C.M.: The preprojective algebra of a modulated graph. In: Representation Theory II, Springer Lecture Notes 832 (1980), 216-231.

[3] Faith, C.: The Basis theorem for modules. A brief survey and a look to the future. In: Ring theory. Marcel Dekker (1978), 9-23.

[4] Fuller, K.R.: Weakly symmetric rings of distributive module type. Comm. Alg. 5 (1977), 997-1008.

[5] Gabriel, P.: Auslander-Reiten sequences and representation finite algebras. In: Representation Theory I. Springer Lecture Notes 831 (1980), 1-71.

[6] Bongartz, K.; Gabriel, P.: Covering spaces in representation theory. To appear.

[7] Happel, D.; Preiser, U.; Ringel, C.M.: Vinberg's characterization of Dynkin diagrams using subadditive functions with application to DTr-periodic modules. In: Representation Theory II, Springer Lecture Notes 832 (1980), 280-294.

[8] Happel, D.; Ringel, C.M.: Tilted algebras. Trans. Amer. Math. Soc. (to appear).

[9] Janusz, G.: Indecomposable modules for finite groups. Ann. Math. 89 (1969), 209-241.

[10] Jøndrup, S.; Ringel, C.M.: Remarks on a paper by Skornjakov concerning rings for which every module is a direct sum of left ideals. Archiv Math. 31 (1978), 329-331.

[11] Köthe, G.: Verallgemeinerte abelsche Gruppen mit hyperkomplexem Operatorring. Math. Z. 39 (1935), 31-44.

[12] Kupisch, H.: Projective Moduln endlicher Gruppen mit zyklischer p-Sylow-Gruppe, J. Algebra 10 (1968), 1-7.

[13] Kupisch, H.: Unzerlegbare Moduln endlicher Gruppen mit zyklischer p- Sylow-Gruppe. Math. Z. 108 (1969), 77-104.

[14] Kupisch, H.: Symmetrische Algebra mit endlich vielen unzerlegbaren Darstellungen II. J. Reine Angew. Math. 245 (1970), 1-14.

[15] Ringel, C.M.: Report on the Brauer-Thrall conjectures. In: Representation theory I, Springer Lecture Notes 831 (1980), 104-136.

[16] Ringel, C.M.: Tame algebras. In: Representation theory I. Springer Lecture Notes 831 (1980), 137-287.

[17] Ringel, C.M.; Tachikawa, H.: QF-3 rings. J. Reine Angew. Math. 272 (1975), 49-72.

[18] Rojter, A.V.: The unboundedness of the dimensions of the indecomposable representations of algebras that have an infinite number of indecomposable representations. Izv. Acad. Nauk SSR 32 (1968), 1275-1282.